APPLIED OPTICS

and

OPTICAL ENGINEERING

VOLUME VI

Contributors to This Volume

N. Balasubramanian

Donald E. Bode

David C. Brown

Jack F. Butler

David Casasent

J. M. Forsyth

D. B. Keck

Adrianus Korpel

Robert D. Leighty

R. E. Love

Gerald F. Marshall

Herbert K. Pollehn

B. J. Thompson

J. Wilson

APPLIED OPTICS
and
OPTICAL ENGINEERING

EDITED BY

RUDOLF KINGSLAKE

Institute of Optics
University of Rochester
Rochester, New York

BRIAN J. THOMPSON

College of Engineering and Applied Sciences
University of Rochester
Rochester, New York

VOLUME VI

1980

ACADEMIC PRESS
A Subsidiary of Harcourt Brace Jovanovich, Publishers
New York London Toronto Sydney San Francisco

ACADEMIC PRESS, INC.
111 Fifth Avenue, New York, New York 10003

United Kingdom Edition published by
ACADEMIC PRESS, INC. (LONDON) LTD.
24/28 Oval Road, London NW1 7DX

Library of Congress Cataloging in Publication Data

Kingslake, Rudolf, ed.
Applied optics and optical engineering.

Vol. 6–
"Cumulative index": v. 5, p. 358–382.
Includes bibliographical references.
PARTIAL CONTENTS:––v. 1. Light, its generation and modification.––v. 2. The detection of light and infrared radiation––v. 3. Optical components.––v. 4–5. Optical instruments.
1. Optical instruments. 2. Optics.
I. Title.
QC371.K5 621.36 65–17761
ISBN 0–12–408606–3 (v. 6)

PRINTED IN THE UNITED STATES OF AMERICA

80 81 82 83 9 8 7 6 5 4 3 2 1

Contents

CHAPTER 4

Acousto-Optics

Adrianus Korpel

CHAPTER 5

Coherent Light Valves

David Casasent

CHAPTER 6

Scanning Devices and Systems

Gerald F. Marshall

CHAPTER 7

Coherent Optical Processing in Mapping

N. Balasubramanian and Robert D. Leighty

CHAPTER 8

Infrared Detectors

Donald E. Bode

CHAPTER 9

Principles and Applications of Optical Holography

B. J. Thompson

CHAPTER 10

Image Intensifiers

Herbert K. Pollehn

CHAPTER 11

Fiber Optics for Communications

D.B. Keck and R. E. Love

List of Contributors

Numbers in parentheses indicate the page on which the authors' contributions begin.

N. Balasubramanian, *Route 62, Cupertino, California 95014 (263)*

Donald E. Bode, *Santa Barbara Research Center, Goleta, California 93017 (323)*

David C. Brown, *Laboratory for Laser Energetics, University of Rochester, Rochester, New York 14627 (1)*

Jack F. Butler, *Laser Analytics, Inc., Lexington, Massachusetts 02173 (53)*

David Casasent, *Department of Electrical Engineering, Carnegie–Mellon University, Pittsburgh, Pennsylvania 15213 (143)*

J. M. Forsyth, *Laboratory for Laser Energetics, University of Rochester, Rochester, New York 14627 (29)*

D. B. Keck, *Corning Glass Works, Corning, New York 14830 (439)*

Adrianus Korpel, *Department of Electrical and Computer Engineering, University of Iowa, Iowa City, Iowa 52242 (89)*

Robert D. Leighty, *Center for Coherent Optics, USAETL, Fort Belvoir, Virginia 22060 (263)*

R. E. Love, *Corning Glass Works, Corning, New York 14830 (439)*

Gerald F. Marshall, *Department of Optics, Energy Conservation Devices, Incorporated, Troy, Michigan 48084 (203)*

Herbert K. Pollehn, *U.S. Army Night Vision and Electro-Optics Laboratory, Fort Belvoir, Virginia 22060 (393)*

B. J. Thompson, *College of Engineering and Applied Science, University of Rochester, Rochester, New York 14627 (357)*

J. Wilson, *Laboratory for Laser Energetics, University of Rochester, Rochester, New York 14627 (29)*

Preface

The preface to the first five volumes of Applied Optics and Optical Engineering stated that "There are, of course, numerous aspects of applied optics and optical engineering that will not have been covered in these volumes." It is to cover some of these "numerous aspects" that Volume VI is offered. This volume is devoted to coherent optical devices and systems.

In recent years, applied optics and optical engineering have continued to show strength in the traditional areas but have been expanded to include whole new areas that were not even thought of in 1965, when Volume I of this series was published. Coherent optical science and technology have developed as important branches of applied optics and optical engineering. The stimulus for this has been the rapid development and exploitation of the laser as a general purpose coherent light source.

What is coherent optical engineering? It is that special area that is concerned with the practical application of the unique properties of coherent light. Coherent light as obtained from a laser is highly spatially coherent, highly temporally coherent (narrow spectral profile), highly directional, and highly energetic. Spatial coherence allows the classical diffraction phenomena to be produced readily and used in a wide variety of measurement and pattern recognition procedures which are made especially viable by the advances in detector technology and minicomputers. Temporal coherence allows interferometry to be carried out with large path differences between the interfering beams; thus conventional interferometry can be extended. The spatial and temporal properties of the light are the properties that make holography work. A hologram is the recorded intensity distribution in the interference pattern produced by the light diffracted (or scattered) from an object and a known or reproducible reference or background beam. In its turn, holography has made possible very interesting new methods of interferometry.

Diffraction combined with spatial filters, and particularly holographic filters, has formed the basis for image and signal processing methods which have become interesting alternatives to digital image processing techniques. This is particularly true today because of the development of light valves and spatial light modulators.

The directionality of the laser beam means that it can be focused to a very small highly energetic spot. This has revolutionized and revitalized optical scanning systems for reading, recording, and display purposes. Well-known acousto-optic and electro-optic effects can be effectively used to control both the direction and intensity of a coherent beam. Who would have guessed in 1965, for example, that a laser scanning system would be on line at the checkout points in major supermarkets throughout the country?

The ability to encode information on an optical beam and transmit that information by sending the beam down an optical fiber is already having a dramatic effect in the communications field.

These are merely examples of coherent optical engineering.

There are, of course, many others. The eleven chapters of this volume cover some of the devices and systems that have made coherent optics an exciting field in which to work, and one that has already produced a useful harvest of engineered devices and systems.

The basic philosophy of this volume is no different from that of the previous five. We sincerely hope that these chapters will be useful to the applied scientist or engineer when he conceptualizes and designs new instruments and systems based on this body of knowledge of coherent optics.

B. J. THOMPSON
R. KINGSLAKE

Rochester, 1979

Contents of Other Volumes

Volume IV: Optical Instruments—Part I

Edited by *Rudolf Kingslake*

Volume V: Optical Instuments—Part II

Edited by *Rudolf Kingslake*

Volume VII

Edited by *Robert R. Shannon and James C. Wyant*

Volume VIII

Edited by *Robert R. Shannon and James C. Wyant*

List of Acronyms, Abbreviations, and Initials

ABC	Automatic brightness control
ADP	Ammonium dihydrogen phosphate
AGC	Automatic gain control
AOPSA	Advanced optical power spectrum analyzer
APD	Avalanche photodiode
AR	Antireflection (coating)
ASE	Amplified spontaneous emission
BLIP	Background limited infrared photoconductor
CATV	Cable television
CCD	Charge coupled device
CID	Compositional interdiffusion technique
CMU	Carnegie Mellon University
DH	Double heterostructure (laser)
DKDP	Deuterated potassium dihydrogen phosphate
EALM	Electronically addressed light modulator
EBI	Equivalent background input (of photocathode)
ERIM	Enivronmental Research Institute of Michigan
FET	Field effect transistor
FLIR	Forward looking infrared
FLN	Fluorescence line narrowing
FOM	Figure of merit
FOV	Field of view
FPA	Focal plane array (of detectors)
FPC	Frequency plane correlator
FSBW	Frame space–bandwith product
FSK	Frequency shift keying
FWHM	Full width at half maximum
GDL	Glass development laser system
G.E.C.	General Electric Company
HOC	Heterodyne optical correlator
IMF	Image matched filter (correlator)
JTC	Joint transform correlator
HOALM	Holographic optically addressed light modulator
KDP	Potassium dihydrogen phosphate
LED	Light emitting diode
LEP	Laboratoire d'Electronique Physica Appliqué
LIDAR	Light radar
LLE	Laboratory for Laser Energetics (at University of Rochester)
LSI	Large scale integration
MCP	Microchannel plate
MFPA	Monolithic focal plane array
MIT	Massachusetts Institute of Technology
MRTR	Maximum readout transfer ratio
MTF	Modulation transfer function
NASA	National Aeronautics and Space Administration
NEP	Noise equivalent power
OALM	Optically addressed light modulator
OEM	Original equipment manufacturer
OPS	Optical power spectrum
OPSA	Optical power spectrum analysis
OTTO	Optical-to-optical converter
PIN	*p*-intrinsic-*n* (diode)
PMT	Photomultiplier tube
POS	Point of sale
PROM	Pockels readout optical modulator
ROSA	Recording optical spectrum analyser
RMS	Root-mean-square
SAR	Synthetic aperture radar
SBRC	Santa Barbara Research Center
SBW	Space–bandwidth product
SH	Single heterostructure (laser)
SLAR	Side-looking airborn radar
SNR	Signal-to-noise ratio
SOALM	Scanned optically addressed light modulator
TDI	Time delay and integration
TE	Transverse electric (mode)
TEM	Transverse electromagnetic (laser fields)

TM	Transverse magnetic (mode)
TOPR	Thermoplastic optical phase recorder
UPC	Universal product code
USAETL	United States Army Engineering Topographic Laboratory
WKB	Wentzel–Kramers–Brillouin
YAG	Yttrium aluminum garnet

CHAPTER 1

Solid State Lasers

DAVID C. BROWN

Laboratory for Laser Energetics, University of Rochester, Rochester, New York

I. INTRODUCTION

In recent years, those of us fortunate to be working in the field of solid state laser physics have witnessed rapid advances in the quality and variety of materials available, the understanding of how rare-earth ions interact with various hosts, and in the design and operation of sophisticated laser systems for a myriad of applications. Much of the stimulus for these developments has come from a desire to use well-characterized Nd:glass lasers operating at a wavelength of 1.06 μm to investigate the physics of inertial confinement or laser fusion. Military applications such as designators, illuminators, and rangers have spawned a large variety of research areas concerned with oscillator and material development for smaller devices.

The solid state laser has become important and even indispensable in some applications, primarily because of its necessity and reliability. Energy, military, scientific, and industrial research would in many cases be impossible or in others surely impaired without it. It is likely that in the future many additional areas of endeavor will be affected by its development.

ISBN 0-12-408606-3

It is the intent of this chapter to provide an overview of solid state lasers to the reader with some training in optical physics. Although the ruby and Nd:YAG laser are treated in what follows, the majority of space has been allotted to the Nd:glass laser which has assumed the most scientific importance. Section II provides the reader with a general review of the physics relevant to the solid state laser. Building upon that, Section III covers the types of solid state materials available and the detailed physics of ruby, Nd:YAG, and Nd:glass. In Section IV an overview is provided of the basic types of devices now in common use in most laboratories which employ Nd:glass laser systems. Finally, in Section V two current applications of solid state lasers are reviewed.

II. SOLID STATE LASER PHYSICS

A. Interaction of Radiation with Matter—Stimulated Emission

The fundamental effect which determines the properties of a laser is that of stimulated emission, a process first described by Albert Einstein to explain the Planck law which describes the distribution of radiation with frequency of a thermodynamic equilibrium state of radiation with matter. In particular, referring to Fig. 1, three distinct processes are possible:

(a) An atom in the upper level m may spontaneously decay to the lower level n, with the emission of a photon of energy $h\nu$. The rate is not dependent upon radiation density and may be described by a spontaneous emission coefficient A_{mn} which when multiplied by the number

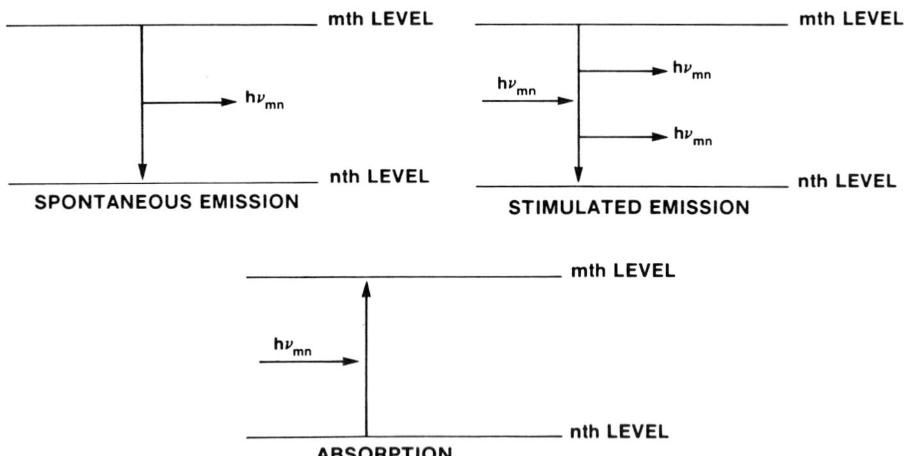

FIG. 1. Emission and absorption of radiation.

of atoms in the mth state N_m represents the rate at which atoms from the mth level decay to the nth level.

(b) Atoms in the mth level may also decay by the process of stimulated emission in which an initial photon of energy $h\nu_{mn}$ induces the atom in the mth level to decay to the nth level with the simultaneous emission of a second photon with energy $h\nu_{mn}$. In the process the direction, phase, polarization, and wavelength of the two resulting photons are exactly the same. In this case, however, the transition probability is dependent upon the radiation density ρ_{mn} at frequency ν_{mn} and the rate given by $N_m B_{mn} \rho(\nu_{mn})$ where B_{mn} is the stimulated emission coefficient.

(c) Finally, the lower level n may absorb radiation of frequency ν_{mn}, the inverse process of (b) above, and the rate is given by $N_n B_{nm} \rho(\nu_{mn})$.

In thermodynamic equilibrium, the principle of detailed balancing requires:

$$N_m A_{mn} + N_m B_{mn} \rho(\nu_{mn}) = N_n B_{nm} \rho(\nu_{mn}) \tag{2.1}$$

For a system in thermodynamic equilibrium, the usual Boltzmann distribution

$$N_m/N_n = (g_m/g_n) \exp[-h\nu_{mn}/kT] \tag{2.2}$$

connecting the populations N_m and N_n may also be used. Here g_m and g_n are the statistical weights of the two levels, k the Boltzmann constant, and T the absolute temperature. Using Eqs. (2.1) and (2.2), one can derive

$$B_{mn}/B_{nm} = g_n/g_m \tag{2.3}$$

and

$$B_{mn}/A_{mn} = c^3(8\pi h \nu_{mn}^3)^{-1} \tag{2.4}$$

equalities which determine the relationship of B_{nm} and B_{mn} to A_{mn}.

These equalities have been derived using radiation which is isotropic and of radiative frequency ν_{mn}. In real situations, the atomic response to the radiative field is not constant, but described by a response or linewidth function.

B. Atomic Absorption and Gain

It can be shown that the increase in intensity ΔI of radiation of frequency ν_{mn} propagating a distance dz through an atomic medium is given by

$$\Delta I = (B_{mn} N_m - B_{nm} N_n)(h\nu/c) I f(\nu, \nu_{mn})\, dz \tag{2.5}$$

where $f(\nu, \nu_{mn})$ is the normalized atomic response function and ν_{mn} the atomic resonance frequency. Using Eqs. (2.3) and (2.4), we have

$$\frac{1}{I}\frac{dI}{dz} = \frac{A_{mn} c^2}{8\pi \nu^2} f(\nu, \nu_{mn}) \left[N_m - \frac{g_m}{g_n} N_n \right] = \alpha(\nu) \tag{2.6}$$

or integrating,

$$I = I_0 \exp[\alpha(\nu)z] \tag{2.7}$$

where

$$\alpha(\nu) = \frac{A_{mn}c^2}{8\pi\nu^2} f(\nu, \nu_{mn})\left[N_m - \frac{g_m}{g_n} N_n\right] \tag{2.8}$$

Note that if $N_m > (g_m/g_n)N_n$, $\alpha(\nu) > 0$ and a population inversion is said to exist between the two states, leading to an increase in intensity (gain). If, however, $N_m < (g_m/g_n)N_n$, the intensity decreases. Equation (2.8) may be rewritten, and evaluated at the resonant frequency ν_{mn}

$$\alpha(\nu_{mn}) = (\sigma/h\nu)h\nu[N_m - (g_m/g_n)N_n] = \alpha \tag{2.9}$$

or

$$\alpha = \alpha_0 E_s \tag{2.10}$$

where

$$\alpha_0 = \sigma/h\nu = A_{mn}c^2/8\pi h\nu^3 \tag{2.11}$$

and

$$E_s = h\nu[N_m - (g_m/g_n)N_n] \tag{2.12}$$

Here $f(\nu_{mn}, \nu_{mn}) = 1$, α_0 is the specific gain coefficient, and E_s the stored energy density. The cross section σ is defined at the center of the atomic transition. In practice, there is usually an additional loss which must be taken into account which represents loss of intensity from host absorption, scattering, or other processes, and Eq. (2.10) must be rewritten

$$\alpha = \alpha_0 \mathrm{E}_s - \alpha_L \tag{2.13}$$

where α_L is the so-called passive loss coefficient. Equation (2.7) represents the small signal regime in which amplification is truly exponential. In most amplifiers, however, intensities increase to the level where significant energy extraction takes place and the process must be described by saturated gain equations.

C. Three- and Four-Level Lasers

With the sole exception of ruby, which is a three-level system, most solid state lasers of any scientific and commercial importance are four-level systems. That type of physical system occurs again and again in laser physics and we shall spend some time discussing it here.

Referring the reader to Fig. 2, such a system consists of a ground state (level 0), a terminal level (level 1), a metastable level (level 2), and a number of pump bands. During optical pumping, atoms are raised from the ground state

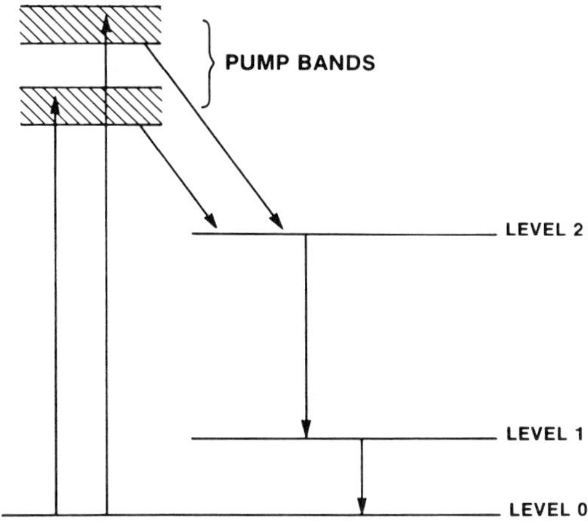

FIG. 2. A typical 4-level laser system.

to the pump bands at a rate R_{ij}; atoms in the pump bands relax quickly (compared to the lifetime of the metastable level) to level 2 with some characteristic rate W_{ij}. Such relaxation may be radiative or nonradiative. In the nonradiative process multiphonon emission takes place, thus the energy difference between the pump photon and the laser emission photon goes into heating the crystal lattice or glass. The rate of relaxation of the pump levels is determined by the energy gap to the next lowest level as well as the phonon spectrum of the specific material. Such effects have been studied extensively by Layne.[1] The metastable level is long lived since electric dipole transitions to lower energy levels are forbidden by quantum mechanical (LaPorte's) selection rules. In addition, the decay is virtually all radiative since multiphonon transitions across such a large energy gap are normally not probable due to the low phonon density of states at that energy. The terminal level is important in determining the threshold and dynamical behavior of any four-level system. Typically, the lifetime is short compared to that of level 2, thus making it relatively easy to maintain a population inversion between levels 2 and 1. The terminal level relaxation is also principally via multiphonon transitions. The location of the terminal level above ground state determines what its population is at ambient temperatures. Indeed, we want the thermal population to be small, and can calculate the number N_1, of atoms in state 1 from the number N_0 in the ground state using the Boltzmann relationship, Eq. (2.2). The energy separation ΔE between levels (0) and (1) must satisfy ($\Delta E \gg kT$) so that $N_1/N_0 \ll 1$.

[1] C. B. Layne, Multiphonon Relaxation and Excitation Transfer in Rare-Earth Doped Glasses. PhD thesis, UCRL-51862, Univ. of California.

If the population of the terminal level can be considered empty during a certain process, the four-level laser essentially degenerates into a three-level one and can be described using only a ground state (level 0), metastable state (level 1), and pump bands (level 2). The threshold properties of a true three-level system are, however, considerably different from a degenerate four-level system. It is necessary in a three-level system, to remove at least half of the ground state population to achieve threshold, thus typically much larger pump powers are necessary than for a four-level system. We will not discuss threshold properties of three- or four-level lasers here but refer the reader to the excellent recent book by Koechner.[2]

In real lasers, such as Nd:glass, it is possible to use rate equations to describe the amplification process; modeling using such equations has been found to describe accurately the amplification of pulses in Nd:glass and Nd:YAG systems. For the levels (1) and (2) we write

$$dN_2/dt = N_2\omega_{21} - W(N_2 - N_1) + R_{32} + R_{02} \tag{2.14}$$

$$dN_1/dt = N_2\omega_{21} + W(N_2 - N_1) + R_{02} - w_{10}N_1 + R_e \tag{2.15}$$

As described before, the ω_{ij} and R_{ij} are the relaxation rates and pumping rates from the ith to the jth level. R_e is the pumping of the terminal level to some possible excited state while W, the rate of stimulated emission, is given by

$$W = (\sigma/h\nu)P(t) \tag{2.16}$$

where, as before, σ is the stimulated emission cross section, $h\nu$ the photon energy, and $P(t)$ is the power.

If we consider pulse amplification in a solid state laser, it is possible to neglect the ω_{ij} and R_{ij} during the process (but not the terminal level relaxation); hence we have from Eqs. (2.14) and (2.15)

$$dN_2/dt = -W(N_2 - N_1) \tag{2.17}$$

$$dN_1/dt = W(N_2 - N_1) - \omega_{10}N_1 \tag{2.18}$$

We now consider two limiting cases where the pulse (FWHM)τ_p is either much shorter or much longer than the terminal level lifetime τ.

Case 1. $\tau \ll \tau_1$: *Three-Level System*

Here we can assume $\omega_{10} = 0$, and Eqs. (2.17) and (2.18) may be combined to give

$$(d/dt)(N_2 - N_1) = -2W(N_2 - N_1) \tag{2.19}$$

---

[2] W. Koechner, "Solid State Laser Engineering." Springer-Verlag, Berlin and New York, 1976.

Defining $E_s = h\nu(N_2 - N_1)$ and recalling Eq. (2.13) with $\alpha_L = 0$ gives

$$d\alpha/dt = -2W\alpha \tag{2.20}$$

Note that we have assumed that the degeneracy factors are equal for simplicity ($g_1 = g_2 = 1$). Integrating Eq. (2.20) gives

$$\alpha = \alpha' \exp\left[-2\frac{\sigma}{h\nu}\int P(t)\,dt\right] \tag{2.21}$$

or letting $E = \int P(t)\,dt$ and defining

$$E_{sat} = h\nu/2\sigma \tag{2.22}$$

we have

$$\alpha = \alpha' \exp(-E/E_{sat}) \tag{2.23}$$

The significance of E_{sat} is that it is the energy density where the gain has been reduced to e^{-1} of its initial value. It is called the saturation energy density and is related to the specific gain coefficient α_0, Eq. (2.11), through

$$E_{sat} = h\nu/2\sigma = 1/2\alpha_0 \tag{2.24}$$

Since we have assumed that during the pulse duration τ the relaxation of the terminal level was not important ($\omega_{10} = 0$), the system is a three-level one.

Case 2. $\tau \gg \tau_1$: Four-Level System

If the pulse duration is much larger than the terminal level lifetime we can assume $\omega_1 = 0$ ($d\omega_1/dt = 0$), hence we have from Eqs. (2.17) and (2.18)

$$dN_2/dt = -WN_2 \tag{2.25}$$

or since here $E_s = h\nu N_2$ we have

$$d\alpha/dt = -W\alpha \tag{2.26}$$

Following the treatment in case 1 above, the solution to Eq. (2.26) becomes identical with that in Eq. (2.23), but now

$$E_{sat} = h\nu/\sigma = 1/\alpha_0 \tag{2.27}$$

As before, E_{sat} is the energy density required to deplete the gain to e^{-1} of its initial value but the essential difference between the three- and four-level systems for pulse amplitudes is that the three-level system requires twice the energy density before saturation occurs. Thus, long pulses (compared with τ_1) will contain higher energy before the gain in an amplifier saturates.

We have examined the gain at a specific spatial location in a solid state laser, and seen how saturation affects gain amplification when terminal level lifetime effects are taken into account. Pulses propagating in a real solid state laser along a direction x may be shown to obey

$$\text{(three-level)}\ (\tau \ll \tau_1) \qquad (dE/dx) = \tfrac{1}{2}E_s(1 - e^{E/E_{sat}}) - \alpha_L E \tag{2.28}$$

$$\text{(four-level)}\ (\tau \gg \tau_1) \qquad (dE/dx) = E_s(1 - e^{E/E_{sat}}) - \alpha_L E \tag{2.29}$$

In Eq. (2.28), $E_{sat} = h\nu/2\sigma = 1/2\alpha_0$ and in (2.29) $E_{sat} = h\nu/\sigma = 1/\alpha_0$. When the passive loss coefficient $\alpha_L \neq 0$, these equations are not integrable in closed form. Since $\alpha_L \neq 0$ in most situations of practical interest, the propagation of pulses in solid state lasers must be modeled by solving Eqs. (2.28)–(2.29) numerically, using sophisticated digital computers. Closed form solutions may be found in the book by Koechner.[2]

III. SOLID STATE LASER MATERIALS

A. Types of Solid State Materials

Having concluded a brief discussion of the physics of typical solid state lasers, we now turn to the consideration of specific materials employed in most systems now in use. Although Nd:YAG and Nd:glass have found the most widespread use in applications, we include ruby as an illustration of the three-level system.

The first laser action was demonstrated by Maiman[3] in crystalline ruby ($Cr^{3+}:Al_2O_3$). Subsequently, solid state lasers have been operated using a number of trivalent rare earths (Nd^{3+}, Eu^{3+}, Pr^{3+}, Er^{3+}, Ho^{3+}, Tm^{3+}, Gd^{3+}, Yb^{3+}), divalent rare earths (Sm^{2+}, Dy^{2+}, Tm^{2+}), some transition metals, and U^{3+} in various host materials. The rare earths have become particularly important. In the solid form they possess absorption spectra similar to the free ions which overlap commonly available optical pump sources and lase in the visible to near infrared region of the spectrum. Since 1961 when Snitzer[4] first demonstrated a Nd:glass laser, the Nd^{3+} ion has become the most important among the variety of materials available. The reasons for this are many and would include the following.

(a) The absorption of Nd^{3+} extends from $\simeq$3500 Å to $\simeq$9000 Å, overlapping well with high brightness pump sources such as Xe or Kr flashlamps.

(b) The output wavelength in the vicinity of 1.06 μm is of interest in laser inertial confinement experiments and although shorter wavelengths are desirable, leads to reasonable plasma coupling efficiencies.

---

[3] T. H. Maiman, *Nature* (*London*) **187**, 493 (1960).

[4] E. Snitzer, *Phys. Rev. Lett.* **7**, 444 (1961).

(c) Nd^{3+} can be incorporated into a variety of glasses, a technology which is relatively well understood, and which can provide large pieces of good optical quality suitable for scaled-up systems.
(d) Due to the relatively long fluorescence lifetime of the $^4F_{3/2}$ metastable level, the laser transition of most interest at 1.06 μm is capable of large energy storage.
(e) The stimulated emission coefficient for Nd^{3+} is in the intermediate regime, large enough to provide gain in reasonably sized amplifiers, but not so large as to make the problem of amplified spontaneous emission severe.
(f) The spectroscopy of Nd^{3+} is well understood, leading to the capability to predict optical and nonlinear properties.

Due to the above combination of favorable properties and knowledge, Nd^{3+} has become the preeminent solid state laser material, and devices have progressed to an advanced state (see Section IV below). Difficulties in applying Nd^{3+} to any future power systems include low efficiency and a currently low repetition rate, both of which will be discussed in Section III,D.

The physics of the specific ion and the way in which it interacts with the crystal lattice or glass determines macroscopic optical properties of most interest, for instance specific gain, nonlinear index, and terminal level lifetime. For many applications, however, physical as well as optical properties must be taken into account. Such properties are characteristic of the host, and may include hardness, thermal conductivity, specific heat, rupture strength, etc. A figure-of-merit approach to evaluating laser glasses for various applications has been discussed by Brown *et al.*[5] and will be investigated in Section III,D.

B. Ruby ($Cr^{3+}:Al_2O_3$)

Ruby is a textbook example of the three-level laser system, and we review its spectroscopic and laser properties here. In Fig. 3 we show the energy level structure of Cr^{3+} in ruby; it consists of a ground state ($^4A_2$) with a degeneracy of $4(g_1 = 4)$, two excited state or pump bands ($^4F_1$, $^4F_2$) located at approximately 4000 Å and 5555 Å, respectively, and the metastable levels ($2\bar{A}$, $\bar{E}$) which have degeneracy of $2[g(R_2) = g(R_1) = 2]$. The pump bands ($^4F_1$, $^4F_2$) relax quickly compared to the metastable levels which have a typical lifetime of 3 msec at room temperature. The transitions ($\bar{E} \rightarrow {}^4A_2$, $2\bar{A} \rightarrow {}^4A_2$) are both observed in fluorescence, corresponding to the wavelengths (6943 Å, 6929 Å) in the red, and are known as the R_1 and R_2 lines. Laser output from the ruby

[5] D. C. Brown, S. D. Jacobs, J. A. Abate, O. Lewis and J. Rinefierd, Laser-Induced Damage in Optical Materials. National Bureau of Standards Special Publ. 509 (December 1977).

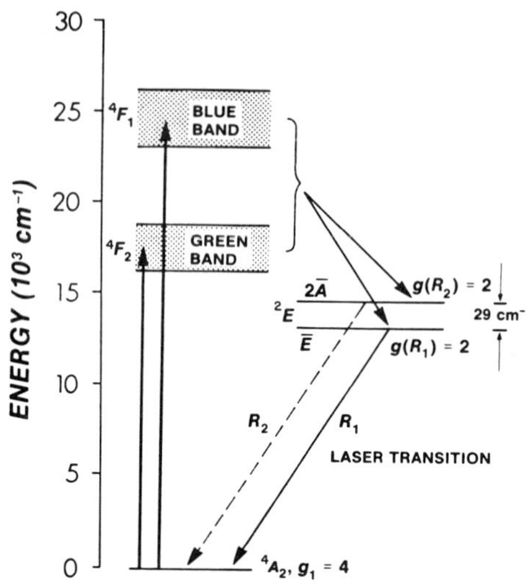

FIG. 3. Energy levels in ruby.

laser, however, consists of only a single line (6943 Å) corresponding to the ($\bar{\mathrm{E}} \rightarrow {}^4\mathrm{A}_2$) transition, primarily due to the higher inversion of the Ē relative to the 2Ā level. Considerable cross relaxation between the Ē and 2Ā levels can take place in a time scale of the order of a nanosecond, and for pulses long compared with that time virtually all of the populations of both states is channeled through emission at 6943 Å.

Because ruby can be described by only three levels (ground state, excited state, metastable state), it can be classified as a three-level laser and, of course, displays the threshold property of such a system. In particular, if n_2 is the number of atoms in the Ē state and n_1 the number in the ground state, then taking degeneracy factors into account $n_2 \geq \frac{1}{2}n_1$ must be satisfied for laser action to occur, or the Ē level must have at least half the population of the ground state.

The absorption spectrum of ruby displays two broad peaks located in the visible and ultraviolet region and which efficiently overlap the output spectra of common Xe and Kr flashlamps used as pump sources. The output of the ruby laser is in the visible, unlike most four-level lasers which radiate in the near infrared.

As a material, sapphire (Al_2O_3) displays unusually attractive properties. Among them are good durability, hardness, and a thermal conductivity which is high compared to most glasses. Ruby crystals are available in large sizes and display excellent optical quality. For more details concerning the properties of ruby, the reader is encouraged to see Koechner.[2]

C. Nd : YAG

The Nd:YAG laser, a four-level one, consists of the rare-earth ion Nd^{3+} incorporated into a YAG host, with typical doping level being 1 wt %. Unlike Nd:glass lasers, the absorption and emission bands are relatively narrow. A typical Nd:YAG absorption spectrum is shown in Fig. 4 and consists of a series of bands which range from ≃3500 Å in the ultraviolet to ≃8900 Å in the near infrared. The bands are partially resolved into discrete lines at 300°K but narrow and become completely resolved at lower temperatures. In Fig. 5

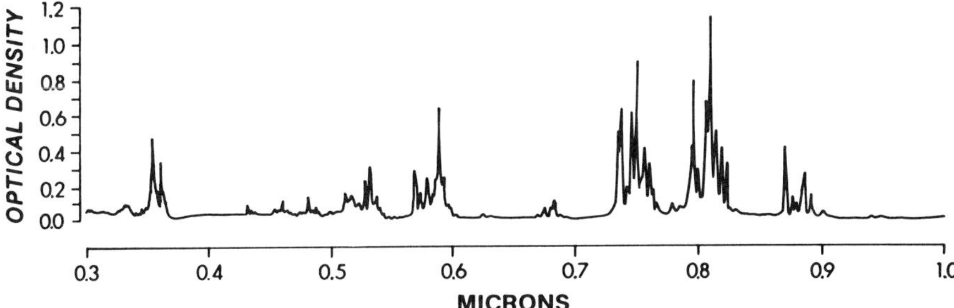

Fig. 4. A typical Nd:YAG absorption spectrum. YAG—5 mm/1 % wt at room temperature.

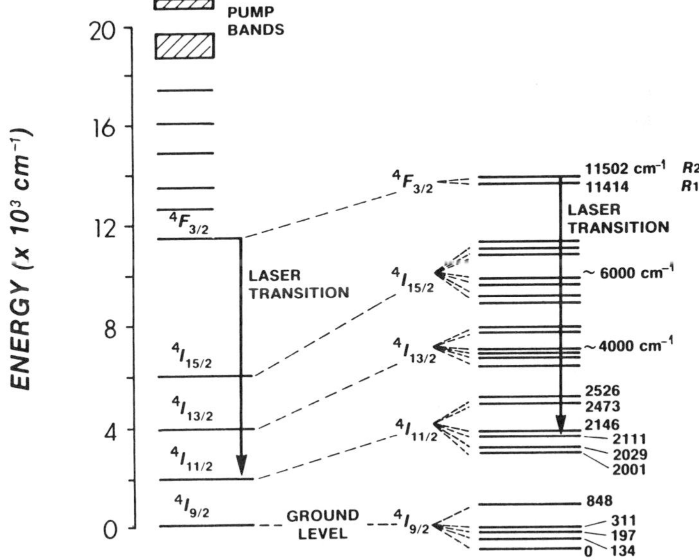

Fig. 5. Energy levels of Nd^{3+}.

the reader can see the four-energy-level structure of Nd^{3+}. Absorption corresponds to transitions from the $^4I_{9/2}$ ground state to various excited states or pump bands. With the exception of the ($^4D_{3/2} \rightarrow {}^4F_{3/2}$) transition at 3530 Å which decays radiatively and nonradiatively, all transitions decay quickly via multiphonon emission to the $^4F_{3/2}$ metastable level. The metastable level is resolved into two (R_1, R_2) separate lines in YAG, and has a fluorescence lifetime of $\simeq$230 μsec at room temperature. As may be seen from Fig. 5, the laser transition at 1.064 μm occurs between the R_2 line of the $^4F_{3/2}$ level at 11414 cm^{-1}, and the Y_3 component of the $^4I_{11/2}$ level at 2111 cm^{-1}. For a 1 wt % doping and room temperature operation, the fluorescence linewidth of the ($^4F_{3/2} \rightarrow {}^4I_{9/2}$) transition is $\simeq$4 Å, resulting in high gain (stimulated emission cross section) relative to Nd:glass. The 1.064 μm transition linewidth is homogeneously broadened by thermal lattice vibrations. The metastable level ($^4F_{3/2}$) has a quantum efficiency of $\geq$99.5 %,[6] while the branching rate for the transitions ($^4F_{3/2} \rightarrow {}^4I_{11/2}$), ($^4F_{3/2} \rightarrow {}^4I_{9/2}$), ($^4F_{3/2} \rightarrow {}^4I_{13/2}$) is (0.60; 0.25; 0.14).[6]

Relative to other materials, YAG offers a low threshold for laser action on the 1.064 μm line, principally due to the high stimulated emission cross section, but also because the $^4I_{11/2}$ level is not thermally populated at room temperature. By inserting a prism, etalons, or other dispersive elements into the laser cavity, it is possible to force the laser to oscillate on one of the twenty or more transitions observed in Nd:YAG. The exact transitions, their cross sections, and relative threshold for continuous (cw) operation have been tabulated elsewhere.[2]

D. Nd : Glass

As mentioned previously, in recent years Nd:glass has been favored as the material for most systems built for high power applications. The reasons are many, including the manufacturing scalability of glass to large sizes, a process very difficult for a crystal such as YAG, the smaller stimulated emission cross section for glass which helps to avoid problems such as parasitic oscillations and amplified spontaneous emission (ASE), and the ability of glass to convert a larger fraction of the pump light into useful emission than YAG due to the spectrally wider transitions.

Unlike a crystal in which the local site of a Nd^{3+} ion is regular, in glass a distribution of sites is obtained which leads to inhomogeneous broadening of the pump and laser transitions. Thus compared to YAG and other similar crystals, the absorption spectrum of Nd:glass yields much broader lines which are more efficient in converting pump radiation into useful inversion. In Figs. 6 and 7 we see typical absorption spectra of a silicate based (ED-2) laser glass and a phosphate one (LHG-5). The bands in both figures are much broader

---

[6] Lawrence Livermore Laboratory, Nd Laser Glasses Data Sheets (1977).

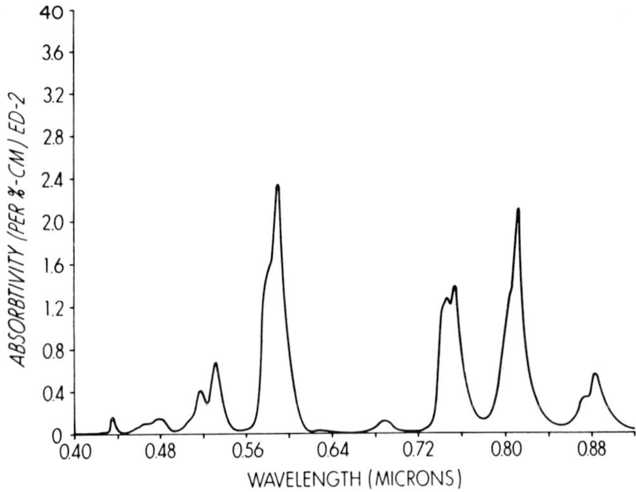

FIG. 6. Absorption spectrum of silicate glass ED-2.

than in Nd:YAG (Fig. 4), but the phosphate composition displays narrower but more intense absorption than the silicate. The site-to-site variation in the local environment has recently been probed by Brecker *et al.*[7] using fluorescence line narrowing (FLN) techniques; they obtained wide variations in the lifetime, quantum efficiency, and branching ratios. Thus in describing these quantities

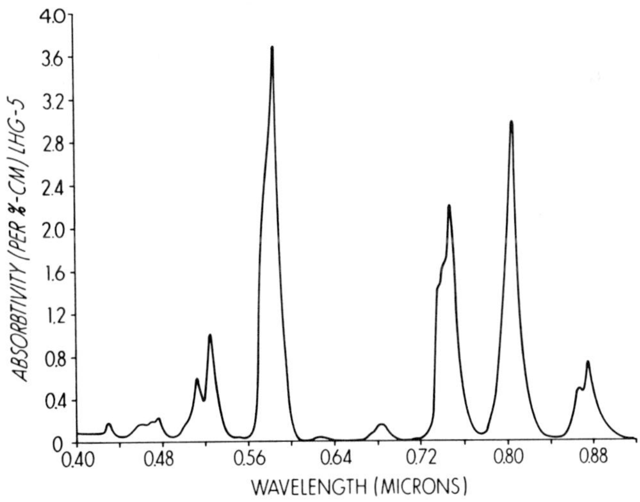

FIG. 7. Absorption spectrum of phosphate glass LHG-5.

---

[7] C. Brecker, L. A. Riseberg, and M. J. Weber, *Appl. Phys. Lett.* **30**, 475 (1977).

for a particular glass we really refer to the statistical average over the distribution of sites.

The absorption spectra of laser glasses may be utilized to study a number of problems inherent in the design of large laser systems. Among these are the prediction of the stored energy density and heat density profiles in amplifiers, and the thresholds for parasitic oscillations and ASE in such devices. It may be shown[8] that the inversion density $E(v)$ and heat density $H(v)$ are described by

$$E(v) = c_1 \int_{\lambda_1}^{\lambda_2} \alpha_\lambda' I_{0\lambda} \left(\frac{\lambda}{\lambda_L}\right) e^{-\alpha_\lambda' v}\, d\lambda \tag{3.1}$$

$$H(v) = c_2 \int_{\lambda_1}^{\lambda_2} \alpha_\lambda' I_{0\lambda} \left(1 - \frac{\lambda}{\lambda_L}\right) e^{-\alpha_\lambda' v}\, d\lambda \tag{3.2}$$

where c_1 and c_2 are constants, λ_1 and λ_2 the wavelength limits, λ_L the laser wavelength, $I_{0\lambda}$ the spectral emittance of the flashlamp, v the optical thickness (weight percent centimeter), and α_λ' the absorptivity per weight percent [(weight percent centimeter)$^{-1}$]. To generate such curves, computer programs have been developed such as GENEFF at Lawrence Livermore Laboratory and FLASH-STORE at the Laboratory for Laser Energetics which numerically integrate Eqs. (3.1) and (3.2). The α_λ' are obtained by digitizing absorption spectra such as those shown previously, while the spectral emittance $I_{0\lambda}$ is computer generated. In Fig. 8 we show curves obtained by this method for typical commercially available silicate (ED-2) and phosphate (EV-2, LHG-5) laser glasses. The predictions of such models have been tested experimentally recently[9] and found to be very accurate. The form of the inversion curves has also been found to be important in determining where the threshold for parasitic oscillations occurs in different glasses.[8]

In considering the utility of a laser glass for a given application, it has been shown that the specific gain coefficient is important in a pump-limited case, but not in a parasitic-limited one. Another quantity of interest is the ratio (n_2/n_0) of nonlinear and linear refractive index. In lasers in which high intensities ($\gtrsim 5 \times 10^9$ W/cm^2) are propagated, intensity-dependent refractive index changes become important and vary according to

$$n = n_0 + \gamma I \tag{3.3}$$

where the nonlinear coefficient γ is related to n_2 by

$$\gamma = c(n_2/n_0) \tag{3.4}$$

---

[8] D. C. Brown, S. D. Jacobs, and N. Nee, *Appl. Opt.* **17**, 211 (1978).

[9] J. A. Abate, D. C. Brown, C. Cromer, S. D. Jacobs, J. Kelly, and J. Rinefierd, Laser-Induced Damage in Optical Materials. National Bureau of Standards Special Publ. 509, (December 1977).

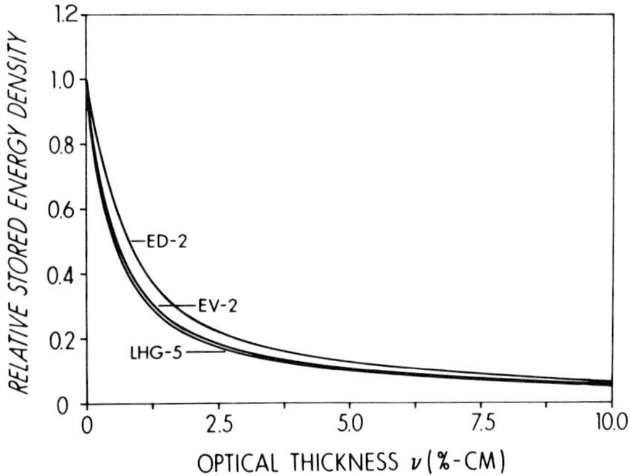

FIG. 8. Stored energy in three laser glasses.

Here c is a constant. The variation of refractive index with intensity has led to a variety of nonlinear effects which include small scale self-focusing and whole beam self-focusing.[10] These effects limit the achievable power from present solid state glass lasers, and can be minimized by minimizing the ratio (n_2/n_0). Recently through the work of colleagues at Lawrence Livermore Laboratory it has become possible to relate nonlinear index n_2 to the Abbe value (dispersion) and linear index at the sodium D lines (n_d) of the glass through the expression (3.5)

$$n_2(10^{-13}\,\text{esu}) = \frac{68(n_d - 1)(n_d^2 + 2)^2}{v_d\{1.517 + [(n_d^2 + 2)(n_d + 1)v_d]/6n_d\}^{1/2}} \tag{3.5}$$

Thus low index, high dispersion materials have been sought which minimize n_2/n_0, and in recent years phosphate, fluorophosphate, and fluoroberylate glasses have been widely investigated to achieve that end. Figures of merit for evaluating the optical properties of laser glasses for high power lasers were chosen based upon gain and nonlinear considerations and an equation

$$\eta = \alpha_0(n_2/n_0) \tag{3.6}$$

used for the pump-limited, and

$$\eta = (n_2/n_0) \tag{3.7}$$

for the parasitic-limited figures of merit (η). More recently, Brown *et al.*[5] have given a more general figure of merit which includes the optical properties, Eqs.

[10] Lawrence Livermore Laboratory, Laser Program Annual Rep. UCRL-50021-74 (March 1975).

(3.6) and (3.7), as a special case and also includes thermal, mechanical, and photoelastic effects. This general figure of merit is formally defined by

$$\eta = \prod_{i=1}^{n} (J_i)^{W_i} \tag{3.8}$$

where the individual J_i factors are defined for rod amplifier geometries and short pulses, and the W_i are weighting factors ($0 \leq W_i \leq 1$) which enable one to either emphasize or de-emphasize a particular attribute for a given application.

The individual J factors that have been identified as being important are as follows.

J factor	Origin
$J_1 = n/n_2$	(minimize B integral)
$J_2 = \alpha_0$	(maximize specific gain)
$J_3 = (E/H)_0$	(maximize inversion/heat)
$J_4 = \eta_0$	(maximize quantum efficiency)
$J_5 = k/\rho c$	(minimize thermal decay time constant)
$J_6 = \rho c$	(minimize induced temperature change)
$J_7 = \lvert \Delta P_r + \Delta P_\theta \rvert^{-1}$	(minimize average change in optical pathlength)
$J_8 = \lvert \Delta P_r - \Delta P_\theta \rvert^{-1}$	(minimize stress-induced birefringence)

An abbreviated description of the significance of each factor is given below.

J_1: The ratio of glass refractive index n to the nonlinear refractive index n_2 should be maximized to minimize the accumulation of B integral in a glass laser system.

J_2: It is of interest to maximize the gain per unit length in a laser glass. Small signal and saturated gain considerations lead to a desire to maximize the specific gain coefficient α_0.

J_3: This is the ratio of flashlamp pump energy converted to useful inversion, E, over that converted into unwanted heating of the laser glass, H, evaluated for zero optical thickness.

J_4: To maximize the obtained stored energy density E, one would like to maximize the quantum efficiency η_0 (assumed wavelength independent).

J_5: For a rod geometry and an assumed parabolic temperature profile (center to edge), a characteristic thermal decay time constant can be derived which gives a measure of the time required for a glass to dissipate absorbed heat. To minimize this thermal decay time constant we want to maximize the glass thermal diffusivity, defined as thermal conductivity k, divided by the product of glass density ρ, and specific heat c.

J_6: One wants to minimize the induced temperature change in a volume of glass due to heat input H by maximizing the product of glass density and specific heat.

J_7: In the absence of stress, the change in optical path length in a rod of length L for a given uniform temperature increase ΔT of the glass is given as

$$\Delta P' = L[\beta + \alpha(n - 1)]\,\Delta T \tag{3.9}$$

where $\beta = dn/dT$ or the temperature coefficient of refractive index and α is the linear coefficient of thermal expansion. The quantity in brackets is referred to as the thermo-optic coefficient. In the presence of stress caused by an assumed radial temperature gradient, one can derive a comparable expression in terms of the change in optical path length per unit length per unit temperature for light polarized radially (ΔP_r) or tangentically (ΔP_θ) within a glass rod. The average of the optical pathlength change for r and θ polarization with thermally induced strain present is

$$\Delta P_r + \Delta P_\theta = 2\beta + [\alpha Y/(1 - p)](3B_\perp + B_\parallel) \tag{3.10}$$

where Y is Young's modulus, p is Poisson's ratio, $B_\perp$ and $B_\parallel$ are photoelastic constants for light polarized perpendicular or parallel to the direction of internal strain. One wants to minimize the amount of thermo-optic distortion, and thus J_7 is expressed as the inverse Eq. (3.10).

J_8: The difference between ΔP_r and ΔP_θ can be shown to give

$$\Delta P_r - \Delta P_\theta = \frac{\alpha Y}{(1 - p)}(B_\perp - B_\parallel) = \frac{\alpha Y B}{(1 - p)} \tag{3.11}$$

where B is the stress–optic coefficient. Thus J_8 is related to the susceptibility of the glass to stress-induced birefringence, being large for glasses possessing a small stress–optic coefficient.

It is, in general, difficult to specify the weighting factors W_i. There are, however, a number of limiting cases where certain W_i may be specified. If only optical effects are considered to be important, then by setting $W_3 = W_4 = W_5 = W_6 = W_7 = W_8 = 0$ and $W_1 = W_2 = 1$, the FOM reduces to $n\alpha_0/n_2$, the familiar pump-limited case. Likewise, one can obtain the parasitic-limited FOM, $\eta = n/n_2$, by setting $W_2 = 0$.

Tabulations of physical properties data for silicate (ED-2) and phosphate (EV-2, LHG-5, LHG-7, Q-88) glasses have recently been given,[5] and are reproduced in Table I.

The individual J factor components for the FOM, calculated from the data in Table I, are given in Table II. Results are normalized relative to the silicate laser glass ED-2. The last column gives an average component FOM for the four phosphates evaluated. Based on a comparison of these averages with ED-2 one concludes the following.

J_1, J_2: Clearly the phosphates offer higher gain/lower n_2 performance advantages over the silicates.

J_8: Thermally induced birefringence is less severe for phosphates versus silicates.

TABLE I
GLASS PHYSICAL PROPERTIES DATA

Property	ED-2	EV-2	LHG-5	LHG-7	Q-88
Refractive index, $n_{1.06\,\mu m}$	1.5554	1.5037	1.5291	1.5050	1.5310
Nonlinear index, $n_2(10^{-13}$ esu)	1.4	1.0	1.2	1.0	1.2
Specific gain, α_0 (cm^2/J)	0.16	0.25	0.21	0.20	0.23
Conductivity, k(cal/cm sec °C(10^{-3})	3.00	1.08	1.75	1.46	1.76
Density, ρ (gm/cc)	2.51	2.72	2.64	2.60	2.67
Specific heat, c (cal/gm °C)	0.22	0.15	0.17	0.17	0.21
Expansion, α (10^{-7}/°C)	77	141	77	91	90
Temp. coefficient refractive index (10^{-7}/°C)	31	−100	−2	−26	−16
Thermo-optic coefficient, $\beta + \alpha(n - 1)(10^{-7}/°C)$	74	−28	39	20	32
Young's modulus, Y (kg/mm^2)	9620	4258	7258	6049	7123
Poisson's ratio, p	0.24	0.24	0.19	0.24	0.24
Stress–optic coefficient, B (nm/cm/kg/cm^2)	1.92	1.45	2.17	2.14	2.07
Photoelastic constant, $B_\perp$ (nm/cm/kg/cm^2)	2.6	4.9	3.5	3.5	2.9
Photoelastic constant, $B_\parallel$ (nm/cm/kg/cm^2)	−0.2	3.1	2.2	2.0	0.0

TABLE II
COMPONENT FIGURES OF MERIT

J factor	Glass: silicate ED-2	Phosphate EV-2	LHG-5	LHG-7	Q-88	Phosphate average
n/n_2, J_1	1	1.4	1.2	1.4	1.2	1.3
α_0, J_2	1	1.6	1.3	1.3	1.4	1.4
$(E/H)_0$, J_3	1	1	1	1	1	1
η_0, J_4	1	1	1	1	1	1
$k/\rho c$, J_5	1	0.5	0.7	0.6	0.6	0.6
ρc, J_6	1	0.7	0.8	0.8	1.0	0.8
$(\Delta P_r + \Delta P_\theta)^{-1}$, J_7	1	−2.3	1.6	3.5	3.3	2.7
$\lvert\Delta P_r - \Delta P_\theta\rvert^{-1}$, J_8	1	1.6	1.3	1.2	1.1	1.3

J_7: Thermally induced wave front distortion is much smaller for phosphates than silicates. This may be attributed to the negative value of temperature coefficient of refractive index for these glasses. In fact, dn/dT is so negative for EV-2 that one can predict convergence for a beam propagating through an amplifier rod of this composition (18) versus predicted divergence for all other glasses in Table II.

J_5, J_6: The one disadvantage for phosphates is their tendency to heat to a higher level under flashlamp pumping, and their diminished capacity to rapidly dissipate heat deposited in them. In practice, however, these somewhat unfavorable thermal properties do not prevent 90 mm diameter by 370 mm long laser rods from being fired in amplifier heads at a repetition rate of twice per hour in the Glass Development Laser (GDL) system at the Laboratory for Laser Energetics (LLE).

Recently, a voluminous amount of information covering the optical properties of silicate, phosphate, fluorophosphate, and fluoroberylate glasses has become available from Lawrence Livermore Laboratory.[6] A similar comparison of the thermal, mechanical, and photoelastic properties of the fluorophosphate and fluoroberylate composition is still lacking.

The fluorescence lifetime of the ($^4F_{3/2} \rightarrow {}^4I_{11/2}$) transition at 1.06 μm in Nd^{3+} is of great interest because it can affect the achievable energy storage in a laser amplifier. Before the advent of the fluorescence line narrowing (FLN) techniques[7] to investigate site-to-site variations in the Nd^{3+} environment, it had been noticed that the decay is in fact nonexponential,[10] even for very low dopings where ion–ion interactions are absent. It had been conjectured that such nonexponential decay was due to the site-to-site variations in the radiative decay rate. FLN experiments have demonstrated that if a particular spectrally narrow subset of the available sites is probed, purely exponential decay results.[7]

At higher dopings, ion–ion interactions result which reduce the decay rate; the phenomenon is referred to as "concentration quenching" and is present to some extent in all glasses. Recently, Nd pentaphosphate crystals have been discovered[11] in which concentration quenching is to a larger degree absent up to large weight percent dopings. Unlike glass in which Nd^{3+} is an interstitial, in pentaphosphate crystals it is part of the lattice, and shielding prevents ion–ion interactions from becoming very important. An excellent discussion of relaxation in laser glasses and its importance in laser design may be found in the thesis by Layne.[1]

IV. SOLID STATE DEVICES

There is not room in a chapter such as this to discuss the operation and design of a large number of solid state laser devices. That is particularly true of oscillators which may operate in any number of distinct modes (cw, normal mode, *Q*-switched, mode-locked). For that, the reader must consult another reference[2] which details the theory of such devices or discusses specific experimental techniques. Here, we will discuss three separate types of laser amplifiers which have found common use in solid state laser systems, summarizing their salient features, operating characteristics, advantages, and disadvantages.

A. Rod Amplifiers

The first type of amplifier we investigate is the rod amplifier, shown schematically in Fig. 9. It consists of a cylinder of Nd:glass or other material, surrounded by a water cooling jacket and an array of Xe flashlamps which are normally water cooled. The end faces of the rod may be antireflection (AR) coated, or wedged at a small angle to avoid back reflections in a large system. Such amplifiers have been operated routinely up to 90 mm in diameter; various Nd:glass laser systems at the Laboratory for Laser Energetics at the University of Rochester include amplifiers of diameter 16, 30, 40, 64, and 90 mm. The design of rod amplifiers includes a consideration of the parasitic oscillator problem, the doping of the glass, and the size and number of flashlamps used. Rod amplifiers, due to the absorption of the flashlamp radiation, display radial stored energy density or gain profiles which significantly affect the propagation of pulses through large laser systems.

Rod amplifiers possess a number of advantages, among them simplicity of design, the propagation of pulses with circular polarization which leads to a

---

[11] H. G. Danilmeir and H. P. Weber, *IEEE J. Quantum Electron.* **QE-8**, 805 (1972).

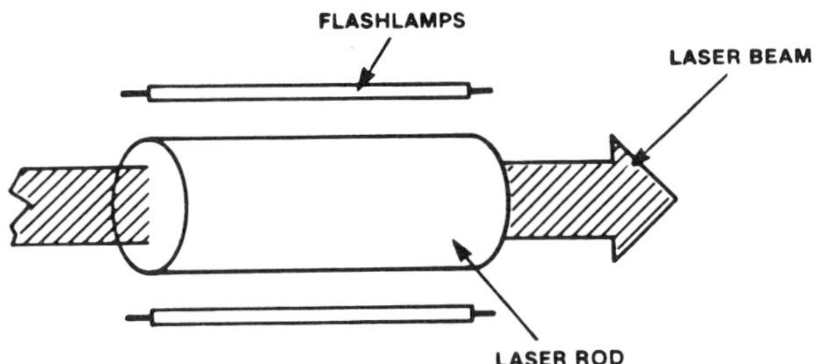

FIG. 9. Rod amplifier.

lower nonlinear index,[12] short thermal cycling times typically of the order of 20–30 min for large diameters, and low cost. Among the disadvantages are the large volume of optical quality glass required, the difficulty in achieving reasonable stored energy density at large diameters, and the necessity of having unpumped ends on the amplifier for holding which reduces the ultimate power handling capability of the device.[10] A rod amplifier system (GDL) has recently been described which produces pulses in excess of 750 GW from a 90-mm diameter.[13]

B. DISK AMPLIFIERS

A commonly used alternative to the rod amplifier is the disk amplifier, shown in Fig. 10. This device, first demonstrated by the General Electric Company, has been intensively developed at the Lawrence Livermore Laboratory for a number of years. It consists of disks mounted at Brewster's angle which are pumped by an array of Xe flashlamps wrapped in a circle around the cavity. Devices have been operated up to 30 cm in diameter; the design has been discussed in detail[10,14] and involves, like rod amplifiers, consideration of disk doping, size and number of flashlamps, etc. A particularly serious problem for disk amplifiers is parasitic oscillations and has led to the development of solid and liquid edge claddings to suppress edge reflections.[15–17] The onset of parasitic

[12] M. J. Weber, D. Milam, and W. L. Smith, Lawrence Livermore Laboratory preprint UCRL-80779 (February 1978).

[13] J. Soures, W. Seka, D. C. Brown, S. Kumpan, and J. Bunkenburg, Topical Meeting on Inertial Confinement Fusion, San Diego, California, paper TuB7 (February 1978). [Abstract in *J. Opt. Soc. Am.* **68**, 542, (1978)].

[14] Lawrence Livermore Laboratory, Laser Program Annual Rep. UCRL-50021-75 (March 1976).

[15] G. Dubé and N. L. Boling, *Appl. Opt.* **13**, 699 (1974).

[16] J. A. Glaze, S. Guch, and J. B. Trenholme, *Appl. Opt.* **13**, 2808 (1974).

[17] R. B. Bennett, K. R. Shillito, and G. L. Linford, Laser-Induced Damage in Optical Materials, National Bureau of Standards Special Publ. 509 (December 1977).

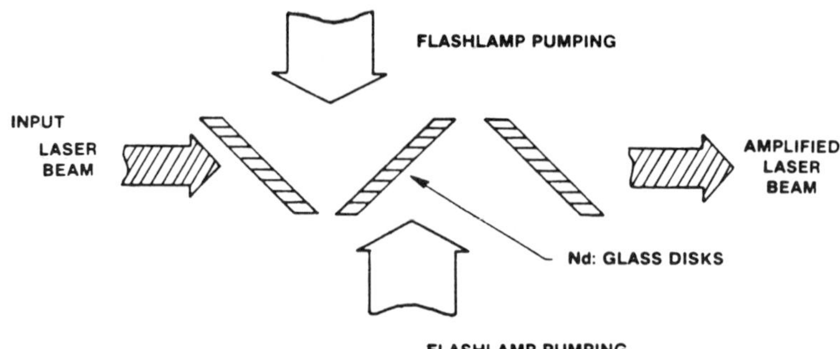

FIG. 10. Brewster disk amplifier.

oscillations leads to a clamping of the gain; further pumping results in no increase in gain. It has been shown that the type of glass composition used also affects the onset of parasitic oscillations.[8] Disk amplifiers fitted with silicate glass have achieved large gain/length and have an excellent power handling capability. The Argus system at Lawrence Livermore Laboratory has produced pulses in excess of 2 TW/beamline.[18] Disk amplifiers also have a 2× area advantage over rod amplifiers due to the Brewster configuration, leading to an energy handling capability which is higher by a factor of 2 than rods. The disadvantages of disk amplifiers are that they are complex and expensive compared to rod amplifiers, cannot propagate circular polarization, have long thermal recycling times (typically $\gtrsim 2$ hr), and can lead to astigmatism on the beam.

C. ACTIVE-MIRROR AMPLIFIERS

The active-mirror amplifier has recently been developed into a reliable, medium-repetition-rate amplifier for large laser systems.[19] and is shown in Fig. 11.

It consists of a laser disk which is antireflection coated on one side at 1.06 μm, while the other uses a special multilayer dielectric coating which transmits the flashlamp radiation in the region of the Nd^{+3} pump bands, but is also a high reflector (HR) for 1.06 μm light. The disk is pumped by an array of Xe flashlamps; a cooling solution, usually ethylene glycol and water, is circulated across the face near the lamps for cooling. At the disk edge, glass beads are used which strongly absorb 1.06 μm light and at the same time support the disk. The cooling liquid is also flowed around the entire edge of the disk for cooling, and also index matches the disk to the glass beads. The absorbing glass beads and index

[18] W. W. Simmons, D. R. Speck, and J. T. Hunt, *Appl. Opt.* **17**, 999 (1978).

[19] O. Lewis *et al.*, *Topical Meeting on Inertial Confinement Fusion, San Diego, California* paper TuBll (February 1978) [Abstract in *J. Opt. Soc. Am.* **68**, 542 (1978)].

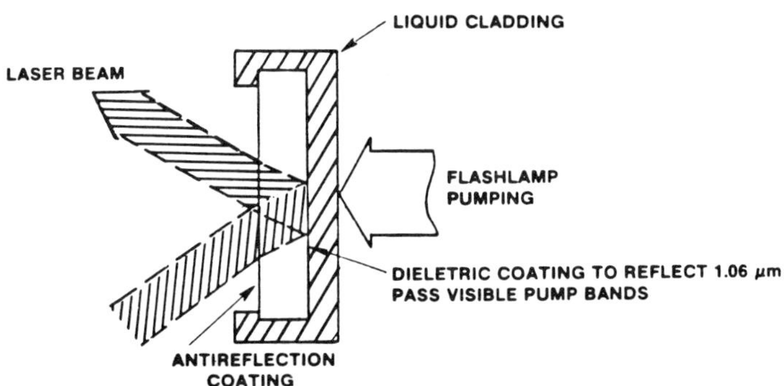

FIG. 11. Active mirror amplifier.

matching are necessary to avoid parasitic oscillations. The front and rear faces are placed at a wedge angle with respect to one another to avoid back reflections being propagated in a system. The edge is beveled to geometrically discriminate against parasitic oscillations. A beam incident on the mirror is supplied in the small signal regime with a gain coefficient twice that of a single disk due to the double pass.

Active mirrors possess a number of advantages over disk amplifiers; in particular the firing rate is once every fifteen minutes, the storage efficiency is approximately equal to a factor of 2 times that of a disk, greater energy extraction results from the double pass, and it is able to propagate circular polarization. Coating damage has not been a problem because the most susceptible one near the flashlamp array is actively cooled. Active mirror devices are considerably more complicated than rods because of the required coating technology and the mechanical problems associated with mounting the disk in a stress-free manner. During the pump pulse thermal distortion is apparent which gives the mirror a finite curvature due to the preferential heating of the disk face nearest the flashlamp array. In a large laser system this leads to a change in the focal distance in the final focusing lens; it has been found, however, that it is easily possible to compensate for the shift since its magnitude is reproducible from shot to shot.

D. EFFICIENT, HIGH-REPETITION-RATE DEVICES

As can be surmized from the above, most large laser systems have low repetition rates, once per half hour for rods and active mirrors, and once $\simeq 2$ hr for disk amplifiers. The principal reasons for this are the small value of the thermal conductivity for glass and the large amount of heat generated during the pump pulse. Heat is deposited into the glass by background absorption of the ultraviolet output of Xe flashlamps as well as the multiphonon emission in Nd:glass and other hosts. The difference (quantum defect) between an absorbed pump photon and the laser photon (1.06 μm) is converted directly into heat. Due to the fact that a Xe flashlamp emits broadband, Nd^{+3} has a finite number of absorption bands, and the quantum defect, low overall efficiency results.

For fairly small amplifier and oscillator sizes, a repetition rate of 1–10 Hz is not uncommon. For oscillators, the large thermal conductivity of Nd:YAG is usually preferred and 10-Hz systems common.[20] From a laser fusion standpoint, a glass laser system with a repetition rate of 10 Hz and overall efficiency of $\gtrsim 10\%$ is desirable.[21] Although we are now at least a factor of 10 off the efficiency value for most amplifiers, there have been a number of schemes to improve the situation. The first is the use of semiconductor converters[22] or organic dyes as wavelenth shifters[23] to provide a better spectral match between the Xe flashlamp and Nd:glass. Analysis has shown,[24] however, that such schemes are unlikely to produce an efficient glass laser. Another method involves the use of light-emitting diodes (LEDs) or injection lasers to pump the Nd^{3+} band at 0.81 or 0.88 μm. LEDs are intrinsically very efficient devices, can easily be made to overlap a Nd^{3+} band, and minimize the quantum defect or heat deposition in the glass. Early analysis has shown the technique to be cost prohibitive[25] but it is likely that the situation will change favorably. An analysis of the system shows that overall efficiencies of 10–15% are possible.

Recently, excimer pumping of rare-earth vapors has been considered by Jacobs *et al.*[26]; the use of excimer laser pumps is indicated because of the high potential efficiency of such lasers, and their general unsuitability for use in a fusion laser system. Wilson and Brown[27] have pumped $Nd{:}P_5O_{14}$ in an oscillator configuration using a XeF laser. It is possible that efficiencies $\gtrsim 3\%$ may result by choosing a different rare earth which minimizes the quantum defect.

It is likely that the major advances in solid state lasers to be made in the next few years will be in the areas of efficiency and minimization of thermal effects, leading to a high repetition rate.

---

[20] R. M. Fogan and T. G. Crow, *Appl. Opt.* **17**, 927 (1978).

[21] J. Wilson and D. Ham, *Laser Focus* **12**, 38 (1976).

[22] Z. I. Alferov *et al.*, *Sov. J. Quantum Electron.* **6**, 734 (1976).

[23] W. W. Morey, General Electric Technical Rep. ECOM-0055-F (August 1972). Available from NTIS, Arlington, Virginia.

[24] D. C. Brown, unpublished results.

[25] S. Warshaw, General Electric TIS Rep. 77CRC098 (May 1977).

[26] R. R. Jacobs and W. F. Krupke, *Appl. Phys. Lett.* **32**, 31 (1978).

[27] J. Wilson and D. C. Brown, *Appl. Phys. Lett.* **33**, 614 (1978).

V. APPLICATIONS

Progress in solid state lasers has been rapid in recent years due to the driving force of laser-induced fusion. The applications of such lasers has proliferated and range from laser welders to sophisticated military devices. In this section, an attempt is made to familiarize the reader with at least two current areas of interest. They are laser fusion systems and systems to be used in current and planned NASA missions.

A. High Peak Power Nd : Glass Systems

We have discussed in Section III,D some of the properties of Nd:glass materials which are important in various applications. In short pulse (50–150 ps) high-peak-power lasers, by far the most important mechanism limiting output is the nonlinear index of refraction n_2, which leads to small-scale self-focusing and various related nonlinear effects. To see how such effects are controlled, we refer the reader to Fig. 12, where we show a schematic design of a typical laser system. A mode-locked oscillator injects a pulse into the system which is approximately Gaussian temporally and spatially. An apodizer is normally placed before the first amplifier; its function is to provide a "soft" cutoff at the edges of the beam to minimize the propagation of diffraction rings through the system components. Spatial filters, which consist of a lens pair and a pinhole which is placed in a vacuum, are used at every stage in the system to remove the most damaging spatial frequencies. Small-scale amplitude and phase perturbations, if left uncurtailed, would also self-focus and cause irreversible damage to the optics and laser glass. Since noise is present to some degree in every system, arising from such sources as dust and scratches on surface, air turbulence, etc., spatial filters are necessary and force the growth of the noise to re-exponentiate at strategic locations. The whole beam phase, or B integral, defined by

$$B = \frac{2\pi}{\lambda} \int_0^L \frac{n_2}{n_0} I(z)\, dz \tag{5.1}$$

has become a standard way of analyzing the nonlinear design of high-peak-power laser systems. It can be shown that the maximum spatial frequency growth rate proceeds according to $\exp(B)$,[10] hence it is of interest in minimizing the value of B in any system. Whole beam phase effects, such as zooming in spatial filters, are also alleviated by reducing B. The tolerable phase accumulation ΔB between the location of spatial filters determines the staging of the system; the exact value of ΔB is determined by a consideration of the magnitude and spectrum of the noise available in the stage. Spatial filters also serve as beam expanders at strategic locations and as devices for implementing "imaging" through the entire system. It has been shown recently that imaging leads to

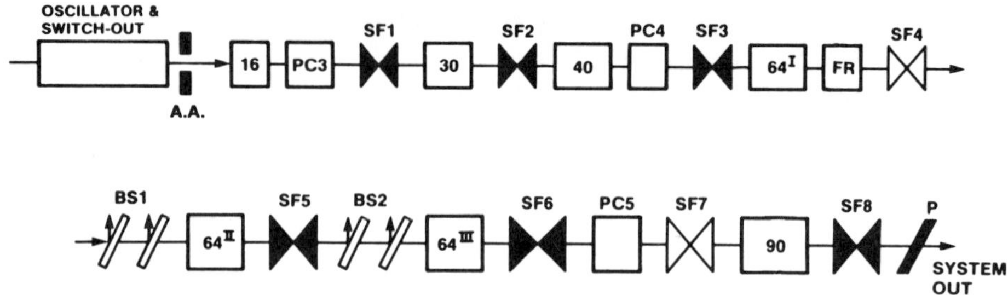

FIG. 12. A typical glass laser system.

some alleviation of the problems of diffraction and whole beam self-focusing.[18] By keeping a beam constantly in the near field, it is much easier to control the radial beam profile, and thus achieve a large fill factor in the final stage; power output is proportional to the fill factor available there.[10] In the imaging scheme, the initial apodizer is imaged through the entire system to the location of the target chamber. Results on the ARGUS system at Lawrence Livermore Laboratory using partial imaging have appeared;[18] the GDL system at LLE has operated similarly.[13] In designing high-peak-power systems to be used for target experiments, it is necessary to take into account possible back reflection. Since only part of the available energy has been extracted by a pulse, a large net gain still exists which can amplify a back reflection to the extent where damage can occur to optical components. As shown in Fig. 12, Faraday rotators and Pockels cells are normally used to provide sufficient isolation. Pockels cells are used exclusively in some systems to isolate the target from amplified spontaneous emission (ASE) which often destroys targets before the arrival of the main pulse. In regions of the system where circular polarization may be used to reduce the nonlinear index, antireflection-coated quarter wave plates convert linear to circular polarization, and circular to linear where necessary.

The development of new laser materials, the use of imaging and spatial filters, and increased understanding of the physics of high-peak-power glass lasers has led to impressive performance from such devices. ARGUS has produced over 2 TW/beamline from a 20-cm clear aperture[18] and the GDL system at LLE 0.75 GW from a 9 cm one.[13] We may confidently expect further significant advances in the near future.

B. Scientific Applications—NASA

Laser radar (LIDAR) systems are commonly used today to perform a variety of atomospheric measurements of pollutants and physical quantities important in understanding its dynamics. The evaluation of LIDAR systems and the power of the approach have led NASA to consider the inclusion of a versatile LIDAR package aboard Space Shuttle/Skylab flights in the 1980s. Indeed, a Request for Proposal was issued in March, 1978 for a system definition study. Of importance here is that the LIDAR system initially will include the use of a solid state YAG laser. Indeed, the YAG laser is the only one that seems capable at present of meeting all of the system requirements, which follow.[28]

Nd:YAG LIDAR Requirements

Energy	1 J/pulse
Pulsewidth	15 nsec
Pulse repetition rate	10 Hz
Linewidth	<1 Å
Mode	TEM_{00}
Amplitude stability	5% p–p

Due to the relatively high thermal conductivity and gain of YAG, it is not difficult to implement a system which operates at 10 Hz; to obtain 1 J/pulse one must use at least two amplifiers in addition to the oscillator. Such systems are now fairly common, one being described recently in the literature[20] while another is available commercially from Quanta-Ray. An important feature of the LIDAR system is the ability to frequency double, triple, and quadruple using various nonlinear crystals. Of the 26 proposed experiments for the system, most may be performed with the Nd:YAG system or a derivation thereof. The attractiveness of the basic Nd:YAG laser is enhanced if the capability to pump a tunable dye laser is also utilized. In such a system, for example, the frequency doubled Nd:YAG output at 5300 Å pumps a Hänsch-type dye laser as shown in Fig. 13. The frequency doubled beam is focused into a line on a dye cell where dye concentration is adjusted to approximately match the width of the line. A beam expander and grating allow spectral narrowing; further reduction in bandwidth may be provided by using an intracavity etalon. Linewidths down to $\simeq 10^{-3}$ Å are required for some measurements, and may be achieved using this scheme. The versatility and flexibility of this system are made possible by the high energy output, typically good beam quality leading to respectable harmonic conversion efficiencies, available repetition rate, and well developed technology.

Typical measurements will likely include the distribution of atmospheric particulates, ozone profile in the stratosphere, wind speed and direction in the troposphere and stratosphere, and the measurement of Na, Mg^+, OH, H_2O,

[28] NASA Request for Proposal 1-106-8000-3018, Atmospheric Lidar Multi-User Instrument System Identification Study (April 1978).

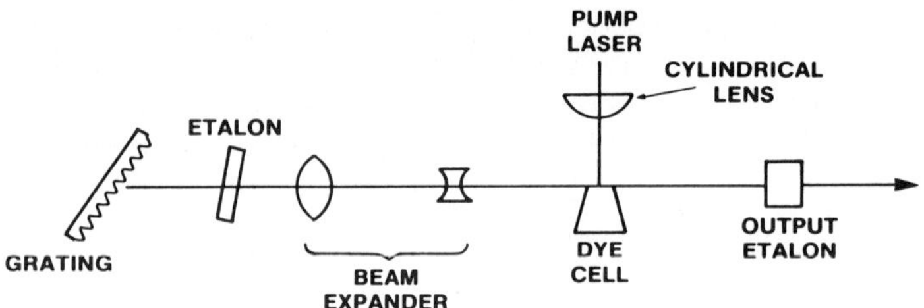

FIG. 13. Dye laser pumped by a Nd:YAG laser.

NO_2, and other species in various parts of the atmosphere. We may expect that a successful mission using this and other systems will add to our detailed understanding of the atmosphere as well as lead to improved modeling and predictive capabilities.

CHAPTER 2

Gas Lasers

J. M. FORSYTH and J. WILSON

Laboratory for Laser Energetics, University of Rochester, Rochester, New York

I. INTRODUCTION: GENERAL CHARACTERISTICS OF GAS LASERS

The optical engineer considering the use of a laser in some application will want to know whether there exists a laser with the power level required, at the right frequency and divergence, and whether it will prove cost effective. There is almost certain to be a gas laser able to fit the bill, since there exists an amazing range of gas lasers. For example, cw power levels range from a milliwatt for visible helium–neon lasers, to about 100 kW from infrared carbon dioxide lasers. The only area in which gas lasers do not excel is in wavelength tunability, for which one must turn to dye lasers. This chapter will describe the properties of gas lasers in order to give the optical engineer the basis on which to make his selection. Gas lasers possess high spectral purity, high power capability, and high efficiency, though not necessarily all three combined.

The output power, efficiency, and lasing wavelength of a gas laser may be attributed primarily to the physical conditions and constituents of the lasing gas. On the other hand, the spectral and spatial purity of the output beam are dominated by the properties of the mirror structure which supplies optical

ISBN 0-12-408606-3

feedback to the laser, i.e., by the optical cavity. The gaseous media in lasers are considerably more transparent near the laser frequency than are liquids or solids; they exhibit an index of refraction close to unity. Thus the optical properties of the feedback system tend to be far less influenced by conditions of uneven gas heating, uneven or turbulent flow, or other effects compared to liquid or solid lasers. This results in a capability for achieving high spectral and spatial purity from gas lasers, a quality which we will now address.

A. High Spectral Purity

The output from a gas laser is usually composed of several discrete frequency components, each component having a very high degree of monochromaticity. The ultimate in spectral purity is achieved when the laser is designed to be operated on one component to give a single frequency output. In a given laser system this requires carefully tailoring the characteristics of the optical cavity to the frequency and bandwidth of the amplifying transition. After single frequency operation is obtained, further increases in spectral purity may be obtained by rigorous control of the mechanical and thermal stability of the optical cavity.

Each step in engineering a gas laser to give very high spectral purity increases the complexity and cost of the laser. In some commerical gas lasers the degree of spectral purity may be controlled by the customer (to some degree) by specifying certain accessories for the laser. Some of these accessories must be specified at the time of purchase—they cannot be added on later. Therefore, the optical engineer must be able to understand and specify the spectral purity required in his application before selecting a suitable gas laser. High spectral purity is often required in applications such as interferometry, holography, or optical mixing, while it is unimportant in such applications as materials working.

In interferometry and holography it is often convenient to discuss the properties of a light source by the concept of the "coherence length" since this concept translates directly into the maximum usable optical path difference in the application. However, this concept must be applied with some caution to the output from a gas laser when the laser produces a number of discrete frequency components. If the output from such a laser is analyzed in a Michelson interferometer, it is found that the interference fringes disappear when the path difference in the two arms of the interferometer is equal to the length of the laser cavity. Now, if the path difference is further increased, the interference fringes return. This variation in fringe visibility with path difference may be repeated, at integral multiples of the laser cavity length, out to very large path differences. A practical ambiguity in the operational definition of coherence length has developed in this case; it is usually resolved by taking the smallest path difference for which the fringes disappear—i.e., the laser cavity length.

For the optical engineer with moderate requirements for coherence length, it may be adequate to select a laser with a cavity which is slightly longer than the required coherence length (provided that this length yields adequate output power, of course).

In more critical applications, such as long path length interferometry or optical heterodyning, it is usually necessary to specify single frequency operation of the laser. Some gas lasers, such as the argon ion laser, are readily operated in a single frequency configuration. Such lasers tend to exhibit a broadening of the laser transition which is dominated by collisions of the gas or plasma particles. Other lasers, such as the helium–neon laser, are difficult to coax into single frequency operation. The laser transition in these devices tends to be Doppler broadened.

Since we have asserted that the spectral and spatial characteristics of gas lasers are dominated by the properties of the optical cavity, we should briefly explain this. The explanation is most conveniently made using the concept of modes.

A mode of a laser cavity is a stationary configuration of the electromagnetic field which satisfies the boundary conditions imposed by the mirrors. Field configurations which do not satisfy these boundary conditions are not retained for long within the cavity, but are quickly lost due to absorption, diffraction, etc. Mode fields, on the other hand, may be retained for a sufficient time within the cavity to interact strongly with an amplifying medium placed within the cavity. Cavity oscillation fields always exhibit the characteristic mode structure imposed by the cavity mirror geometry.

An optical resonator of reasonable dimensions will be very much larger than a wavelength. An open-sided structure, consisting of two opposed mirrors, can support oscillation in modes with low diffraction loss. Because of the relationship between the resonator dimensions and the wavelength, a microwave cavity is often excited on the so-called fundamental mode, in which a half wavelength of the electromagnetic field fits exactly inside the structure. By contrast, an optical cavity is inevitably excited on a very high overtone. Many of the overtone resonances typically fall within the bandwidth of a gas laser amplifier (Fig. 1). By resorting to a structure with open sides, the losses for most of these overtone resonances can be made so high that oscillation will be possible on only a few modes. Nevertheless, gas lasers which employ a single pair of mirrors for feedback are multimode devices—the output exhibits several frequency components.

To restrict laser oscillations to a single frequency, additional boundary conditions on the mode structure are usually imposed by placing sharp spectral filters such as etalons inside the cavity or, in stubborn cases, resorting to multiple mirror reflectors which resemble one or more of the familiar optical interferometers. It is always better to control the gas laser frequency spectrum from within the laser cavity than to try to filter it externally because of the

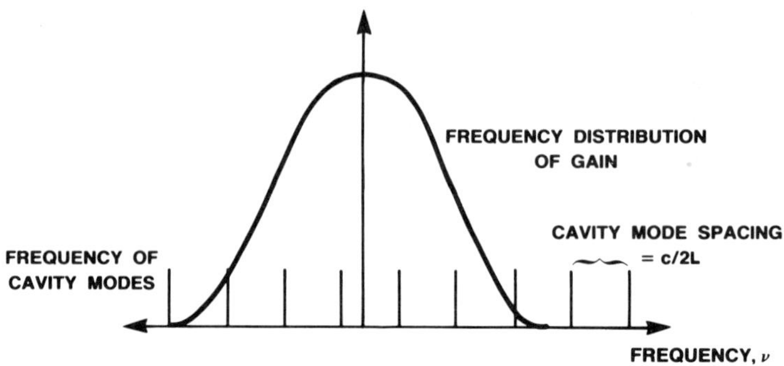

FIG. 1. Possible distribution of cavity modes and gain distribution as a function of frequency, showing several modes with gain inside the bandwidth of the laser.

difficulty of matching the center of the filter passband to a particular mode frequency and because higher single-frequency power results if the laser amplifier medium interacts with a single cavity mode only.

In contrast to the multimode laser where the coherence length is effectively the length of the cavity, coherence lengths of tens of meters are readily available from single-frequency gas lasers of reasonable dimensions. With special efforts on the dimensional stability of the cavity the coherence length may be increased to tens of kilometers. Gas lasers have been employed as optical frequency standards, and the development work in this area continues.

B. HIGH SPATIAL PURITY

Specification of either the required or the actual spatial purity in a gas laser is often a difficult task. Terms such as "divergence," "focusability," and even "diffraction limited" may be misapplied or misunderstood even among knowledgable engineers. In most applications it is required to be able to pass a certain flux through an aperture of specified cross-sectional area using a given set of optical components. This will usually constitute a practical and unambiguous test of the suitability of a particular gas laser in a given application. Laser manufacturers, however, do not usually include such information in their technical data because of the endless variety of specific applications. Instead, the spatial mode structure of the laser output is often specified. In order to match the characteristics of a gas laser to the particular application, the optical engineer should be familiar with the propagation characteristics of fields derived from the modes of a laser cavity.

In order to minimize diffraction losses, the mirrors which compose a gas laser resonator are slightly curved. The curvatures are chosen so that a certain set of rays will be retained within a finite aperture even after any number of passes between the mirrors. The modes of such structures have been extensively analyzed and a large literature exists on their properties. We will attempt to highlight only the most important properties of these mode fields here.

In open-sided resonators where the mirror apertures are small compared to their separation, the mode fields are composed of oppositely directed (i.e., traveling wave) components which propagate between the mirrors essentially as they would in free space. Thus the fields are designated transverse electromagnetic, or TEM. We will use the indices (m, n, q) to identify the spatial field configuration of the modes in a rectangular coordinate system.

The index q measures the number of half wavelengths between the mirrors along the optical axis. The indices m and n measure variations in the standing wavefront structure transverse to the optical axis; in particular, the number of phase reversals which occur across the wave front. The complete designation of the mode would be written TEM_{mnq} but this is usually abbreviated to TEM_{mn} because the index q is usually a very large number whose precise value is not important. The terms "low order" or "high order" modes are commonly used with reference to the indices m and n only.

The transverse structure of a mode is found to be the same in all planes normal to the optical axis of the cavity, apart from an overall change of scale. Since the output of a laser consists of one of the traveling wave components of the mode field, we may readily appreciate that the characteristic structure of a laser beam will remain unaltered, apart from a change in scale, when processed by an ordinary optical system (provided the system aperture is large compared to the transverse extent of the beam). In most applications it is desirable for this structure to be as simple and uniform as possible. Therefore, it is normal practice in gas laser design to arrange for operation in the lowest order, TEM_{00} mode. In fact, it can be shown that while a given laser system and cavity will deliver more total power into higher order modes, when the output from such a laser is focused by a lens, the illumination in the focal plane (or any other plane) is independent of the order number of the mode.

Since the TEM_{00} mode has the smallest transverse spatial extent in a given cavity, it is normal practice to aperture the laser cavity to increase the losses for high order modes and favor oscillation on the TEM_{00} mode. Often the laser amplifier geometry will accomplish this directly with the appropriate cavity design.

The scalar amplitude distribution in the TEM_{00} mode in rectangular coordinates may be represented in any x–y plane (normal to the cavity axis) by the expression

$$E(x, y) = E_0 e^{-(x^2+y^2)/w^2} e^{i\omega t} \tag{1.1}$$

i.e., by a Gaussian amplitude distribution. The parameter w measures the radial distance from the cavity axis at which the amplitude has dropped to $1/e$ of its (maximum) on-axis value. Since the appearance of such a beam when it strikes a screen is that of a somewhat diffuse spot, the parameter w is sometimes called the "spot radius" or "spot size" of the beam. Its value depends on the geometry of the laser cavity and on the distance of the plane of observation from the cavity. In a focusing system there will be some point at which a minimum value w_0 is reached. This minimum spot size is called the "beam waist" of the system.

If we take the direction of propagation to be aligned with the z axis, then the variation of the spot size is described by

$$w^2(z) = w_0^2[1 + (\lambda z/\pi w_0^2)^2] \tag{1.2}$$

where we take the origin of coordinates to be at the beam waist (Fig. 2). This expression applies to the mode field inside the laser cavity as well as to the external field. Of course, each time the beam passes into an optical system with focusing properties one must find the new location of the beam waist before applying Eq. (1.2) in the system. We will show how to do this shortly.

Equation (1.2) describes one aspect of the divergence of a gas laser beam. With reference to the coordinate system established above, the radius of curvature of a wave front at some point z is

$$R(z) = z[1 + (\pi w_0^2/\lambda z)^2] \tag{1.3}$$

For a gas laser operating in the TEM_{00} mode, it is tempting to regard the output beam as a plane wave disturbance. Equations (1.2) and (1.3) tell us that this is wrong and that we must use some care in discussing the propagation of a laser beam through an optical system. To see an important difference between laser light and light from a classical or ideal "point source," we recall that if the point source were located at $z = 0$ then the radius of the emitted wave fronts at the point z would be $R(x) = z$. Equation (1.3) shows a more complicated behavior, especially in the vicinity of the beam waist, i.e., at $z = 0$.

If a gas laser beam falls on a thin lens of focal length f such that the radius of the incident wave front at the lens is R, then the emerging wave front will have a radius R' according to

$$1/R + 1/R' = 1/f \tag{1.4}$$

From our experience with ordinary optical systems we would expect to find the laser beam brought to a focus at a distance R' from the lens. However, Eqs. (1.2) and (1.3) tell us that the laser beam energy is most tightly concentrated

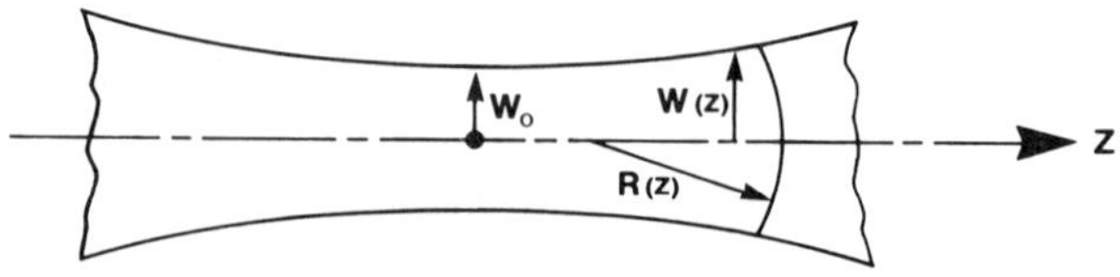

FIG. 2. Coordinate system of mode geometry.

in the plane of the beam waist where $R' \to \infty$, and that this point is not at a distance R' from the lens. If the spot size of the incident beam is w_a in the aperture of the lens, then we may invert Eqs. (1.2) and (1.3) to find that the beam waist is

$$w_0{}^2 = \frac{w_a{}^2}{1 + (\pi w_a{}^2/\lambda R')^2} \tag{1.5}$$

and that it will be found at the point

$$Z = \frac{R'}{1 + (\lambda R'/\pi w_a{}^2)^2} \tag{1.6}$$

from the lens.

Before elaborating on the difference between laser beam focusing and conventional geometrical optics, it is useful to introduce the notion of beam divergence. This is conveniently done by following the growth or shrinkage of the spot size as the beam propagates, using Eq. (1.2). With our point of measure beginning at the beam waist, the half angle θ of the beam can be defined by $\tan\theta = w(z)/z$. If the beam waist is substantially larger than the wavelength itself, then the divergence angle will be small. In this case an observer some distance from the laser will measure the (far field) half angle to be

$$\theta = \lambda/\pi w_0 \tag{1.7}$$

To see when Eq. (1.7) reduces to ordinary geometrical optics, namely, when $R' = z$, we may combine Eqs. (1.5) and (1.7) to give

$$\theta z/w_0 = \pi w_a{}^2/\lambda R' \tag{1.8}$$

Thus if

$$\theta \gg w_0/Z \tag{1.9}$$

Eq. (1.6) reduces to the classical result. It is sometimes said that geometrical optics, Eq. (1.9), applies to laser focusing problems in which the convergence or divergence angle is "large." More specifically, we require that the input spot size w_a be large compared to the beam waist w_0 for Eq. (1.9) to hold.

The effective spot radius of the higher order modes is larger than that of the TEM_{00} mode by a factor which is roughly the square root of the mode order number. Thus the volume of active medium which interacts with a high order mode is larger and, with a given cavity, more total output power is generated. However, we have already seen that this increased power does not result directly in increased illumination in the focal plane of an optical system. Consequently, a great deal of research has been performed in an attempt to efficiently employ the output from a laser having a relatively large cross section. We will briefly discuss the principal approaches to this problem.

In principle a high order mode field could be passed through a filter which would yield an output wave front of uniform phase, which would then have

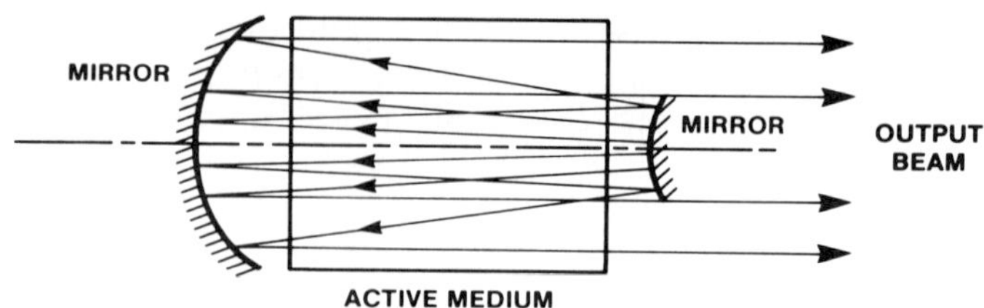

FIG. 3. Unstable cavity for achieving large mode volume.

improved focusing characteristics. The filter used would be specific to the mode field to be focused. Unfortunately, lasers do not operate stably in a particular high order mode configuration. This is because of the great similarity in the intensity distributions of high order modes of nearly identical order number. Even the most minute perturbations of the laser cavity or of the optical properties of the amplifier may cause the oscillation to shift from one mode to another. So far, it has not proved practical to build active compensation filters which can follow such "mode hopping" effectively, although research in this field continues at an active pace.

Another approach to efficient use of gas lasers of large cross-sectional area has been to employ unstable resonators, i.e., resonators whose geometry does not result in permanently containing a set of rays within a finite aperture. While such resonators are characterized by high diffraction losses, these "losses" may be taken in the form of directed output in some configurations. Figure 3 shows one arrangement of an unstable resonator which gives a parallel output beam of annular shape.

Unstable resonators can be specified for some models of commercially available lasers, particularly for carbon dioxide lasers. By using unstable resonators, much higher focusable powers can be obtained than would otherwise be possible from laser amplifiers of similar dimensions. For example, by changing the output mirror of a commercially available rare-gas halide laser to a lens, the cavity is converted to an unstable cavity, and the intensity at focus increases, although the total power is reduced.[1]

C. High Average Power

Gas lasers possess a capability for high average power due to the fact that a gas can be made to flow at a high rate. The average laser power available from any laser is limited by the ability of the laser medium to reject the pumping power that is not converted into laser power.

[1] D. L. Barker and T. R. Loree, *Appl. Opt.* **16**, 1792 (1977).

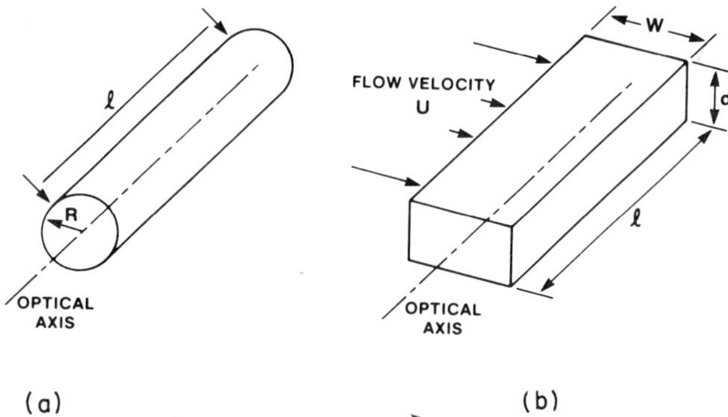

FIG. 4. Geometrics of (a) cylindrical laser without flow, (b) laser with flow

If P_{in}, P_{laser}, and P_{reject} are the input power, laser power, and rejected power, respectively, then

$$P_{in} = P_{laser} + P_{reject}$$

If $P_{laser} = \eta P_{in}$, thereby defining the laser efficiency, rearrangement gives

$$P_{laser} = [\eta/(1 - \eta)]P_{reject}$$

In a conventional laser which does not use flow, the rate of rejecting waste heat is limited by thermal conductivity. Given a cylindrical laser of length l and radius R (Fig. 4a), with a mean free path λ and average molecular velocity $\bar{c}$, the characteristic time for heat rejection is $\tau = R^2/\lambda\bar{c}$. If the gas has an allowable temperature rise ΔT, and density N molecules/cc, the rate of heat rejection is

$$P_{reject} = \tfrac{3}{2}(k\,\Delta T\lambda\bar{c}N/R^2)\pi R^2 l$$

in which k is Boltzmann's constant. Thus the laser power is

$$P_{laser} = [\eta/(1 - \eta)]\tfrac{3}{2}\pi k\,\Delta T\,\bar{c}(N\lambda)l$$

The quantity $N\lambda$ is independent of pressure, so we see that the maximum laser power is determined solely by the laser length, and not by the pressure or radius. Evaluating the expression, one finds that the power is of the order of 200 W/m. In practice, values around 100 W/m, i.e., of the same order, are found.

However, in a fluid the rejected heat can be removed by convection. The heat rejection time is the flow time through the cavity, i.e.,

$$\tau = W/U$$

(see Fig. 4b), in which W is the dimension of the cavity in the flow direction, and U is the flow velocity. For this case the rate of heat rejection becomes

$$P_{reject} = \tfrac{3}{2}(k\,\Delta T\,NWld)/(W/U)$$

and hence the laser power is

$$P_{laser} = [\eta/(1 - \eta)]\tfrac{3}{2}k\,\Delta T\,NU\,ld$$

Now, the laser power increases with flow velocity, gas pressure, and laser height as well as length. For this reason very high-power lasers employ a flowing active medium.

Not all gas lasers are suitable for convective heat removal. If we consider a gas laser transition which terminates on an excited state, then the steady output power is directly proportional to the rate of decay of the lower laser level. If the lower laser level lifetime is shorter than a practically realizable heat rejection time defined above, then convective flow will not increase the power output. For example, the lower laser level in the argon ion laser has a decay time shorter than 1 nsec. A plasma flow velocity approaching 10^9 cm/sec would be required to effect useful convection, too high a value to warrant practical consideration!

D. Efficiency

Lasers may operate on electronic, vibrational, or rotational transitions. Visible and near infrared lasers are generated only by electronic transitions. Infrared lasers (2–20 μm approximately) are generated most efficiently by vibrational transitions; far infrared lasers by rotational transitions. Only a few far infrared lasers are on the market, and they will not be discussed further here. Information on them may be found in a review article.[2]

The efficiency of a laser can be broken down into three parts, the pumping efficiency η_{pump}, the quantum efficiency η_{quantum}, and the extraction efficiency η_{extract}, such that

$$\eta_{\text{laser}} = \eta_{\text{pump}}\eta_{\text{quantum}}\eta_{\text{extract}}$$

The pumping efficiency is the efficiency with which the basic source of energy is converted into energy in the upper laser level of the atom or molecule forming the laser medium. The quantum efficiency is the ratio of the energy of a laser photon to the energy of the upper laser level. Of all the atoms or molecules that achieve excitation to the upper laser level, only a fraction are stimulated to give off laser light. This fraction is the extraction efficiency. In principle, at least, by clever design of the laser, the pumping efficiency and extraction efficiency could be made close to unity. However, the quantum efficiency is a property of the atom or molecule involved, and so represents an ultimate efficiency for a particular medium.

The energy level diagrams of three laser media are given in Fig. 5. Helium–neon was the first visible gaseous laser. The upper laser level in neon (3*S*) is at quite high energy (166,500 cm^{-1}) and the quantum efficiency is poor, 9.5%. The vibrational energy levels in carbon dioxide, Fig. 5b, are much better spaced, giving a quantum efficiency of 41%. As a result, carbon dioxide lasers can have

[2] T. Y. Chang, *IEEE Trans. Microwave Theory Technol.*, **MTT-22**, 983 (1974).

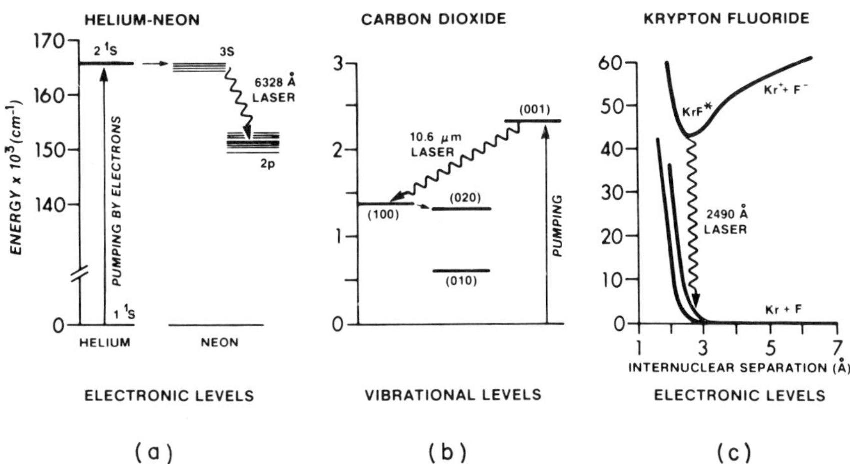

FIG. 5. The important energy levels in three laser systems. (a) Helium–neon, (b) carbon dioxide, and (c) krypton fluoride.

overall efficiencies as high as 20%. For some time it was thought that since electronic levels are typically high compared with the energy of visible photons, efficient visible lasers would be impossible. However, recently, laser action in rare-gas halides has been discovered. In these systems, Fig. 5c, the excited upper laser level is a stable excited state and the lower level is the ground state which is an unstable or very weakly bound state. The quantum efficiency is 50% for krypton fluoride. These systems promise to provide efficient visible gaseous lasers in the near future.

II. DESCRIPTIONS OF IMPORTANT LASER SYSTEMS

A. HELIUM–NEON

Helium–neon lasers operate at a wavelength of 6328 Å in the red region of the visible spectrum, and 1.15 μm and 3.3 μm in the infrared. They are available with between 1 and 10 mW of cw power output. The beam has good coherence properties and will easily demonstrate interference effects. The beam is about 1 mm in diameter, with a 1 mrad divergence. This makes the helium–neon laser an excellent alignment device. Helium–neon lasers can be cheap, costing about $100, or even less in OEM quantities. The medium is pumped by a dc wall-stabilized discharge. The units themselves are small, about 2 in. × 2 in. × 12 in. for the laser head, and simple, requiring only to be plugged into a wall outlet and switched on to operate. If the infrared line is desired, different mirrors are used, so this should be stated when ordering a unit. Helium–neon lasers are not efficient, 0.02% overall efficiency being typical.

B. Argon Ion

Argon ion lasers are low pressure discharge pumped devices. They may be pulsed or cw. Operation is at several wavelengths between $\gtrsim$3550 Å and 5285 Å. Argon ion lasers are available up to cw power levels of 40 W and pulsed energies of 4 J, and require either water cooling or forced air cooling. Because the upper laser level is an excited state of the argon ion, the quantum efficiency (referred to the ground state atom) is low, and overall efficiencies of 0.05 % are typical. A well-rated source of three-phase power is usually required to operate a cw ion laser yielding several watts or more of laser power.

A related laser, the krypton ion laser, is also available, generally being supplied in the same housing and plasma tube as the argon ion laser. The output wavelengths of the krypton ion laser span the near uv (3650 Å) to the red (6471 Å) and the multicolored visible outputs (e.g., blue, green, yellow, red) are visually spectacular. However, the overall output power is significantly less than from a similar plasma tube filled with argon. The strongest output from krypton ion lasers is available from the red transition (6471 Å). To obtain this transition it must be specified when ordering the laser since a somewhat higher filling pressure is required for this transition to yield optimum output than is required for the other transitions. In this configuration a krypton ion laser will yield up to 40 % of the power achievable in an argon ion laser of the same size.

Since argon and krypton are compatible gases in an electric discharge it is possible to construct a so-called "mixed gas" ion laser. Such a system can be tailored to produce comparable power simultaneously in red, green, and blue. However, the color balance of the system changes as the plasma tube ages and this option is rarely specified.

C. Carbon Dioxide

The carbon dioxide laser is important in that it is the highest powered cw laser avaiable. However, this power is in the infrared region, at a wavelength of 10.6 μm. At this wavelength special materials must be used for transmission of the beam; zinc selenide, germanium, salt, and potassium bromide are commonly used. Another consequence of this long wavelength is that for the same beam diameter the diffraction angle is higher than for visible wavelengths.

Carbon dioxide lasers have been pumped by almost every technique available; electric discharge, photon pumping, thermal pumping (gas-dynamic laser), neutron pumping, and chemical pumping. However, all commercially available lasers are discharge pumped. The highest cw power reported is of the order of 100 kW, from a gas-dynamic laser. Gas-dynamic lasers suffer from inherently low efficiency, less than 2 %, whereas discharge lasers may have as high as 20 % efficiency. The highest power available commercially is 20 kW.

FIG. 6. The Avco carbon dioxide HPL industrial laser with 20 kW output, shown welding $\frac{1}{2}$-in. thick stainlesss steel plate at 25 in./min, using standard milling machine for work traverse. The beam is directed down at the work through the output telescope in the foreground. The dome in the background at left houses the electron beam generator for the cavity. The flexible duct vents the work region shielded by Plexiglas panels. An alternate work station is shown in the background at far right. Photograph courtesy Avco Everett Research Laboratory Inc. (HPL is a registered trademark of Avco Everett Research Laboratory Inc).

These high-power lasers use considerable flow of the medium for the reason explained above, and are not simple devices. Figure 6 is a photograph of an AVCO 20-kW unit giving some idea of its size.

Carbon dioxide lasers can be made to operate at a gas pressure of 1 atm or higher. By tuning the laser cavity with a grating, the laser can be forced to operate on one of many different rotational lines, covering the wavelength space of 9.17 to 9.44 μm and 10.09 to 10.44 μm. At about 10 atm, the rotational lines become so broad that they are smeared together, and continuous tuning is possible.[3]

Pulsed operation of carbon dioxide lasers is possible in addition to cw operation, and very high pulse power levels are possible; 10^8 W is available from a commercial unit.

---

[3] A. J. Alcock, K. Leopold, and M. C. Richardson, *Appl. Phys. Lett.* **23**, 562 (1973).

D. Nitrogen

The nitrogen laser is a pulsed laser with pulse lengths of about 20 nsec, operating in the ultraviolet region of the spectrum at 3371 Å. It is debatable whether it really deserves the name "laser" since it does not produce a coherent beam. Due to its extremely high gain it operates by amplified spontaneous emission, producing output within the solid angle subtended by the discharge, and not in a cavity mode. Nevertheless, the nitrogen laser is of importance because of the high pulsed power available in the ultraviolet region of the spectrum. It is not very efficient—typically $<0.1\%$.

E. Other Lasers

Other gas laser systems of importance are (1) the hydrogen-fluoride chemical laser, (2) the iodine laser, and (3) the rare-gas halide lasers. There is only one commercially available hydrogen fluoride laser, and three rare-gas fluoride lasers. No iodine laser is available commercially at this writing.

The hydrogen fluoride laser has produced laser energy of 4 kJ with a 20-nsec pulse duration in a research device.[4] Laser action results from a reaction between hydrogen and fluorine, with hydrogen fluoride as the product. The difficulty of handling these gases precludes commercial application. The laser wavelength is 2.7 μm in the infrared region.

The iodine laser has produced laser pulses of 1 kJ in a 1-nsec pulse.[5] The efficiency is low, around 0.2%, so that despite its high power it has limited application. The wavelength is 1.3 μm. Recently,[6] the iodine laser was successfully pumped in a discharge, which may open the way to commercial operation.

The rare-gas halide lasers are a family of lasers having as the lasing species an excited molecule formed from an atom of one of the rare gases and an atom of one of the halides. Wavelengths run from 1930 Å for argon chloride to 3550 Å for xenon fluoride. They can operate in a discharge, and efficiencies as high as 24% have been predicted.[7]

III. COMMERCIAL AVAILABILITY

The properties of commercially available lasers have been listed in Table I for cw lasers, and Table II for pulsed lasers. No attempt has been made to list every single laser manufactured, as such a list would soon become obsolete

[4] R. A. Gerber and E. L. Patterson, *J. Appl. Phys.* **47**, 3524 (1976).

[5] G. Brederlow, K. J. Witte, E. Fill, K. Hohla, and R. Volk, *IEEE J. Quantum Electron.* **QE-12**, 152 (1976).

[6] M. C. Wong and R. E. Beverly, *Opt. Commun.* **20**, 19 (1977).

[7] C. A. Brau and J. J. Ewing, "Electronic Transition Lasers" (J. I. Steinfield, ed.), p. 195. MIT Press, Cambridge, Massachusetts, 1976.

TABLE I
CONTINUOUS WAVE GAS LASERS

Laser type	Wavelength (μm)	Power (W)	Divergence (m rad)	Beam diameter (mm)	Price range ($)	Manufacturers
1. Helium–neon	0.6328	0.0001–0.05	0.5–6	0.5–15	55–8500	CILAS
						Coherent
						C.W. Radiation
						Ealing
						Hughes
						International Res. & Dev.
						Jodon
						Lansing Research
						Laser Inst. Co. India
						Liconix
						Metrologic
						Newport Research
						Nippon Electric
						Scientific Radiation
						Selectro Scientific
						Siemens
						Spectra-Physics
						Tropel
						W.E.C. Engineering
	1.15	0.001–0.015	0.8–2	0.8–2	700–8600	Scientific Radiation
	3.39	0.0007–0.003	3–6	0.8–2	700–9000	Jodon
						Spectra-Physics
2. Helium–cadmium	0.325	0.001–0.010	0.5–1.5	0.7–2.0	3500–9000	Liconix
						Nippon Electric
						RCA
	0.4416	0.008–0.05	0.6–1.5	0.6–1.5	3000–9000	Liconix
						Nippon Electric
						RCA

TABLE I (continued)

Laser type	Wavelength (μm)	Power (W)	Divergence (m rad)	Beam diameter (mm)	Price-range ($)	Manufacturers
3. Argon ion	0.33–0.53	0.015–40	0.5–1.5	0.7–1.6	3000–45,000	American Laser Corp.
						Coherent
						Control Laser Corp.
						Lexel
						Nippon Electric
						Spectra-Physics
4. Krypton	0.3375–0.86	0.25–7	0.5–1.3	1–2	7000–30,000	Coherent
						Control Laser
						Interactive Radiation
						Spectra-Physics
5. Carbon dioxide	10.6	1–100	1–10	1.4–10	4000–19,000	Advanced Kinetic
						Apollo Lasers
						CILAS
						Coherent
						Edinburgh Instr.
						G.T.E. Sylvania
						Hughes
						Israel El-Op Ind.
						Kristallopt. Laserbau
						Lasag
						Laser Technique
						Nippon Electric
						Photon Sources
						Systems EIKO

	10.6	100–1000	1–5	5–25	20,000–56,000	CILAS
						Coherent
						Garching Instr.
						Kristallopt. Laserbau
						Lasag
						Laser Inc.
						Laser Technique
						Nippon Electric
						Photon Sources
	10.6	1000–20000	1–5	10–75	60,000–400,000	Avco–Everett
						B.O.C.
						CILAS
						Garching Instrument
						G.T.E. Sylvania
						United Technologies
6. Hydrogen cyanide	311 and 337	10^{-2}–1	40	10	14,000	Advanced Kinetics
7. Methyl fluoride	96–1800	10^{-5}–10^{-3}	40	1–3	11,000	Advanced Kinetics
8. Water	118, 78, 28	10^{-3}–10^{-2}	10	5	16,000	Advanced Kinetics
9. Hydrogen fluoride	2.58–3.05	2–35	1	1.5	17,000–22,000	Helios

TABLE II
PULSED GAS LASERS

Laser type	Wavelength (μm)	Pulse energy (J)	Pulse repetition rate	Pulse length (μsec)	Divergence (m rad)	Beam diameter (mm)	Price range ($)	Manufacturers
1. Argon ion	0.4579–0.5285	5×10^{-8}–3×10^{-5}	10–60	0.0003–6	0.7–5	1.2–2.5	4000–19,000	Holotron Spectra-Physics
2. Carbon dioxide	10.6	3×10^{-5}–75	10^{-1}–10^{5}	0.04–10^{6}	0.6–10	2–90	4500–45,000	Advanced Kinetics Apollo Gen-tec GTE Sylvania Israel El-Op Ind. Jodon Lambda Physik Lasag Laser Inc. Laser Technique Lumonics Maxwell Labs Systems, Science & Software Tachisto, Inc. United Technologies
		10^{2}–10^{3}	0.02–10^{4}	0.1–10^{6}	0.5–2	8–200	50,000–110,000	CILAS Coherent Laser Technique Lumonics Systems, Science & Software
		10^{3}–10^{4}	0.02–10^{2}	2–100	0.3–0.5	200–350	Not available	Coherent Systems, Science & Software

3.	Carbon monoxide	5–6	8×10^{-3}–2.2	1–10^2	0.07–10^3	0.6	5–32	17,500–24,000	Advanced Kinetics Israel El-Op Ind. Lumonics
4.	Copper vapor	0.5106	$(1–2) \times 10^{-4}$	$(2–5) \times 10^3$	0.02	1	15–35	20,000–26,000	Laser Consultants
5.	Deuterium fluoride	3.5–4	0.04–0.7	1–25	0.1–0.5	0.5–0.75	10–30	16,000–19,000	Lumonics United Technologies
6.	Hydrogen cyanide	311, 337	10^{-3}	1–100	30	40	10	15,000	Advanced Kinetics
7.	Hydrogen fluoride	2.8–3.2	0.005–1	1–3	0.5–1	0.3–4	25–30	19,000	Israel El-Op Ind. Lumonics
8.	Krypton	0.458–0.647	5×10^{-8}–10^{-6}	10^6–10^8	$(0.03 - 1.5) \times 10^{-2}$	0.7	1.2	18,600	Spectra-Physics
9.	Nitrogen	0.3371	5×10^{-6}–10^{-1}	0.3–500	4×10^{-4}–10^{-2}	0.1–10	$1–6 \times 32$ (rectangular)	250–29,500	Garching Instr. Interactive Radiation Lambda Physik Laser Energy Lumonics Molectron National Research Phase-R Sopra Systems, Science & Software
10.	Nitrous oxide	5–6	0.3–2.2	1 10^2	0.1–60	0.6	30	19,000–24,000	Lumonics
11.	Water	118, 78, 28	10^{-5}	1–10	30	15	8	15,000	Advanced Kinetics
12.	Xenon fluoride	0.355	0.05–0.1	Up to 10	0.01–0.02	4	4×15 (rectangular) –20	14,000–29,500	Lambda Physik Lumonics
13.	Argon fluoride	0.193	0.04–0.1	Up to 10	0.01–0.02	4	4×15 –20	14,000–29,500	Lambda Physik Lumonics
14.	Krypton fluoride	0.249	3×10^{-5}–0.25	Up to 10	0.01–0.02	3–4	3–20	2000–29,500	Tachisto Lambda Physik Laser Energy Lumonics Tachisto

and would undoubtedly be incomplete. Instead, groups of lasers are given, so the engineer can get an approximate idea of lasers available for his application, and manufacturers to contact for detailed information. An excellent list of all lasers currently available is produced annually by *Laser Focus Journal*, and this should also be consulted.

IV. APPLICATIONS

Applications of gaseous lasers are described below. This list is not intended to be exhaustive, but rather to illustrate the useful features of lasers for these applications. The engineer can then compare his intended application with this list to determine approximately the kind of laser he requires.

A. Materials Working

Materials working can be divided into machining, welding, and heat treating. For these applications the laser is a means of providing very high power densities.

1. *Machining*

CO_2 lasers can cut various materials at low powers (less than 1 kW) with oxygen assist, or at high powers without it (see Table III). One major application is cutting materials that are difficult to work by ordinary techniques, e.g., glass, ceramics, cloths, and rubber.[8,9] The laser leaves a clean edge that often requires no further finishing. The work piece is usually placed so that the focal point of the laser beam is slightly below the surface. There is no tool erosion, and the process is well suited to computer control. As a result, there can be significant cost savings due to the elimination of tool changing and layout.

In machining, the laser power is sufficiently high that the irradiated region vaporises before heat can be conducted into the bulk of the material. It is sometimes necessary to remove this vaporized material with a stream of inert gas.

2. *Welding*

If the laser intensity is somewhat lower than that used for cutting, heat is conducted further into the body of the material, and the surface temperature is

[8] O. O. Yessick and D. J. Schmate, Tech. Paper MR74-962, Soc. Mfg. Eng. (1974).

[9] D. A. Belforte, *Electro-Opt. Syst. Design*, 7, 18 (1975).

TABLE III
TYPICAL VALUES USED IN LASER MATERIALS WORKING

Cutting				
Material	Power (W)	Rate (in./min)	Material thickness (in.)	Reference
Polyester	400	60	0.130	8
Steel	400	240	0.030	8
Steel	400	60	0.075	8
Stainless steel	500	45	14 gauge	9
Plywood, Plexiglas	8000	60	1	10
Stainless steel	20,000	50	0.187	10

Welding				
Material	Power (W)	Rate (in./min)	Material thickness (in.)	Reference
Titanium	500	180	0.024	11
Stainless steel	1500	140	0.05	12
Metal	1000	70	0.05	8
Car underbody	6000	450	0.035	8
Stainless steel	8000	30	0.35	10
Stainless steel	20,000	100	0.5	10

Heat treating ($15\ kW/mm^2$ max, 0.020 in. depth)			
Material	Power (W)	Area affected/unit time ($in.^2$/min)	Reference
Cast iron	400	0.6	8
Steel	10,000	60	13

kept below vaporization. The surface melts and so welding can occur. The laser power required depends on the thermal conductivity, welding speed, weld dimensions, and melting temperature, and shows the same correlation between these parameters as does electronbeam welding.[10] For a given power, the welding speed is inversely proportional to weld depth.[11,12]

[10] E. V. Locke, E. D. Hoag, and R. A. Hella, *IEEE J. Quantum Electron.* **QE-8**, 132 (1972).

[11] M. Pasturel and R. Saunders, *Electro-Opt. Syst. Design*, **8**, 23 (1976).

[12] S. L. Engel, *Laser Focus* **12**, 44, (1976).

The advantages of laser welding are that dissimilar materials can be welded; welding is possible in any inert atmosphere; a minimum of heat can be applied, thereby minimizing distortions; and it does not produce x rays.

3. *Heat Treating*

At still lower intensities than those used for welding, the surface is only heated and not melted. If the laser is then removed, the surface is rapidly quenched by heat conduction. In some metals this can lead to surface hardening. Alternatively, if a compound is placed on the surface which can diffuse into it, the surface can be laser heated to promote this and thereby cause chemical hardening. For example, steel has been surface hardened with chromium carbide by this technique.[13]

Laser heat treating has the advantage of being very selective in its application, and can reach into places difficult to access mechanically.

B. Other Applications

Some applications merely require a laser in order to define a straight line. Here low power is acceptable and low cost advantageous, making the helium–neon laser perfect for this application. Examples of this application in the construction industry are sewer pipe alignment, grading, and levelling. The helium–neon laser has also been used in aircraft manufacturing to define a reference line for wing assembly.

Good coherence properties can be used to focus to a small area or to generate interference fringes. This property is used in scanning systems, again at low power, and mainly using the helium–neon laser, for example, in supermarket label readers, security devices, fingerprint comparators.

The high spatial and spectral purity of gas lasers has led to a greatly expanded field for optical interferometry. Its numerous applications include optical surface testing, distance measuring, optical component alignment and assembly, holography, and holographic interferometry, to name a few. The high brightness of the interference patterns made possible even with a lower power helium–neon laser make it possible to detect the patterns electronically and to interface the signal with high precision electromechanical positioning devices.

Large screen displays and holographic displays use either helium–neon lasers, or when more power is required, argon and/or krypton lasers.

Argon lasers have been used in medical applications as photocoagulators and in dermatology. The carbon dioxide laser has found application in medicine as a scalpel.

---

[13] E. V. Locke, D. Ghanamuthu, and R. A. Hella, Tech. Paper MR74-706, Soc. Mfg. Eng. (1974).

Carbon dioxide and nitrogen lasers have been used for pollution monitoring. In scientific work, the carbon dioxide laser is being used for isotope separation, which could become an industry by itself in the future.

V. ECONOMIC CONSIDERATIONS

"Is the use of a laser cost effective?" should be of concern to the engineer. The cost will depend on initial cost, operating cost, and reliability (or lifetime).

The purchase cost of several cw lasers has been plotted in Fig. 7 against the laser power output. There is no dependence on wavelength, as can be seen by the fact that the visible argon ion laser lies between the infrared hydrogen cyanide laser and the infrared carbon dioxide laser. The cost does depend on output power and can be represented approximately by

$$(\text{cost in dollars}) = 6000(\text{power in watts})^{1/2}$$

However, if the laser required has special features, as, for example, high stability, it may cost considerably more than the price indicated by this simple formula. The class of helium–cadmium lasers is priced a factor of 10 above the formula due to the intricacies of the design.

No such simple rule applies to pulsed lasers, whose cost depends on energy as well as power output. A high energy output needs a physically large active volume, which is expensive; a high energy, low repetition rate laser may cost more than a low energy, high repetition rate laser with higher average power.

All commercially available lasers are electrically powered, so the operating cost will be determined primarily by the electrical input. Some gas lasers employ

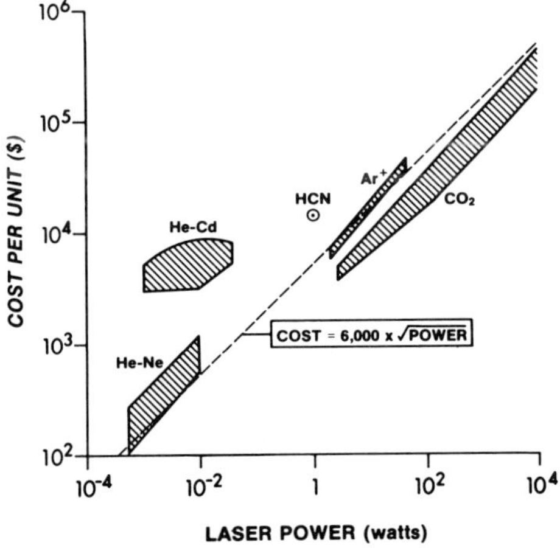

FIG. 7. Approximate cost of cw lasers as a function of laser power.

a flowing gas, which will increase the operating cost if it is open cycle. This is particularly true of the xenon and krypton fluoride lasers, since xenon and krypton are expensive gases.

It must be borne in mind that all lasers have a finite lifetime. As an example of laser lifetime, RCA quotes about 5000 hr for its helium–cadmium laser to drop to 50% of its original output. However, lifetimes vary enormously from laser system to laser system and no universal statements about them can be made.

CHAPTER 3

Semiconductor Diode Lasers

JACK F. BUTLER

Laser Analytics, Inc., Lexington, Massachusetts

I. INTRODUCTION

Semiconductor diode lasers possess a number of features not generally exhibited by other types of lasers, including extremely small size, high quantum efficiency, operating simplicity, ease of modulation to high data rates, wide spectral tunability during operation, and coverage of a broad spectral range from the visible into the far infrared regions. These and other features form the basis for a variety of important applications in optical communications, laser spectroscopy, ranging, and other areas. While the potential for such

ISBN 0-12-408606-3

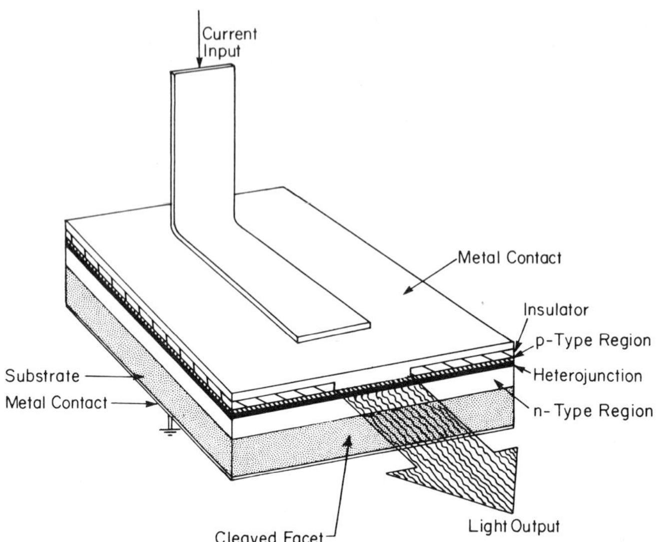

FIG. 1. Schematic diagram of a semiconductor diode laser. Laser emission occurs when the bias current exceeds a threshold value.

applications has been recognized since the first development of diode lasers,[1-3] realization has, until lately, been severely hindered by low operating temperatures, poor lifetime, and other problems. Recent technological advances have led to major improvements in these areas, and to a corresponding increase in the interest in and utilization of diode lasers.

The essential features of a semiconductor diode laser are diagrammed in Fig. 1. The *p–n* junction region is adjacent to one crystal surface and generally incorporates a heterojunction structure for improved performance. Radiation may be confined to a stripe at the center of the crystal by limiting the current distribution, as shown, or by other techniques. Two planar, parallel, and partially reflecting faces perpendicular to the junction, form the Fabry–Perot laser cavity. Linear dimensions are generally a few hundred micrometers on the longest side. Laser emission occurs when the forward bias current exceeds a threshold value, typically in the order of 100 mA for cw operation.

This chapter presents a broad review of diode lasers and their uses, with emphasis being given to underlying principles. Sections II and III are concerned with various aspects of solid state theory[4] and emission processes in semi-

---

[1] R. N. Hall, G. E. Fenner, J. D. Kingsley, T. J. Soltys, and R. O. Carlson, *Phys. Rev. Lett.* **9**, 366 (1962).

[2] M. J. Nathan, W. P. Dumke, G. Burns, F. H. Dill, Jr., and G. Lasher, *Appl. Phys. Lett.* **1**, 62 (1962).

[3] T. M. Quist *et al.*, *Appl. Phys. Lett.* **1**, 91 (1962).

[4] See, for example, S. M. Sze, "Physics of Semiconductor Devices," Parts I and II. Wiley, New York, 1969.

conductors which are particularly relevant to these devices. Section IV considers operating principles of diode lasers and discusses specific properties of various types and configurations. The final section reviews the major applications of diode lasers and considers their prospects for the future.

II. SOLID STATE THEORY

A. Energy Bands and Electrical Conductivity in Semiconductors

Electron energies in single-crystalline solids occur in series of allowed bands separated by energy gaps and occupied with electrons up to a maximum level (the Fermi level) determined by the total numbers of electrons and states available. In a pure semiconductor the Fermi level lies in the gap between two allowed bands, the nearly filled valence band lying below and nearly empty conduction band lying above.

Neither completely filled nor completely empty bands contribute to electrical current flow. Electrical conductivity in semiconductors results from thermal excitation of electrons across the gap between the valence and conduction bands (the bandgap) and from thermal ionization of impurities and defects in the crystal. These processes produce electrons in the conduction band and unfilled states (holes) in the valence band, both of which are mobile and contribute to the conduction process. The electrical conductivity is

$$\sigma = (n\mu_e + p\mu_h)e \tag{2.1}$$

where n and p are the steady state concentrations of conduction band electrons and valence band holes, respectively, and μ_e and μ_h their mobilities. A semiconductor is n-type or p-type, respectively, according to whether electrons or holes dominate the conduction process.

B. Direct and Indirect Bandgaps

As a result of the periodicity of the crystal lattice, electronic wave functions in a semiconductor may be expressed as Bloch functions, which are of the form

$$\psi_{n,\mathbf{k}}(\mathbf{r}) = \exp(i\mathbf{k}\cdot\mathbf{r})u_{\mathbf{k}}(\mathbf{r}) \tag{2.2}$$

where $u_{n,\mathbf{k}}(\mathbf{r}) = u_{n,\mathbf{k}}(\mathbf{r}+\mathbf{a})$, $\mathbf{a}$ is a lattice vector, and n enumerates the energy bands. The reduced wave vector $\mathbf{k}$ lies within the first Brillouin zone which is a primitive cell in the crystal's reciprocal lattice defined by the relation

$$\mathbf{k}\cdot\mathbf{G} = \pi|\mathbf{G}|^2 \tag{2.3}$$

where **G** represents the set of reciprocal lattice vectors of minimum length measured from a given reciprocal lattice point. Wave vectors **k** which satisfy Eq. (2.3) are Bragg-reflected by the crystal. It can be shown that the matrix element for optical transitions between Bloch functions in the valence and conduction bands,

$$M_{cv} = i\hbar(\psi_{c,\mathbf{k}}|\nabla|\psi_{v,\mathbf{k}'}) \tag{2.4a}$$

$$= 0, \qquad \text{when} \quad \mathbf{k} \neq \mathbf{k}' \tag{2.4b}$$

In Eq. (2.4) it is assumed that the wave vector of the photons involved in the transition is negligible compared with that of the electron. This result, the "momentum conservative rule," is of fundamental importance for diode laser technology because mobile electrons and holes reside in the narrow regions in the k space corresponding to the bottom of the conduction band and top of the valence band, respectively. Equation (2.4) thus implies that efficient radiative recombination across the bandgap is possible only when these extreme points coincide; for this case the semiconductor is said to have direct bandgap. Hence, semiconductors can generally be classified as having direct or indirect bandgaps, diode laser action being possible only in the first class. Figure 2 shows examples of energy band diagrams for indirect and direct bandgap semiconductors.

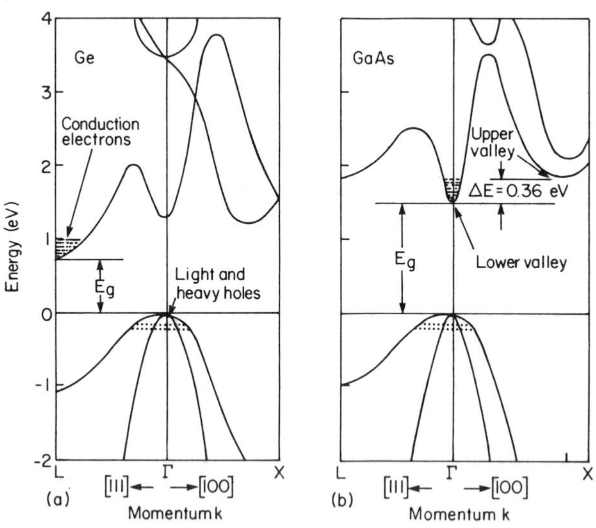

FIG. 2. Energy bands along two directions in $\bar{k}$ space for (a) Ge, an indirect bandgap semiconductor, and (b) GaAs, a direct bandgap semiconductor. Occupation of the band extrema with appropriate charge carriers is indicated. [M. L. Cohen and T. K. Bergstresser, *Phys. Rev.* **141**, 789 (1966).]

C. The Effective Mass

Being restricted to small regions in k space, electrons and holes can generally be treated by an effective mass approximation in which the charge carriers are considered to be free particles, with their lattice interaction being accounted for by replacing the true electron mass with an effective mass m^* given by

$$\left(\frac{1}{m^*}\right)_{ij} = \left|\frac{1}{\hbar^2}\frac{d^2E}{dk_i\,dk_j}\right| \tag{2.5}$$

The subscripts i and j in Eq. (2.5) represent Cartesian coordinates. In general, m^* is a tensor quantity, but may often be approximated as a scalar by using appropriate averages.

For the region in which the effective mass approximation is valid, the bands are parabolic in the sense that the dependence of the kinetic energy of a charge carrier on wave vector is similar to that of a free particle of mass m^*. That is, $E = |\hbar^2k^2/2m^*|$, where E is measured upward from the bottom of the conduction band for electrons and downward from the top of the valence band for holes.

D. Dopants in Semiconductors

In a pure semiconductor without lattice imperfections, the conductivity is determined by thermal excitation of electron–hole pairs across the bandgap and is said to be intrinsic. Conduction band electrons and valence band holes are also derived from ionized dopants, which are foreign impurities or defects in the lattice. Donors, n-type dopants, are easily ionized in the crystal lattice environment to produce electrons in the conduction band, while p-type dopants (acceptors) capture electrons from the valence band, leaving holes. A semiconductor whose charge carriers are derived mainly from the ionization of dopants is said to be extrinsic.

The ionization energy E_I of a dopant center is manifested as an allowed energy level within the forbidden bandgap. It can often be approximated as the ionization energy of a hydrogen atom immersed in a medium having the static dielectric constant ε_s of the crystal, and with the electron mass replaced by the effective mass, i.e.,

$$E_I = \left(\frac{\varepsilon_0}{\varepsilon_s}\right)^2 \frac{m^*}{m_0}(13.6) \qquad \text{eV}$$

As examples, Eq. (2.6) predicts donor ionization energies of 0.025 eV for Si, 0.007 eV for GaAs, and 1.4×10^{-6} eV for PbTe.

However, at the high doping concentrations generally used in diode lasers, the ionization levels are actually broadened by interaction into a band that merges continuously into the nearby conduction or valence band, Under these conditions, the dopant ionization levels are zero and the dopants are completely ionized at all temperatures. The continuum of states thus extending from the edge of the allowed band into the bandgap is termed a conduction or valence band tail.

Dopant states within the bandgap play an important role in emission processes by providing an alternate, and in some instances, highly efficient mechanism for radiative recombination. Donor level-to-acceptor level, donor level-to-valence band, and conduction band-to-acceptor level, are examples of radiative transitions involving dopant levels. Certain dopants may, on the other hand, provide an efficient pathway for nonradiative recombination and thereby prevent the emission of radiation.

E. Charge Carrier Statistics

The energy distribution of electrons in a semiconductor at equilibrium is described by the Fermi distribution function,

$$f(E, E_F) = \{\exp[(E - E_F)/kT] + 1\}^{-1} \tag{2.6}$$

where E_F is the Fermi energy. It is apparent that the distribution function for holes is $1 - f(E, E_F)$. The total number of electrons in the conduction band is then given by

$$n_0 = \int_{E_c}^{\infty} g(E) f(E, E_F)\, dE \tag{2.7}$$

where E_c is the energy at the bottom of the conduction band or merged donor impurity band and $g(E)$ is an appropriate density-of-states function. It is assumed that the top of the band is far above E_F. A similar relation can be written for holes in the valence band. At a given temperature, the position of the Fermi level is a sensitive function of the nature and density of dopants. In an extrinsic n-type semiconductor, E_F lies near the conduction band edge, hence there exists an appreciable density of electrons in the conduction band and a much smaller density of holes in the valence band. In this case, the electrons are said to be the majority and holes the minority carriers. Conversely, in a p-type extrinsic semiconductor, E_F lies near the valence band and holes are the majority carriers. When E_F lies substantially within an allowed band, the majority carrier concentration is degenerate.

Nonequilibrium densities of charge carriers Δn and Δp, may be generated in the respective bands by irradiating the semiconductor with short wavelength light, injection across a p–n junction (described in Section III), and other processes. In general, these excess carriers recombine by both radiative and nonradiative mechanisms. The overall recombination process can often be characterized by a simple lifetime τ. Generally, τ is long enough for the carriers to come into equilibrium within their respective bands before recombining. When this is the case, quasi-Fermi levels, F_n for electrons and F_p for holes, can be defined such that

$$n + \Delta n = \int_E^\infty g(E) f(E, F_n)\, dE \tag{2.8}$$

and a similar equation relates Δp to F_p. The validity of the quasi-Fermi level concept may sometimes be questionable in a diode laser operating above threshold because the lifetime, being reduced by stimulated emission, may become comparable to the intraband relaxation times.

F. The *p–n* Junction

A *p–n* junction consists of a more-or-less abrupt transition from *p*- to *n*-type conductivity within a single semiconductor crystal due to a change of dopant type from net acceptor to net donor. In the neighborhood of the junction, electrons from donors transfer to nearby acceptors, and a dipole layer consisting of positively ionized donors on the *n* side and negatively ionized acceptors on the *p* side is formed. For an abrupt change from *p*- to *n*-type, the dipole layer extends from one side of the junction to the other a distance

$$d = \frac{1}{e}\left(\frac{2\varepsilon_s e V_d}{N_a + N_d}\right)^{1/2}\left[\left(\frac{N_a}{N_d}\right)^{1/2} + \left(\frac{N_d}{N_a}\right)^{1/2}\right] \tag{2.9}$$

where V_d is the double layer potential, and N_a and N_d the net acceptor and donor concentrations, respectively. V_d is just the potential difference needed to equalize the Fermi levels in the *p*- and *n*-type regions of a crystal in equilibrium. When a voltage V is applied across the junction with positive polarity to the *p* side (forward bias), V_d in Eq. (2.9) is replaced by $(V_d - V)$. Figure 3 shows a diagram of energy levels versus distance near a *p–n* junction. Note that the quasi-Fermi levels are split by an amount $\Delta F = eV$.

According to the diffusion theory of current flow in a *p–n* junction diode, the application of a forward bias causes a lowering of the barrier. This allows electrons to flow into the *p* region and holes into the *n* region where they establish nonequilibrium population densities. The diffusion into the bulk regions and recombination of these excess carriers constitutes the forward current flow. This process, in which majority carriers from one side of the junction are made to diffuse into the opposite side where they become excess minority carriers, is termed *injection*. The injection and subsequent radiative recombination of excess minority carriers forms the basis for diode laser action. (In reverse bias, the barrier becomes higher, suppressing all current

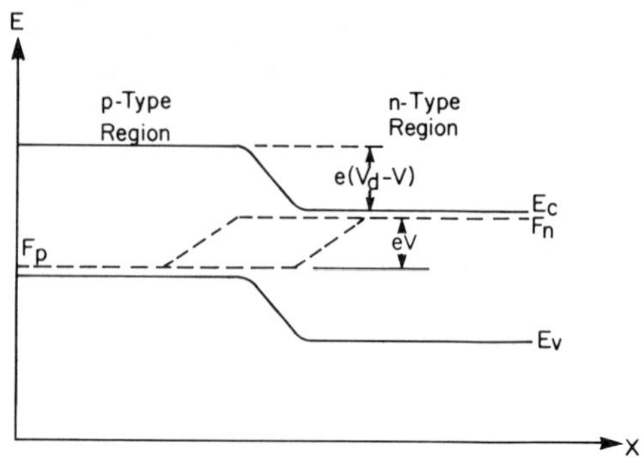

FIG. 3. Energy band diagram of a p–n junction with a forward bias voltage V. Splitting of the quasi-Fermi levels is shown.

flow except for a small saturation current resulting from the thermal excitation of electron–hole pairs.)

From the above described picture, the following relation between current density J and voltage V for a semiconductor diode can be derived:

$$J = e\left(\frac{D_{\mathrm{e}} n_p}{L_{\mathrm{e}}} + \frac{D_{\mathrm{h}} p_n}{L_{\mathrm{h}}}\right)\left[\exp\left(\frac{eV}{kT}\right) - 1\right] \tag{2.10}$$

where D is the diffusion coefficient, L the diffusion length ,and the subscripts e and h denote minority electron and holes, respectively; n_p and p_n are, respectively, minority electron and hole densities. In real semiconductor diodes, the functional form of the I–V dependence may be complicated by the occurrence of tunneling through the barrier, recombination within the space charge region, surface effects, and other factors.

G. Injection Efficiency and Heterojunctions

According to the derivation of Eq. (2.10), the two current components, $J_{\mathrm{e}} = eD_{\mathrm{e}}n_p/L_{\mathrm{e}}$ and $J_{\mathrm{h}} = eD_{\mathrm{h}}p_n/L_{\mathrm{h}}$ represent injection of carriers in opposite directions across the p–n junction. In diode lasers, it is advantageous to make the injection process unidirectional by suppressing one current component in favor of the other. It is thus desired that the injection efficiency, defined as

$$\gamma_{\mathrm{e}} = I_{\mathrm{e}}/(I_{\mathrm{h}} + I_{\mathrm{e}}) \tag{2.11a}$$

for electrons, or

$$\gamma_{\mathrm{h}} = I_{\mathrm{h}}/(I_{\mathrm{h}} + I_{\mathrm{e}}) \tag{2.11b}$$

for holes, be near unity. From Eq. (2.10) it is apparent that one current component can be made to dominate by designing in a higher carrier concentration and/or mobility on one side of the junction.

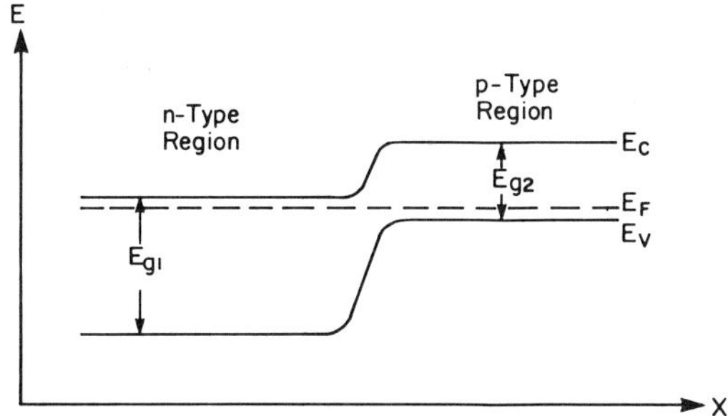

FIG. 4. Energy band diagram of a single heterojunction diode. The difference in bandgaps, $\Delta E = E_{g1} - E_{g2}$ leads to essentially 100% injection efficiency for electrons injected into the p-type region.

The injection efficiency can be made essentially unity by using the heterojunction construction diagrammed in Fig. 4. In this configuration, the two sides of the p–n junction have different bandgaps, a condition which may be achieved by composition control with certain alloy semiconductors. It may be shown that one current component is thereby suppressed by the factor $\exp(-\Delta E/kT)$, where ΔE is the difference in bandgaps.

III. EMISSION PROCESSES

A. SPONTANEOUS AND STIMULATED EMISSION RATES

Emission of optical energy from semiconductors results from the recombination of excess carriers produced by injection across a p–n junction, as described in Section II, or other processes. Optical transitions are either spontaneous or stimulated, which may be considered as two distinct processes operating in parallel. The spontaneous emission rate depends on the densities of filled upper level states and empty lower level states, and on the transition probabilities of the particular states involved. The stimulated rate depends on these factors and the density of radiation of the same frequency already present. Stimulated emission provides the linear, phase-coherent amplification mechanisms for laser action.

The overall rate at which photons of energy $\hbar\omega$ are emitted can be expressed as[5]

$$r(\hbar\omega) = r_{sp}(\hbar\omega) + N_p(\hbar\omega)r_{st}(\hbar\omega) \tag{3.1}$$

[5] G. J. Lasher and F. Stern, *Phys. Rev.* **A 133**, 533 (1964).

where $r_{sp}(\hbar\omega)$ is the rate of spontaneous downward transitions and $N_p(\hbar\omega)r_{st}(\hbar\omega)$ is the difference between the stimulated rates of downward and upward transitions. $N_p(\hbar\omega)$ is the number of photons per mode which, in thermal equilibrium, is given by the Bose–Einstein distribution function

$$N_{p0}(\hbar\omega) = [\exp(\hbar\omega/kT) - 1]^{-1} \tag{3.2}$$

The transition rates can be related to the properties of the semiconductor by the following equations

$$r_{sp}(\hbar\omega)\, d(\hbar\omega) = \sum Bf_u(1 - f_l) \tag{3.3a}$$

$$r_{st}(\hbar\omega)\, d(\hbar\omega) = \sum B(f_u - f_l) \tag{3.3b}$$

Here, f_u and f_l are the probabilities that the upper level and lower level states involved in the transitions are occupied and B is given by

$$B = (4ve^2\omega/m^2\hbar c^3)|M|^2$$

where v is the refractive index and M the transition matrix element averaged over all photon polarizations. The summations in Eq. (3.3) are taken over all pairs of states per unit volume having energy differences between $\hbar\omega$ and $\hbar\omega + d(\hbar\omega)$.

When the quasi-Fermi level approximation is valid, it follows from Eq. (3.3) that

$$r_{st}(\hbar\omega) = r_{sp}(\hbar\omega)\left[1 - \exp\left(\frac{\hbar\omega - \Delta F}{kT}\right)\right] \tag{3.4}$$

where $\Delta F = F_n - F_p$. We note from Eqs. (3.1), (3.2) and (3.4) that in thermal equilibrium ($\Delta F = 0$), the spontaneous emission rate in each energy interval just balances the stimulated absorption rate, as expected.

The absorption coefficient of the semiconductor, $\alpha(\hbar\omega)$ is related to the stimulated emission rate by[6]

$$\alpha(\hbar\omega) = -(\pi c^2 h/v^2\omega^2)r_{st}(\hbar\omega) \tag{3.5}$$

where the minus sign arises because $r_{st}(\hbar\omega)$ is defined to be positive when radiation is emitted, whereas $\alpha(\hbar\omega)$ is positive when radiation is absorbed. Equations (3.4) and (3.5) can be combined to yield

$$r_{sp}(\hbar\omega) = \frac{v^2\omega^2 m\alpha(\hbar\omega)}{\pi^2 c^2 h\{\exp[(\hbar\omega - \Delta F)/kT] - 1\}} \tag{3.6}$$

which reduces to the well known van Roosbroeck–Shockley relation for the case of thermal equilibrium.[7]

---

[6] F. Stern, *Solid State Phys.* **15**, 364 (1963).

[7] W. van Roosbroeck and W. Shockley, *Phys. Rev.* **94**, 1558 (1954).

B. Spontaneous Emission Spectra

The spectral distribution of spontaneous emission intensity emitted within a semiconductor can, in principle, be calculated from Eq. (3.3a). As an illustrative, if idealized, example, we consider spontaneous emission for the simple parabolic band model of Fig. 5 and assume the momentum conservation rule to be valid. The summation in Eq. (3.3a) can then be replaced by the density-of-states function[4]

$$\rho = \frac{1}{4\pi^2}\left(\frac{2m_r}{\hbar^2}\right)^{3/2}(\hbar\omega - Eg) \tag{3.7}$$

where $m_r = m_c m_r/(m_c + m_v)$ is the reduced effective mass; m_c and m_v are the conduction band and valence band effective masses, respectively. It follows that the spontaneous emission function is given by

$$r_{sp}(\hbar\omega) = \frac{2ve^2\omega(2m_r)^{3/2}}{m^2\hbar^4c^3\pi^2}(\hbar\omega - Eg)^{1/2}f_u(1 - f_l) \tag{3.8}$$

for this case. Equation (3.8) should be summed over spins, taking spin conservation into account.

Equation (3.8) describes the spontaneous emission spectrum in the absence of reabsorption within the semiconductor. A spectrum calculated from Eq. (3.8), assuming both quasi-Fermi levels to be situated deep within their respective bands, is shown in Fig. 6.

Spontaneous emission spectra of narrow bandgap lead salt semiconductors when corrected for reabsorption, have been found to be in reasonable agreement

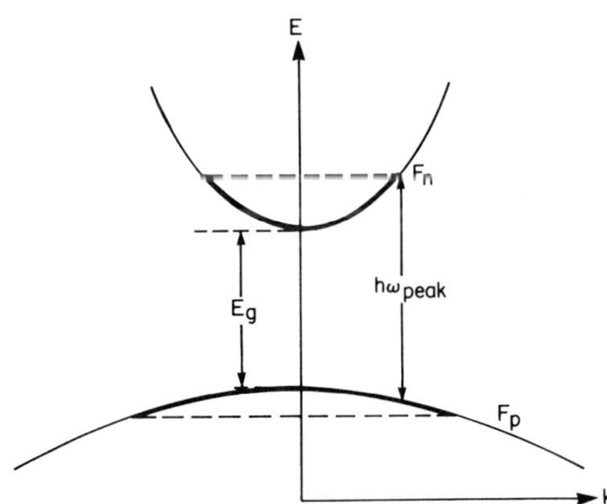

FIG. 5. Energy band diagram in k space for a semiconductor with simple parabolic valence and conduction bands. The position of the quasi-Fermi levels correspond to both electron and hole degeneracy.

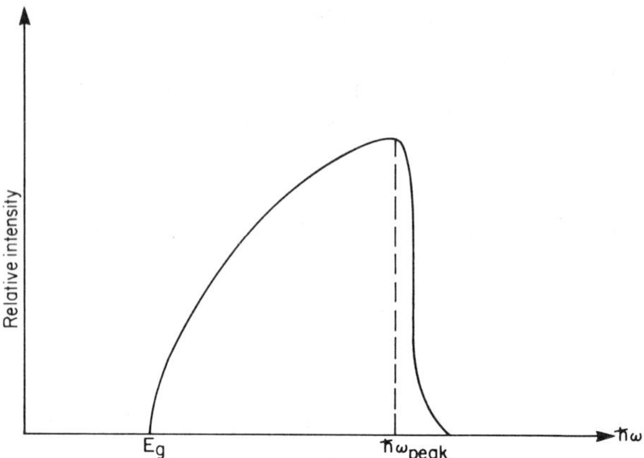

FIG. 6. Schematic diagram of the spontaneous emission spectrum for the band model of Fig. 5. Reabsorption within the crystal is not taken into account.

with Eq. (3.8),[8] This result is not surprising in view of the simple, essentially parabolic form of the lead salt band edges,[9] and the negligible dopant ionization energies predicted by Eq. (2.6). (In fact, theory indicates that dopant ionization energies in lead salts may actually be negative and lie within the allowed bands.[10])

For comparison, Fig. 7 shows spontaneous emission spectra from forward biased GaAs diodes. The spectra exhibit a pronounced tail extending far into the long wavelength region, due to the existence of tail states. Furthermore, with increasing current, the emission peak shifts to shorter wavelengths suggesting that the tail states are filled to progressively higher energies as the density of injected carriers increases. Photon-assisted tunneling may also play a role in the shift of the peak to shorter wavelengths.[11] It is believed that k conservation is not applicable in the vicinity of the band edges in GaAs because of electron–dopant and electron–electron interaction.

C. Laser Action in Semiconductors

A basic requirement for laser action to occur at a frequency ω is that the rate of stimulated downward transitions must exceed the rate of upward

[8] E. R. Washwell and K. F. Cuff, *in* "Radiative Recombination in Semiconductors" (M. Hulin, ed.). (Proc. 7th Int. Conf. on Physics of Semiconductors, 1964, Vol. 4.) Dunod, France.

[9] J. O. Dimmock, *in* "Semimetals and Narrow Gap Semiconductors" (D. L. Carter and R. T. Bate, eds.), p. 319. Pergamon, Oxford, 1971.

[10] G. W. Pratt, Jr., *in* "Physics of IV–VI Compounds and Alloys" (S. Rabii, ed.), p. 85. Gordon and Breach, New York, 1974.

[11] J. I. Pankove, *Phys. Rev. Lett.* **9**, 283 (1962).

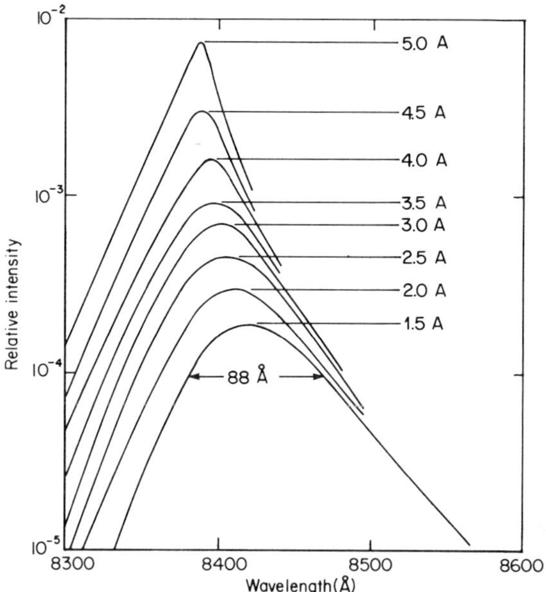

FIG. 7. Spontaneous emission spectra from a GaAs diode for several values of forward bias current. A shift in the peak position and decrease in spectral width with current are evident. [R. N. Hall, *Solid State Electron.* **6**, 405 (1963).]

transitions at that frequency, i.e., $r_{st}(\hbar\omega) > 0$. From Eq. (3.3b) this condition is seen to be satisfied in a semiconductor when $f_u > f_l$, or, if the quasi-Fermi level approximation is valid, when

$$\hbar\omega < F_n - F_c \tag{3.9}$$

Equation (3.9) is a necessary condition for laser action.[12]

Additional characteristics of laser action can be derived by assuming the region of inverted population to uniformly fill the space between two parallel, planar end faces of reflectivity R separated by a distance L; it is also assumed that the laser radiation is in the form of plane waves travelling along the laser's longitudinal axis. The requirement for self-sustained oscillations is then that a wave front make a full round trip within the cavity without attentuation and arrive at the starting point with the same phase. This requirement can be expressed as

$$R \exp\left\{\left[\frac{2\pi i v}{\lambda} - (\alpha(\hbar\omega) + \alpha')\right]2L\right\} = 1 \tag{3.10a}$$

$$= \exp(2\pi q i) \tag{3.10b}$$

[12] M. G. A. Bernard and B. Duraffourg, *Phys. Status Solidi* **1**, 699 (1961).

where q is an integer, $\alpha(\hbar\omega)$ is the gain due to the inverted population, and α' includes distributed losses due to free carrier absorption, interband absorption, and other factors.

The spectrum of the laser emission can now be deduced by equating the imaginary parts of Eq. (3.10), from which it follows that

$$\lambda_q = 2\nu L/q \tag{3.11}$$

where λ_q is the wavelength of the qth longitudinal mode. The separation between spectral modes can be then derived from Eq. (3.11), giving in wavelength units,

$$\Delta\lambda = \lambda^2 \bigg/ 2\nu L\left(1 - \frac{\lambda}{\nu}\frac{d\nu}{d\lambda}\right) \tag{3.12a}$$

and in frequency units

$$\Delta\omega = 2\pi c \bigg/ 2\nu L\left(1 + \frac{\omega}{\nu}\frac{d\nu}{d\omega}\right) \tag{3.12b}$$

Thus, the laser radiation exists in a ladder of longitudinal modes with separation given by Eq. (3.12) and which span the spectral region near the peak of the underlying spontaneous emission function. The spectral width of an individual mode is extremely small, typically much less than 10^{-5} cm^{-1}. Figure 8 shows a typical diode laser mode spectrum. The mode separation of

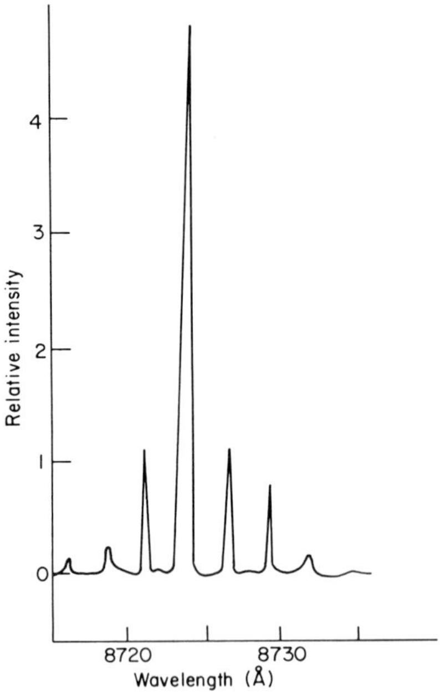

FIG. 8. Emission spectrum of a GaAs laser operating well above threshold. A well-defined longitudinal mode structure is exhibited. [T. L. Paoli, *J. Quantum Electron.* **QE-9**, 267 (1973).]

about 3.0 Å and reported cavity length of 380 μm yield from Eq. (3.12a) $\nu[1 - (\lambda/\nu)(d\nu/d\lambda)] = 3.4$. As will be apparent in Section IV the model of a uniform lasing region and plane wave modes is only approximately valid for diode lasers and, unless special precautions are taken, conditions will exist for the generation of families of transverse modes.

Threshold conditions for the stimulated emission rate can be derived from Eq. (3.10) by equating its real parts. Using Eq. (3.5), the following condition is found

$$r_{st}(\hbar\omega) = \frac{\nu^2\omega^2}{\pi c^2 \hbar}\left[\alpha' + \frac{1}{L}\ln\left(\frac{1}{R}\right)\right] \tag{3.13}$$

D. Nonradiative Recombination and Efficiency

Electron–hole pairs can recombine by a variety of processes that do not involve the emission of radiation. The existence of diode lasers, has, in fact, been made possible in large part by the development of materials processing techniques for the various direct bandgap semiconductors which yield high-purity crystals in which nonradiative recombination processes are minimized. Several important nonradiative recombination mechanisms are as follows.[13]

Point defects in a crystal lattice include vacancies, interstitials, and foreign impurity atoms at either substitutional or interstitial positions. Nonradiative recombination at such defects can provide highly effective sites for the recombination of electron–hole pairs in which the recombination energy is dissipated in the lattice as heat.

Macroscopic defects in a crystal include precipitated material and gross crystal imperfections such as grain boundaries. They generally provide a medium for recombination through a continuum of states without the emission of radiation. A uniform distribution of small precipitates may have a much stronger adverse effect on emission efficiency than a few large ones.

In addition to recombining within the bulk of a semiconductor, electrons and holes may diffuse to the surface where they recombine nonradiatively. The surface recombination rate depends on many factors and under some conditions may be much higher than the bulk rate. Laser crystals are subjected to many different processing steps that affect the chemistry of the surface, and are generally exposed to an atmosphere with constituents such as H_2O and O_2 which react with the surface. Various thermal cycling and polishing operations may also result in damaged regions on the crystal surface which increase the recombination rate.

---

[13] For a general discussion, see J. Pankove, "Optical Processes in Semiconductors," Chapter 7, p. 160. Prentice-Hall, Englewood Cliffs, New Jersey, 1971.

In the Auger effect, the energy given off in the recombination process is absorbed as kinetic energy by another electron or hole and then dissipated as heat into the lattice as the third particle relaxes into equilibrium within its band. Various Auger-effect processes are possible, involving interaction with electrons, holes, or impurity levels. Auger-effect recombination is the inverse of avalanche multiplication, in which the kinetic energy of a hot carrier excites an electron–hole pair into existence.

The efficiency of the radiative process in semiconductor diodes may be characterized by various quantum efficiencies. The internal quantum efficiency is the ratio of the number of photons emitted within the crystal to the number of electrons passing into the crystal in the same time period,

$$\eta_i = eP_i/\hbar\omega I \tag{3.14}$$

where P_i is the power emitted internally and I is the diode current. The external quantum efficiency is then defined as

$$\eta_e = eP_e/\hbar\omega I \tag{3.15}$$

where P_e is the power emitted external to the crystal; $P_e < P_i$, hence $\eta_e < \eta_i$, because of reabsorption. The power efficiency is defined as

$$\eta_P = P_e/IV \tag{3.16}$$

where $\eta_P < \eta_e$ because of resistive losses in the diode and other factors. The differential quantum efficiency is

$$\eta_d = (e/\hbar\omega)(dP_e/dI) \tag{3.17}$$

IV. DIODE LASERS

A. Threshold Currents

An expression for the threshold current in a diode laser can be derived from the following considerations.[5] At a current level just below the onset of laser action, the photon density N_P is very small, hence nearly all of radiative recombination of injected carriers derives from the spontaneous term in Eq. (3.1). Defining this current as threshold,

$$I_t = I_t/A = \left[eU\int r_{sp}(\hbar\omega)\,d(\hbar\omega)\right]\Big/\eta_i A \tag{4.1a}$$

$$= (e\gamma' t/\eta_i) r_{sp}(\hbar\omega_m)\Delta(\hbar\omega) \tag{4.1b}$$

where A, U, and t are the area, volume, and thickness, respectively, of the active region and $r_{sp}\hbar\omega_m$ is the spontaneous rate at the peak. It has been assumed that the spontaneous emission is characterized by a linewidth $\Delta(\hbar\omega)$ and γ' has been introduced as a line shape correction factor. Substituting the stimulated rate from Eq. (3.4),

$$J_t = \frac{e\gamma' t}{\eta_i} \frac{r_{sp}\hbar\omega_m \Delta(\hbar\omega)}{\{1 - \exp[(\hbar\omega - \Delta F)/kT]\}} \tag{4.2}$$

Combining Eq. (4.2) with the threshold condition of Eq. (3.13) then gives

$$J_t = \frac{\gamma e t v^2 \omega^2 \Delta(\hbar\omega)}{\eta_i \pi^2 c^2 \hbar} \left[\alpha' + \frac{1}{L}\ln\left(\frac{1}{R}\right)\right] \tag{4.3a}$$

$$\equiv \frac{1}{\beta}\left[\alpha' + \frac{1}{L}\ln\left(\frac{1}{R}\right)\right] \tag{4.3b}$$

where $\gamma = \gamma'/\{1 - \exp[(\hbar\omega - \Delta F)/kT]$. The "demerit" factor γ is close to unity at low temperatures and increases with temperature. Values of γ near 10 have been calculated for GaAs diode lasers at room temperature.[14]

In the derivation of Eq. (4.3), the volume of the active recombination region was assumed to be $U = At$. However, as will be discussed below, the laser radiation in a semiconductor laser extends beyond the active region, possessing a width, say, $t' > t$ (see Fig. 12). The injection current must therefore be increased by the factor t'/t to achieve a given level of gain. In other words, Eq. (4.3) is correct for the case where the radiation extends beyond the active region provided t is redefined as the effective width of the total region encompassed by the radiation.

Equation (4.3) can be compared with experiment, and values for β and α deduced by observing the dependence of J_1 on laser cavity length. Figure 9, for example, shows experimental results at four different temperatures for a set of diode lasers fabricated from the same substrate, but with different cavity lengths. Sample heating at high current densities was avoided by using pulsed bias current with short duty cycles. As can be seen, the predicted linear relation holds at all temperatures up to room temperature.

Figure 9 also shows that for these early homostructure lasers, the threshold current increased strongly with temperature. For the 100-μm device, for example, J_t increased from 3×10^3 A/cm^2 at 4.2 K to about 3×10^5 A/cm^2 at 296 K. The corresponding gain factor β decreased from 5.1×10^{-2} to 4.9×10^{-4} cm/A, while the loss factor α increased from 1 to 30 cm^{-1}.

B. Gain Saturation and Power Output

As the injection current of a diode laser is increased through threshold, the photon density in the cavity builds up rapidly to high levels. The corresponding

[14] F. Stern, *Phys. Rev.* **148**, 186 (1966).

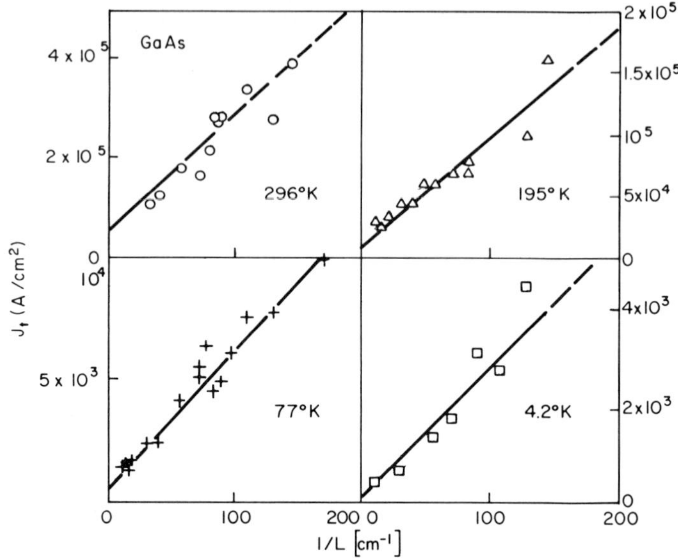

FIG. 9. Dependence of threshold current density J_t on cavity length L for a set of GaAs diode lasers at various temperatures. Currents were applied in 27 nsec pulses at repetition rates of 10–50 pps. [M. Pilkuhn, H. Rupprecht, and S. Blum, *Solid State Electron.* **7**, 905 (1964).]

large increase in the stimulated emission rate ensures that the pumping level, which would otherwise increase until the gain greatly exceeds the losses, stabilizes at a rate which just balances the supply of injected carriers. Thus, the gain and injected carrier concentrations both saturate at a current slightly above threshold. According to the considerations relating to Eq. (2.8), saturation of the injected carrier density implies a pinning of the quasi-Fermi levels to a fixed separation, which should lead to a corresponding voltage saturation in the diode volt–ampere characteristics. Such an effect is commonly observed experimentally in lead salt diode lasers.[15] Figure 10, for example, shows V–I curves for a $Pb_{1-x}Sn_xSe$ diode laser at two temperatures. The slope discontinuities occur at threshold (as separately determined from optical measurements) and near voltages V_t such that $eV_t \approx \hbar\omega_t$, where ω_t is the threshold emission frequency. The finite slope above threshold is due to bulk and contact series resistances.

When gain saturation occurs, the power emitted by stimulated emission within the cavity becomes directly proportional to the current above threshold,

$$P_i = [(I - I_t)\eta_i/e]\hbar\omega \tag{4.4}$$

[15] J. N. Walpole, A. R. Calawa, R. W. Ralston, T. C. Harman, and J. P. McVittie, *Appl. Phys. Lett.* **14**, 333 (1969).

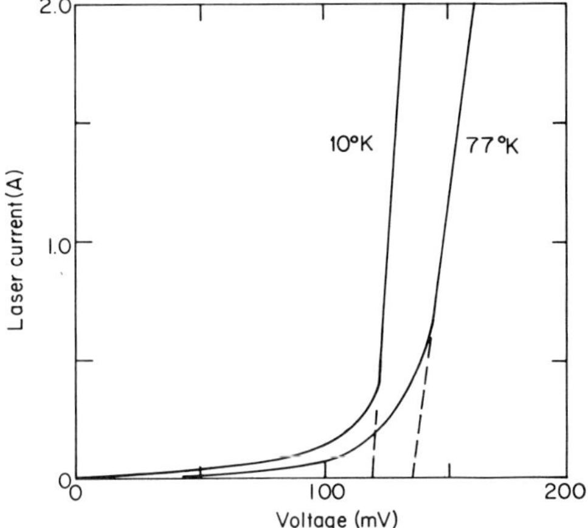

FIG. 10. Forward volt–ampere characteristics for a $Pb_{0.97}Sn_{0.03}Se$ diode laser at two temperatures. Extrapolated voltage intercepts correspond closely to the measured threshold frequencies [K. J. Linden, K. W. Nill, and J. F. Butler, *J. Quantum Electron.* **QE-13**, 720 (1977).]

A portion of this power is dissipated inside the laser cavity and the rest transmitted through the end reflectors. These two components of the total emitted power are proportional to α' and $(1/L)\ln(1/R)$, respectively; hence the power emitted from the laser can be written as

$$P_e = \frac{(I - I_t)\eta_i \hbar\omega}{e} \frac{(1/L)\ln(1/R)}{\alpha' + (1/L)\ln(1/R)} \tag{4.5}$$

Using Eq. (3.17), then, the differential quantum efficiency is

$$\eta_d = \eta_i \frac{(1/L)\ln(1/R)}{\alpha' + (1/L)\ln(1/R)} \tag{4.6}$$

while, from Eq. (3.16), the power efficiency is

$$\eta_p = \frac{I - I_t}{I} \frac{\hbar\omega}{eV} \eta_i \frac{(1/L)\ln(1/R)}{\alpha' + (1/L)\ln(1/R)} \tag{4.7}$$

In practice, $eV \approx \hbar\omega$ and $(1/L)\ln(1/R) \gg \alpha$. Therefore, well above threshold η_e approaches η_i. Since η_i is generally near unity, diode lasers can possess very high power efficiencies.

C. Carrier Confinement

As discussed in the derivation of Eq. (4.3), the thickness of both the active gain region and the region encompassed by the laser radiation play important roles in determining threshold currents in diode lasers. In this section we consider the factors establishing the active gain region thickness and how this thickness may be reduced by the use of heterostructures. Section V will consider the confinement of laser radiation to the active gain region to achieve the desired condition of $t'/t \approx 1$.

Under the free-diffusion conditions assumed in the derivation of Eq. (4.3), the spatial distribution of excess electrons injected across a p–n junction is described by a diffusion length, L_D, as

$$\delta n(x) = \frac{JL_D}{eD} \exp\left(-\frac{x}{L_D}\right) \tag{4.8}$$

where D is the diffusion coefficient and J the electron current density. A similar relation holds for an excess of holes $\delta p(x)$. If $\delta n'$ is the excess electron density necessary for an inverted population, the thickness of the active recombination region is, then

$$t = L_D \left| \frac{eD\delta n'}{J_t L_D} \right| \tag{4.9}$$

and the fraction of electrons recombining in the active region is

$$\gamma^* = 1 - \exp\left(\frac{t}{L_D}\right) \tag{4.10}$$

Electrons recombining outside of the active region do not contribute to laser action.

The fraction γ^* can be increased by halting the free diffusion of carriers with the reflecting heterobarriers diagrammed in Fig. 11. In this configuration, carriers injected across the p–n junction are confined to the region of width d, which may be much less than L_D. Using this technique, it is quite feasible to achieve values of γ^* close to unity. An additional advantage of the heterobarrier

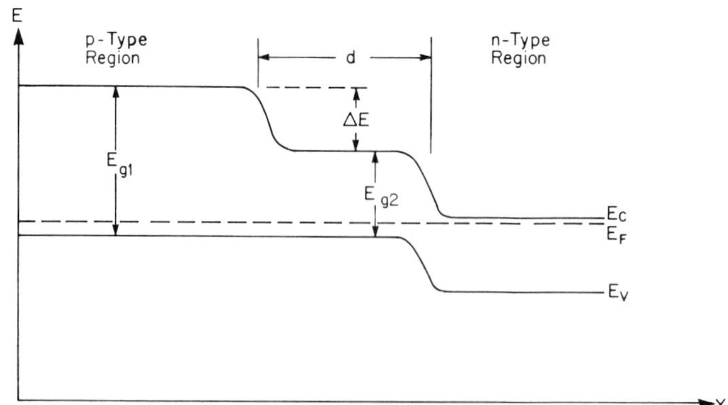

FIG. 11. Energy band diagram of a single-sided heterostructure. Electrons injected into the narrow bandgap p-type region are confined by the heterobarrier.

configuration is that the density of excess injected carriers is uniformly distributed throughout the gain region rather than varying with distance as described by Eq. (4.8).

D. Optical Confinement

Electromagnetic radiation in a diode laser is always confined to the junction region to some degree by an effective dielectric waveguide along the plane of the junction. In homojunction lasers the waveguiding effect is due to a small increase (much less than 1 %) in the dielectric constant in the junction region related to changes in the dopant concentrations and plasma oscillation frequency, the existence of gain, and other factors. This waveguiding effect, and the degree of optical confinement, can be greatly enhanced by using a heterostructure configuration in which the bandgap is decreased, hence the refractive index is increased in the active region relative to that in adjacent regions. Dielectric constant increments of from 5 to 10 % are thereby achievable. Figure 12 illustrates schematically the confinement of laser radiation to the junction region.[16]

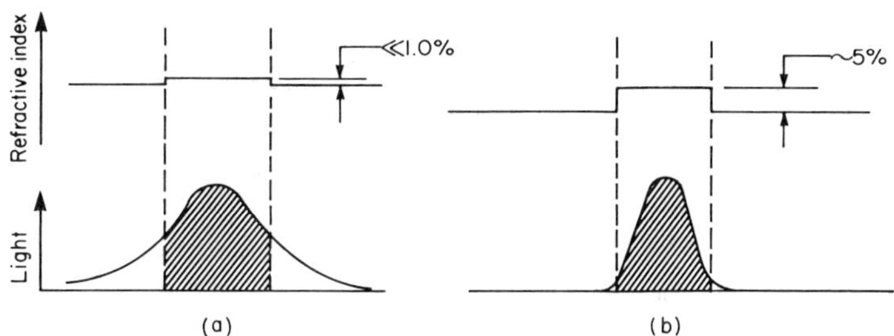

Fig. 12. Schematic representation of the refractive index changes and optical field distribution in (a) a homostructure and (b) a double heterostructure diode laser.

The essential features of mode confinement in a dielectric waveguide can be derived from the simplified model of Fig. 13. The dielectric constant of the central slab ε_1 is greater than that of the infinitely extending outer regions ε_2 by an amount $\delta\varepsilon = \varepsilon_1 - \varepsilon_2$. It is assumed that there is no variation normal to the plane of the figure (y direction). Solutions to Maxwell's equations for this model can be decoupled into TE and TM modes with electric and magnetic field vectors, respectively, restricted to the y direction.[17] Using the appropriate boundary conditions, solutions are found to be in the form of discrete propagating modes within region 1, with evanescent components extending into regions 2. The electric field components for even TE modes, for example, are

[16] I. Hayashi, M. B. Panish, and F. K. Reinhart, *J. Appl. Phys.* **42**, 1929 (1971).

[17] A. Yariv, *in* "Introduction to Optical Electronics," p. 40. Holt, New York, 1971.

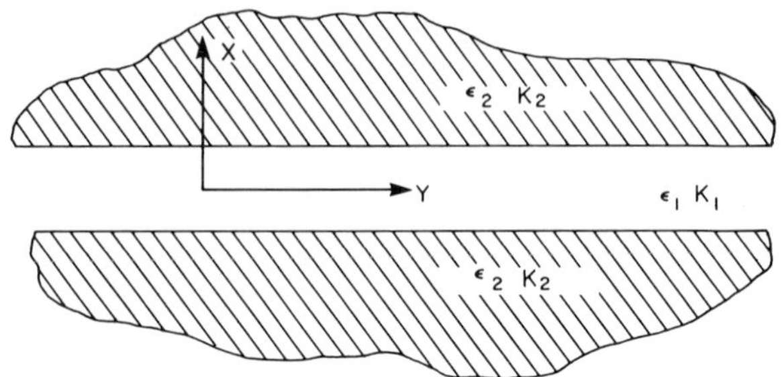

FIG. 13. Model of refractive index variation leading to optical confinement. $\varepsilon_1 > \varepsilon_2$.

given by

$$E_y = E_0 \exp[-p(|x|\tfrac{1}{2}d) - i\beta z] \qquad |x| \geq \tfrac{1}{2}d \tag{4.11a}$$

and

$$E_y = \frac{E_0 \cos(hx)\exp(-i\beta z)}{\cos(h\frac{1}{2}d)} \qquad |x| \leq \tfrac{1}{2}d \tag{4.11b}$$

where p and h are evaluated from the equations

$$pd = hd\tan(h\tfrac{1}{2}d) \tag{4.12a}$$

and

$$(pd)^2 + (hd)^2 = (\delta\varepsilon/\varepsilon_1)(k_1 d)^2 \tag{4.12b}$$

Solutions representing propagation exist where the function $p = h\tan(hd/2)$ intersects the family of circles defined by Eq. (4.12). When the "radius" $(\delta\varepsilon/\varepsilon_1)^{1/2}k_1 d$ is small, such that $0 < (\delta\varepsilon/\varepsilon_1)^{1/2}k_1 d < 2\pi$, only one confined mode exists and is designated as TE_0. With the radius increased to the range $2\pi < (\delta\varepsilon/\varepsilon_1)^{1/2}k_1\, d < 4\pi$, we obtain two intersections of Eq. (4.12), giving the TE_0 mode again and the first order even TE_1 mode. (The TE_0 mode has no zero crossings while the TE_1 mode has one zero crossing in the region $|x| < d/2$.) Further increase in the radius brings higher order even modes into propagation. In addition to the even TE modes, there exist families of odd TE modes, and even and odd TM modes. With fixed frequency and dielectric properties, the number of possible propagating modes increases with d. Thus, for large d, many possible modes correspond to the same frequency.

The evanescent wave extends into region 2 a distance $1/p$, which can be calculated from Eq. (4.12) if the dielectric mismatch, emission frequency, and region 1 thickness are known. Since this wave is not amplified in the lasing process and propagates in a lossy region, it is desirable to confine the radiation to region 1 as much as possible. The effectiveness of confinement is described by a parameter Γ, which is essentially the fraction of the radiation flux restricted to the region of inverted population. If g_0 is the laser gain for radiation com-

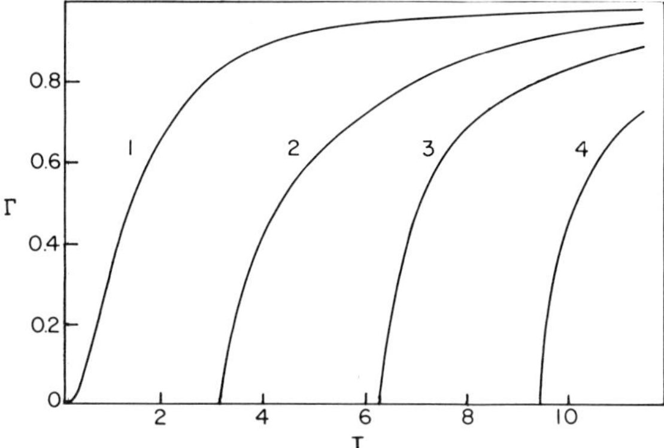

FIG. 14. Dependence of the optical confinement parameter Γ on the reduced active layer thickness T for lower order TE modes in a symmetric waveguide. (P. G. Eliseev, Preprint No. 33 [in Russian], Physics Institute, Academy of Sciences of the USSR, Moscow (1970); [in English] *Sov. J. Quantum Electron.* **2**, 505 (1973).)

pletely confined to the active region, then the effective gain when confinement is not complete is $g = g_0\Gamma$. The confinement parameter Γ is given by $2(t + 1/p)$ if the active region is restricted to and uniformly fills region 1. Figure 14 shows Γ calculated for this case for several low order TE modes as a function of the reduced thickness, $T = (\delta\varepsilon/\varepsilon_1)(k_1 d)^2$.

E. DIODE LASER STRUCTURES

Diode lasers have been made from a wide variety of direct bandgap semiconducting compounds. Methods of materials preparation and device fabrication vary widely from one compound to another and with composition for a given alloy system. For this reason, specific fabrication details will only be alluded to briefly here.

Figure 1 showed schematically general features common to all diode lasers. The two end faces forming the laser cavity are usually uncoated since the relatively high dielectric constants of the semiconductors involved provide sufficient reflectivity. The p–n junction region may incorporate a compositional variation to form an appropriate heterostructure and is often in the form of a stripe down the center of the crystal (stripe geometry) to eliminate side bounce and internal circulating modes. Ideally, the stripe width is minimized to eliminate all but the lowest order transverse modes in one dimension.

The following paragraphs consider in somewhat more detail the various structural configurations and how they relate to emission characteristics.

1. *Homojunctions*

A homojunction laser is a simple p–n junction diode with a uniform bandgap throughout. The first GaAs diode lasers were of this type as are some types of

Pb–salt diode lasers currently available. Optical confinement in homojunction lasers is relatively poor and the electromagnetic energy spreads extensively into the inactive region. Carriers are also not well confined in a homojunction laser, the active region width being defined by the minority carrier diffusion length. An increase with temperature of both the optical losses in the inactive region and the diffusion length leads to high threshold currents at elevated temperatures. As a result, cw operation in homojunction lasers has been achieved only at cryogenic temperature. Because of the weak waveguiding effect in homojunction lasers, the field distribution of the laser energy can be well approximated by TEM modes. Correspondingly, the output of a homojunction diode laser may exhibit a single uniform spatial lobe and poorly defined polarization characteristics.

2. *Single Heterostructures* (*SH*)

The next step in complexity is the SH laser, usually consisting of a homojunction separated by a small distance d from a heterobarrier as shown in Fig. 11. The distance d is much less than a diffusion length, hence, this structure provides strong carrier confinement. The barrier also provides partial optical confinement. Threshold currents in a SH laser decrease with d, as expected from Eq. (4.3), but reach a minimum and increase rapidly for widths below a critical value d_c. The increase in threshold for $d < d_c$ is thought to be due to a decrease in injection efficiency across the p–n junction at the high carrier densities involved.[16]

SH lasers were used in III–V compounds before the successful development of double heterostructure devices. High performance lead-salt lasers have been made using SH structures formed by the compositional interdiffusion (CID) technique.[18] This method relies on the selective out-diffusion of one constituent from the crystal surface to produce the heterobarrier.

3. *Double Heterostructures* (*DH*)

Generally, the DH structure is preferred for diode lasers. This structure is defined by two heterobarriers as diagrammed in Fig. 15, with one of them being also the injecting p–n junction. This structure provides both strong carrier and optical confinement; the injection efficiency is essentially unity and nearly independent of d. The confinement parameter Γ in a properly designed DH laser is near unity, which results in greatly improved performance at elevated temperatures over that of homojunction and SH lasers. The cw operation of diode lasers up to room temperature has been possible only since the development of DH laser fabrication methods in GaAs and some other III–V compounds.

---

[18] E. D. Hinkley, R. T. Ku, K. W. Nill, and J. F. Butler, *Appl. Opt.* **15**, 1653 (1976).

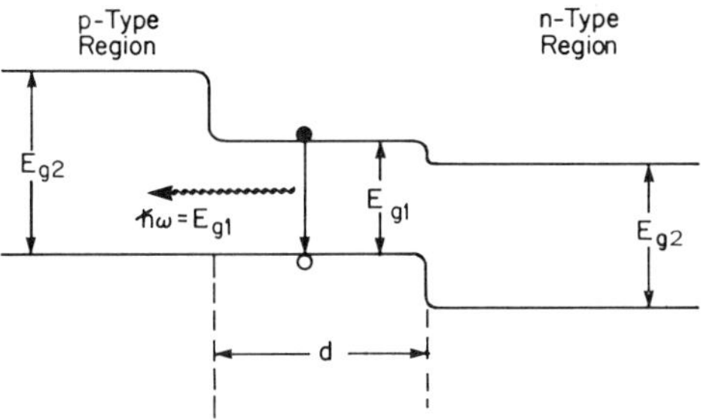

FIG. 15. Energy band diagrams of a double heterojunction. Radiation is generated in the narrow bandgap region.

The emission output from DH lasers is generally completely polarized parallel to the plane of the junction, indicating the predominance of TE over TM modes. Also, unless special precautions are taken, the spatial characteristics of the output of DH lasers often show a multilobed structure indicative of higher order modes. These observations can be qualitatively understood by considering the reflectivity of the laser end faces, as follows. The field distribution within the active region can be decomposed into two plane waves propagating at the same angle to the plane of the junction but with components of the propagation vector normal to the junction being in opposite directions. As is well known, the reflection coefficient of a plane wave at a planar dielectric interface depends on the angle of incidence and the polarization, being larger than that of a normally incident wave when the E vector is normal to the plane of incidence (case of TE modes) and smaller if the E vector is in the plane of incidence (case of TM modes). Higher order TE modes predominate because they have larger angles of incidence, and thus larger reflection coefficients. As shown previously, higher order modes can, in principle, always be suppressed by making d small enough so that $(\delta\varepsilon/\varepsilon)^{1/2}k_1 d < 2\pi$.

The problem of lattice parameter matching is of considerable importance to heterojunction technology. A 1% lattice mismatch at an interface will result in a dislocation being generated at approximately every 100 atom planes. Since a dislocation core is generally an effective site for nonradiative recombination, such a high dislocation density will degrade the internal quantum efficiency. Furthermore, dislocations can propagate through the multilayer structure, causing eventual complete failure.[19] Therefore, in designing heterostructures, lattice-matched systems must be chosen or else dislocations removed by compositional grading. It is found experimentally in III–V compound lasers that a lattice mismatch of as much as 0.1% is acceptable.[20]

[19] P. M. Petroff and L. C. Kimerling, *Appl. Phys. Lett.* **29**, 461 (1976).

[20] C. J. Nuese, Private communication (1979).

F. Diode Laser Materials

1. *III–V Compounds*

The most highly engineered and tested of the diode lasers are probably DH devices derived from the ternary alloy system $Al_xGa_{1-x}As$. Thin layers (<0.1 μm) of high quality can be sequentially grown with a high degree of precision by epitaxial growth on GaAs substrates. The success of the $Al_xGa_{1-x}As$ system is due in part to the fact that the lattice parameter varies only by 0.14% as x goes from zero to unity, and hence strain-induced defects are negligible. As an example of attainable performance levels in these materials, commercially available, off-the-shelf GaAs DH lasers are specified by the manufacturer[21] to provide cw operation to 27°C, cw power outputs of 20 mW, and room temperature threshold currents as low as 100 mA. On a pulse basis, peak power outputs of 20 W are attainable with individual diode lasers and more than a kilowatt of peak pulsed power with laser arrays. Beam patterns of individual lasers operating in the cw mode are, according to the manufacturer, uniform and single lobed, with a width of about 56° in one direction and 4° in the other.

By increasing the Al composition of the active region in $Al_{1-x}Ga_xAs$ DH lasers, emission wavelengths as short as 0.74 μm have been achieved, with room temperature cw operation.[22] Using other III–V compounds, cw room temperature operation has been extended to wavelengths as long as 1.12 μm.[23,24] In general, the various physical processes related to laser action in semiconductors become more favorable at lower temperatures. Thus, operation at cryogenic temperatures greatly increases the range of materials available for diode lasers. Using various III–V compounds operating at cryogenic temperatures and possibly on a pulse basis, diode laser action has been achieved at wavelengths as short as 0.58 μm[25] and as long as 5.2 μm.[26] As discussed in Section IV, F, 2, much longer wavelengths can be achieved with the lead salt semiconductors.

2. *The Lead Salts*

The lead salt semiconductors include PbTe, PbSe, and PbS, and alloys of these compounds amongst themselves and with SnSe, SnTe, CdS, and other materials. The lead salts provide a class of diode lasers completely spanning the

---

[21] For example, Laser Diode Laboratories, Inc., Metuchen, New Jersey, 08840.

[22] H. Kressel and F. Z. Hawreylo, *Appl. Phys. Lett.* **28**, 598 (1977).

[23] J. J. Hsieh, J. A. Rossi, and J. P. Donnelly, *Appl. Phys. Lett.* **28**, 709 (1976).

[24] C. J. Nuese, G. H. Olsen, M. Ettenberg, J. J. Gannon, and T. J. Zamerowski, *Appl. Phys. Lett.* **29**, 807 (1976).

[25] W. R. Hitchens, N. Holonyak, Jr., P. D. Wright, and J. J. Coleman, *Appl. Phys. Lett.* **27**, 245 (1975).

[26] R. J. Phelan, Jr., A. R. Calawa, R. H. Rediker, R. J. Keyes, and B. Lax, *Appl. Phys. Lett.* **3**, 143 (1963).

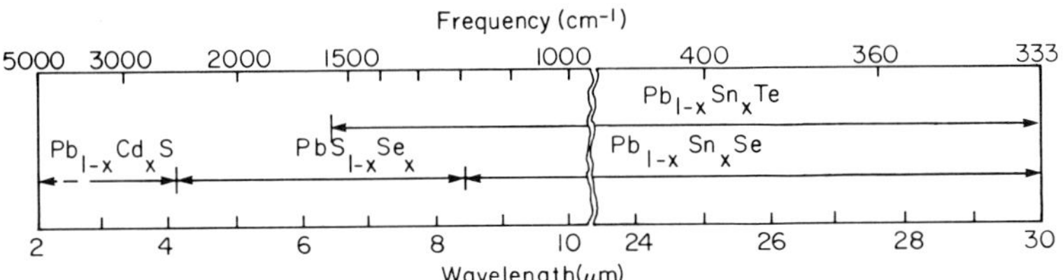

FIG. 16. Dependence of emission frequency on composition for Pb salt diode lasers.

infrared spectrum from about 2.7 to over 30 μm (see Fig. 16). This spectral region includes absorption bands of nearly all molecular compounds, and much of the interest in the lead salt lasers relates to their use in spectroscopy for diverse applications.

The lead salts are strikingly different from the III–V compounds in several respects. For example: their crystal structure is that of rock salt (hence their common name) instead of the zinc blende structure possessed by the III–V semiconductors; their bandgaps generally increase with increasing temperature, whereas III–V compound bandgaps decrease; their bandgaps occur at ⟨111⟩ edges of the Brillouin zone instead of at the center; they can be doped strongly *n*- or *p*-type by adding a stoichiometric excess of the metal or nonmetal constituent, respectively, without adding impurities. Until recently, only homojunction lead salt diode lasers were available. Reliable SH lasers are now possible for some compositions, and limited success has been achieved in fabricating DH lead salt lasers using liquid-phase epitaxy[27] and molecular-beam-epitaxy growth techniques.[15]

Lead salt lasers operate at cryogenic temperatures. Operation at 77 K is generally possible for wavelengths as long as 10 μm with even lower temperatures required beyond. Power outputs as high as 330 mW cw and 2.9 W on a pulse basis have been observed from experimental lead salt lasers.[28] Typical power levels for commercially available devices are a few milliwatts out to 8.5 μm and a few hundred microwatts or less at longer wavelengths.

V. APPLICATIONS OF DIODE LASERS

A. LIGHTWAVE COMMUNICATIONS

Modern interest in lightwave communication began with the first development of the laser in 1960. Since, in principle, the information-carrying capacity

[27] K. J. Linden, K. W. Nill, and J. F. Butler, *IEEE J. Quantum Electron.* **QE-13**, 720 (1977).

[28] R. W. Ralston, J. N. Walpole, and T. C. Harman, Solid State Rep., Lincoln Laboratory, MIT, p. 1. Massachusetts Institute of Technology, Cambridge, Massachusetts, 1974.

of a communications system increases directly with frequency, the laser, operating in the 100 THz region offered the promise of increasing this capacity more than 4 orders of magnitude over that of the then existing microwave systems. Diode lasers, when they appeared, seemed particularly suitable for this application because of their small size, ease of modulation, and low power requirements. However, realization of even a small part of this potential has remained elusive until quite recently because of the cumbersomeness and unreliability of available lasers, atmospheric propagation problems, and other factors.

Several recent developments have brought practical lightwave communications systems close to reality. First, the development and refinement of heterostructure lasers made room temperature operation possible.[29] Then, understanding and eliminating degradation mechanisms in the highly developed $Al_{1-x}Ga_xAs$ system has led to diode laser lifetimes for this material of more than a year, with indications that 100 year lifetimes are feasible.[30] Finally, highly transparent, low cost fiber lightguides have become available for long distance transmission. It is fortuitous that emission wavelengths of $Al_{1-x}Ga_xAs$ DH lasers can be tailored to be close to the wavelength of minimum transmission loss of the lightguide (see Fig. 17). Considerable research and development activity is presently aimed at achieving wavelengths even closer to the minimum by using other materials, such as $In_{1-x}Ga_xAs$, $In_{1-x}Ga_xP$, or quaternary materials such as $In_xGa_{1-x}As_yP_{1-y}$. These rather elaborate material combinations are needed for lattice matching.

Figure 18 shows schematically a lightwave communications system for long distance transmission. Diode lasers are employed as the modulated source and in the repeater stations. Silicon avalanche photodiodes (ADPs) provide the high-speed detection function. Digital coding is used to carry information because of its signal-to-noise ratio advantages in this type of system. For fast pulse modulation the diode laser is biased with a direct current just below threshold and pulsed to an appropriate level above threshold. This eliminates lasing delay related to the spontaneous emission lifetimes and allows high data rates to be achieved.

Light emitting diodes (LEDs) may in some instances be used in place of diode lasers for optical communication. LEDs have the advantage of a simple construction and a smaller temperature dependence of the emitted power. In addition, reliable LEDs emitting nearer the 1.1 μm transmission maximum of fiber optics can be fabricated in materials such as $In_{1-x}Ga_xAs$. However, the spectral bandwidth of the LED is typically 1 to 2 orders of magnitude broader than that of the diode lasers (all modes); because of spectral dispersion in the fibers, this may limit their application. Furthermore, the modulation speed of

[29] J. Hayashi, M. B. Panish, P. W. Foy, and S. Sumski, *Appl. Phys. Lett.* **17**, 109 (1970).

[30] R. L. Hartman, Private communication (1977).

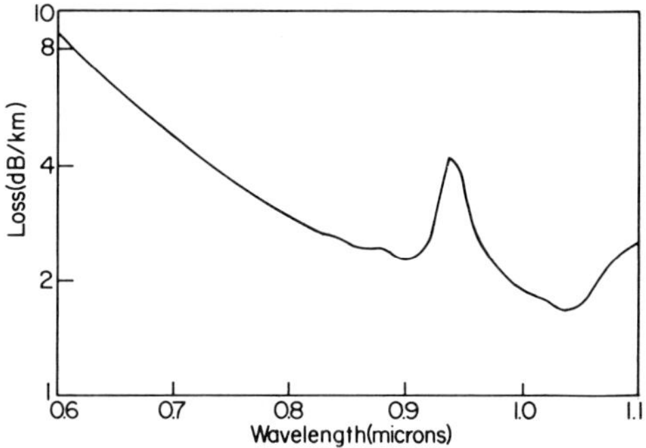

FIG. 17. An absorption spectrum typical of glass fibers used for lightguides. The peak is due to the OH group. [G. W. Tasker and W. G. French, *Proc. IEEE* **62**, 1282 (1974).]

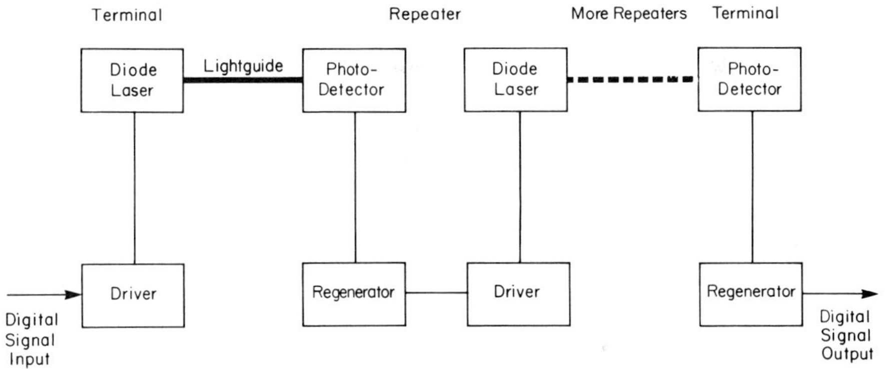

FIG. 18. Schematic diagram of an optical communications system. Diode lasers are used as transmitters in the input terminal and repeaters.

LEDs is limited by the spontaneous lifetime, and the coupling efficiency of LEDs into low-numerical-aperture fibers is much lower than that of diode lasers.

Properties of the optical fibers themselves play an important role in the performance of lightwave communication systems.[31] Optical fibers are often characterized by the maximum distance allowed between repeaters, assuming a repeater gain of 50 dB. At high frequencies, dispersion in the fibers is the dominant loss mechanism. For multimode fibers with a simple refractive index step, modal dispersion limits the frequency to about 20 MHz km (that is, a 5-km repeater spacing at 4 MHz). By contrast, transmission in a single-mode fiber with a specially profiled index may be limited only by material dispersion, giving a quality factor of 50 GHz km.

[31] A. G. Chynoweth, *Phys. Today* **29**, 28 (1976).

Advantages of lightwave communication systems over conventional systems based on transmission through copper wires or cables include much smaller size and weight for a given high rate of data transmission, freedom from electrical interference, weak dependence of transmission loss on frequency (which simplifies receiver electronics), and, ultimately, lower cost. Possible near term benefits resulting from these advantages include increased capacity of existing urban underground telephone ducts, and decreased size and weight of aircraft or satellite guidance systems.[32] The Bell System is, as of the time of this writing, operationally testing a lightwave system which carries voice, data, and video signals through 1.5 mi of underground urban cable.[33] The lightguide cable is only 0.5 in. in diameter and contains 24 fibers, each having an information capacity of 44.7 MHz. This provides a cable capacity equivalent to more than 8000 simultaneous two-way conversations.

Major new developments in lightwave communication are predicted to result from research efforts presently under way in integrated optics.[33] This technology incorporates techniques for performing a variety of functions at optical frequencies, such as transmission, modulation, switching, mixing, and upconversion, all on a single substrate or chip. The size of the components involved are potentially in the order of the wavelength of light, in one and possibly two dimensions. Thus an overall integrated optical circuit may be quite small, having linear dimensions less than a centimeter.

In general, lightwave communications appears to be an important emerging technology on the verge of becoming economically feasible. As its application spreads and new developments arise, it is expected to have a major impact in all areas of communications. Economic forecasts for the decade of the 1980s project a multibillion dollar market for optical communications systems and components.[34] It is claimed that optical communications will have an impact on society similar to that of the transistor in the 1950s and the integrated circuit in the 1970s.

B. Spectroscopic Applications (Tunable Diode Lasers)

Diode lasers have a number of unique features, including broad tunability, spectral purity, widely spaced spectral modes, ease of modulation, and operating simplicity, which give them important applications in spectral measurement and analysis. Instruments employing diode lasers as the source of infrared radiation have measurement capabilities in some cases many orders of magnitude better than conventional instruments using thermal emitters.

---

[32] Discussed in W. S. Boyle, *Sci. Am.* **237**, 40 (August 1977).

[33] E. M. Conwell, *Phys. Today* **29**, 48 (1976).

[34] J. D. Montgomery, *Fiber Integrated Opt.* **1**, 101 (1977).

Generally, these advanced spectroscopic applications would not be possible without the broad tuning ranges uniquely attainable with semiconductor lasers. In the first place, Fig. 16 shows that diode lasers can be "compositionally tuned," i.e., their emission wavelengths selected by controlling the composition of the active region during manufacture. An emission wavelength at any preselected value from less than 0.6 μm to beyond 30 μm is, in principle, attainable by this method. Beyond this, however, diode lasers can be tuned during operation by varying one of a number of parameters which affect the bandgap and refractive index. These include pressure, magnetic field, and temperature, variation of the latter being at present the most used in practice. Bias-current variation provides a fine tuning adjustment based on the small but significant I^2R heating.

Using temperature variation, individual lasers have been tuned over 400 cm^{-1}, covering the complete 9–12 μm atmospheric window, for example.[16,27] Figure 19 shows an example of bias-current tuning and illustrates the continuous tuning behavior of individual spectral modes. The limited tuning range of each mode results from the slow variation of the refractive index compared with the variation of the underlying gain peak. These continuous ranges of a few reciprocal centimeters are more than sufficient for most uses, which involve the measurement of spectral features 10^{-4}–10^{-1} cm^{-1} wide, and are far greater than attainable by other types of lasers.

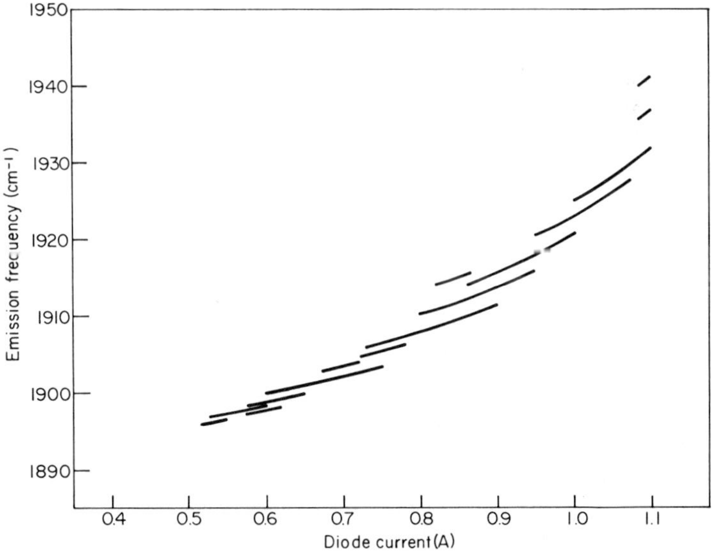

FIG. 19. Current tuning characteristics of a $PbS_{0.65}Se_{0.35}$ diode laser [J. F. Butler, Final Report, NASA Contract NASI-12979 (1975).]

When used in spectroscopic applications, these devices are termed tunable diode lasers. Most tunable diode laser applications so far have utilized the lead salt family of semiconductors, which span a spectral region that includes spectral absorption features of nearly every molecular compound. The need for cryogenic operating temperatures of the lead salt lasers, while undesirable, is not a serious barrier to their practical utilization. Reliable, low-maintenance, mechanical refrigeration systems with the required precision of temperature stability and control are available for cooling and temperature tuning the lasers without liquid cryogens.

The remainder of this section will discuss demonstrated applications of tunable diode lasers to high resolution spectral absorption measurements and the potential applications to infrared analytical instrumentation and heterodyne radiometry.

1. *High Resolution Spectrometry*

Infrared spectrometry has been an important and widely used tool in many branches of science and technology for many years. Conventional spectrometers are designed to use a heated element to produce broadband infrared radiation, which is dispersed by a prism or grating to provide a source of narrow-band energy. This narrow band of radiation may then be tuned in frequency by mechanical adjustment of the dispersing element. In this type of instrument, as the bandwidth is narrowed to increase the resolution of the measurement, the amount of available energy is correspondingly decreased, making its detection increasingly more difficult. Signal processing techniques are employed which generally lead to a trade-off between resolution and measurement time. With high quality commercial dispersion instruments, resolutions near $0.1\ cm^{-1}$ in the infrared region are achievable. Using specially constructed, very expensive instruments in a laboratory environment and very long integration times, this may be reduced in practice to the order of $0.005\ cm^{-1}$.

By contrast, in tunable diode laser spectrometry, all of the output energy is contained in a few discrete emission lines which may be individually isolated to provide a source of extremely narrow ($<0.0001\ cm^{-1}$), intense radiation which is varied in frequency by tuning the laser. The measurement resolution, being governed by the laser linewidth, is orders of magnitude better than that of conventional instruments. Furthermore, as a result of the high intensity of the laser lines, scan times may be very short, allowing, for example, real-time display of high resolution spectra on an oscilloscope. This vastly improved capability allows measurements to be made with relative ease which are completely beyond the limit of dispersion instruments. For example, the strengths and fully resolved line shapes of absorption features in gas phase samples at pressures corresponding to Doppler broadening, can be measured with relative ease using a tunable diode laser spectrometer.

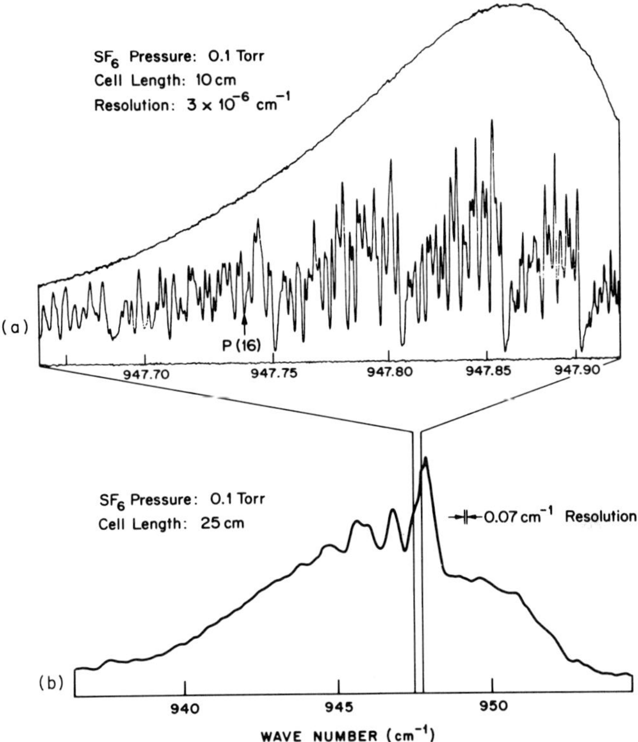

FIG. 20. Spectrum of the ν_3 band of SF_6 measured with (a) a diode laser and, for comparison, (b) a high quality grating spectrometer (Brunet and Perez, 1969). The rich structure brought out by the diode laser scan is completely lost with the conventional measurement. [E. D. Hinkley and P. L. Kelley, *Science* **171**, 635 (1970).]

As an example, Fig. 20 shows a diode laser scan of low pressure SF_6 gas compared with a spectrum measured with a high quality conventional grating spectrometer. This scan was published in 1970 and was the first reported tunable diode laser spectrum. The laser scan is fully resolved and shows a great deal of rich structure which is completely lost in the grating spectrometer measurement. These early results have been followed by many publications describing a wide range of tunable diode laser spectral measurements. Figure 21 shows, as an example, spectra of supercooled CO_2 in the 15 μm Q branch. Complete systems for carrying out tunable diode laser spectrometry are now available commercially.

2. *Analytical Instruments*

Infrared analytical instruments have long been used for air pollution monitoring, toxic gas monitoring, process control, and many other applications

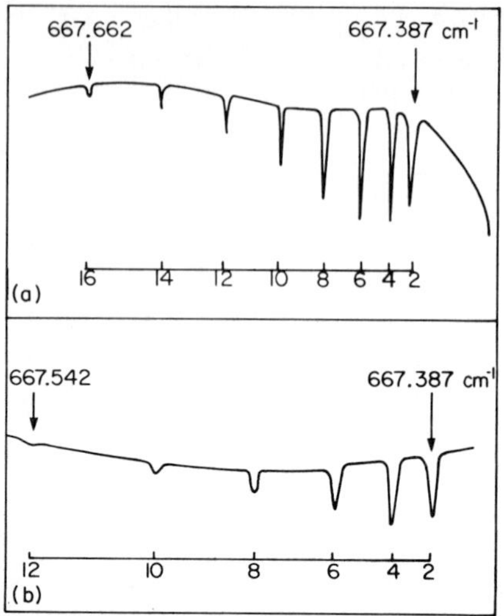

FIG. 21. Absorption spectrum of expansion-cooled CO_2 near 12 μm. (a) Expansion of pure CO_2, $T_{rot} = 32$ K, (b) expansion of 10% CO_2 in He, $T_{rot} = 18$ K. [D. N. Travis, J. C. McGurk, D. McKeown, and R. G. Denning, *Chem. Phys. Lett.* **45**, 287 (1977).]

where trace quantities of gases are detected by their infrared signatures. Conventional infrared instruments use heated elements as the source of infrared radiation and dispersion elements or filters to select desired spectral intervals. Infrared instruments of this class have, in many instances, been developed to a high degree of sophistication.

The use of a tunable diode laser as the source of radiation offers the prospect of vastly improved performance in many types of infrared analytical instruments and of measurement functions never before possible. Some of the advantages provided by tunable diode lasers are as follows.

(a) The high spectral purity of the emission makes it possible to select one or more absorption lines in a spectral region containing lines of interfering constituents. This is important, for example, where relatively large and variable quantities of atmospheric water vapor, CO_2, etc., effectively mask the spectrum of the trace gas to be detected except for a few narrow windows of good transmission.
(b) The high intensity in conjunction with the coherence of the emission makes measurements over path lengths as long as a kilometer relatively straightforward.
(c) The ease of tuning makes possible differential absorption, spectroscopy, and other techniques which utilize the information contained in absorption line shapes to enhance discrimination against background absorption.

(d) The ease of rapid modulation makes novel signal processing techniques possible and provides a means of greatly reducing effects of atmospheric turbulence on long-path measurement.

So far, demonstrated applications of tunable diode lasers to monitoring have been primarily in pollution measurements carried out during the past few years at MIT Lincoln Laboratory.[35] Tunable diode lasers have been applied to the detection of ethylene and nitric oxide in automobile exhausts, in stack monitoring of sulfur dioxide, and in a long-path (up to 1 km) ambient air monitoring system used for carbon monoxide, nitric oxide, ammonia, ozone, and other pollutants.

The increasing concern for the environment and worker exposure to hazardous substances, along with the growth in complexity of nearly all types of industrial processing and control, form the basis of a real and expanding need for advanced analytical instrument such as those made possible with tunable diode lasers. The market for such instruments appears to be potentially very large.

C. Infrared Heterodyne Radiometry

In heterodyne detection the diode laser is used as a local oscillator to increase a detector's effective sensitivity by a large amount. In the heterodyne radiometer configuration, the incoming optical signal is broadband, sunlight, for example, and the diode laser only interacts with a small spectral interval defined by its own emission frequency and the bandwidth of the detection electronics. If an absorbing gas is interposed between the broadband source and detector, its spectrum can be measured with high resolution and sensitivity by then tuning the local oscillator through the absorbtion lines.

This general approach has potential applications of considerable significance, although it is still in the research and development stage at the time of this writing. The technique was, for example, used to measure a spectrum of ozone in the 10 μm region in the upper atmosphere. Since the spectrum obtained was fully resolved, one could, in principle, calculate the concentration profile of atmospheric ozone from the results.[36] This offers the interesting possibility of ground-based monitors to track ozone concentration on a global basis.

Other potential applications of heterodyne radiometry include remote stack gas monitors, plume detection, and various process control uses.

---

[35] E. D. Hinkley, *Opt. Quantum Electron.* **8**, 155 (1976).

[36] M. A. Frerking and D. J. Muehlner, *Appl. Opt.* **16**, 526 (1977).

D. Additional Application Areas

Diode lasers are being used in many applications besides communications and spectroscopy. A few of these uses are described in the following paragraphs and generally employ GaAs or other III–V compound lasers operating at room temperature.[37]

a. Range Finding. In the range finder, the laser transmits a narrow pulse that is reflected from the objects of interest and returned to a sensitive receiver in the instrument. An electronic circuit measures the round trip time of the pulse, multiplies by half of the speed of light, and gives a readout of distance. Diode laser range finders are generally less accurate and have less range capability than range finders employing high power lasers, but are compact, inexpensive, and easy to operate.

b. Aids for the Blind. The range finding concept is also used as a basis for aids for the blind. These handheld instruments, of which there are a number of configurations, provide a signal in the form of a vibration or audible tone warning of obstacles or dropoffs.

c. Fuzing. Another application of range finding of considerable interest to the military, is in fuzing bombs or missile warheads. Built-in range finders employing inexpensive diode lasers cause detonation at preset heights.

d. Illuminators. Covert illuminators use diode laser arrays to produce infrared power in the kilowatt range to illuminate wide areas of terrain. Infrared detection systems then view the activity in the illuminated region. Illuminators are of interest to both the military and civilian police forces.

e. Intrusion Alarms. Intrusion alarms using visible light beams were in use for many years before the laser. Replacing the visible light source with a diode laser operating in the infrared makes the beam invisible to the intruder, and coverage of much longer paths possible.

f. Laser Pump Source. Arrays of diode lasers can be used to optically pump YAG or other lasers with relatively high efficiency. The emission wavelength can be compositionally tuned to match the minimum required absorption energy in the pumped laser. Therefore, very little energy is lost in heating.

[37] R. W. Campbell and F. M. Mims, III, "Semiconductor Diode Lasers." Sams and Co., Bobbs-Merrill Co., Indianapolis, Indiana, 1972.

CHAPTER 4

Acousto-Optics

ADRIANUS KORPEL

Department of Electrical and Computer Engineering, University of Iowa, Iowa City, Iowa

I. INTRODUCTION

The word "acousto-optics" describes the field of light–sound interaction phenomena. That such interactions exist is due to the fact that a sound wave induces localized refractive index variations in the medium through which it travels. That the interactions are not just a curiosity but form the basis of a whole range of practical devices (modulators, beam deflectors, frequency shifters, etc.) follows from the fact that it is simple to generate and control such sound waves. Moreover, the effect is relatively strong, and important device parameters such as amount of light deflected, angle of deflection, and frequency shift are directly related to sound power and sound frequency; hence they can be controlled electronically.

Before discussing in some detail the physics and engineering of acousto-optics, it is useful to discuss a simple experiment that shows in rudimentary form most of the features to be found in modern acousto-optic devices. The experimental setup is shown in Fig. 1. It consists of a water-filled glass tank into which a sound wave frequency f_s is launched by means of a transducer.

ISBN 0-12-408606-3

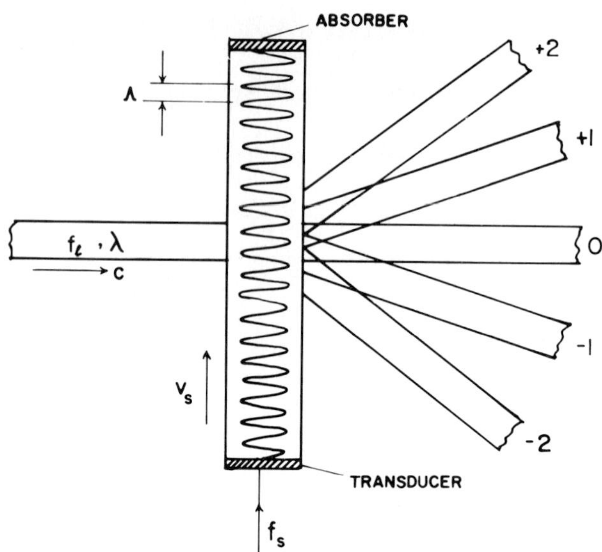

FIG. 1. Diffraction of a light beam into separate orders by interaction with a high-frequency sound wave.

The far end of the tank is lined with a sound absorbing material in order to prevent reflections. When a collimated beam of light is made to impinge on the tank from the left, we observe that the light leaving the tank is split into many orders, labeled -2, -1, 0, $+1$, $+2$, etc., in the drawing. Upon increasing the sound power the number of orders visible is seen to increase gradually, although any one of them appears to go through maxima and minima for specific sound levels (Fig. 10). The angle between the various orders does not depend on the sound amplitude. However, if we change the sound frequency, the deflection angles change, and careful measurement reveals that the nth order is characterized by an angle Θ_n, where $\sin \Theta_n = n\lambda/\Lambda$, λ and Λ denoting the wavelengths of light and sound, respectively.

It thus appears that the sound wave acts like a thin phase grating with a grating period equal to the sound wavelength and a refractive index variation Δn, proportional to the sound level. From what was said before, this seems plausible enough: the rarefactions and compressions in the sound wave induce a periodic change in the refractive index of the medium.

What about the fact that the sound wave (and hence the induced grating) is moving upward with velocity V_s? This should give a Doppler shift which, for the nth order, is given by $\Delta f = f_l(V_s \sin \Theta_n)/c = nf_s$, where f_l and c denote the frequency and velocity of the light. Frequency measurements on the various orders do, in fact, indicate that the nth order is upshifted (or downshifted when n is negative) by n times the frequency of the sound.

If a standing wave is used in the experiment, each of the orders will be observed to be *modulated*, rather than frequency shifted, at a frequency f_s. The sound wave may now be thought of as inducing a *stationary* grating with a refractive index varying in time only.

The simple experiment described above has revealed some very interesting phenomena which may well be put to use in a practical device: modulation, deflection, and frequency shifting. The engineering aspects of acousto-optics are essentially concerned with optimizing these effects for specific applications.

To achieve this it is necessary to analyze in greater detail the effect of certain parameters such as, for instance, the width of the sound beam (i.e., thickness of the induced grating). As we will see later, increasing this width leads to a mode of operation more analogous to X-ray diffraction in crystals than diffraction by a thin phase grating. Consequently, useful effects occur only at certain critical angles (Bragg angles); this in turn has an important bearing on the design of devices, specifically as regards efficiency and bandwidth.

Another important parameter is the width of the incident beam of light relative to the wavelength of sound. If this width is small, the concept of diffraction loses much of its attraction; it is then more convenient to analyze the phenomenon in terms of a local time-varying gradient and curvature of refractive index (Fig. 2). Rather than splitting up in various orders, the beam as a whole will now be deflected and defocused periodically. Such low-frequency phenomena have been used, for instance, in a traveling lens application. However,

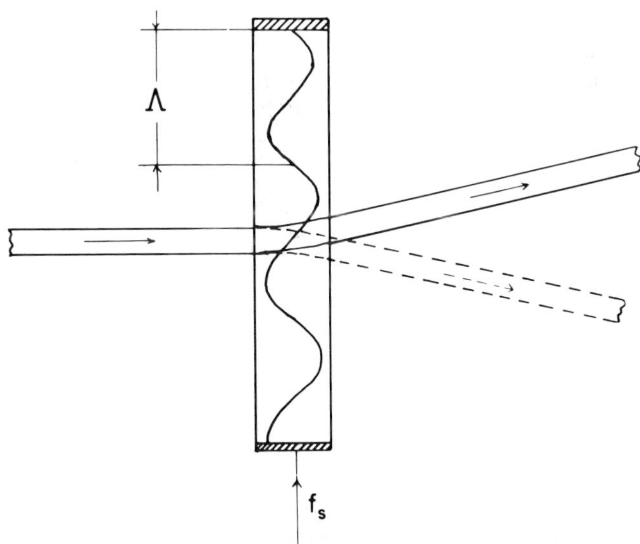

FIG. 2. Deflection of a light beam by interaction with a low-frequency sound wave.

most modern acousto-optic devices use the high-frequency diffraction effect in some form or other, generally in combination with a laser.

Prior to the invention of the laser, practical applications of acousto-optics had been limited to relatively few areas such as low-frequency light modulation, schlieren-type studies of acoustic fields, measurement of acoustic phase velocities etc.[1,2] The availability of coherent light sources stimulated a new interest in this field and gave rise to many new applications which had previously been impossible to implement. To name but a few, lasers were used to generate sound[3,4], to investigate thermal phonon processes,[5] and to probe acoustic fields[6,7]. Conversely, sound cells were used to frequency shift,[8] deflect[9,10] and modulate[11] coherent light for purposes of signal processing[12–14] and display.[15] Reviews of these and similar applications may be found in Adler,[2] Gordon,[16] Quate *et al.*,[17] Korpel,[18–19] Chang,[20] Schmidt,[21] and Stegemann.[22]

The underlying theory most often used in the past is that of Raman and Nath.[23–28] It is restricted to a plane wave of light traversing a rectangular

---

[1] L. Bergmann, "Der Ultraschall," Chapter 3. Hirzel, Stuttgart, 1954.

[2] R. Adler, *IEEE Spectrum* **4**, 42 (May 1967).

[3] R. Y. Chiao, C. H. Townes, and B. P. Stoicheff, *Phys. Rev. Lett.* **12**, 592 (1964).

[4] A. Korpel, R. Adler, and B. Alpiner, *Appl. Phys. Lett.* **5**, 86 (1964).

[5] G. Benedek and T. Greytak, *Proc. IEEE* **53**, 1623 (1965).

[6] M. G. Cohen and E. I. Gordon, *Bell Syst. Tech. J.* **44**, 693 (1965).

[7] A. Korpel and L. W. Kessler, *Acoust. Hologr.* **2**, Chapter 9 (1970).

[8] A. E. Siegman, C. F. Quate, J. Bjorkholm, and G. Francois, *Appl. Phys. Lett.* **5**, 1 (1964).

[9] A. Korpel, R. Adler, P. Desmares, and T. M. Smith, *IEEE J. Quantum Electron.* **QE-1**, 60 (1965).

[10] H. A. Heynan and G. M. Barnard, *Proc. Elec. Opt. Syst. Design Conf.*, p. 640, (1970).

[11] W. H. Watson and G. W. Hrbek, *Proc. Elec. Opt. Syst. Design Conf.* p. 630 (1970).

[12] R. L. Whitman, A. Korpel, and S. N. Lotsoff, *Proc. Symp. Mod. Opt.* p. 243. Polytechnic Press, Brooklyn, New York, 1967.

[13] D. H. McMahon, *Proc. IEEE* **55**, 1602 (1967).

[14] M. King, W. R. Bennett, L. B. Lambert, and M. Arm. *Appl. Opt.* **6**, 1367 (1967).

[15] A. Korpel, R. Adler, P. Desmares, and W. Watson, *Proc. IEEE* **54**, 1429 (1966).

[16] E. I. Gordon, *Proc. IEEE* **54**, 1391 (1966).

[17] C. F. Quate, C. D. W. Wilkinson, and D. K. Winslow, *Proc. IEEE* **53**, 1604 (1965).

[18] A. Korpel, *Appl. Solid State Sci.* **3**, 72 (1972).

[19] A. Korpel, *in* "Optical Information Processing" (Y. E. Nesterikhin, G. W. Stoke, and W. E. Kock, eds.), p. 171. Plenum Press, New York, 1976.

[20] I. C. Chang, *IEEE Trans. Sonics Ultrason.* **SU-23**, 2 (1976).

[21] R. V. Schmidt, *IEEE Trans. Sonics Ultrason.* **SU-23**, 22 (1976).

[22] G. I. Stegemann, *IEEE Trans. Sonics Ultrason.* **SU-23**, 33 (1976).

[23] C. V. Raman and N. S. N. Nath, *Proc. Indian Acad. Sci.* **2**, 406 (1935).

[24] Lord Rayleigh, "The Theory of Sound," Vol. 2, Section 272a. Dover, New York, 1945.

[25] C. V. Raman and N. S. N. Nath, *Proc. Indian Acad. Sci.* **2**, 413 (1935).

[26] C. V. Raman and N. S. N. Nath, *Proc. Indian Acad. Sci.* **3**, 75 (1936).

[27] C. V. Raman and N. S. N. Nath, *Proc. Indian Acad. Sci.* **3**, 119 (1936).

[28] C. V. Raman and N. S. N. Nath, *Proc. Indian Acad. Sci.* **3**, 459 (1936).

sound column. Modern treatments that allow light and sound fields of less restricted shape are to be found in Gordon[16] and Korpel.[18] The method adopted here combines both aspects. It follows the treatment in a previous article[18] but leaves out those detailed mathematical derivations that are of little relevance to engineering applications.

The point of departure for the understanding of device physics is the plane wave interaction analysis. This will be discussed in some detail form a heuristic point of view. Although at first glance this analysis may appear to be of somewhat academic interest, it is, in fact, of great benefit in trying to understand such diverse items as the difference between Raman–Nath interaction and Bragg diffraction, acousto-optics of birefringent media, bandwidth of modulators, etc.

Next, the Raman–Nath equations will be discussed; again the derivation is treated heuristically and most of the mathematics is concerned with applying these equations to actual devices.

A section on acousto-optic materials will acquaint the reader with current terminology and measurement techniques, as well as provide some insight into basic physical mechanisms.

The part dealing with applications is limited to devices and techniques that are of current general practical interest. Thus Bragg diffraction imaging, sound field sampling, etc., have been left out. The interested reader will find such special applications discussed in some detail in Ref. 18.

A. Heuristic Background

The interaction of plane waves of light and sound is conveniently illustrated by a momentum or wave vector diagram. In this diagram, the interacting plane waves are considered to represent photons and phonons having a completely specified momentum and, consequently (through the uncertainty relation), a completely indeterminate location.[29] The interaction between these plane waves is then depicted as a collision process between photons and phonons in which the total momentum must be conserved. Thus, if the incident plane waves are characterized by a wave vector $\mathbf{k}$ for the light, and $\mathbf{K}$ for the sound, the corresponding momentum vectors are (according to the wave–particle equivalence of quantum mechanics) given by $\hbar\mathbf{k}$ and $\hbar\mathbf{K}$, where $\hbar = h/2\pi$ and h denotes Planck's constant.

Two kinds of collision processes are possible, corresponding to the quantum-mechanical model of absorption and stimulated emission, respectively.[30] In the first one, that of absorption, one phonon is absorbed for every scattered

[29] R. W. Ditchburn, "Light," Chapter 18. Wiley (Interscience), New York, 1963.

[30] R. W. Ditchburn, "Light," Appendix 19C. Wiley (Interscience), New York, 1963.

photon generated. Denoting the wave vector of the scattered photon by $\mathbf{k}_+$ we may write the condition for conservation of momentum

$$\hbar\mathbf{k} + \hbar\mathbf{K} = \hbar\mathbf{k}_+ \tag{1.1}$$

In the second process the incident phonon stimulates the emission of a photon plus another phonon. This emitted phonon has the same momentum as the incident one. If we denote the wave vector of the scattered photon by $\mathbf{k}_-$

$$\hbar\mathbf{k} + \hbar\mathbf{K} = \hbar\mathbf{k}_- + \hbar\mathbf{K} + \hbar\mathbf{K} \tag{1.2}$$

Dividing by $\hbar$, (1.1) and (1.2) may be more conveniently written

$$\mathbf{k} + \mathbf{K} = \mathbf{k}_+ \tag{1.3}$$

$$\mathbf{k} - \mathbf{K} = \mathbf{k}_- \tag{1.4}$$

The two-dimensional wave vector diagrams illustrating (1.3) and (1.4) are shown in Fig. 3. In three dimensions the interaction is best illustrated by vector cones obtained by rotating the triangles of Fig. 3 about any of the three vectors. Thus a particular propagating vector of scattered light is contributed to by a cone of vectors of sound and incident light.

In addition to the momentum, the energy must be conserved in the process. If ω, Ω, ω_+, and ω_- are the frequencies of the incident light, sound, and scattered light [ω_+ for (1.3), ω_- for (1.4)] then the energy of the corresponding particles is given by $\hbar\omega$, $\hbar\Omega$, $\hbar\omega_+$, and $\hbar\omega_-$. For the two cases just discussed the law of energy conservation may then be written (after division by $\hbar$)

$$\omega + \Omega = \omega_+ \tag{1.5}$$

$$\omega - \Omega - \omega_- \tag{1.6}$$

where (1.5) refers to (1.3) and (1.6) to (1.4). The light of frequency ω_+ will hereafter be referred to as the upshifted light, that of frequency ω_- as the downshifted light.

Classically, Eqs. (1.3)–(1.6) may be derived by considering the interaction to be a parametric process in which the sound wave perturbs the dielectric constant (refractive index, polarizability) of the medium according to a plane

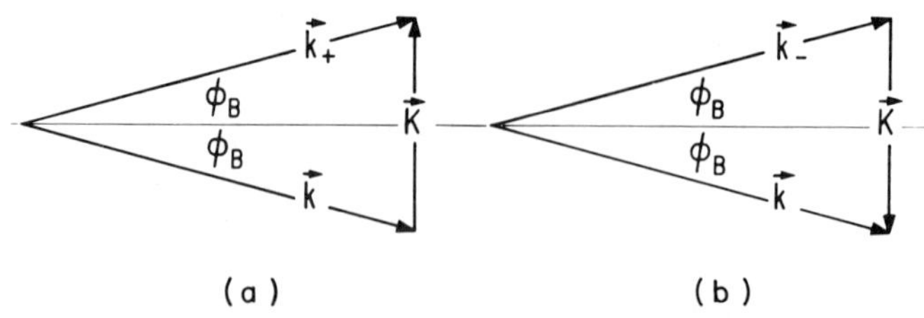

FIG. 3. Wave vector interaction diagrams. (a) Upshifted interaction. (b) Downshifted interaction. Arrows over symbols indicate vector quantities.

wave dependence $S\cos(\Omega t - \mathbf{K}\cdot\mathbf{r})$. The incident light $E\cos(\omega t - \mathbf{k}\cdot\mathbf{r})$ then induces a spatiotemporal polarization of the form $ES\cos(\Omega t - \mathbf{K}\cdot\mathbf{r}) \times \cos(\omega t - \mathbf{k}\cdot\mathbf{r})$. This polarization field may be considered to consist of terms $\frac{1}{2}ES\cos[(\omega + \Omega)t - (\mathbf{K} + \mathbf{k})\cdot\mathbf{r}]$ and $\frac{1}{2}ES\cos[(\omega - \Omega)t - (\mathbf{k} - \mathbf{K})\cdot\mathbf{r}]$. If one now calculates the electric field generated by this polarization by integrating over the interaction volume, it is found that essentially two plane waves of scattered light exist with frequencies and wave vectors given by (1.3)–(1.6). The above method was first used by Brillouin[31,32] whose work implicitly contains the relations (1.3)–(1.6). In essence, then, the conditions (1.3) and (1.4) represent conditions of phase synchronism which implicitly specify propagation directions for which mutual, spatially extended, cumulative interaction is possible.

It should be noted that in the foregoing it is assumed that the induced polarization is caused by the incident field only, i.e., the effect of the polarization induced by the scattered light (rescattering) is ignored. This is the so-called weak interaction case.

In the usual experimental conditions, the sound frequency is very small compared with the light frequency

$$\Omega \ll \omega, \tag{1.7}$$

and consequently

$$k_+ = \frac{\omega_+}{v} = \frac{\omega + \Omega}{v} \approx \frac{\omega}{v} = k \tag{1.8}$$

$$k_- = \frac{\omega_-}{v} = \frac{\omega - \Omega}{v} \approx \frac{\omega}{v} = k \tag{1.9}$$

where v and k denote the light velocity and the propagation constant in the medium of interaction. Hence the wave vector diagram illustrating the vector relations (1.3) and (1.4) is an isosceles triangle ($k_+ = k$, $k_- = k$) as shown in Figs. 3a and 3b, respectively. The angle ϕ_B between the direction of incident (or diffracted) light and the normal to the sound propagation vector (i.e., between the incident light and the sound wave fronts) is given by

$$\sin\phi_\mathrm{B} = K/2k = \lambda/2\Lambda \tag{1.10}$$

where λ and Λ are the wavelengths of light and sound, respectively. This angle is called the Bragg angle in analogy to X-ray diffraction where similar processes take place.[33] (Instead of sound wave fronts one considers atomic planes.) In the usual acoustic Bragg diffraction experiments the Bragg angle is small, typically smaller than 10°, even for sound frequencies as high as 1 GHz.

---

[31] L. Brillouin, *Ann. Phys.* (*Paris*) **17**, 88 (1922).

[32] L. Brillouin, *Actual. Sci. Ind.* **59** (1933).

[33] W. H. Zachariasen, "Theory of X-Ray Diffraction in Crystals." Dover, New York, 1967.

It may be shown[18] that in two dimensions, at least, the plane wave amplitude $E_+(\mathbf{k}_+)$ of the upshifted light is related to that of the incident light $E(\mathbf{k})$ and the sound $S(\mathbf{K})$ in the following way

$$E_+(\mathbf{k}_+) \propto \frac{p}{\lambda \cos \phi_B} E(\mathbf{k})S(\mathbf{K}) \tag{1.11}$$

In this relation S denotes sound strain or dilatation and p is the so-called strain–optic coefficient of the medium

$$p = \frac{\Delta(1/n^2)}{S} \tag{1.12}$$

where n is the refractive index. For the downshifted light a similar formula applies but with S replaced by its complex conjugate S^*

$$E_-(\mathbf{k}_-) \propto \frac{p}{\lambda \cos \phi_B} E(\mathbf{k})S^*(\mathbf{K}) \tag{1.13}$$

The heuristic plane wave interaction model is of great usefulness in preliminary analysis of experimental configurations. As an example, consider the experiment shown in Fig. 4, where a plane wave of light is incident at an angle ϕ on a sound column of width L. Assume first that L is large enough so that the sound field may be thought of as a plane wave. It is then clear from Fig. 3 that no interaction will take place at all unless $\phi = \pm\phi_B$. In that case, just one order of diffracted light will be generated. What happens when we decrease L? The sound column will act less and less like a single plane wave and, in fact, it is now more appropriate to consider it an angular spectrum of plane waves.

Figure 5 shows such a spectrum pertaining to a sound column of width L and wavelength Λ. Its shape is the familiar $\sin X/X$ curve with the first zero at an angle which is approximately equal to Λ/L (when $\Lambda/L \ll 1$) and the peak amplitude proportional to L. The plane wave of incident light is indicated on the same diagram by an arrow, labeled 0, at the Bragg angle ($\phi \approx \frac{1}{2}\lambda/\Lambda$). This plane wave will interact with a sound plane wave at $\phi = 0$ to create a downshifted plane wave of light (arrow labeled -1 at $\phi \approx -\frac{1}{2}\lambda/\Lambda$). Because the

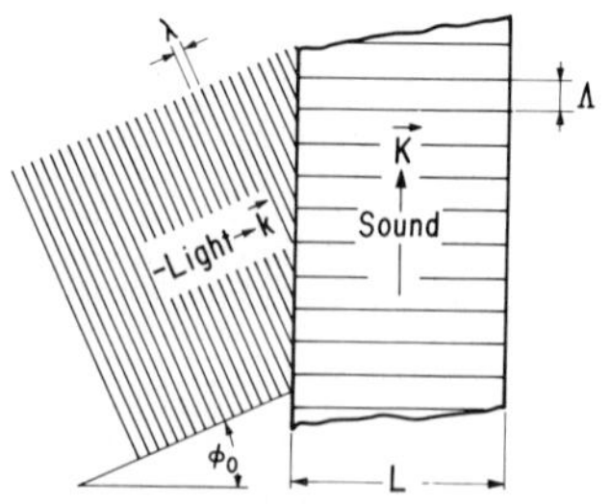

FIG. 4. Conventional light–sound interaction experiment with plane wave of light incident under an angle ϕ on sound column of width L.

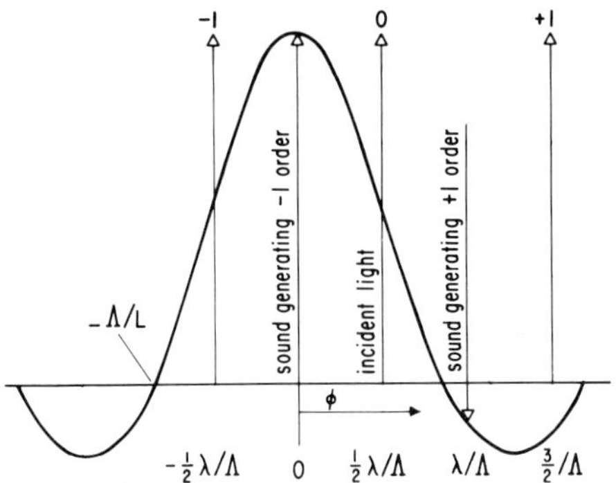

FIG. 5. Angular plane wave spectrum of sound column with arrows indicating the various interacting plane waves.

amplitude of the plane wave of sound at $\phi = 0$ is proportional to L, the same will apply to the diffracted light. Hence, to a first approximation (i.e., that of weak interaction) *the amplitude of the diffracted order will increase linearly with the interaction length.* (The weak interaction assumption in this case means that we ignore rescattering from the minus-first order back into the zeroth order. If this is taken into account we find that eventually when all the power is in the minus-first order, it will start flowing back into the zeroth order. In common with a large class of similar coupled wave phenomena, the process is characterized by a periodic transfer of power from one mode to the other. We will come back to this later.)

Inspection of Fig. 5 shows that, in addition to the minus-first order, a small amount of plus-first order will be generated. The plane wave of sound responsible for this is indicated by the arrow at $\phi \approx \lambda/\Lambda$. With respect to this plane wave, the light is incident at an angle $\phi_B = -\frac{1}{2}\lambda/\Lambda$, a condition appropriate to the generation of the plus-first order. The latter has been shown in the diagram by the arrow labeled $+1$ at $\phi = +1\frac{1}{2}\lambda/\Lambda$. Because the sound wave at $\phi - \lambda/\Lambda$ is relatively weak, the amount of plus-first order is relatively small. It will be clear, however, that when L decreases, the relative strength of the two orders tends to become equal. On the basis of this it is feasible to define a minimum length L such that it is possible to predominantly generate one order only, whenever L is larger than this minimum length. The region so defined is called the Bragg region. If, as an arbitrary basis, we use the criterion that the sound wave responsible for the generation of the residual order ($+1$ in the example used) must fall well outside the main lobe of the sin X/X curve we find, from Fig 5,

$$\lambda/\Lambda \gg \Lambda/L$$

or

$$L \gg \Lambda^2/\lambda \tag{1.14}$$

as the defining definition for the Bragg region. Later on this same definition will be derived from mathematical considerations regarding phase synchronous interaction.

With the heuristic model developed so far, it is relatively simple to explain a number of features characteristic of acousto-optic modulators and deflectors.[34] Also, an important experimental observation by Cohen and Gordon[6] can be explained readily: *When, in weak two-dimensional Bragg interaction* (1.14), *the angle of the incident light is varied, the strength of the diffracted light as a function of this angle will trace out the radiation pattern of the transducer.* It will be clear that, physically, this phenomenon is a direct result of the one-to-one correspondence between plane waves of sound and of scattered light as illustrated by Figs. 3a and 3b. A more far-reaching consequence of this same principle is that, *when a wide uniform angular spectrum of light is incident on a sound field, the angular spectrum of scattered light mirrors that of the sound, i.e., the scattered light carries a spatial image of the sound field.* This principle which was first postulated and experimentally verified by the present author[35] is the basis of the various techniques called Bragg diffraction imaging. It should be stressed again that this simple plane wave correspondence principle only applies in two dimensions. In three dimensions the situation is more complicated because, as said before, one scattered plane wave vector may have been contributed to by a whole cone of incident light and sound vectors.

Originally developed in connection with imaging, one other heuristic conception is quite useful: that of the ray diagram. If, in the limit, all wavelengths involved are assumed to be infinitely small, even a very thin pencil of light or sound (a ray) may be considered to represent a plane wave. The concept of Bragg interaction then refers to localized conditions of rays meeting at the proper angles, and the wave vector diagrams of Fig. 3 are replaced by the ray diagrams of Fig. 6. As shown there, A or B now signify points in real space. For a given distribution of sound and light rays they form a locus of ray crossings from which diffracted rays originate. The eikonal or ray theory developed recently[36] allows calculation of the diffracted ray amplitude.

The inverse relation to (1.14), i.e.,

$$L \ll \Lambda^2/\lambda \tag{1.15}$$

defines what is known as the Raman–Nath or Debye–Sears region. In this region, for weak interaction, two orders are generated simultaneously. For strong interaction many orders exist because plane waves of sound are available at the various angles required for rescattering. The principle of the generation of many orders by rescattering is illustrated in Fig. 7a, while Fig. 7b shows the

[34] E. I. Gordon, *Proc. IEEE* **54**, 1391 (1966).

[35] A. Korpel, *Appl. Phys. Lett.* **9**, 425 (1966).

[36] A. Korpel, *Acoust. Hologr.* **2**, Chapter 4 (1970).

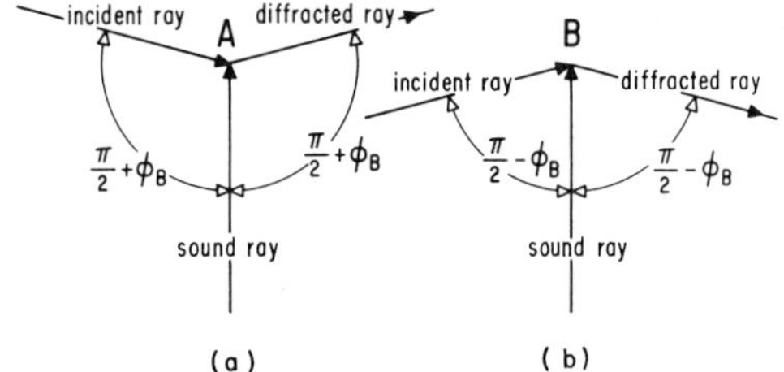

FIG. 6. Ray interaction diagrams. (a) Upshifted interaction. (b) Downshifted interaction (after Korpel[36]).

various plane waves in the sound spectrum that take part in the interaction. It is evident from Fig. 7a that the angle ϕ_n between the nth order and the zeroth order is given by

$$\phi_n - \phi_0 = 2n\phi_{\mathrm{B}} \tag{1.16}$$

In the usual derivation of the Debye–Sears phenomena a slightly different relation is obtained

$$\sin \phi_n - \sin \phi_0 = nK/k = 2n \sin \phi_{\mathrm{B}} \tag{1.17}$$

This is because the sharply bounded nonspreading sound column, which is the usual physical model, cannot be decomposed into plane waves with a unique magnitude of the K vector. The difference is slight and not really important for small Bragg angles but it may lead to confusion if the somewhat artificial nature of the bounded column is not realized. The calculation of the strength of the various orders in strong interaction, on the basis of plane wave theory, has not been attempted to the author's knowledge. The usual analysis, which

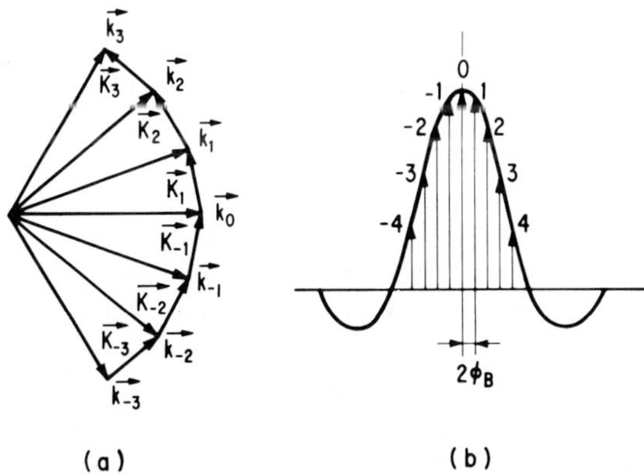

FIG. 7. Multiple order generation by rescattering. (a) Wave vector diagram. (b) Magnitude of interacting acoustic plane waves.

is the one followed in this article, employs the Raman–Nath equations and adopts the bounded sound column model because it provides simple boundary conditions.

II. DIFFRACTION BY A SOUND COLUMN

In this section a brief derivation will be given of the operating equations describing specific sound and light field configurations.

It might, at first glance, be thought that such derivations may well be left out because the final operating equations are all that is needed for engineering applications. In the author's experience this is not the case, however. In working with acousto-optic devices the practicing engineer may be confronted, at first glance, with many, rather puzzling effects. It is true that sometimes the operational equations predict these effects, provided these equations are applicable for the working range of the device. In many cases, however, the equations are insufficient, i.e., no exact solutions have been worked out.

A typical example would be a device working between the Raman–Nath region (thin phase grating) and the Bragg region (thick phase grating). In using such a device the engineer may notice that undesirable orders appear which he had not counted on. At present, there are no explicit equations available to him describing this quantitatively. If, however, he had strengthened his physical intuition by understanding the derivation of the Bragg region equation, he would have been able to make qualitative predictions based on such physical concepts as phase synchronism, coupled modes, etc. Moreover, he may already be very familiar with such concepts from his work in other areas, for instance, microwaves. To obtain quantitative information he may now refer back to less restrictive equations, make approximations appropriate to his own specific conditions, and then derive computer solutions.

Thus the most important benefit of understanding the derivations may well be the insight gained into the applicability of the operating equations and the ability to make predictions even when conditions differ from those in textbook examples.

In a number of cases, the experimental configuration is characterized by sound propagating in one direction only, with the incident light similarly restricted to one angle of incidence. For low sound frequencies the bounded column configuration shown in Fig. 8a is most common; for very high frequencies the plane wave of finite length, illustrated in Fig. 8b, sometimes applies. For most of the current applications Fig. 8a applies; consequently, we shall restrict ourselves to this case. The interested reader will find applications characterized by Fig. 8b in Quate *et al.*[17] and Korpel.[18]

As pointed out before, the angle ϕ_n of orders diffracted by a column of sound

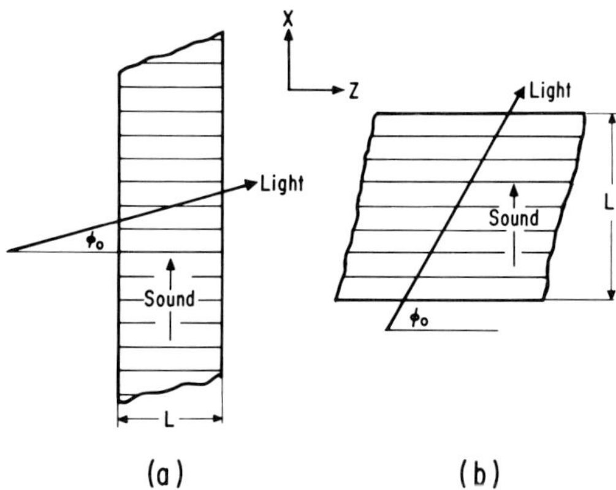

FIG. 8. Conventional interaction configurations. (a) Small Bragg angle scattering. (b) Large Bragg angle scattering.

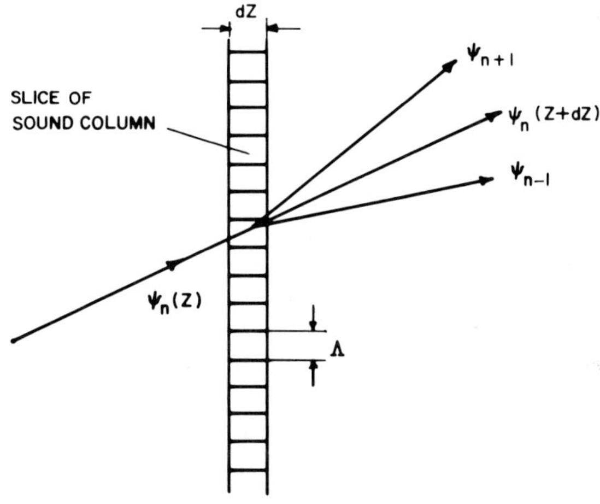

FIG. 9. Generation of new orders by thin slice of sound column.

is given by Eq. (1.17) which, in essence, is just the grating equation of optics.[37] We shall represent the various orders as plane waves

$$E_n(z, x) = \psi_n(z) \exp[-jkz \cos \phi_n - jkx \sin \phi_n] \tag{2.1}$$

A convenient way to analyze the interaction between these orders is by dividing the sound beam into infinitesimally thin slices dz as illustrated in Fig. 9. Because

[37] E. B. Brown, "Modern Optics," Chapter 8. Van Nostrand-Reinhold, Princeton, New Jersey, 1965.

each slice represents an infinitesimally thin phase grating, a particular order ψ_n incident on it will give rise to only two additional diffracted orders. It is then easy to see that this leads mathematically to an infinite set of differential equations expressing the contribution $d\psi_n$ to a particular order in terms of the orders ψ_{n-1} and ψ_{n+1} which coexist with ψ_n at the point under consideration (i.e., z) in the sound beam. This method of analysis was first used by Van Cittert[38] and leads immediately to the Raman–Nath equations discussed below.

A. Raman–Nath Equations

Van Cittert's analysis leads to the following equations

$$\begin{aligned}\frac{d\psi_n}{dz} &= \frac{-jkCS}{4\cos\phi_n}\psi_{n-1}\exp[-jkz(\cos\phi_{n-1}-\cos\phi_n)] \\ &\quad + \frac{-jkCS^*}{4\cos\phi_n}\psi_{n+1}\exp[-jkz(\cos\phi_{n+1}-\cos\phi_n)]\end{aligned} \tag{2.2}$$

where

$$C = -n^2 p \tag{2.3}$$

The boundary conditions are

$$\psi_n = 0 \qquad \text{for} \quad z < 0 \qquad n \neq 0 \tag{2.4}$$

$$\psi_0 = E_0 \qquad \text{for} \quad z < 0 \tag{2.5}$$

Equation (2.2) is one variant of the celebrated Raman–Nath equations.[35] They are usually stated in a slightly different fashion because of a definition different from (2.1) and one which does not represent plane waves with propagation constant k. We shall, however, use (2.2), as this formulation is both simpler and more readily understood from the physical point of view.

As stated previously, each order ψ_n is being contributed to by adjacent orders ψ_{n+1} and ψ_{n-1}. Generally, the phase of the contributions varies with z, and the exponents $k(\cos\phi_{n-1} - \cos\phi_n)$ and $k(\cos\phi_{n+1} - \cos\phi_n)$ denote the lack of phase synchronism in this process. Because of this synchronism there will be little net interaction in the general case. Broadly speaking, there are only two experimental configurations in which strong effects occur.

(a) The interaction length L is short enough so that the accumulated degree of phase mismatch is small. (A more precise criterion will be given later.) This is called the Raman–Nath or Debye–Sears region

[38] P. H. van Cittert, *Physica* **4**, 590 (1937).

and, for strong interaction, is characterized by the simultaneous existence of many orders.

(b) There exists phase synchronism between the orders 0 and -1, or 0 and $+1$ by virtue of the condition

$$\phi_{\pm 1} = -\phi_0 \tag{2.6}$$

Condition (2.6) implies that the zeroth order and diffracted order propagate symmetrically with respect to the sound wave fronts. The angle of incidence of the zeroth order, according to (2.6) and (1.17), is given by

$$\sin \phi_0 = \sin \phi_B = K/2k \qquad \text{for generation of the } -1 \text{ order} \tag{2.7}$$

$$\sin \phi_0 = -\sin \phi_B = -K/2k \qquad \text{for generation of the } +1 \text{ order} \tag{2.8}$$

This region, characterized by the existence of two orders only, is called the Bragg region; the angle ϕ_B defined in (2.7) is called the Bragg angle. For the generation of only the $+1$ or -1 order to be possible it is necessary to choose L sufficiently large so that a large degree of phase mismatch exists for all other orders.

B. Raman–Nath Diffraction

Assuming small values of K/k the phase-asynchronism terms $(\cos \phi_{n+1} - \cos \phi_n)$ and $(\cos \phi_{n-1} - \cos \phi_n)$ may be expressed in a power series by using (1.17)

$$\cos \phi_{n+1} - \cos \phi_n = -\frac{K}{k} \tan \phi_0 - \left(n + \frac{1}{2}\right)\left(\frac{K}{k}\right)^2 \frac{1}{\cos \phi_0} + \cdots \tag{2.9}$$

$$\cos \phi_{n-1} - \cos \phi_n = \frac{K}{k} \tan \phi_0 + \left(n - \frac{1}{2}\right)\left(\frac{K}{k}\right)^2 \frac{1}{\cos \phi_0} + \cdots \tag{2.10}$$

For $\phi_0 = 0$, the accumulated degree of phase mismatch for order n is negligible if

$$n(K^2/k)L \ll 1 \tag{2.11}$$

where L is the interaction length as shown in Fig. 8a. Condition (2.11) is the criterion for operation in the Raman–Nath or Debye–Sears region with the configuration of Fig. 8a. It is usually stated with $n = 1$, although, strictly, the condition is more stringent the more orders are considered. In the subsequent calculations we shall assume that (2.11) is satisfied and neglect the $(K/k)^2$ term in (2.9) and (2.10) for all n.

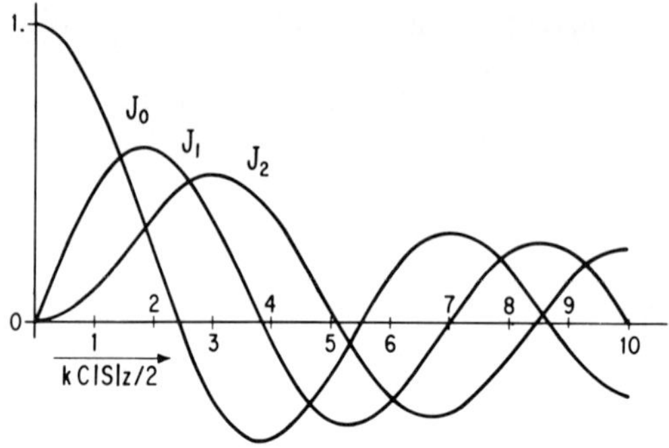

FIG. 10. Amplitude of scattered orders in Debye–Sears interaction as a function of peak phase delay $\hat{\alpha}_z = kC|S|z/2$.

1. *Perpendicular Incidence*

Historically, the first case treated is the one for perpendicular incidence, i.e., $\phi_0 = 0$. With (2.9) and (2.10) we may write for (2.2)

$$\frac{d\psi_n}{dz} = \frac{kC|S|}{4}[\psi_{n-1} - \psi_{n+1}] \qquad \phi_n \ll 1 \tag{2.12}$$

where, for simplicity we have chosen S to equal $j|S|$. Using the functional relation for Bessel functions[39]

$$dJ_n/dz = \tfrac{1}{2}J_{n-1} - \tfrac{1}{2}J_{n+1} \tag{2.13}$$

it follows immediately, with (2.5), that inside the sound field

$$\psi_n = E_0 J_n(kC|S|z/2) \tag{2.14}$$

This behavior is illustrated in Fig. 10 for various values of n.

2. *Oblique Incidence*

For oblique incidence ($\phi_0 \neq 0$) we may write for (2.2), with (2.9) and (2.10),

$$\frac{d\psi_n}{dz} = \frac{kC|S|}{4\cos\phi_0}[\psi_{n-1}e^{-jKz\tan\phi_0} - \psi_{n-1}e^{+jKz\tan\phi_0}], \tag{2.15}$$

where, in the denominator, we have approximated $\cos\phi_n$ by $\cos\phi_0$.

Applying (2.13), together with a second recurrence formula[39]

$$(2n/z)J_n = J_{n-1} + J_{n+1} \tag{2.16}$$

[39] I. A. Stegun, *in* "Handbook of Mathematical Functions" (M. Abramowitz and I. A. Stegun, eds.), Chapter 9. Dover, New York, 1965.

to a newly defined function

$$I_n = \exp(jnbz)J_n[(\alpha \sin bz)/b] \tag{2.17}$$

it is easy to prove that

$$dI_n/dz = \tfrac{1}{2}\alpha \exp(2jbz)I_{n-1} - \tfrac{1}{2}\alpha \exp(-2jbz)I_{n+1} \tag{2.18}$$

Comparing (2.18) with (2.15) we find immediately

$$\alpha = kC|S|/2 \cos \phi_0 \qquad \text{and} \qquad b = (-K/2) \tan \phi_0 \tag{2.19}$$

and, with (2.5),

$$\psi_n = E_0 \exp(-jn\tfrac{1}{2}Kz \tan \phi_0)J_n\left[\frac{kC|S|z}{2 \cos \phi_0} \frac{\sin(\frac{1}{2}Kz \tan \phi_0)}{\frac{1}{2}Kz \tan \phi_0}\right] \tag{2.20}$$

With (1.12) and (2.3) it is easily shown that

$$\hat{\alpha}_z = \Delta(kz) \tag{2.21}$$

Thus $\hat{\alpha}_z$ represents the accumulated peak phase delay over a distance z. It is to be noted that, in the case of oblique incidence (2.20) the phase modulation index is multiplied by a factor

$$\frac{1}{\cos \phi_0} \frac{\sin(\frac{1}{2}Kz \tan \phi_0)}{\frac{1}{2}Kz \tan \phi_0}$$

This factor represents the decrease in integrated phase delay due to the fact that rays now no longer propagate parallel to the sound wave front. Indeed for an angle of incidence given by

$$\tan \phi_0 = m\Lambda/z = 2\pi m/Kz \tag{2.22}$$

the rays have traversed all possible phases of the sound and as a consequence the integrated phase delay equals zero. This is shown in Fig. 11.

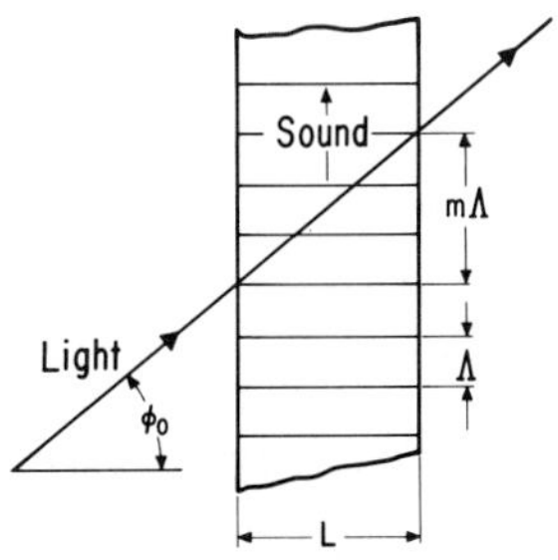

FIG. 11. Phase cancellation along incident ray for angle of incidence ϕ_0 such that $\tan \phi_0 = m\Lambda/L$.

C. Bragg Diffraction

As pointed out before, this region is characterized by the phase-synchronous interaction of two orders (0 and $+1$ or 0 and -1). The interaction length is assumed to be long enough so that the accumulated phase mismatch for the other orders is sufficiently large that their contribution may be neglected. The criterion for this is the opposite of that which defines the Debye–Sears region (2.11) and may be written

$$(K^2/k)L \gg 1 \tag{2.23}$$

where the strongest condition is implied (i.e., $n = 1$).

1. *Bragg Angle Incidence*

We shall now calculate the interaction between orders 0 and -1 (the other case proceeds quite similarly) with the assumption that $\psi_n = 0$ for all n other than 0 and -1. With (2.6) it follows from (2.2) that

$$\frac{d\psi_0}{dz} = \frac{-jkCS}{4 \cos \phi_B} \psi_{-1} \tag{2.24}$$

$$\frac{d\psi_{-1}}{dz} = \frac{-jkCS^*}{4 \cos \phi_B} \psi_0 \tag{2.25}$$

Equations (2.24) and (2.25) are the standard expressions for coupled modes; their solution, taking the proper boundary value (2.5) into account, is given by

$$\psi_0 = E_0 \cos(kC|S|z/4 \cos \phi_B) \tag{2.26}$$

$$\psi_{-1} = -j(S^*/|S|)E_0 \sin(kC|S|z/4 \cos \phi_B) \tag{2.27}$$

As illustrated in Fig. 12, the interaction is characterized by a periodic exchange of power from one order into the other, the first total power transfer being reached when the accumulated peak phase delay ($\alpha_z = kC|S|z/2 \cos \phi_B$) equals π radians.

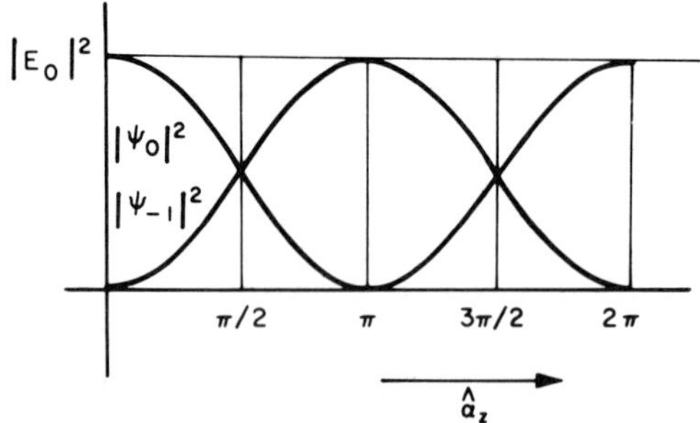

Fig. 12. Amplitude of orders 0 and -1 in small Bragg angle interaction as a function of peak phase delay $\hat{\alpha}_z = kC|S|z/2 \cos \phi_B$.

2. *Deviations from Bragg Angle Incidence*

It is sometimes of interest to consider small deviations from phase synchronism due to the fact that the incident angle no longer satisfies the exact condition (2.7). Let us write

$$\phi_0 = +\phi_B + \Delta\phi \tag{2.28}$$

where $\Delta\phi$ is a small angle such that $\cos\Delta\phi \approx 1$ and $\sin\Delta\phi \approx \Delta\phi$. From (1.17) it is then shown readily that (for small Bragg angles)

$$\phi_{-1} \approx -\phi_B + \Delta\phi \tag{2.29}$$

whence

$$\cos\phi_{-1} - \cos\phi_0 \approx 2\Delta\phi \sin\phi_B = (K/k)\,\Delta\phi \tag{2.30}$$

With these modifications it follows from (2.2) that

$$\frac{d\psi_0}{dz} = \frac{-jkCS}{4\cos\phi_B}\psi_{-1}\exp(-jKz\,\Delta\phi) \tag{2.31}$$

$$\frac{d\psi_{-1}}{dz} = \frac{-jkCS^*}{4\cos\phi_B}\psi_0\exp(+jKz\,\Delta\phi) \tag{2.32}$$

where the denominator has been approximated by $\cos\phi_B \approx \cos\phi_0 \approx \cos\phi_1$. The following solution for ψ_{-1} may be readily derived from (2.31) and (2.32)

$$\psi_{-1} = -j\frac{S^*(kC|S|/4\cos\phi B)E_0}{|S|[(kC|S|/4\cos\phi_B)^2 + (K\,\Delta\phi/2)^2]^{1/2}}\exp[jK\,\Delta\phi\, z/2]$$
$$\times \sin\{[(kC|S|/4\cos\phi_B)^2 + (K\,\Delta\phi/2)^2]^{1/2}z\} \tag{2.33}$$

Inspection of (2.33) shows that there still exists a periodic, although incomplete, transfer of power. This is illustrated in Fig. 13.

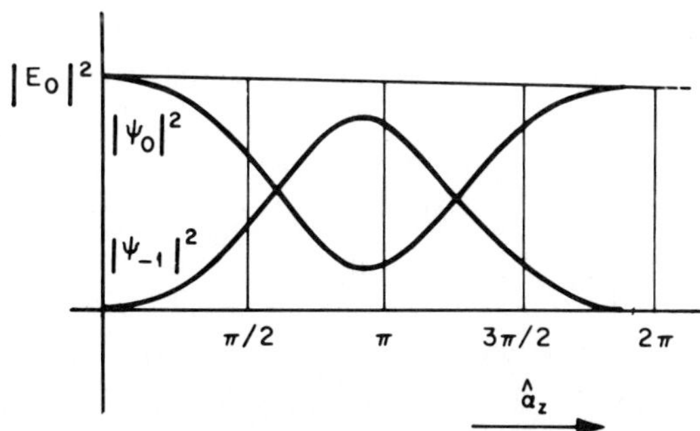

FIG. 13. Amplitude of orders 0 and -1 in near synchronous small Bragg angle interaction as a function of peak phase delay $\hat{\alpha}_z = kC|S|z/2\cos\phi_B$.

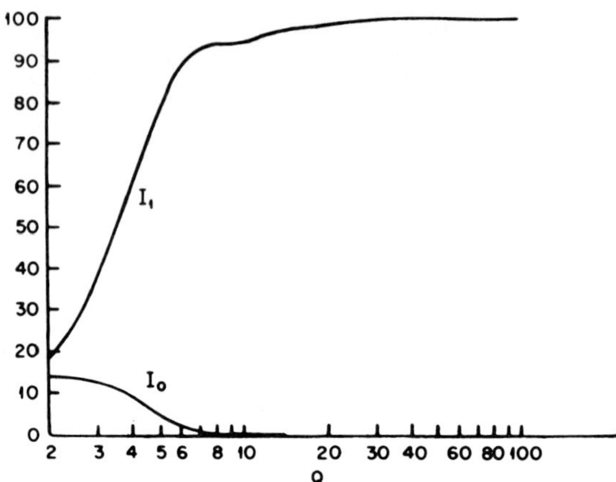

FIG. 14. Intensity of orders 0 and 1 in near Bragg region interaction as a function of the parameter $Q = K^2L/k$. The peak phase delay is assumed to equal 180° (after Klein and Cook[40]).

To explicitly show the effect of $\Delta\phi$ it is useful to write down the weak interaction ($S \to 0$) version of (2.33)

$$\underset{S\to 0}{\psi_{-1}(z)} \propto (kCSz/4\cos\phi_B)E_0 \exp(jK\,\Delta\phi\,z/2)\sin(Kz\,\Delta\phi/2)/(Kz\,\Delta\phi/2) \quad (2.34)$$

It should be noted that the sin X/X term in (2.34) describes the angular plane wave spectrum of a transducer with width z. Thus, if the power in the scattered light is measured and plotted as a function of the incident angle $\Delta\phi$, the radiation pattern of the transducer will be traced out. This was already briefly mentioned in Section I, and was seen to be a consequence of the one-to-one correspondence in plane wave interaction.

3. *Near Bragg Region*

The condition (2.23) for operation in the Bragg region is necessarily rather vague. Often it is of interest to be able to define this region more precisely by, for instance, specifying the maximum percentage of light that can be diffracted in a Bragg interaction of the kind described by (2.26) and (2.27). This problem has been tackled by Klein and Cook[40] who calculated numerically the amount of light diffracted as a function of the parameter.

$$Q = K^2L/k \quad (2.35)$$

which forms the left-hand side of condition (2.23). They assumed that the value of the argument $(kC|S|z/4\cos\phi_B)$ in (2.26) equalled $\pi/2$ so that for the ideal Bragg case described by (2.26) and (2.27) all the light would have been diffracted. Figure 14 shows a plot of their results. It is seen from the graph that

[40] W. R. Klein and B. D. Cook, *IEEE Trans. Sonics Ultrason.* **SU-14**, 123 (1967).

for $Q \to \infty$, $I_1 \to 100\%$ as is to be expected. Note that, in general, $I_1 + I_0 \neq 100\%$ due to the generation of other orders. If the Bragg region is defined arbitrarily by the condition $I_1 > 90\%$, then a more quantitative restatement of (2.23) is given by

$$Q = K^2 L/k > 7 \tag{2.36}$$

III. ACOUSTO-OPTIC MATERIALS

A. Macroscopic Description

The strain-optic or elasto-optic coefficients p_{ij} describe the change in orientation and size of the optical indicatrix[41]

$$\frac{x_1^2}{n_1^2} + \frac{x_2^2}{n_2^2} + \frac{x_3^2}{n_3^2} + \frac{2x_2x_3}{n_4^2} + \frac{2x_3x_1}{n_5^2} + \frac{2x_1x_2}{n_6^2} = 1 \tag{3.1}$$

brought about by application of the mechanical strains[41,42] S_j ($j = 1$–6)

$$\Delta(1/n_i^2) = p_{ij} S_j \tag{3.2}$$

The stress-optic or piezo-optic constants q_{ij} refer to the application of stresses T_j rather than strains

$$\Delta(1/n_i^2) = q_{ij} T_j \tag{3.3}$$

The constants p and q are related as follows

$$p_{ij} = q_{ik} c_{jk} \tag{3.4}$$

$$q_{ij} = p_{ik} s_{jk} \tag{3.5}$$

where c_{jk} and s_{jk} are the elastic stiffness constants and elastic compliance constants, respectively.

The relations used here are based on the definitions of the 1949 IRE Standards.[42] According to these standards extensional strain and tensile stress are taken as positive. The abbreviated notation S_j or T_j ($j = 1$–6) is derived from S_{kl} ($k, l = 1$–3) by the following rules:

$$\begin{aligned} S_{11} &= S_1 & T_{11} &= T_1 \\ S_{22} &= S_2 & T_{22} &= T_2 \\ S_{33} &= S_3 & T_{33} &= T_3 \\ 2S_{23} &= S_4 & T_{23} &= T_4 \\ 2S_{31} &= S_5 & T_{31} &= T_5 \\ 2S_{12} &= S_6 & T_{12} &= T_6 \end{aligned}$$

---

[41] J. F. Nye, "Physical Properties of Crystals." Oxford Univ. Press (Clarendon), London and New York, 1960.

[42] Standards on piezoelectric crystals, *Proc. IRE* **37**, 1391 (1949).

Notice the additional factor 2 in the abbreviation of what are often called "engineering shear strains" $S_4 \to S_6$. They are also used in the well-known book by Nye.[41] The conventional strains S_{kl} are defined by the relations

$$S_{kl} = \tfrac{1}{2}(u_{kl} + u_{lk}) = \tfrac{1}{2}(\partial u_k/\partial x_l + \partial u_l/\partial x_k) \tag{3.6}$$

where $\partial u_k/\partial x_l$ is the gradient in the direction x_l of the displacement u_k along x_k. In a nonpiezoelectric crystal the relation between stresses and strains is given by

$$S_i = s_{ij} T_j \tag{3.7}$$

$$T_i = c_{ij} S_j \tag{3.8}$$

where s_{ij} and c_{ij} are the same constants as used in (3.4) and (3.5). In some older definitions where compressional stress is taken as positive, relations (3.3) are often preceded by a minus sign so that the same values of q apply as in the present case. The notation π_{ij} instead of q_{ij} is followed by some authors.[41] The number of photoelastic (p's or q's) constants varies from 36 in a triclinic crystal to three or four in a cubic crystal. In an isotropic solid only two coefficients are different

$$p_{11} = p_{22} = p_{33}, \qquad p_{12} = p_{21} = p_{13} = p_{23} = p_{32}$$
$$p_{44} = p_{55} = p_{66} = \tfrac{1}{2}(p_{11} - p_{12})$$

all other coefficients are zero.

It has recently been shown by Nelson and Lax[43,44] that the formulation (3.2) is not sufficiently general to describe all acousto-optical phenomena. Specifically, the case of shear waves in anisotropic media is not covered by the conventional description. This is so because a shear wave not only causes a shear strain but also a physical rotation of the infinitesimal volume elements.[45] The rotation R_{kl} may be expressed in the displacement gradients u_{mn} in the following way

$$R_{mn} = \tfrac{1}{2}(u_{mn} - u_{nm}) \tag{3.9}$$

with the abbreviated notation R_k ($k = 4$–6), where $23 \to 4$, $31 \to 5$, $12 \to 6$. The three rotations defined by (3.9) cause a change in the local refractive index if the medium is optically anisotropic. To take this into account a second term has to be added to the right-hand side of (3.2)

$$\Delta(1/n_i^2) = p_{ij} S_j + \bar{p}_{ik} R_k, \qquad i = 1 \to 6, \quad k = 4 \to 6 \tag{3.10}$$

This introduces a maximum of 18 new constants $\bar{p}_{ik}$. As they relate to a simple rotation of the crystal they may be calculated directly from the optical indicatrix. Table I lists the values of $\bar{p}_{ik}$ calculated according to a formula given by

[43] D. F. Nelson and M. Lax, *Phys. Rev. Lett.* **24**, 379 (1970).

[44] D. F. Nelson and M. Lax, *Phys. Rev. B* **3**, 2778 (1971).

[45] H. Kolsky, "Stress Waves in Solids," Chapter 2. Dover, New York, 1963.

TABLE I

ACOUSTO-OPTIC COEFFICIENTS FOR ROTATION

$\bar{p}_{ik}$	$k = 4$	$k = 5$	$k = 6$
$i = 1$	0	$-2n_5^{-2}$	$2n_6^{-2}$
$i = 2$	$2n_4^{-2}$	0	$-2n_6^{-2}$
$i = 3$	$-2n_4^{-2}$	$2n_5^{-2}$	0
$i = 4$	$(n_3^{-2} - n_2^{-2})$	n_6^{-2}	$-n_5^{-2}$
$i = 5$	$-n_6^{-2}$	$(n_1^{-2} - n_3^{-2})$	n_4^{-2}
$i = 6$	n_5^{-2}	$-n_4^{-2}$	$(n_2^{-2} - n_1^{-2})$

Nelson and Lax.[43] As seen from Table I the number of new nonzero constants introduced ranges from 0 for a cubic crystal ($n_1 = n_2 = n_3, n_4 = n_5 = n_6 = \infty$) to 15 for a triclinic crystal (all n_k's exist and are different).

Another formulation of (3.10) is possible by expressing the change in the optical indicatrix directly in terms of the displacement gradients:

$$(1/n_i^2) = p'_{ikl} u_{kl} \tag{3.11}$$

An abbreviated form for p'_{ikl} has been suggested by Nelson and Lax based on contracting the subscripts kl in the following way: $11 \to 1$, $22 \to 2$, $33 \to 3$, $23 \to 4$, $32 \to \bar{4}$, $31 \to 5$, $12 \to 6$, $21 \to \bar{6}$. Using this notation it can be shown that

$$\begin{aligned} p'_{ik} &= p_{ik} \qquad k \leq 3 \\ p'_{ik} &= p_{ik} + \tfrac{1}{2}\bar{p}_{ik} \qquad k > 3 \\ p'_{i\bar{k}} &= p_{ik} - \tfrac{1}{2}\bar{p}_{ik} \end{aligned} \tag{3.12}$$

For acousto-optic applications it is useful to define a figure of merit M that would relate the amount of light diffracted (P_d) per unit sound intensity (I_s) to the properties of the medium. As seen before, for small sound intensities we may write

$$P_d \propto (\Delta n)^2 \tag{3.13}$$

Neglecting tensor properties it follows from (3.2) that

$$(\Delta n)^2 \propto n^6 p^2 S^2 \tag{3.14}$$

It may be shown readily[46] that the following relation holds for an isotropic medium

$$I_s = \tfrac{1}{2} S^2 \rho V^3 \tag{3.15}$$

[46] A. Wood, "Acoustics," Sect. 11.3. Wiley (Interscience), New York, 1941.

where ρ is the density of the medium and V the sound velocity. From (3.13)–(3.15) it follows that

$$P_d/I_s \propto n^6 p^2/\rho V^3 \tag{3.16}$$

On the basis of this, Smith and Korpel[47] defined a figure of merit

$$M = n^6 p^2/\rho V^3 \tag{3.17}$$

Later investigators[34,48–50] defined additional figures of merit

$$M_1 = MnV^2 \tag{3.18}$$

$$M_3 = MnV \tag{3.19}$$

$$M_4 = M(nV^2)^2 \tag{3.20}$$

which apply to the optimization of some special interaction configurations such as used in light deflectors. We will return to this later in Section IV,B,4. In order to avoid confusion the constant M (3.17) is often referred to as M_2 in the literature. The figure of merit as defined in (3.17)–(3.20) must, of course, be specified with respect to the particular mode of operation for which it applies. This determines which ρ and V are appropriate.

B. Microscopic Description

Under certain simplifying assumptions, the index of refraction n in cubic or isotropic materials is related to the molecular polarizability α by the Lorentz–Lorenz equation[51]

$$(n^2 - 1)/(n^2 + 2) = AN\alpha \tag{3.21}$$

where A is a constant and N the number of dipoles per unit volume. Following Pinnow[52] we find, by differentiation of (3.21),

$$\rho\, dn/d\rho = [(n^2 - 1)(n^2 + 2)/6n](1 - \Lambda_0) \tag{3.22}$$

where

$$\Lambda_0 = -(\rho/\alpha)\, d\alpha/d\rho \tag{3.23}$$

and ρ is the density of the material.

---

[47] T. M. Smith and A. Korpel, *IEEE J. Quantum Electron.* **QE-1**, 283 (1965).

[48] R. W. Dixon, *J. Appl. Phys.* **38**, 5149 (1967).

[49] E. I. Gordon, *IEEE J. Quantum Electron.* **QE-2**, 104 (1966).

[50] I. C. Chang, *l.c.*[20]

[51] A. R. von Hippel, "Dielectrics and Waves," Chapter 2. Wiley, New York, 1954.

[52] D. A. Pinnow, *IEEE J. Quantum Electron.* **QE-6**, 223 (1970).

In hydrostatic compression, all longitudinal strain components are equal; hence

$$S_1 = S_2 = S_3 = -\tfrac{1}{3}\Delta\rho/\rho \tag{3.24}$$

From (3.22), (3.24), and (3.2) it follows readily that, for an isotropic material,

$$p_{\text{av}} = (p_{11} + p_{12} + p_{13})/3 = [(n^2 - 1)(n^2 + 2)/3n^4](1 - \Lambda_0) \tag{3.25}$$

From (3.25) it is clear that the change in refractive index is brought about both by a change in the density of elementary oscillators and by a change in their intrinsic polarizability. The first factor is identified by setting $\Lambda_0 = 0$ in (3.25). It results in a value of p_{av} close to $\frac{1}{3}$ for the usual values of n. The second factor is represented by Λ_0. As pointed out by Pinnow, this factor can be both negative or positive, depending on the crystal. Most materials, however, appear to exhibit a positive value of Λ_0. Thus for most alkali halides $\Lambda_0 \approx 0.50$–0.70. For YAG the peculiar situation obtains where $\Lambda_0 \approx 0.99$, so that the resulting p_{av} is very small. For lead nitrate $\Lambda_0 \approx -0.25$, resulting in an increase in p_{av}. For a wide range of materials p_{av} appears to be between 0.05 and 0.60. The older, very high values for lead nitrate, sometimes found in the literature[53] appear to disagree with more recent measurements[47] which give only moderate values. Although semiempirical theories exist[54] for the calculation of those photoelastic constants that refer to compression, no comprehensive theories exist for constants such as p_{44}, p_{55}, etc., that refer to distortion only. Usually, however, these constants are $\frac{1}{2}$ to 1 order of magnitude smaller than the first mentioned ones.

C. Measurements

In earlier static methods of measurement, compression was applied to a crystal by means of a lever arrangement and the induced birefringence was measured by means of a Babinet compensator.[55] With this technique the piezo-optic constants q rather than the elasto optic constants p are the primary measured quantities. The p's are obtained by using relations (3.4), a procedure which adversely affects the final accuracy because of the many separate uncertainties involved. The first dynamic measurements of photoelastic properties were proposed by Bergmann and Fues[56] and by Mueller.[57] As an accurate measurement of sound intensity is extremely difficult to perform, the quantities measured by this method are ratios of certain elasto-optic coefficients rather

[53] S. Bhagavantam and K. V. Krishna Rao, *Acta Crystallogr.* **6**, 799 (1953).

[54] E. Burstein and P. L. Smith, *Proc. Indian Acad. Sci.* **A28**, 377 (1948).

[55] S. Bhagavantam and D. Suryanarayana, *Proc. Indian Acad. Sci.* **A26**, 97 (1947).

[56] L. Bergmann and E. Fues, *Naturw. Bd.* **24**, 492 (1936).

[57] H. Mueller, *Z. Kristallogr.* **99**, 122 (1938).

than absolute values.[58] Thus, with longitudinal waves in amorphous solids, one can measure the ratio p_{12}/p_{11} by observing the change in the amount of diffracted light occurring when the polarization of incident light is rotated by 90°.

More recent methods measure the figure of merit M_2 of a material with respect to a standard reference material, usually water or fused quartz. Knowing the other parameters appearing in M_2 (i.e., n, ρ, V) one is then able to derive a value for ρ in terms of the standard. The first such technique was proposed and evaluated by Smith and Korpel[47] who used the configuration shown in Fig. 15. Pulsed, high frequency sound waves are launched by the transducer into the reference liquid. The solid under investigation is immersed in this liquid with appropriate orientation of the crystal axes. Three separate measurements of Bragg diffracted light are performed with the incident laser beam at positions A, B, and C such that B is in the center of the solid and A and C are in the liquid, equidistant from B. If the sound intensities at A, B, and C are denoted by I_A, I_B, and I_C, respectively, the amounts of diffracted light P_a, P_b, and P_c may be expressed as

$$P_a = DM_{2\mathrm{r}} I_a P_0 \tag{3.26}$$

$$P_b = DM_{2\mathrm{s}} I_b P_0 \tag{3.27}$$

$$P_c = DM_{2\mathrm{r}} I_c P_0 \tag{3.28}$$

where D is a constant, P_0 the power of the incident light, $M_{2\mathrm{r}}$ the figure of merit of the reference liquid, $M_{2\mathrm{s}}$ that of the solid under investigation. With the location of A, B, and C chosen as described, it is readily shown that independent of sound attenuation in the solid and liquid, and independent of the sound reflection at the solid–liquid interface, the following relation holds

$$I_b = (I_a I_c)^{1/2} \tag{3.29}$$

Substituting (3.29) into (3.26)–(3.28) it follows that

$$M_{2\mathrm{s}}/M_{2\mathrm{r}} = P_b/(P_a P_c)^{1/2} \tag{3.30}$$

The use of pulsed sound has the advantage that resonances in the solid can be avoided, so that the expression on the right-hand side of (3.30) is simply evaluated by collecting the diffracted light with a photodetector and measuring the height of the generated current pulses. Moreover, the relative time of occurrence of the various pulses (including echos from the crystal faces) serves as a precise means to determine the exact spatial location of A, B, and C.

---

[58] H. F. Gates and E. A. Hiedemann, *J. Acoust. Soc. Am.* **28**, 1222 (1950).

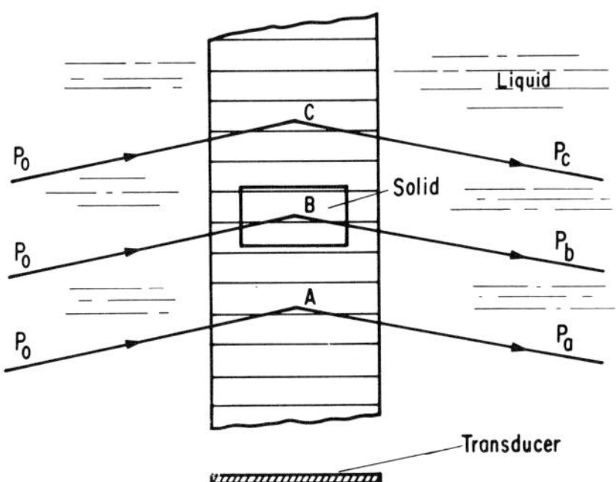

FIG. 15. Technique for measuring figure of merit M of a solid with respect to a liquid (after Smith and Korpel[47]).

Although the technique of Fig. 15 is simple and makes it possible to check various crystal orientations quickly, it does not lend itself readily to the evaluation of photoelastic constants pertaining to shear. To that purpose the method has been modified by Dixon and Cohen[59] who use a quartz buffer rod rather than a liquid medium as a reference. An extensive list of constants so determined has been given by Dixon.[48] This list is included in Table II together with data on some newer materials such as TeO_2,[60,61] $PbMoO_4$,[62] $Pb_5(GeO_4)(VO_4)_2$,[63] BGO, BSO, and SBN,[64] and chalcogenide glass ($Ge_{33}As_{12}Se_{55}$).[65] The last entries contain results of early measurements[47] on some glasses, plastics, and the infrared material KRS-5. More information may be found in Chang,[20] Uchida and Niizeki,[66] Pinnow,[67] and Gottlieb *et al.*[68]

It should be noted that with the dynamic methods discussed here, it is generally not possible to determine the sign of the photoelastic constants. In order to measure this the older static techniques have to be used or a more recently developed refractive method.[69]

---

[59] R. W. Dixon and M. G. Cohen, *Appl. Phys. Lett.* **8**, 205 (1966).

[60] G. Arlt, J. Liebertz, and H. Schweppe, *Int. Congr. Acoust., 6th, Tokyo* Paper H-3-3 (1968).

[61] N. Uchida and Y. Ohmachi, *J. Appl. Phys.* **40**, 4592 (1969).

[62] D. A. Pinnow, L. C. van Uitert, A. W. Warner, and W. A. Bonner, *Appl. Phys. Lett.* **15**, 83 (1969).

[63] T. Yano, Y. Nabita, and A. Watanabe, *Appl. Phys. Lett.* **18**, 570 (1971).

[64] E. L. Venturini, E. G. Spencer, and A. A. Ballman, *J. Appl. Phys.* **40**, 1622 (1969).

[65] J. T. Krause, C. R. Kurkjian, D. A. Pinnow, and E. A. Sigety, *Appl. Phys. Lett.* **17**, 367 (1970).

[66] N. Uchida and N. Niizeki, *Proc. IEEE* **61**, 1073 (1973).

[67] D. A. Pinnow, *in* "CRC Handbook of Lasers" (R. J. Pressley, ed.). Chemical Rubber Company, Cleveland, Ohio, 1971.

[68] M. Gottlieb *et al.*, *J. Appl. Phys.* **45**, 5145 (1974).

[69] D. K. Biegelsen, *Appl. Phys. Lett.* **22**, 221 (1973).

TABLE II
PROPERTIES OF ACOUSTO-OPTIC MATERIALS

Material	λ (μm)	n	ρ (g/cm^3)	Acoustic wave polarization and direction	V (10^5 cm/sec)	Optical wave[a] polarization and direction	$M_1(n^7p^2/\rho V)$[b]	$M_2(n^6p^2/\rho V^3)$	$M_3(n^7p^2/\rho V^3)$
Fused quartz	0.63	1.46	2.2	Long.	5.95	$\perp$	7.89×10^{-7}	1.51×10^{-18}	1.29×10^{-12}
Fused quartz	0.63			Trans.	3.76	$\parallel$ or $\perp$	0.963	0.467	0.256
GaP	0.63	3.31	4.13	Long. in [110]	6.32	$\parallel$	590	44.6	93.5
GaP	0.63			Trans. in [100]	4.13	$\parallel$ or $\perp$ in [010]	137	24.1	33.1
GaAs	1.15	3.37	5.34	Long. in [110]	5.15	$\parallel$	925	104	179
GaAs	1.15			Trans. in [100]	3.32	$\parallel$ or $\perp$ in [010]	155	46.3	49.2
TiO_2	0.63	2.58	4.6	Long. in [11$\bar{2}$0]	7.86	$\perp$ in [001]	62.5	3.93	7.97
$LiNbO_3$	0.63	2.20	4.7	Long. in [11$\bar{2}$0]	6.57	c	66.5	6.99	10.1
YAG	0.63	1.83	4.2	Long. in [100]	8.53	$\parallel$	0.16	0.012	0.019
YAG	0.63			Long. in [110]	8.60	$\perp$	0.98	0.073	0.114
YAG	1.15	2.22	5.17	Long. in [100]	7.21	$\perp$	3.94	0.33	0.53
$LiTaO_3$	0.63	2.18	7.45	Long. in [001]	6.19	$\parallel$	11.4	1.37	1.84
As_2S_3	0.63	2.61	3.20	Long.	2.6	$\perp$	762	433	293
As_2S_3	1.15	2.46		Long.		$\parallel$	619	347	236
SF4	0.63	1.616	3.59	Long.	3.63	$\perp$	1.83	1.51	3.97
β-ZnS	0.63	2.35	4.10	Long. in [110]	5.51	$\parallel$ in [001]	24.3	3.41	4.41
β-ZnS	0.63			Trans. in [110]	2.165	$\parallel$ or $\perp$ in [001]	10.6	0.57	4.9
α-Al_2O_3	0.63	1.76	4.0	Long. in [001]	11.15	$\parallel$ in [11$\bar{2}$0]	7.32	0.34	0.66
CdS	0.63	2.44	4.82	Long. in [11$\bar{2}$0]	4.17		51.8	12.1	12.4
ADP	0.63	1.58	1.803	Long. in [100]	6.15	$\parallel$ in [010]	16.0	2.78	2.62
ADP	0.63			Trans. in [100]	1.83	$\parallel$ or $\perp$ in [001]	3.34	6.43	1.83
KDP	0.63	1.51	2.34	Long. in [100]	5.50	$\parallel$ in [010]	8.72	1.91	1.45
KDP	0.63			Trans. in [100]		$\parallel$ or $\perp$ in [001]	1.57	3.83	0.95
H_2O	0.63	1.33	1.0	Long.	1.5		4.36	160	29.1
Te	10.6	4.8	6.24	Long. in [11$\bar{2}$0]	2.2	$\parallel$ in [0001]	10.200	4400	4640
TeO_2	0.63		5.99	Long. in [100]	2.98	[100] in [010]	0.097	0.048	
TeO_2	0.63			Long. in [100]		[001] in [010]	22.9	10.6	

TeO_2	0.63			Long. in [001]	4.26	[100] in [010]	142	34.5	
TeO_2	0.63			Long. in [001]		[001] in [010]	113	25.6	
TeO_2	0.63			[010] in [100]	3.04	arbitrary in [001]	3.70	1.76	
TeO_2	0.63			Long. in [110]	4.21	[110] in $[\bar{1}10]$	323	0.802	
TeO_2	0.63			Long. in [110]		[001] in $[\bar{1}10]$	16.2	3.77	
TeO_2	0.63			Long. in [101]	3.64	[010] in $[\bar{1}01]$	101	33.4	
TeO_2	0.63			Long. in [010]	2.98	[101] in $[\bar{1}01]$	42.6	20.4	
TeO_2	0.63			$[\bar{1}10]$ in [110]	0.617	arbitrary in [001]	68.6	793	
TeO_2	0.63			$[\bar{1}01]$ in [101]	2.08	[100] in [010]	76.4	77	
$PbMoO_4$	0.63		6.95	Long. in [100]	3.98	[001] in [010]		7.5	
$PbMoO_4$	0.63			Long. in [100]	3.98	[100] in [010]		24	
$PbMoO_4$	0.63			Long. in [100]	3.98	[010] in [001]		24	
$PbMoO_4$	0.63			[100] in [010]	2.20	[100] in [001]		12.8	
$PbMoO_4$	0.63			[100] in [001]	1.99				
$PbMoO_4$	0.63			Long. in [001]	3.75	[100] in [010]		35.6	
$PbMoO_4$	0.63			Long. in [001]	3.75	[001] in [010]		35.6	
$PbMoO_4$	0.52			Long. in [001]	3.75	[100] in [010]		43.4	
$PbMoO_4$	0.49			Long. in [001]	3.75	[100] in [010]		56.1	
$Pb_5(GeO_4)(VO_4)_2$	0.63	2.281	7.15	Long. in *y*	3.11	*y* in *x*	78.6	35.6	
	0.63			Long. in *y*		*x* in *z*	57.4	26.0	
	0.63	2.275		Long. in *y*		*z* in *x*	29.5	13.4	
	0.63			Long. in *z*	3.45	*x* in *y*	54.9	20.2	
	0.63			Long. in *z*		*z* in *y*	137.0	50.6	
	0.63			Long. in *y*		*x–y* in *z*	73.9	33.5	
	0.63			*z* in *y*	1.49	arbitrary in *x*		0.03	
	0.63			*x* in *y*	1.66	arbitrary in *z*	0.83	1.32	
$Bi_{12}GeO_{20}$(BGO)	0.63	2.25 at 51 μm	9.2	Long. in [110]	3.42	arbitrary pol.[d]	29.5	9.91	8.64
$Bi_{12}SiO_{20}$(BSO)	0.63			Long. in [100]	3.83	arbitrary pol.	33.8	9.02	8.83
$Sr_{0.75}Ba_{0.25}Nb_2O_6$	0.63	2.30		Long. in [001]		[001] pol.	26.8	38.6	48.8
(SBN)		2.31		Long. in [100]		[100] pol.	26.9	2.66	4.08
$Sr_{0.50}Ba_{0.50}Nb_2O_6$	0.63	2.27		Long. in [001]		[001] pol.	59.3	8.62	10.8
(SBN)		2.31		Long. in [001]		[100] pol.	36.4	5.19	6.62
$Ge_{33}As_{12}Se_{55}$ glass	1.06	2.55	4.0	Long.	2.5			246	

TABLE II (continued)

Material	λ (μm)	n	ρ (g/cm^3)	Acoustic wave polarization and direction	V (10^5 cm/sec)	Optical wave[a] polarization and direction	$M_1(n^7p^2/\rho V)$[b]	$M_2(n^6p^2/\rho V^3)$	$M_3(n^7p^2/\rho V^3)$
Clear lead glass	0.63	1.72	4.8	Long.	3.80	$\perp$		10	
				Long.		$\parallel$		8	
Yellow lead glass	0.63	1.96	6.3	Long.	3.10	$\perp$		16	
				Long.		$\parallel$		20	
Lucite	0.63	1.55	1.18	Long.	2.68	$\perp$		50	
				Long.		$\parallel$		56	
Polystyrene	0.63	1.59	1.06	Long.	2.35	$\perp$		127	
				Long.		$\parallel$		113	
KRS-5	0.63	2.6	7.37	Long.	2.11	$\perp$		180	
				Long		$\parallel$		254	
$PbNO_3$	0.63	1.78	4.7	Long.	2.62	$\perp$		26	
				Long.		$\parallel$		16	
m-Xylene	0.63	1.50	0.86	Long.	1.30	$\perp$ or $\parallel$		40	

[a] The symbols $\parallel$ and $\perp$ refer to polarization directions parallel or perpendicular to the plane of light and sound propagation vectors.
[b] Figures of merit in cgs units.
[c] Dixon and Cohen.[59]
[d] Due to strong optical activity of BGO and BSO the diffracted beam intensity is independent of polarization.

IV. APPLICATIONS

A. Light Modulators

The very first light modulators[2,70] made use of the Debye–Sears effect at relatively low frequencies below 10 MHz. Figure 16a illustrates the principle of operation. The information-bearing signal modulates the rf sound carrier and hence also the amount of diffracted light. The zeroth order light is obstructed by a suitably located stop. A glance at Fig. 10 indicates that reasonably linear operation may be achieved with a carrier amplitude corresponding to less than 2.4-rad peak phase shift of the light. Although in Fig. 16a the diffracted light is used, it will be clear that the zeroth order may be employed also. In that case, the stop is replaced by an aperture and the resulting light modulation is the inverse of the modulation of the sound beam.

A disadvantage of this type of modulator is the small interaction length permitted by the Debye–Sears criteria. According to (2.11) we find that for 10 MHz in glass the interaction length is limited to about 5 cm which is still a reasonable value. However, at 50 MHz the maximum allowable length decreases to approximately 2 mm. As we shall see, with such a narrow sound column, a great deal of sound power is required for effective operation. Thus the Debye–Sears type modulator is limited to relatively low carrier frequencies and, consequently, limited bandwidths. For higher frequencies and bandwidths, the Bragg type modulator shown in Fig. 16b, is to be preferred. As before, either the diffracted or the undiffracted light may be used.

1. *Efficiency*

An important parameter is the acoustic power required for 100% modulation, i.e., for total conversion of the incident light. From (2.27) it follows that the condition for this is given by

$$|kCSL/4 \cos \phi_B| = \pi/2 \tag{4.1}$$

where L is the interaction length. With (2.3) and assuming $\cos \phi_B \simeq 1$, it follows that

$$|S| = 2\pi(kn^2pL)^{-1} \tag{4.2}$$

Substituting (3.15) we find for the required acoustical intensity

$$I_s = \lambda_v^2/2ML^2 \tag{4.3}$$

where M is the figure of merit defined in (3.17) and λ_v is the vacuum wavelength of the light. If the height of the transducer is denoted by h then the total required acoustic power P_s is determined by the following relation

$$P_s = I_s hL = (\lambda_v^2/2M)h/L \tag{4.4}$$

[70] L. Bergmann, "Der Ultraschall," Chapter 5. Hirzel, Stuttgart, 1954.

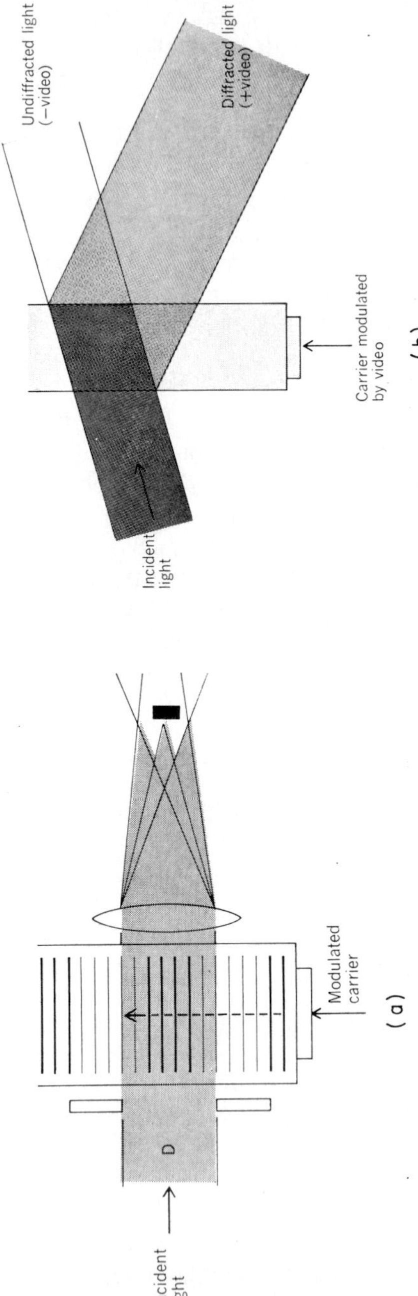

FIG. 16. Acousto-optic light modulators. (a) Debye–Sears operation. (b) Bragg operation (after Adler[2]).

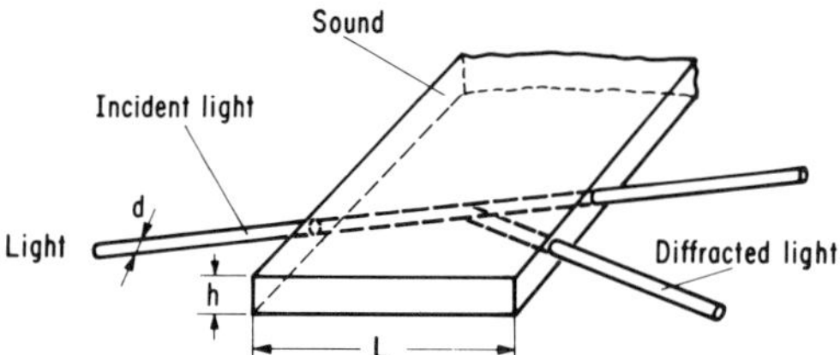

FIG. 17. Efficient Bragg modulator configuration with sheet beam of sound.

It is seen from (4.4) that the aspect ratio h/L should be small for efficient operation. This leads to a sheet beam configuration for the sound as illustrated in Fig. 17. For convenience we may write (4.4) in the form

$$P_s = 1.25 M_w^{-1}(h/L)(\lambda/\lambda_r)^2 \qquad \mathrm{W} \tag{4.5}$$

where M_w is the figure of merit relative to water; λ_r equals 6328 Å, the wavelength of the He–Ne laser.

A typical modulator is shown in Fig. 18. It uses glass as a medium ($M_w \simeq 0.04$) and operates at 40 MHz with a transducer height h of 2 mm and an interaction length of about 50 mm.[11] According to (4.5), the power required for full modulation is 1.25 W, in close agreement with the reported performance.

2. *Bandwidth*

The attainable modulation bandwidth of an acousto-optic modulator is mainly determined by the width d of the light beam.[16] When the modulation frequency increases, the sound intensity across the beam will no longer be uniform and a decrease of modulation depth results. At the point where the modulation frequency has increased so much that a full modulation cycle is

FIG. 18. Typical Bragg modulator operating at 40 MHz (courtesy Intra Action Corp.).

contained within the (rectangular) beam, the resulting effect is just averaged out and the modulation disappears.[71] This frequency is given by

$$f_{m0} = V_s/d = 1/\tau \tag{4.6}$$

where V_s is the sound velocity and τ the sound transit time across the light beam.

It is often thought that the angular dependence on frequency of the diffracted light poses an additional limitation on the bandwidth of acousto-optical modulators. This angular dependence, however, is just another way of explaining this limitation and not a separate effect. For a modulation frequency f_m two sidebands $f_s + f_m$ and $f_s - f_m$ are generated (f_s is the carrier frequency). Because of the nature of acoustical Bragg diffraction, the optical beams corresponding to these sidebands propagate in a direction differing from the carrier beam by angles

$$\pm\delta(\lambda/\Lambda) = \pm(\lambda/V_s)f_m \tag{4.7}$$

The "modulation" of the total beam may be thought of as brought about by the mixing or heterodyning of the carrier beam with the sidebands, as detected, for instance, on the surface of a photodiode. For optimum mixing, the phase fronts must be parallel, i.e., the beams must propagate in the same direction, as is indeed the case for low modulation frequencies. When the modulation frequency increases, however, the angle between the wave fronts increases too and the modulation depth becomes smaller. When the angle reaches a value of approximately λ/d there exists one optical wavelength phase difference across the mixing wave fronts, and the modulation averages out to zero. The frequency for which this happens is given by

$$\pm\delta(\lambda/\Lambda) = \pm(\lambda/V_s)f_{m0} = \pm(\lambda/d) \tag{4.8}$$

or

$$f_{m0} = V_s/d \tag{4.9}$$

It will be seen that this is identical to Eq. (4.6).

More specifically, it may be shown that the dependence of modulation depth m on frequency f_m for a rectangular beam is given by

$$m = m_0 \sin(\pi f_m/f_{m0})/(\pi f_m/f_{m0}) \tag{4.10}$$

where m_0 is the modulation depth at zero modulation frequency. It should be noted, in all this, that for an acousto-optic modulator the concept of modulation depth of the beam must be defined with respect to the demodulator used. Thus (4.10) refers to the conventional case of demodulation by an unapertured

---

[71] H. V. Hance and J. K. Parks, *J. Acoust. Soc. Am.* **38**, 14 (1965).

photodetector. A completely different behavior would follow if spatial filters were used in the demodulation process. As will be clear from the foregoing discussion, this is because the acousto-optical modulator affects the spatial uniformity of the optical beam across the spatial filter and it is primarily this effect that determines the bandwidth.

It is seen from (4.9) that the bandwidth may be increased by reducing the width of the light beam. There is, however, a limit to this procedure set by the increasing angular spread of the incident light. When this becomes larger than the angular spread of the sound, the interaction efficiency decreases rapidly because of the unavailability of plane waves of sound to interact with the tails of the light plane wave distribution. Not only does the efficiency decrease but, at this point, the bandwidth does not increase much any more with further reduction in d. This is because the diffracted beam is now limited in angular spread, and hence in size, by the spread in the sound beam. (The effect is already beginning to show up in the configuration of Fig. 16b.) Mathematically, the d in (4.8) refers to the diffracted beam rather than the incident one, and although the latter may decrease in size the former tends to remain constant. More detailed analyses may be found in Gordon,[16] Chang,[20] and Maydan.[72]

Thus, for efficient interaction, $\lambda/d < \Lambda/L$ or

$$d > \lambda L/\Lambda \tag{4.11}$$

Once a certain d has been chosen on the basis of a desired modulation frequency, L should be made as large as possible within the constraints set by (4.11). This ensures maximum overall device efficiency [see (4.4)]. As an example, in a typical modulator[72] operating at 350 MHz with As_2S_3 as the interaction material for 0.63 μm light, the following values apply: $\lambda \simeq 0.24$ μm (inside the material), $L = 0.69$ mm, $\Lambda \simeq 7.5$ μm. From (4.11) we find $d > 22$ μm and from (4.6) it follows that $f_{m0} \simeq 120$ MHz. The actual device uses a Gaussian beam with 19 μm waist and achieves a rise time of about 6 nsec.

With the devices discussed so far, the emerging light beam is modulated at the modulation frequency of the carrier. It is sometimes useful to have the light modulated at a much higher frequency, for instance at the carrier frequency itself. There are various ways to bring this about. All of them introduce a second light beam which coincides with the first diffracted beam but has a frequency different by an amount equal to the carrier frequency. In the simplest method the sound cell is made resonant. The resulting standing wave inside may be thought of either as a fixed phase grating with time-varying phase delay or as a combination of two traveling waves running in opposite direction.[73] In either case, the emerging diffracted light can be shown to be 100% modulated

[72] D. Maydan, *IEEE J. Quantam Electron.* **QE-6**, 15 (1970).

[73] L. E. Hargrave, *J. Acoust. Soc. Am.* **34**, 1547 (1962).

at twice the sound frequency. The disadvantage of this system is its relative inflexibility; because of the sharp resonance condition it is not of much use in optical communication applications. This is not the case in a second technique in which the light incident on the modulator cell consists of two beams.[74] These beams are angularly offset by twice the Bragg angle so that the diffracted order of one coincides with the zeroth order of the other one and vice versa. Consequently, each of the two sets of beams emerging from the modulator appears to be modulated at the sound frequency. If the latter itself is modulated in phase, then the carrier frequency variations of the output light will also be so modulated. Such a system may be well suited to so-called subcarrier FM methods of optical communication.[75] Various modifications of this scheme appear possible by choosing different means of generating the two input beams. One method employing two cascade Debye–Sears cells has been used in an experimental television link.[76]

3. *Availability*

At the time of writing, approximately nine companies have available off-the-shelf acoustic-optic modulators. Modulation frequencies are limited to 50 MHz with carrier frequencies ranging from 40 to 250 MHz. Spectral range is typically from 0.4 to 0.7 μm, although some infrared and 10.6 μm devices are available. A number of companies have developed special modulators at 1.06 μm to be used as Q switches. Nominal aperture dimensions are a few millimeters and efficiency is of the order of 70–90%. The most commonly used materials are glass, lead molybdate, fused silica, tellurium dioxide, and germanium. Prices range from $\simeq$\$1300—for a 5 MHz modulator (with electronics) to $\simeq$\$5000—for a 50 MHz model.

B. Light Deflectors

1. *Resolution*

Basically, acousto-optic light deflectors operate in the same way as Bragg diffraction modulators, the essential difference being that the frequency rather than the amplitude of the sound carrier is varied. Some of the pertinent parameters of a deflector are illustrated in Fig. 19. The change in deflection angle $\Delta(2\phi_B)$ upon a change Δf of the sound frequency is given by

$$\Delta(2\phi_B) = |\Delta(\lambda/\Lambda)| = (\lambda/V_s)\Delta f \tag{4.12}$$

---

[74] R. W. Dixon and E. I. Gordon, *Bell Syst. Tech. J.* **46**, 367 (1967).

[75] F. E. Goodwin, *Proc. IEEE* **58**, 1746 (1970).

[76] G. M. Borsuk and W. J. Thaler, *IEEE Trans. Sonics Ultrason.* **SU-17**, 207 (1970).

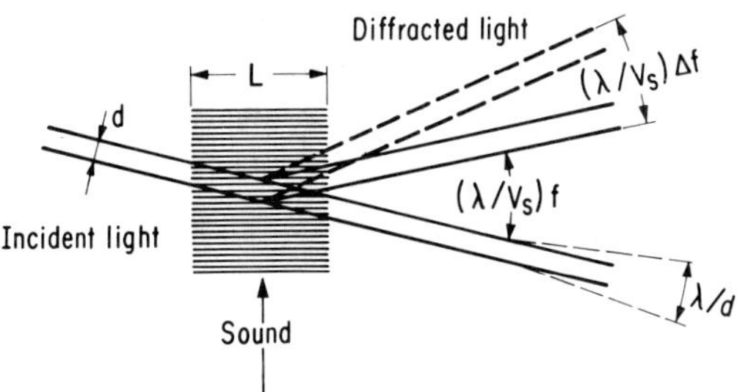

FIG. 19. Incremental acoustic deflection $(\lambda/V_s)\Delta f$ of a light beam of width d with diffraction spread $\simeq \lambda/d$.

The diffraction spread in a light beam of width d is given by[77]

$$\delta\phi \simeq \lambda/d \tag{4.13}$$

Hence the number N of resolvable angles (Rayleigh criterion) for a total frequency change Δf is found to be[9]

$$N = \Delta(2\phi_B)/\delta\phi = (d/V_s)\Delta f = \tau\Delta f \tag{4.14}$$

where τ represents the transit time of the sound through the light beam. Although (4.14) gives a good indication of resolution for a rectangular beam, more general conditions require the calculation of the modulation transfer function of the light deflector.[78] In any case, it is seen that a large aperture (d) is necessary for high resolution. As already discussed in Section IV,A the sound propagates in the shape of a sheet beam for reasons of efficiency. The combination of the two requirements of resolution and efficiency leads to a sheet beam configuration for the light. The entire deflector, including the necessary optics for beam shaping, may typically look as illustrated in Fig. 20.

2. *Bragg Angle Tolerance*

Apart from increasing d (and hence τ), the resolution may also be improved by increasing the frequency deviation Δf. A limit to this is set not only by the transducer characteristics[79] but also by the Bragg angle tolerance. The latter refers to the fact that the proper angle of incidence of the light (the Bragg angle $\phi_B \simeq \lambda/2\Lambda$) is a function of the sound wavelength. For a change δf in frequency this angle changes by

$$\delta\phi_B = (\lambda/2V_s)\delta f \tag{4.15}$$

[77] M. Born and E. Wolf, "Principles of Optics," Section 8.5. Pergamon, Oxford, 1965.

[78] A. Korpel, R. L. Whitman, and M. Odom, *in* "Acoustic Surface Wave and Acousto-Optic Devices." Optosonic Press, New York, 1971.

[79] D. A. Berlincourt, D. R. Curran, and H. Jaffe, *Phys. Acoust.* **IA**, Chapter 3 (1964).

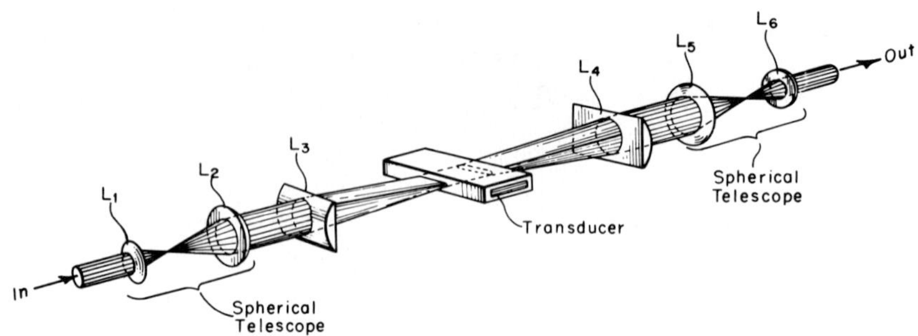

FIG. 20. Optics for acousto-optic deflector (after Korpel *et al.*[78]).

The incident light, however, always propagates in the same direction and if only one plane wave each of sound and light were available, the Bragg angle condition could be satisfied at one frequency only, i.e., $\Delta f = 0$. However, the sound originates from a transducer of finite length L and, as pointed out in Section I, such a sound beam exhibits a plane wave spectrum of angular width $\simeq \Lambda/L$. Hence, upon a change in frequency δf, the interaction will shift to another plane wave in the sound spectrum so as to satisfy the new Bragg angle condition. The limit of this process is reached when no more plane waves of sound are available at the angles required. Hence, for efficient interaction, $\delta\phi_B < \Lambda/L$ or, with (4.15), the total change Δf is limited to

$$\Delta f/f_{s0} = 2\Lambda_0{}^2/\lambda L \tag{4.16}$$

where f_{s0} and Λ_0 are the nominal sound frequency and wavelength. From (4.14) and (4.16) it follows that there exists a limit on the interaction length L if a given resolution is to be achieved for fixed τ. On the other hand, according to (4.5), a long interaction length means increased efficiency. To overcome this dilemma, a beam steering technique has been devised in which the sound beam is made to track the Bragg angle during frequency change.[15] A typical phased array of transducers which accomplishes this is shown in Fig. 21. A number of small transducers are mounted on a stair-step (echelon) type glass block, the difference in height between successive steps being so chosen as to equal one half the sound wavelength at the nominal frequency. The individual transducers are energized with alternating polarity so that at the center frequency the combined wave front propagates perpendicular to the transducer surfaces. When the frequency deviates by δf from the center frequency f_{s0}, a phase delay of $\pi(\delta f/f_{s0})$ radians is introduced between neighboring transducers. This has the effect of tilting the combined wave front by an angle $\delta\phi_t$ which, to a first approximation, is given by

$$\delta\phi_t \simeq (\Lambda_0/2s)\delta f/f_{s0} \tag{4.17}$$

where s is the spacing between transducer elements. In order for the effective

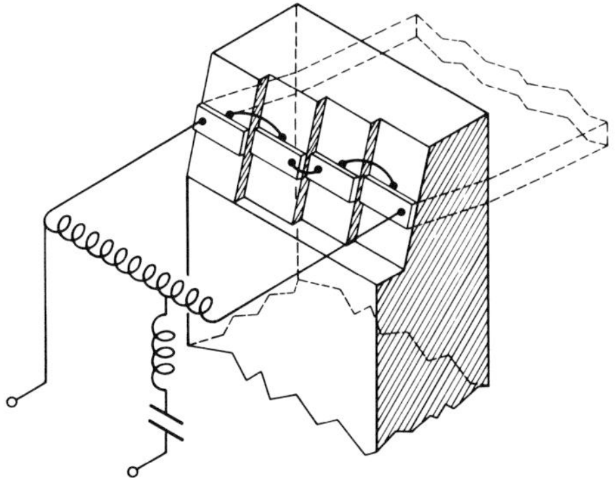

FIG. 21. Phased array of transducers used to track Bragg angle by beam steering (after Korpel *et al.*[15]).

direction $\delta\phi_t$ of the sound wave to track the Bragg angle $\delta\phi_B$, the spacing between transducers must be chosen correctly. Equating (4.17) and (4.15) it follows that

$$s = \Lambda_0{}^2/\lambda \tag{4.18}$$

When (4.18) is satisfied the Bragg angle will be tracked correctly to a first order of δf. It may be shown[16] by expanding (4.17) to include terms in δf^2 that the tolerance Δf, achieved with beam steering, is approximately given by

$$(\Delta f/f_{s0})^2 = 4\Lambda_0{}^2/\lambda L \tag{4.19}$$

The first beam-steering light deflector was used in a laser TV display system.[15] It used water as a medium for the sound and operated over a frequency sweep of 16 MHz around a center frequency of 28 MHz. With an aperture $d = 18$ mm and a transit time $\tau = 12$ μsec, this early deflector achieved about 200 resolvable spots. The stair-step array used four transducers with a spacing s of 6 mm. Figure 22 shows graphs of the diffracted light output as a function of the input angle for this deflector operated (a) with, and (b) without, beam steering.

Beam steering at higher frequencies has been demonstrated as well, although with poor deflection efficiency. An experimental deflector, described by Coquin *et al.*, operates at 300 MHz with 280 MHz frequency swing.[80] It should be stressed, however, that it is not always necessary to use a beam-steering technique if a short interaction length L or a relatively long center wavelength can be tolerated (4.16). Thus the modulator shown in Fig. 18 doubles as a deflector at 40 MHz,[81] having a frequency swing of 20 MHz and achieving 130 resolvable spots with $d = 2.5$ cm and $\tau \simeq 6.5$ μsec. This is made possible because

---

[80] G. A. Coquin, J. P. Griffith, and L. K. Anderson, *IEEE Trans. Sonics Ultrason.* **SU-17**, 34 (1970).

[81] Medium Resolution with an M-40R Acoustic-Optic Modulator. Zenith Radio Corp., Application note.

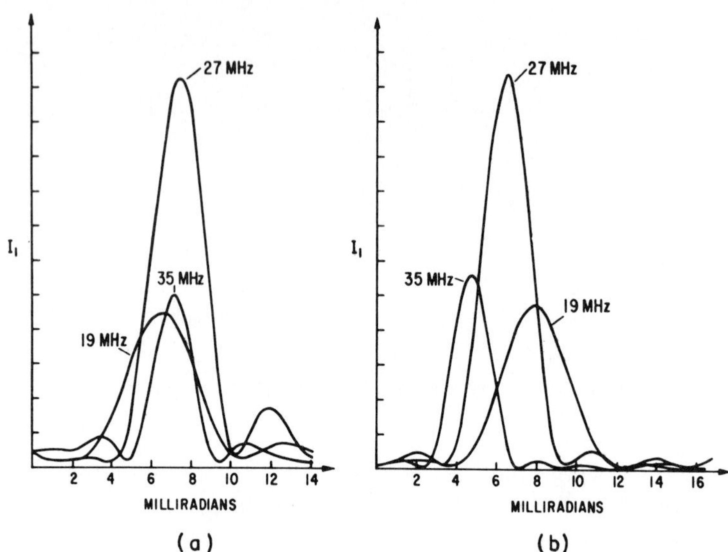

FIG. 22. Intensity of diffracted light as a function of input angle for various sound frequencies (a) With beam steering. (b) Without beam steering (after Korpel *et al.*[15]).

glass is chosen as an interaction medium. The sound velocity is much higher than in water and consequently the wavelength is large enough so that (4.16) is satisfied. For frequencies higher than 40 MHz, a phased array must be used or else the interaction length must be decreased severely. The latter method has been followed[10] in the construction of a 150 MHz deflector with $\Delta f \simeq 80$ MHz, $d = 2.5$ cm, $\tau \simeq 6.5$ μsec, $N \simeq 520$. The interaction length is kept to less than 10 mm. A reasonable efficiency is maintained by the use of $PbMoO_4$ as an interaction medium. The required electrical power equals 4 W for 50% conversion.

3. *Response Time*

Next to the resolution, bandwidth, and efficiency, the response time τ of an acousto-optic deflector is an important parameter. For random positioning applications it represents the minimum access time. From (4.14) it is seen that this access time may be reduced by decreasing the diameter of the light beam, however only at the expense of resolution.

In linear scanning applications, the transit time is relatively unimportant because the linear frequency chirp within the aperture merely acts as an additional cylinder lens.[82] This is illustrated in Fig. 23. In point A of the sound cell the

[82] J. S. Gerig and H. Montague, *Proc. IEEE* **52**, 1753 (1964).

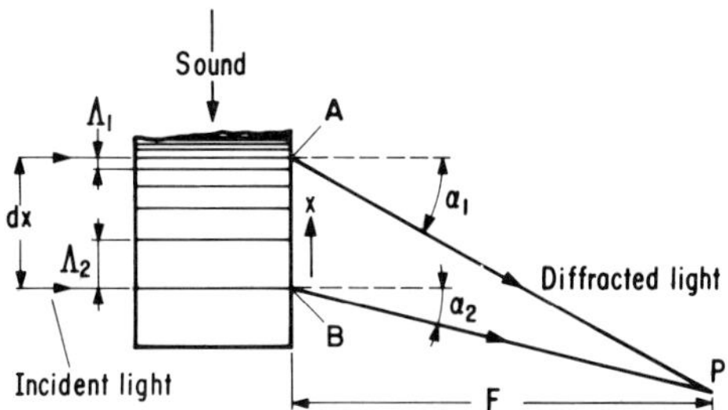

FIG. 23. Cylinder lens effect of acoustic linear frequency sweep. P is the focus of the diffracted light.

incoming rays are diffracted by an angle $\alpha_1 \simeq \lambda/\Lambda_1$, where Λ_1 is the local sound wavelength. In point B the angle equals $\alpha_2 \simeq \lambda/\Lambda_2$. The diffracted rays from A and B cross in point P which is located a distance F from the center of the sound cell. It may be shown that for small angles,

$$F = \frac{dx}{\alpha_1 - \alpha_2} = \frac{V_s}{\lambda}\frac{dx}{f_{s1} - f_{s2}} = \frac{V_s}{\lambda}\left(\frac{df_s}{dx}\right)^{-1} = \frac{V_s^2}{\lambda}\left(\frac{df_s}{dt}\right)^{-1} \tag{4.20}$$

For a linear scan df_s/dt is a constant; hence F is independent of the relative position of A and B and P thus indicates a focus. In operation the cell acts as a cylinder lens whose presence may be readily compensated for.

It is often thought that in actual display applications a single sound cell can double as modulator and deflector. It would appear that for this purpose one need merely modulate the amplitude of the frequency-chirped carrier, used for deflection, with the brightness signal to be displayed. In general, however, this is not possible without affecting either the resolution or the high-frequency response because of the mutually contradictory requirements on beam diameter set by (4.6) and (4.14). It is, however, feasible to electronically process the video signal in such a way that the same sound cell can be used for both modulation and deflection without impairment of image quality.[83]

4. *Figures of Merit*

In Section III,A it was pointed out that several figures of merit are in use for the classification of acousto-optic materials. We shall now discuss their engineering significance.

The acoustic power P_s required for 100% light deflection is minimized by choosing a material with a high figure of merit $M = M_2$. From (4.4)

$$\frac{1}{P_s} = \frac{L}{h}\frac{2M}{\lambda_v^2} \tag{4.21}$$

[83] A. Korpel, S. N. Lotsoff, and R. L. Whitman, *Proc. IEEE* **57**, 160 (1969).

In light deflectors it is sometimes desirable to optimize the efficiency–bandwidth product. From (4.21) and (4.16) we find

$$\frac{\Delta f}{P_s} = \frac{4M_1 f_{s0}^{\ 3}}{\lambda_v^{\ 3} h} \tag{4.22}$$

where $M_1 = MnV_s^2$. Thus maximizing the efficiency–bandwidth product, for a given height h of the sound column, requires a material with high M_1.

In general, [see (4.21)] the efficiency increases with decreasing height h of the sound column. The smallest practical value is given by $h = d$ as seen from Fig. 17. As discussed under (4.11) the optimum value of d is given by $d \simeq \lambda L/\Lambda$. Putting all this together we find with (4.21)

$$\frac{1}{P_s} \simeq \frac{2M_3 f_{s0}^{\ -1}}{\lambda_v^{\ 3}} \tag{4.23}$$

where $M_3 = MnV_s$. Thus maximizing efficiency in the optimum case where $h = d$, requires finding a material with large M_3.

Sometimes transducer power density P_s/hL is a limiting design parameter. With (4.16) and (4.22) we find

$$\frac{(\Delta f)^2}{(P_s/hL)} = \frac{8M_4 f_{s0}^{\ 6}}{\lambda_v^{\ 6}} \tag{4.24}$$

where $M_4 = M(nV_s^2)^2$. In this case, then, M_4 is the relevant figure of merit.

It is worth pointing out that the effective sound beam height h can be made extremely small (of the order of Λ) by using acoustical surface waves rather than bulk waves. The diameter of the light beam also has to be small and this may be achieved by propagating the light in a shallow waveguide. Such planar acousto-optic configurations are characterized by high efficiencies. An excellent review has been given by Schmidt.[21]

5. *Availability*

The same companies (about nine) marketing modulators have deflectors available. The number of resolvable spots range from 100–1000 with prices ranging from ~$1300 to $9500 (including electronics). Aperture dimensions vary from rectangular (1 × 10 mm) to circular (~24 mm). One company manufactures an *X-Y* deflector with a circular 20 mm aperture, capable of generating 800 × 400 resolvable spots. The price of this unit is $16,000.

C. Signal Processors

1. *Image Plane Processing*

The usefulness of acousto-optic devices to signal processing is based on the fact that they make it possible to process a relatively long part of an electrical signal all at once.[18] When an electrical signal $e'(t)$ is modulated upon a carrier $\exp(j\Omega t)$ and the resulting signal $e'(t)\exp(j\Omega t)$ is applied to the transducer of a sound cell, the acoustical signal $S(t)$ in the cell is given by

$$S(t) \propto e(-x + V_s t) \exp j(\Omega t - Kx) \tag{4.25}$$

where

$$e(-x + V_s t) \propto e'(-(x/V_s) + t) \tag{4.26}$$

where x is the position coordinate in the sound cell, V_s the sound velocity, and $K(=\Omega/V_s)$ the acoustic propagation constant. Note that $e'(t)$ is generally a complex number $|e'(t)| \exp j\phi(t)$ where $\phi(t)$ denotes phase modulation of the carrier. When collimated light is directed into the sound cell at the proper angle (we assume Bragg region operation) the spatial pattern of the diffracted light is, for weak scattering, given by

$$E'(x) \propto e(-x + V_s t) \exp j(\omega + \Omega)t \tag{4.27}$$

where ω is the frequency of the incident light and upshifted interaction has been assumed. A common technique is to image this pattern onto a transparency with amplitude transmittance $g(x)$. A 1:1 telescope is used for imaging and the zeroth order light is suppressed by a stop in the common focal plane of the two lenses as shown in Fig. 24. Taking into account the image reversal caused by the telescope, the light amplitude pattern to the immediate right of the transparency is given by

$$E''(x) \propto e(x + V_s t) g(x) \exp j(\omega + \Omega)t \tag{4.28}$$

This light is collected by a third lens so located that the transparency is in its front focal plane. A small aperture on the optical axis is situated in the back focal

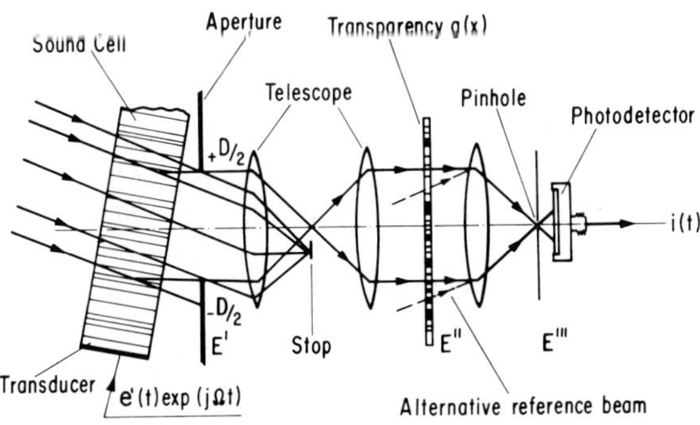

FIG. 24. Bragg diffraction signal processing in image plane. Note correlating mask $g(x)$ in this plane and pinhole aperture in back focal plane of last lens.

plane. Because of the Fourier transform properties of this third lens,[84] the field in the aperture is given by

$$E'''(t) \propto \exp j(\omega + \Omega)t \int_{-D/2}^{+D/2} e(x + V_s t)g(x)\,dx \qquad (4.29)$$

A subsequent photodetector on which E''' is incident generates a current $i(t) \propto |E'''|^2$

$$i(t) \propto \left| \int_{-D/2}^{+D/2} e(x + V_s t)g(x)\,dx \right|^2 \qquad (4.30)$$

From (4.30) it follows that the final signal is proportional to the finite cross correlation between e and g. Note that without the aperture on the optical axis, the current out of the photodiode would simply have been proportional to the total power flux in $E'''(x)$ or

$$i(t) \propto \int_{-D/2}^{+D/2} |e(x + V_s t)g(x)|^2\,dx \qquad \text{(without aperture)} \qquad (4.31)$$

which is basically different from and less useful than (4.30).

With a transparency it is most convenient to keep $g(x)$ a real quantity. If complex values are required, it is sometimes easier to use a reference beam $E_r(x)\exp(j\omega t)$. This is indicated by the dotted lines in Fig. 30. In this mode of operation the aperture is not used and the photodetector again measures the total power P

$$P \propto \int_{-D/2}^{+D/2} < |E_r(x)\exp(j\omega t) + e(x + V_s t)\exp(j\omega t + j\Omega t)|^2\,dx \qquad (4.32)$$

The resulting current has an rf component at Ω which is separated by electronic filtering

$$I \propto \exp(j\Omega t) \int_{-D/2}^{+D/2} E_r{}^*(x)e(x + V_s t)\,dx \qquad (4.33)$$

As seen, the amplitude of the rf current is linearly proportional to the cross correlation of e and $E_r{}^*$. Correlator techniques of the kind described above can also be achieved in the Debye–Sears region provided the proper optical configurations are used. The earliest technique of this kind, devised by Rosenthal, in fact operated in this region using an arc lamp as a light source.[85] Since then, many variations have been devised, mainly for applications in radar processing,

---

[84] L. J. Cutrona, E. N. Leith, C. J. Palermo, and L. J. Porcello, *IEEE Trans. Inf. Theory* **IT-6**, 386 (1960).

[85] A. H. Rosenthal, *IRE Trans. Ultrason. Eng.* **8**, 1 (1961).

such as pulse compression and expansion.[82,86–91] They all have in common the concept of using the sound cell as a moving transparency carrying a picture of the electrical signal. Quite apart from its application to signal processing, however, this idea of a moving image dates back quite far and was used in a very ingenious way in the Scophony system of television projection.[92]

2. *Fourier Plane Processing*

A somewhat different mode of operation follows from the concept of a sound cell as a spectrum analyzer.[12,85,91] This comes about because, as we have seen before, the deflection angle ϕ_d is approximately proportional to the sound frequency

$$\phi_d \simeq (\lambda/V_s)f_s = (1/2\pi)(\lambda/V_s)\Omega \tag{4.34}$$

If, then, acoustically diffracted light is focused by a lens, the position of the spot in the focal plane depends linearly on f_s. This is illustrated in Fig. 25 where l_1 is the focal length of the lens in question. With the usual small angle approximation, the spot position x_s is given by

$$x_s = l_1(\lambda/V_s)f_s = (l_1/2\pi)(\lambda/V_s)\Omega \tag{4.35}$$

The width Δx of the spot depends on the diffraction spread λ/d of the light beam:

$$\Delta x = l_1\lambda/d \tag{4.36}$$

Each frequency f_s thus being characterized by its own spot, it is clear that the device acts as a spectrum analyzer. The frequency resolution f_s follows from (4.34) and (4.35):

$$\delta f_s = (V_s(l_1\lambda)\Delta x = V_s/d = 1/\tau_s \tag{4.37}$$

where τ_s is the transit time of the sound through the light beam. Because in Bragg diffraction not only the amplitude, but also phase and frequency of the sound are carried over to the diffracted light, the plane X of the photodetector in Fig. 24 may be considered to display the true Fourier transform of the signal. It contains simultaneously and in parallel all the frequency components

---

[86] A. Reich and L. Slobodin, *Proc. Nat. Aerospace Electron. Conf.* p. 163 (1961).

[87] L. Slobodin, *Proc. IEEE* **51**, 1782 (1963).

[88] M. Arm, L. B. Lambert, and I. Weissman, *Proc. IEEE* **52**, 842 (1964).

[89] L. B. Lambert, M. Arm, and A. Aimette, *in* "Optical and Electro-Optical Information Processing," Chapter 38. MIT Press, Cambridge, Massachusetts, 1965.

[90] H. R. Carleton, W. T. Maloney, and G. Meltz, *Proc. IEEE* **57**, 769 (1969).

[91] M. Bertolotti, F. Gori, G. Guattari, and B. Daino, *Appl. Opt.* **7**, 1961 (1968).

[92] D. M. Robinson, *Proc. IRE* **27**, 883 (1939).

of the electrical signal within the sound cell, in optical form and up- (or down-) shifted with respect to the optical frequency ω. Neglecting the finite width Δx of the diffracted spot we have (for up-shifted interaction)

$$E(x_s) \propto \tilde{S}(\Omega) \exp j(\omega + \Omega)t \tag{4.38}$$

where $\tilde{S}$ is the Fourier transform of the acoustic signal in the sound cell

$$\tilde{S}(\Omega) \propto \int_{-\tau_s/2}^{+\tau_s/2} S(t) \exp(-j\Omega t)\, dt \tag{4.39}$$

The relation between x_s and Λ in $S(\Lambda)$ is as indicated in (4.35). If now a reference (local oscillator) beam $E_r(x) \exp(j\omega t)$ is made to heterodyne with the signal (4.38) the resulting current out of the photodetector has a component given by[12]

$$I(t) \propto \int E_r^*(\Omega)\tilde{S}(\Omega) \exp(j\Omega t) d\Omega_s \tag{4.40}$$

where the variable x in $E_r(x)$ has been replaced by Ω for simplicity [see Eq. (4.35)]. From (4.40) it is clear that the following holds for the frequency spectrum (Fourier transform) $\tilde{I}(\Omega)$ of $I(t)$

$$\tilde{I}(\Omega) \propto E_r^*(\Omega)\tilde{S}(\Omega) \tag{4.41}$$

Hence the effect is as if S were passed through a filter $E_r^*(\Omega)$. This filter can, within wide limits, be made to have an arbitrary response by the use of a suitable optical system (indicated by dotted lines in Fig. 25) to create the desired phase and amplitude pattern $E_r(x)$. Thus if $E_r(x)$ is a plane wave front $\exp(+j\alpha x)$

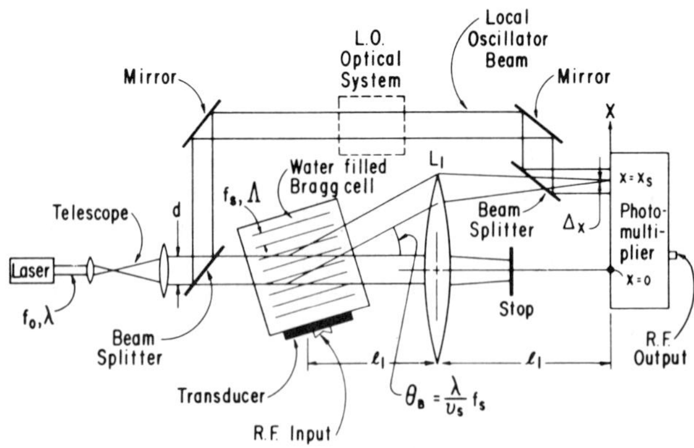

FIG. 25. Bragg diffraction signal processing in Fourier plane. Note the reference beam, called local oscillator beam in drawing (after Whitman *et al.*[12]).

then the corresponding filter $E_r(\Omega)$ is characterized by a linear phase response $\exp(-j\tau\Omega)$ which results in a constant time delay τ for the signal $S(t)$. Whitman *et al.*[12] have simulated a dispersive delay line filter by creating a curved reference wave front (parabolic phase response) through the use of a lens in the reference beam. They point out that a random phase filter is a possibility through the use of a diffused reference beam.

In analogy to the correlation technique discussed before, a transparency may be used instead of a reference beam and the photodetector may be preceded by an aperture on the optical axis.

The basic difference between the techniques illustrated in Figs. 24 and 25 lies in the fact that the processing in Fig. 24 takes place in an image plane and in Fig. 25 in a Fourier transform plane. Although not shown here, it will be clear that the processing could also take place in an arbitrary plane (Fresnel transform plane) with characteristics different from either (4.33) or (4.41). In analyzing such general configurations it is worth noting that the position of the photodetector as such has no effect on the output signal.[93] A photodetector basically detects time variations in the photon flux and, to a first approximation, the shape of these variations is independent of where they are detected along the light path, provided *all* the light is intercepted. Hence, if a reference beam is used in a processing scheme it is most convenient to analyze the heterodyning in a plane where either the reference field or the signal field has a simple character. If necessary, lenses can be added in "gedanken experiments" as long as they process all the light. For instance, inserting a Fourier transform lens between beam splitter and photodetector in Fig. 25 does not change the output, in spite of the fact that the image, rather than the far field of the sound field is now incident on the photodetector, just as in Fig. 24. We realize, of course, that this is because the reference beam is transformed too, but it does point out that the method of Fig. 24 can be made to give results identical to Fig. 25 simply by properly shaping the reference beam to be the spatial Fourier transform of the reference beam used in Fig. 25. As an example, consider the case where the latter is an extended plane wavefront which, as we have seen, results merely in a time delay in the processed signal. The spatial Fourier transform of such a beam shape is a sharp delta-function light spot $\delta(x)$. It is clear that if such a reference light spot is used in Fig. 24 the result will likewise be a simple time delay of the processed signal. Hence to a certain extent the choice between the various processing systems is one of convenience.

Although in all of the above we have referred to correlation operations, it will be clear that reversing the direction of sound travel will result in convolution, the difference being only in the sign of the spatial variable of the processed signal.

[93] A. Korpel and R. L. Whitman, *Appl. Opt.* **8**, 1577 (1969).

V. SPECIAL TOPICS

A. Polarization Effects

Under certain conditions part or all of the diffracted light may have a polarization direction which is orthogonal to that of the incident light. This may happen even in optically isotropic media when $p_{11} \neq p_{12}$. Figure 26 shows a typical experimental configuration which exhibits this "orthogonal" scattering in an isotropic medium. A longitudinal sound wave travels in the X direction and the light is incident in approximately the Z direction (we assume the Bragg angle to be very small in this experiment). It is polarized in the X' direction at 45° to the X and Y axes. Considering the components E_x and E_y it is clear that different strain–optic coefficients must be applied to their respective interactions. From (3.2) it is seen readily that whereas p_{11} applies to E_x, p_{21} applies to E_y. However, since in an isotropic medium $p_{21} = p_{12}$, we will use the latter notation. Assuming weak downshifted Bragg interaction we find from (2.27), with (2.3),

$$E_y' \simeq \tfrac{1}{4} jkS_y{}^* n^2 p_{12} L \tag{5.1}$$

$$E_x' \simeq \tfrac{1}{4} jkS^* E_x n^2 p_{11} L \tag{5.2}$$

where we have replaced the cumbersome notation $\psi_{y(-1)}$ by E_y', etc. Furthermore, L is the interaction length and it has been assumed that $\cos \phi_B \simeq 1$ and that the interaction is sufficiently weak so that the sine function in (2.27) may be replaced by its argument. It is readily seen that with respect to the Y' and X' axes, the components of the scattered light are

$$E_{x'}' = \tfrac{1}{4} jkS^* E_0 n^2 \tfrac{1}{2}(p_{11} + p_{12})L \tag{5.3}$$

$$E_{y'}' = \tfrac{1}{4} jkS^* E_0 n^2 \tfrac{1}{2}(p_{12} - p_{11})L \tag{5.4}$$

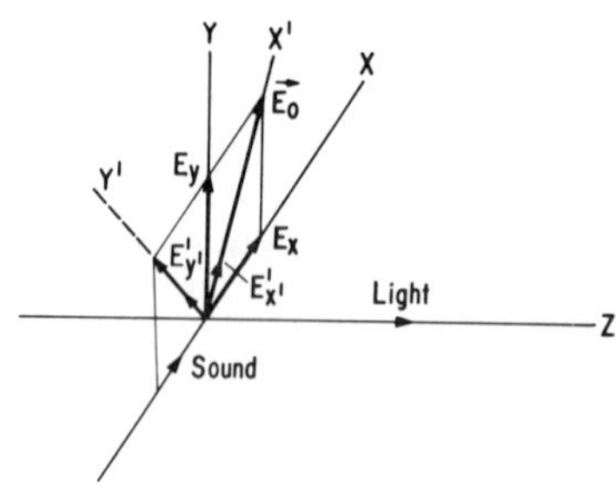

Fig. 26. Coordinate system used in calculation of orthogonal scattering. The incident light E_0 is polarized along X'. Part of the diffracted light is polarized along Y'.

From (5.4) we thus find that "orthogonal" scattering takes place with an effective strain-optic coefficient $\frac{1}{2}(p_{12} - p_{11})$ which, according to Section III,A, is numerically equal to p_{44} in isotropic media. Note that in addition to the orthogonal component, an ordinary component is also present (5.3), and in most isotropic media is much stronger ($|p_{11} + p_{12}| \gg |p_{12} - p_{11}|$). However, if a shear wave is used, the ordinary component may be eliminated. Thus, if in Fig. 26 the sound is in the form of a shear wave with particle motion in the Y direction (S_6) then, according to (3.2), a change in n_6 results and the relevant parameter is p_{66}. From (3.1) it is seen that this introduces an additional term $2x_1x_2/n_6{}^2$ in the description of the optical indicatrix. It is readily shown that this term changes the indicatrix from a sphere (isotropic medium) to an ellipsoid with axes along X', Y', and Z.[94] Whereas the axis along Z is of constant length, the fractional change in length of the other two is given by

$$\Delta n_{x'} = \tfrac{1}{2}n^3\Delta(1/n_6{}^2) = \tfrac{1}{2}n^3 p_{66} S_6 \tag{5.5}$$

$$\Delta n_{y'} = -\tfrac{1}{2}n^3\Delta(1/n_6{}^2) = -\tfrac{1}{2}n^3 p_{66} S_6 \tag{5.6}$$

It is clear that an incident light wave with polarization along X', such as shown in Fig. 26, will not give rise to cross polarization terms in scattering. However, if the incident light is polarized in the X or Y direction, the situation as regards orthogonal scattering is analogous to the longitudinal wave with the light polarized along X'. We find, however, one important difference. Due to the opposite sign of the coefficients in (5.5) and (5.6) the scattered light with polarization in the same direction (say along X) cancels. The orthogonal polarization is again described by an effective strain-optic coefficient which is numerically equal to p_{66} $(=p_{44})$.

An important application of orthogonal scattering in isotropic media is to be found in the field of signal processing. Signal processors, such as described in Section IV,C often operate in the Debye–Sears region. The signal propagating in the sound cell still acts as a transparency, but one which causes spatial phase modulation rather than amplitude modulation. The phase modulation can, however, be changed into amplitude modulation if the phase of the two spatial sidebands (orders $+1$ and -1) is delayed by 90° with respect to the carrier (zeroth order). Normally it is difficult to achieve this. If, however, the polarization of the scattered light (i.e., the sidebands) is orthogonal to that of the incident light, a suitably oriented quarter-wave plate will perform the required 90° differential phase shift. This is shown in Fig. 27.[90] The quarter-wave plate is followed by an analyzer which transmits a common component of the undiffracted and diffracted orders. The final combined light field now represents a moving amplitude image of the sound field within the cell. In the experiment of Fig. 27 this field is correlated with the grating mask $T(y)$ in a manner discussed

[94] M. Born and E. Wolf, "Principles of Optics," Section 14.2. Pergamon, Oxford, 1965.

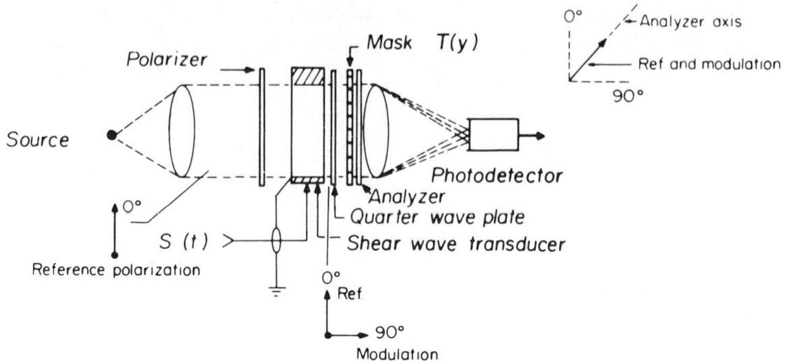

FIG. 27. Use of orthogonal scattering in acousto-optic processing. Note the quarter wave plate–analyzer combination which changes the phase image to an amplitude image (after Carleton *et al.*[90]).

earlier in Section IV,C. It will be clear that this technique, developed by Maloney and Carleton, has certain features in common with the Zernike method of phase contrast imaging.[95] The same technique has been used by Korpel *et al.*[96] in their method of sound field probing by a focused laser beam.

B. BIREFRINGENT MEDIA

An entirely different application of orthogonal scattering concerns optically birefringent media. Here, the magnitudes of the incident and scattered wave vectors are, in general, not equal and hence the momentum triangle is no longer isosceles as it was in Fig. 3. Figure 28 shows some modified triangles for various situations occurring in uniaxial crystals.[97] Particularly interesting is the case of collinear interaction. In ordinary Bragg diffraction this occurs only at very high sound frequencies where $\Lambda = \frac{1}{2}\lambda$ ($\sin \phi_B = 1$). Figure 29 shows the collapsed triangle for this case (a) and also for the equivalent situation in a birefringent medium (b). If the latter is positive uniaxial and the incident wave is ordinarily polarized, then

$$k_{in} = n_0 k \tag{5.7}$$

$$k_{out} = n_e k \tag{5.8}$$

$$K = k_{out} - k_{in} = (n_e - n_o)k \tag{5.9}$$

For small birefringence the required K is thus much smaller than in the case of Fig. 27a (where $K = 2k$) and collinear interaction may occur at low frequencies. Thus, in the case of α-Al_2O_3 it has been observed at about 140 MHz.[97]

---

[95] M. Born and E. Wolf, "Principles of Optics," Section 8.6.3. Pergamon, Oxford, 1965.

[96] A. Korpel, L. W. Kessler, and M. Ahmed, *J. Acoust. Soc. Am.* **51**, 1582 (1972).

[97] R. W. Dixon, *IEEE J. Quantum Electron.* **QE-3**, 85 (1967).

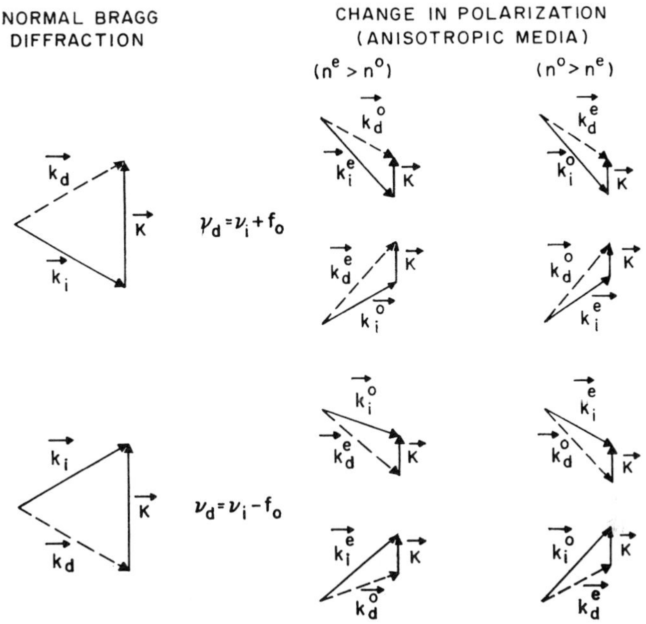

FIG. 28. Wave vector interaction diagrams for uniaxial crystals (after Dixon[97]).

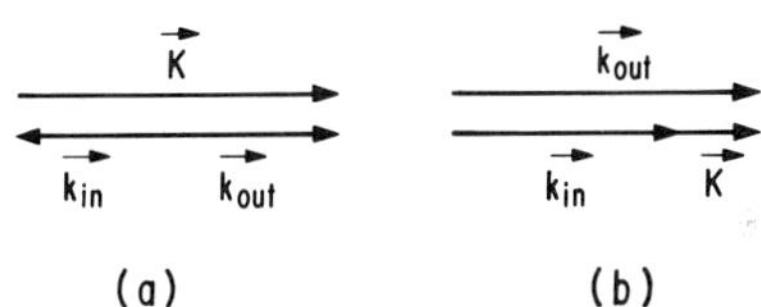

FIG. 29. Wave vector diagrams for collinear upshifted interaction. (a) In isotropic materials. (b) In a positive uniaxial crystal.

Harris *et al.* have made use of polarization-sensitive collinear interaction in a new kind of tunable optical filter[98] shown in Fig. 30. Light at the correct frequency interacts with a shear wave in a $CaMoO_4$ crystal. It is converted into a diffracted wave with orthogonal polarization. Because of birefringence this wave (transmitted signal) leaves the crystal at a slightly different angle, as indicated in the drawing. An analyzer blocks the remaining zero-order light. For the crystal used, the low birefringence ($n_e - n_o \simeq 0.01$) and low shear velocity (2.95×10^3 m/sec) permits operation at rather low sound frequencies in the range of 40–100 MHz. The filter has a half-power bandwidth of approximately 8 Å and achieves 35% transmission at 800 mW continuous acoustic power.

[98] S. E. Harris, S. T. K. Nieh, and R. S. Feigelson, *Appl. Phys. Lett.* **17**, 223 (1970).

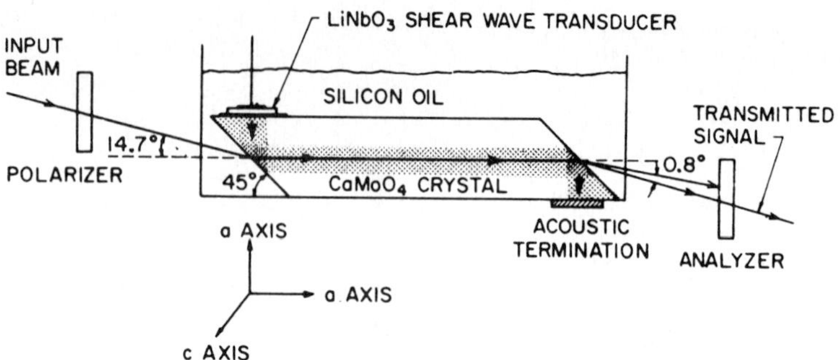

FIG. 30. Electronically tunable optical filter using collinear acoustic Bragg diffraction in $CaMoO_4$ (after Harris *et al.*[98]).

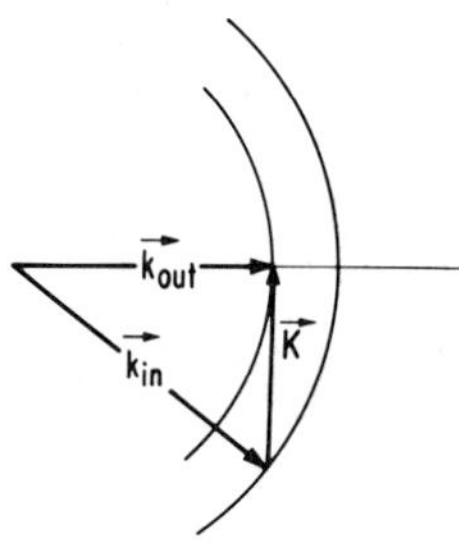

FIG. 31. Wave vector diagram for acousto-optic light deflector using anistropic medium.

A final illustration of the use of birefringent crystals in sound–light interaction is the application to scanning.[99] The useful configuration here is not collinear but orthogonal, i.e., the light is scattered nominally at 90° with respect to the sound. The relevant momentum diagram is shown in Fig. 31. Because the locus of $\mathbf{k}_{out}$ is tangential to the nominal $\mathbf{K}$, there exists a greater than usual tolerance on magnitude of $\mathbf{K}$, i.e., on the frequency variation of the sound. As may be readily seen from the drawing, the orientation of $\mathbf{K}$ relative to the incident light need change but very little when the magnitude K is varied. The phenomenon has been used by Collins *et al.*[100] in a sapphire Bragg diffraction scanner/processor with a bandwidth of 600 MHz, centered at 1560 MHz.

C. MISCELLANEOUS

Some other applications of acousto-optics exist which space does not permit us to discuss in detail. In this section the interested reader will find a brief description together with references for further study.

---

[99] E. G. H. Lean, C. F. Quate, and H. J. Shaw, *Appl. Phys. Lett.* **10**, 48 (1967).
[100] J. H. Collins, E. G. H. Lean, and H. J. Shaw, *Appl. Phys. Lett.* **11**, 240 (1967).

1. *Surface Wave Measurements*

In this application acousto-optics is used to measure or display patterns of acoustic wave motion on a surface.[101] Examples are the monitoring of acoustic surface waves[102–105] and vibrating structures[106] as well as the readout of images in acoustic cameras[107,108] and microscopes.[109]

2. *Refractive Techniques*

When low-frequency sound is used and the diameter of the light beam is smaller than the sound wavelength, it is more convenient to analyze the interaction in terms of refraction rather than diffraction.[110] Low-frequency effects have been used to create a time-varying lens for a Q-switched laser[111] and a traveling lens to increase the resolution of scanners.[112]

3. *Parametric Effects*

When large light powers are used, parametric effects leading to sound amplification or attenuation begin to play a role.[3,4] A review of applications and theory may be found in Quate,[17] Korpel,[18] Tien,[113] and Cassedy and Oliner.[114]

4. *Guided Wave Interaction*

This application was referred to in Section IV,B,4. It involves planar interactions in configurations of integrated optics and acoustical surface waves.[21]

5. *Bragg Diffraction Imaging*

As remarked before, the scattered light field exhibits a plane wave correspondence with the scattering sound field. This property forms the basis of an acoustic imaging technique called Bragg diffraction imaging.[18,115,116]

---

[101] R. L. Whitman and A. Korpel, *Appl. Opt.* **8**, 1567 (1969).
[102] E. P. Ippen, *Proc. IEEE* **55**, 248 (1967).
[103] A. Korpel, L. J. Laub, and H. C. Sievering, *Appl. Phys. Lett.* **10**, 295 (1967).
[104] R. L. Whitman, *Appl. Opt.* **9**, 1375 (1970).
[105] R. Adler, A. Korpel, and P. Desmares, *IEEE Trans. Sonics Ultrason.* **SU-15**, 157 (1968).
[106] R. L. Whitman, L. J. Laub, and W. J. Bates, *IEEE Trans. Sonics Ultrason.* **SU-15**, 186 (1968).
[107] A. Korpel and P. Desmares, *J. Acoust. Soc. Am.* **45**, 881 (1969).
[108] G. A. Massey, *Proc. IEEE* **56**, 2157 (1968).
[109] A. Korpel *et al.*, *Nature* (*London*) **232**, 110 (1971).
[110] R. Lucas and P. Biquard, *J. Phys. Radium* **3**, 464 (1932).
[111] A. J. DeMaria and G. E. Danielson, *IEEE Quantum Electron.* **QE-2**, 157 (1966).
[112] L. C. Foster, C. B. Crumly, and R. L. Cohoon, *Appl. Opt.* **9**, 2154 (1970).
[113] P. K. Tien, *J. Appl. Phys.* **29**, 1347 (1958).
[114] E. S. Cassedy and A. A. Oliner, *Proc. IEEE* **51**, 1342 (1963).
[115] A. Korpel, *Appl. Phys. Lett.* **9**, 425 (1966).
[116] A. Korpel, *J. Acoust. Soc. Am.* **49**, 1059 (1971).

CHAPTER 5

Coherent Light Valves

DAVID CASASENT

Department of Electrical Engineering, Carnegie–Mellon University, Pittsburgh, Pennsylvania

I. INTRODUCTION

The rapidly expanding display market has accelerated the development of many two-dimensional spatial light modulators.[1–4] In these systems the data to be displayed or projected are usually recorded on a transparent material in a two-dimensional format and high-energy light passed through the device to produce the desired display, usually on a large screen. These devices are generally known as light valves. They must typically exhibit TV resolution or better, and be reusable with long lifetime and a cycle time of typically 30 frames/sec. A related class of device is known as a page composer[5] and is used in optical mass memories for the storage of digital data.[6–8]

---

[1] *IEEE Trans. Electron. Dev.* **ED-20**, 917, Special issue on display devices (1973).

[2] *J. Soc. Inf. Disp.* **17**, Special issue on flat panel and large screen displays (1st quarter, p. 1) (1976).

[3] A. Agajanian, *J. Soc. Inf. Disp.* **14**, 76 (1973).

[4] B. Ellis and A. Walton, Royal Aircraft Estab. Tech. Rep. 71,009 (1971).

[5] H. N. Roberts, *Appl. Opt.* **11**, 397 (1972).

[6] O. N. Tufte and D. Chen, *IEEE Spectrum* **10**, 48 (March 1973).

[7] D. Chen and J. D. Zook, *Proc. IEEE* **63**, 1207 (1975).

[8] *Appl. Opt.* **13**, Special issue on optical storage of digital data (April 1974).

ISBN 0-12-408606-3

In this chapter we consider several of these light valve devices. However, the emphasis in this current state-of-the-art survey will be on light valves that are useful in coherent optical data processing. A brief review of the operations possible and general architectures for these coherent optical systems are included in Section II,A for completeness and to place the role of the various devices and the remaining sections in perspective.

The requirements for coherent light valves for coherent optical data processing are more demanding than those for projection or display devices, light valves used in noncoherent processing, etc. The various classes of coherent light valves and the two major modulation mechanisms used are reviewed in Sections II,B and II,C. The performance criteria used for comparing and evaluating the various light valves are then summarized in Section II,D.

No attempt will be made to include every light valve program of the past ten-odd years. Rather, only the six major light valves that are currently being actively used in coherent light are considered. These include the electron-beam addressed thermoplastic and dielectric oil film light valves (Section III), the electron-beam addressed DKDP and optically addressed photo-DKDP devices (Section IV), the optically sensitive PROM (Section V), and hybrid liquid crystal (Section VI). Only two-dimensional light valves are included. One-dimensional modulators such as acousto-optic devices are discussed elsewhere.[9]

II. General Considerations

A. Coherent Optical Data Processing

The classical operations performed in every coherent optical processor include one or more of the following: the Fourier transform, spatial filtering, and correlation. These operations are discussed at length in various books[10,11] and survey articles.[12,13] The present treatment is thus intentionally brief.

The classical frequency-plane correlator (FPC) is shown schematically in Fig. 1. Here P_0 is the input plane whose amplitude transmittance is proportional to the input function $g(x_0, y_0)$. Spherical lens L_1 forms the two-dimensional Fourier transform $G(x_1, y_1)$ of $g(x_0, y_0)$ at P_1. Capital letters are used to denote the Fourier transform of the corresponding lower case functions. The

[9] R. A. Sprague, *Proc. SPIE* **90**, 136 (1976).

[10] J. W. Goodman, "Introduction to Fourier Optics." McGraw-Hill, New York, 1968.

[11] H. J. Caulfield (ed.), "Handbook of Holography." Academic Press, New York, 1979.

[12] A. Vander Lugt, *Proc. IEEE* **62**, 1300 (1974).

[13] J. W. Goodman, *Proc. IEEE* **65**, 29 (1977).

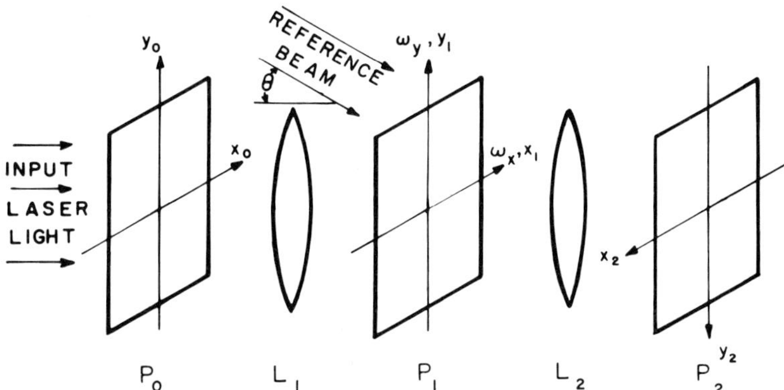

FIG. 1. Schematic of frequency-plane correlator (FPC).

coordinates of P_1 are related to the horizontal f_x and vertical f_y spatial frequencies present in the input by

$$x_1 = \lambda F f_x, \qquad y_1 = \lambda F f_y \tag{2.1}$$

where F is the focal length of L_1 and L_2 and λ is the wavelength of the laser light used. Spatial frequencies and distances, rather than frequency and time coordinates, are used in describing an optical processor, since once the input data or signals are recorded in a given region of P_0, time is converted to distance.

If the interference of the Fourier transform G of g and a plane-wave reference beam $R = A \exp(-j2\pi a y_1)$ at an angle θ, where $a = (\sin\theta)/\lambda$, is recorded at P_1, a holographic matched spatial filter of g results. The term of interest in the subsequent transmittance of P_1 is

$$G_{\mathrm{H}} = (AG^*/\lambda F)\exp(-j2\pi a y_1) \tag{2.2}$$

which contains the amplitude and phase of the conjugate transform of g.

In an optical correlator, one function g_1 is recorded at P_0 and a matched spatial filter with transmittance proportional to $G_1{}^*$ is recorded and stored at P_1 as described above. A second function g_2 is then recorded at P_0, the light distribution incident on P_1 is then proportional to G_2, and the light distribution leaving P_1 is proportional to $G_2 G_1{}^*$. Lens L_2 forms the Fourier transform of this product of two transforms at P_2, where the resultant light distribution at $(x_2, y_2) = (0, -a\lambda F)$ is proportional to the correlation of the input g_1 and reference g_2 functions,

$$g_1 \circledast g_2 = \iint_{-\infty}^{+\infty} g_1(\zeta, \eta) g_2{}^*(\zeta - x, \eta - y + a\lambda F)\, d\zeta\, d\eta \tag{2.3}$$

We will refer to this system as a frequency-plane correlator (FPC) in subsequent sections, and planes P_0, P_1, and P_2 as the input, filter (or transform), and output (or correlation) planes, respectively.

(a)

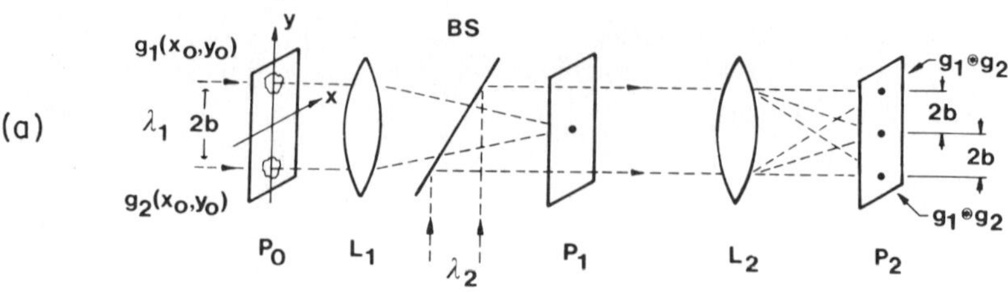

(b)

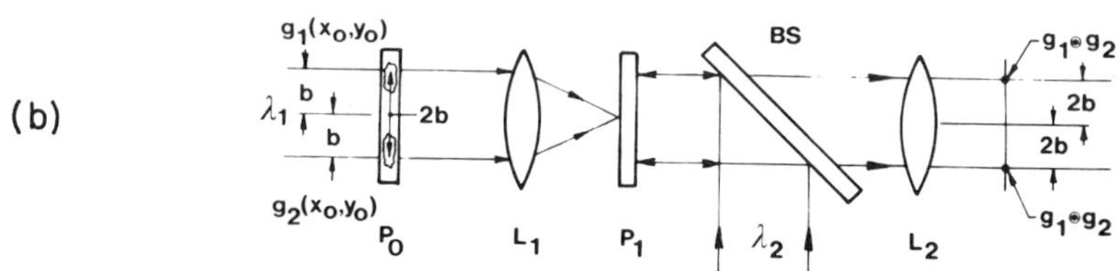

FIG. 2. Schematic of joint transform correlator (JTC). (a) Transmission-mode filter plane light valve, (b) reflex readout filter plane light valve (BS = beam splitter).

Since the light distribution incident on P_1 is a spatial-frequency representation of the input data present at P_0, simple amplitude spatial filters with opaque and transparent areas of various shapes can be placed at various positions in P_1 to alter or remove selected spatial frequencies present in the input. Since L_1 and L_2 constitute an imaging system, the output at P_2 will now be a spatially filtered version of the input function at P_0.

The optical correlation of two functions can also be realized using the joint transform correlator of Fig. 2a. The two spatial functions g_1 and g_2 are now placed side by side at P_0 separated by a center-to-center distance $2b$ (where b is the physical size of each input function). In one dimension, we can describe the transmittance of P_0 by

$$t(x_0, y_0) = g_1(x_0 - b, y_0) + g_2(x_0 + b, y_0) \tag{2.4}$$

The Fourier transform of t is formed by L_1 at P_1 where the modulus squared of the sum of the transforms G_1 and G_2 of g_1 and g_2 is recorded. The term of interest in the subsequent transmittance of P_1 is

$$t(x_1, y_1) = G_1 G_2^* \exp(-j4\pi x_1 b/\lambda F) \tag{2.5}$$

If Plane P_1 is now illuminated by a plane wave, the transform of Eq. (2.5) is produced by L_2 at P_2 or

$$U(x_2, y_2) = g_1 \circledast g_2 * \delta(x_2 - 2b) \tag{2.6}$$

which describes the contents of P_2 as the correlation of g_1 and g_2 centered at $x_2 = 2b$. We will refer to this system as a joint transform correlator (JTC).

Other coherent optical processing configurations exist. The one shown in Fig. 2b is quite practical. It is identical to that of Fig. 2a except that the light valve used at plane P_1 is read in reflection rather than in transmission. Other versions of the FPC or JTC with reflex readout light valves used at P_0 or P_1 are possible and quite practical since most available optically addressed light valves are read in reflection.

To realize the high data rate and parallel processing possible in a coherent optical processor, one must be able to alter the spatial transmittances of P_0 and/or P_1 using real time and reusable materials. It is for these reasons that we consider the use of coherent light valves.

B. Classes of Light Valves

Two general classes of light valves exist. They are distinguished by the manner in which data are recorded on them (optically or electronically). An optically sensitive light valve can be addressed in four ways.

(1) Point by point, by a scanning, modulated, and focused laser beam. We refer to such light valves as scanned optically addressed light modulators (SOALMs);

(2) In parallel, by a spatially modulated noncoherent or coherent light distribution imaged onto the light valve. We refer to such light valves as optical-to-optical converters (OTTOs).

(3) In parallel, by recording a holographic interference pattern on the device as at P_1 in Fig. 1 or 2. We refer to such devices as holographic OALMs (HOALMs).

(4) Line by line in parallel. In this scheme an acousto-optic cell is filled with one line of input data (usually a pulsed radar return or similar signal). The contents of this cell are then imaged onto one line of an OTTO or SOALM, and a vertical deflector is used to select the correct horizontal line. This requires a narrow laser pulse of duration approximately the reciprocal of the bandwidth of the signal in the cell and thus usually a cavity dumped laser. This addressing scheme is thus a modified two-dimensional laser scanning scheme.

SOALM and OTTO light valves are used at the input plane P_0 and HOALMs at plane P_1 of Fig. 1 or 2. The PROM, photo-DKDP, and liquid crystal devices are examples of optically addressed light valves. An image intensifier can also be interfaced to an OALM by fiber-optic bundles; we still refer to such devices as OTTOs.

There are three ways of electronically addressing a light valve.

(1) By a scanning and modulated electron beam.
(2) By an electroded matrix deposited on the surface of the light valve.
(3) By a CCD array interfaced to the active light valve material.

The electrode-addressed devices are typically used as page composers. Because of their limited resolution and considerable cross-talk, these types of light valves are not considered here. The CCD-addressed devices are in a primitive research state but do offer the eventual promise of a highly integrated device. In this present survey, we consider the electron-beam-addressed dielectric oil film, thermoplastic, and DKDP devices, which we refer to as EALMs (electronically addressed light modulators).

The input data for an electron-beam-addressed light valve and SOALM are electronic signals, usually in raster format often from a video pickup camera, although they can also originate from a radar or sonar receiver or a myriad of other transducer sources. The input data for an OTTO are generally available from a CRT or TV monitor, often through an image intensifier. The light valves used at plane P_1 of either optical correlator must be HOALMs, although the use of electron-beam-addressed light valves and SOALMs at P_1 under computer control are also possibilities.

Electron-beam-addressed light valves, OTTOs, and SOALMs can be used in the input plane P_0. Optically addressed light valves generally possess potentially better resolution than electron-beam-addressed units. This is because a laser beam can be focused to a smaller spot size than an electron beam while retaining sufficient energy. In practice, laser beam deflectors and scanning electron beam systems yield essentially the same resolution in TV–rate systems. Since OTTOs must be addressed from a CRT or TV monitor, in practice they exhibit comparable resolution to the SOALM and electron-beam-addressed light valves as well as suffering the limited gray scale and the scan and geometrical nonlinearities of the CRT or TV source used. In the ultimate bandwidth light valve, an acousto-optic line is filled with data which is subsequently imaged onto one line of the light valve. This horizontal acousto-optic line coupled with a vertical deflector such as a rotating mirror or second acousto-optic line used as a deflector provides a fast high bandwidth two-dimensional addressing scheme for a light valve.

For volume storage of multiple holographic matched spatial filters, photorefractive media are the ultimate choice. These materials are reviewed elsewhere.[7,8,14,15] Data stored in these materials are not easily altered. Thus these materials are usually intended for semipermanent storage rather than as easily alterable OALMs. They are also far less available commercially and otherwise than the light valves discussed in this chapter.

C. Modulation Mechanisms

The light valves to be discussed can also be distinguished by the modulation mechanism by which the collimated input laser beam is spatially modulated. In

---

[14] D. Casasent, *Proc. IEEE* **65**, 143 (1977).

[15] G. Knight, Interface devices and memory materials, *in* "Optical Data Processing" (S. H. Lee, ed.). Springer-Verlag, Berlin and New York, 1977.

all cases to be considered, the phase (not the amplitude) of the laser light is modulated by the data recorded on the light valve. The general modulation case is treated first and then the individual cases of deformable and electro-optic modulation. This treatment is contained in this one section to allow a unified and common development and a preparation for more specific light valve discussions.

We describe the input laser beam by a light amplitude A_i (assumed uniform). For simplicity, we ignore the uniform phase of the laser wave front and the time dependence $\exp(j\omega t)$ of the optical wave. After transmission through P_0, the laser's light distribution can be described by

$$A_i \exp(j2\pi nd/\lambda) \tag{2.7}$$

where n and d are the index of refraction and thickness of the material, respectively. In the light valves to be discussed, either d or n is spatially varied proportional to the input data.

In the oil film and thermoplastic light valves (Section III), the charge deposited on the target material's surface by the electron beam causes a physical deformation in the thickness of the target material proportional to the deposited charge. The resultant spatial variations $d(x, y)$ in Eq. (2.7) cause spatial variations in the phase of the output laser beam. These light valves are pure phase modulators. If a sine wave at spatial frequency ω_0 (in line pairs per millimeter) is recorded horizontally, the transmittance in one dimension can be written as

$$t(x_0) = \exp(jm \sin \omega_0 x_0) \tag{2.8}$$

where the modulation index or phase delay amplitude is $m = 2\pi D(n - 1)/\lambda$ in which D is the peak amplitude of the deformation. For simplicity, we ignore effects such as the finite input aperture $W \times L$ described by $\mathrm{rect}(x/W)\mathrm{rect}(y/L)$ multiplying Eq. (2.8). Likewise, effects such as constant phase shifts due to substrate thickness, etc., characterized by an $\exp(jK)$ factor (where K is a constant) multiplying Eq. (2.8) are also ignored for simplicity.

To describe the Fourier transform of a phase modulated light valve, we expand Eq. (2.8) in a Bessel series

$$t(x_0) = \sum_{m=-\infty}^{\infty} J_m[2\pi(n - 1)D/\lambda]\exp(jm\omega_0 x) \tag{2.9}$$

where J_m is the Bessel function of the first kind of order m. From Eq. (2.9), we see that m plane waves each of amplitude J_m emerge from the light valve. A Fourier transform lens will focus each of these m plane waves to separate spots of light in the Fourier transform plane. We are concerned only with the first-order light.

With a single spatial frequency ω_0 present, the Fourier transform of Eq. (2.8) is simply $J_1 \exp(j\omega_0 x_0)$. With P different spatial frequencies $\omega_p (p = 1, 2, \ldots, P)$ present, each spatial frequency ω_p is weighted by a different Bessel function

$J_1[2\pi D_p(2n - 1)/\lambda]$ where D_p is the deformation depth of spatial frequency ω_p. The Fourier transform pattern is thus

$$\sum_{p=1}^{P} J_1(KD_p)\delta(x_1 - K\omega_p) \tag{2.10}$$

It contains P peaks, the displacement of each from the origin of the (x_1, y_1) transform plane being proportional to the individual spatial frequencies ω_p present. With the modulation index restricted to small values, the first-order term $J_1(KD_p)$ describing the amplitude of each peak will be linearly related to D_p for the spatial frequency ω_p under consideration. Since J_1 is linear up to about a 30% amplitude transmittance, the peak diffraction efficiency of such light valves is about 10% for linear operation.

The operation of these devices in coherent light differs from the mode of operation used in noncoherent projection display applications in which a schlieren optical system is used.

In the DKDP and PROM light valves (Sections IV and V), n rather than d is spatially varied by the Pockels electro-optic effect.[16] For these devices as well as the liquid-crystal light valve (Section VI), polarized input light is essential and the orientations of the crystallographic axes of the target crystal, the direction of propagation and the polarization of the input light, and the orientation of the spatial electric field produced across the device must be considered in the modulation analysis.

Crystals exhibiting the Pockels effect have two preferred axes (X and Y) along which the polarized input light waves will propagate. For the devices to be considered here, the orientation shown in Fig. 3 applies. The input light is polarized along x at 45° to X and Y. The components along X and Y are $A_i/(2)^{1/2}$ at the front of the crystal. The crystal's indices of refraction along the crystallographic X and Y axes are equal to n_0 in the absence of a voltage V or field E across the crystal and the crystal is isotropic. In the presence of an E field, the crystal exhibits a birefringence with different indices of refraction along these X and Y axes,

$$n_{1,2} = n_0 \pm aE \tag{2.11}$$

where a is a constant. Since E and the direction of propagation of the input light are collinear, and since n changes linearly with E, the name linear–logitudinal electro-optic or Pockels effect is used to describe the phenomenon.

Using Eqs. (2.8) and (2.11) the components of light along X and Y as they emerge from the crystal of thickness b are described by

$$A_X = [A_i/(2)^{1/2}]\exp(j2\pi n_1 b/\lambda)$$

---

[16] B. H. Billings, *J. Opt. Soc. Am.* **39**, 797 (1949).

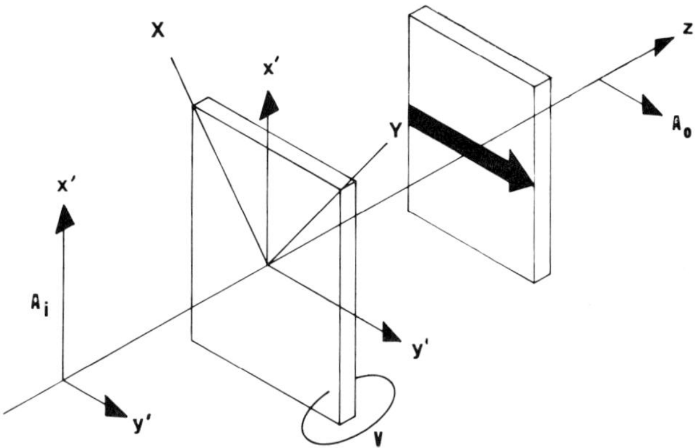

FIG. 3. Schematic representation of electro-optic effect.

and

$$A_Y = [A_i/(2)^{1/2}]\exp(j2\pi n_2 b/\lambda)$$

where all transmission losses of the crystal assembly have been neglected. Summing components along y in Fig. 3 yields

$$\begin{aligned} A_0 &= A_i \sin[K(n_2 - n_1)] \\ &= A_i \sin(\pi V/2V_{1/2}) \end{aligned} \tag{2.12}$$

where K is a constant and $V_{1/2}$ is the crystal's half-wave voltage at which the phase retardation between the components along X and Y equals $\pi/2$.

The light transmitted by a crossed analyzer oriented along y is thus described by Eq. (2.12) where V is the voltage across the crystal. In practice, this is a spatial voltage distribution $V(x, y)$ and spatial modulation of the laser beam results. The output light is linear with voltage up to about 70% modulation. If the polarization of the input light is along X (or Y), there is only one component of the wave in the crystal and output light exhibits pure phase modulation. The output beam is now elliptically polarized and thus has a component along Y (or X) also. The electro-optic effect can thus be viewed either as phase modulation or a rotation of the plane of polarization of the light so that the component along y at the output increases with V. This formulation of the effect of birefringence is particularly useful in describing the operation of the liquid–crystal light valve (Section VI).

D. LIGHT VALVE PERFORMANCE FACTORS

The salient performance factors to be considered in a coherent light valve are listed in Table I. Most are self-explanatory but several deserve expansion.

TABLE I

Coherent Light Valve Performance Factors

Type of addressing	Erase mechanism
Sensitivity	Storage capability
Diffraction efficiency	Nondestructive readout
Wavelengths	Response and cycle times
Resolution and contrast	Complexity of operation
Optical quality	Fabrication ease
Gray scale	Lifetime
Linearity	Special features

These devices are intended to replace photographic film at the input and filter planes and must therefore be reusable and real-time elements of course.

For the optically addressed light valves, the sensitivity (in microjoules per square centimeter) is given. This is the reciprocal of the exposure needed to change the intensity of the transmitted light by $1/e$. Sensitivity should most properly be combined with the resolution of the device and specified as an energy per resolution element. However, for comparison purposes units of microjoules per square centimeter are used. For electron-beam-addressed devices the beam current in microamperes required to produce full modulation is the analogous parameter.

Dynamic range and peak contrast ratio can be obtained from the light valve's sensitometry curve (input intensity or exposure versus output transmittance). However, the device's signal-to-noise ratio (SNR) and optical and cosmetic quality are usually more important, especially in coherent optical data processing applications. Scatter and surface roughness effects destroy the uniformity of the laser's phase front and contribute a spatial modulation to the laser beam, in addition to the intended phase modulation due to the recorded data. An optical flatness of at least $\lambda/2$ is necessary in most coherent optical data processing applications.

The contrast required in display or large-screen projection TV applications of these light valves is often only 10:1. Coherent optical data processing usually requires far higher contrast ratios at P_0 to record the full dynamic range of the input data. Log amplifiers can often be used in this case to compress the dynamic range of the input data. Stringent dynamic range requirements also exist if the light valve is used as a spatial filter plane material. However, the full dynamic range of the Fourier transform of the data must be linearly recorded only if the output is to be an aesthetically pleasing linearly reconstructed image of this stored hologram. If a matched filter for pattern recognition or correlation purposes is recorded, it is conventional to saturate the higher intensity, low spatial frequency data. This enhances the pattern recognition process by basing recognition and correlation on the higher spatial frequency components of the data. This reduces cross correlations and erroneous recognitions based on dc or low spatial frequency data that is common to many images. Thus, in practice, the dynamic range requirements of light valves can often be greatly reduced depending on the application.

In optical mass memory applications, only binary data storage is usually required. However, in coherent optical data processing applications (correlations, pattern recognition, etc.), the light valve must generally exhibit gray-scale recording (not by halftones). The linearity of the light valve's response curve is now important and usually either the light valve's modulation level must be limited to reduce distortions, or external signal correction circuitry employed. Reducing the modulation level reduces the linear dynamic range and diffraction efficiency η (percent of the input light diffracted into the first order) of the light valve. The value of η is usually specified for phase modulators. It is related to the normalized contrast (modulation) for amplitude modulation by $\eta = (\text{modulation})^2/16$.

The resolution and contrast of a light valve may seem like the most obvious parameters. However, the limiting resolution of a light valve at 5% modulation is usually specified together with the peak contrast at low spatial frequency. All potential light valve users should realize that these specified resolution and contrast parameters are not simultaneously obtainable. The light valve's modulation transfer function MTF (a plot of normalized contrast, i.e., modulation versus spatial frequency) is the best descriptor of the resolution and contrast values that are simultaneously obtainable. The light valve's MTF is also often a function of the input exposure, which further complicates analysis. The MTF of all light valves discussed in this chapter will be provided. As a common norm for comparing all light valves, we will specify the device's resolution at 50% modulation and its peak contrast. Device resolution must be specified at a given contrast to be meaningful. Hence both are included as a pair of parameters in Table I. Users of light valves should also be aware that the device's dynamic MTF, rather than its static MTF, is important. Some devices (such as the PROM) have been found to exhibit reciprocity failure at very short exposures.

A clarifying remark on resolution is appropriate for all light valves in which modulation is achieved by a spatial deformation of the target material. The resolution of these devices has a bandpass characteristic centered at about $(2b)^{-1}$, where b is the thickness of the active material. This center frequency and the bandpass of spatial frequencies about it are provided for these cases. In some articles, only the center frequency of the bandpass is provided from which the reader may erroneously assume that the light valve has an ultrahigh resolution (e.g., 2000 line pairs/mm = 2000 lp/mm), whereas the bandwidth about this central spatial frequency may be only 100–200 lp/mm.

Several remarks are also needed concerning the use of these deformable devices in the optical systems described earlier. When used in the input plane, the data must be recorded on a carrier, usually a grid of lines whose spatial frequency equals the central frequency of the device. When used in the matched filter plane, the central frequency of the light valve's response determines the angle θ needed between the input and reference beams. The resultant spatial frequency of the carrier $(\sin \theta)/\lambda$ in line pairs per millimeter must coincide with the center of the light valve's spatial frequency passband. This central spatial frequency of the bandpass similarly determines the spacing required between inputs in the JTC. Comparable filter-plane resolution is required in both correlators.[17]

Several general figures of merit can be defined and used to compare all light valves. However, no one figure of merit is that useful, thus we list several separate light valve parameters, but include two more generally descriptive and comparative parameters, the light valve's space–bandwidth product (SBW) and its frame-space–bandwidth product (FSBW).

The SBW used is the product of the device's resolution in line pairs per millimeter at 50% modulation times the device's active area in square millimeters, i.e., the number of bytes of data it is possible to record. This could be multiplied by the contrast or dynamic range of the light valve to yield the total channel capacity including gray-scale recording. Since the dynamic range requirements of the data vary widely between applications, we choose not to list the product of dynamic range and SBW. The FSBW is the product of SBW and the complete write/read/erase frame cycle rate. This parameter thus includes both the resolution and speed or cycle time of the light valve and is perhaps the most useful parameter when real-time operation is required. This latter parameter (FSBW) is more difficult to determine for parallel addressed OALMs (photo-DKDP and the PROM), a 100 frame/sec rate is thus assumed for these devices although they could be cycled at up to 1000 frames/sec if inputs could be prepared at this rate.

In most OALMs, data are recorded at a write wavelength λ_1 and read at a different read wavelength λ_2. This can cause an undesirable scaling (by λ_2/λ_1) of the pattern that is especially bothersome in correlations and pattern recognition using matched filtering. A resolution loss also occurs in many cases. Another aspect of the spectral sensitivity of these devices is important in any system design using input and filter-plane light valves. The read wavelength of the input light valve must coincide with the write wavelength of the filter-plane light valve to allow synthesis of the filter from the input data. In the FPC system, the read wavelengths of both devices must also coincide to enable correlation of the input and filter-plane patterns to be realized. The read operation must also be nondestructive.

For these and other reasons a system employing an EALM input light valve and HOALM filter-plane light valve in a JTC system (Fig. 2) is one of the more practical total systems. Most input data usually contains from 500 × 500 to

[17] D. Casasent and A. Furman, *Appl. Opt.* **16**, 285 (1977).

1000×1000 points, which is compatible with the resolution easily obtained from an EALM. The light valve used in the filter plane must usually have a far higher spatial frequency resolution. This resolution depends on the angle θ for the FPC system and the spacing between inputs in the JTC system. For example, a resolution of $(\sin \theta)/\lambda = 409$ lp/mm is required for $\theta = 30°$ and $\lambda = 633$ nm.

Both the input and filter-plane materials must also exhibit storage. The storage required depends on the application. Since the Fourier transform of a function is not valid until the entire input pattern has been recorded, EALM or SOALM light valves must have a storage time equal to the write time for an entire frame of data. In some applications, the input data must remain fixed for several frame times, in which case longer input storage of several frames is required. The contents of the filter-plane light valve in the FPC system are usually changed far less frequently than those of the input light valve. Thus these reference filter light valves must generally exhibit even longer storage times of many frames.

In both cases, the light valves must exhibit a nondestructive readout. This requirement is most severe for the filter-plane material in the FPC. These requirements can be relaxed by use of the JTC correlator system in which the contents of the filter plane are updated each time the input is changed. In some instances, the storage time of a light valve in the dark rather than under readout is provided. Such specifications are clearly of minimum use in the correlator systems discussed.

The response time (time between when data is applied and the device's transmittance has changed) for a scanned light valve (EALM or SOALM) must generally be more rapid than that of a parallel addressed light valve (OTTO or HOALM). However, the minimum signal duration to which the device will respond is also important, as noted earlier. For an $m \times n$ resolution pattern, each point on the device will only be addressed for T/mn (where T is the frame assembly time) for a scanned light valve. For real-time applications, the light valve's cycle time is most important. This is defined as the sum of the write time (time to assemble a frame), read time (during which the recorded data is read and during which all processing occurs), and erase time (time to restore the light valve to its original state in which no data are stored).

Since storage (to some degree) is required in all cases, some erase mechanism is essential and the completeness of erasure is most important. Devices in which erasure is by decay of the stored pattern can clearly produce problems in most systems. The television cycle time of 33 msec is generally taken as the norm of real time.

Complexity of operation (noted in Table I) refers to: the need for separate wavelengths for write, read, and erase; the need for unusual wavelengths such as uv; the need for high voltages that must be rapidly switched; and similar aspects of a device that complicate its use and incorporation into a system.

The special features referred to in Table I are operations such as contrast reversal, baseline subtraction, level slicing, and conventional addition and subtraction of images. These are most useful in many applications. Examples of these operations will be presented in Section IV,B,3.

The reader should realize that the six light valves to be described are in various stages of development and refinement. No single light valve is a panacea with ultrahigh resolution, low sensitivity, high bandwidth, fast cycle and response times, permanent storage, and infinite reusability. In most instances, all of these parameters are not required. Many applications (such as on-board missile guidance) require a lifetime of only 5–10 cycles. Some applications do not require storage; whereas in other cases nondestructive readout is vital. In all instances, the specific application determines the required light valve parameters. Several reviews of the applications of coherent optical data processing exist[12,18–20] from which the roles of these real-time devices can be seen.

This somewhat lengthy discussion of the general aspects and features of real-time light valves contains numerous items that are vital to the understanding of these devices; selection of the proper one for a given application, and its incorporation into a system. In the remaining sections, the leading coherent light valves are described and applications of their use in optical data processing presented.

III. DEFORMABLE LIGHT VALVES

Two electron-beam-addressed light valves are described that produce a spatial phase modulation of the input laser beam by controlled spatial deformations of the target material caused by the charge deposited by the electron beam. The General Electric fluid-film light valve and the ERIM (Environmental Research Institute of Michigan) TOPR (thermoplastic optical phase recorder) light valve are the two devices of this type to be discussed. The target material used in both deformable light valves is a thermoplastic; however, the characteristics and operation of both are quite different.

The basic phase modulation mechanism for all deformable devices was described in Section II,C. The following general remarks apply to all deformable light valves.

(1) The spatial frequency response is a bandpass.

(2) The amplitude of the deformations often depends on the spatial frequency recorded and is thus nonlinear.

---

[18] D. Casasent and H. J. Caulfield (eds.), "Applications of Optical Data Processing." Springer-Verlag, Berlin and New York, 1978.

[19] *Proc. IEEE*, **65**, 1, Special Issue on Optical Computing (1977).

[20] *IEEE Trans. Comput.* **C-24**, 337, Special Issue on Optical Computing (April 1975).

(3) Pure phase modulation results.

(4) The optical quality of these light valves after repeated deformations must be considered.

(5) Low modulation depths are required for linear recording.

(6) The lifetime of light valves requiring actual movement of the recording material must be seriously considered.

A. General Electric Fluid-Film Light Valve

1. *Structure and Operation*[21]

A schematic of this coherent light valve is shown in Fig. 4. This device is similar to the Swiss Eidophor and is a version of the company's noncoherent, projection display device modified for use in coherent light. In this modified version of the projection light valve, the schlieren projection optics are removed, higher quality optical windows and target substrate are used, and the maximum target deformations are reduced.

The fluid film is the target for a raster-scanned electron beam which deposits a groove pattern of raster lines on the target. The modulating input signal is superimposed on this constant beam-current raster. The constant beam-current raster defines the location of the groove or scan lines and helps to maintain a uniform thickness of the target fluid. Input modulation is achieved by varying the groove depth of the scan lines. An electron-beam spot whose shape can be varied is used. Input video is impressed on the raster by a dot arrest scheme in which the video is added to the horizontal sweep. The electron beam is actually velocity modulated by the input video data. As a result, this coherent light valve has reached quasi-production status by slight modifications to the company's production facility for the noncoherent projection device. It is 12 in. long, 9 in. in diameter, and weighs about 125 lb.

The reading laser beam passes through an optical window beside the electron gun so that the axes of the optical system and electron beam are nearly concentric. The target fluid is recirculated on a slowly rotating 10-in. diameter glass disk target (1 revolution every 180 sec). A continuously refreshed layer of writing fluid is provided from a reservoir at the bottom of the structure. A small magnetic drive provides the fluid pump and mechanical disk drive with no breaks in the vacuum wall or rotating seals. Considerable attention was also given to the problem of cathode contamination. The company now guarantees a lifetime of over 3000 hr (3×10^8 cycles at TV rates) for the light valve.

As in all deformable light valves, electronic forces cause the target to deform, whereas surface tension and viscosity forces resist these deformations. In this

[21] M. Nobel and W. Penn, *Proc. Elec. Opt. Syst. Design Conf.* p. 301 (1975).

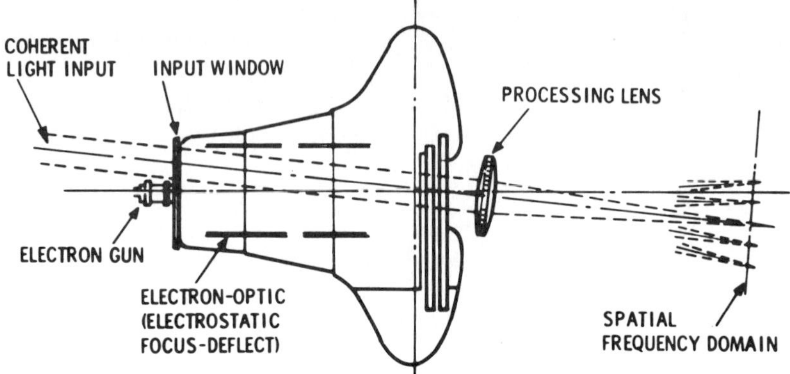

FIG. 4. Schematic of the G.E. fluid-film light valve.[21]

fluid-film light valve, the charge is first deposited on the target at time t_0, and no deformation occurs. The deposited surface charge then starts to decay causing the surface deformations to grow and hence the modulation to increase. At time t_1, the surface tension equals the electronic force and the deformations reach a maximum. The target rapidly reaches these peak deformations. The surface charge now continues to decay causing a slow decrease in the deformations and modulation. The fluid's mechanical properties determine the formation of the grooves and the fluid's electrical properties determine the decay of the charge.

2. *Specifications*

The device's integration time is the elapsed time from the start of deformation until the deformations have decayed to $1/e$ of their peak value. This is one measure of the light valve's cycle time. It is normally adjusted to equal one frame time. However, the lack of a complete erase cycle and storage are some of the more severe limitations of this light valve. The integration time is inversely proportional to the beam current and is a strong function of temperature. The target fluid is normally heated to 50°C and the beam current varied from 0.5 to 4.25 μA to produce integration times from 22 down to 4 msec. Maximum integration time is 300 msec.

For most applications, operation is at a 30-Hz frame rate with a 15.75-KHz horizontal scan rate. The device's bandwidth is usually specified. A 10-MHz signal corresponds to about 500 lp/line = 25 lp/mm; a 20-MHz signal to about 1000 lp/line = 50 lp/mm. By recording the same continuous wave tone on all lines, the two-dimensional optical Fourier transform of the input pattern in the first quadrant consists of a single off-axis peak of light at coordinates proportional to the input spatial frequency recorded. Examples of this two-dimensional

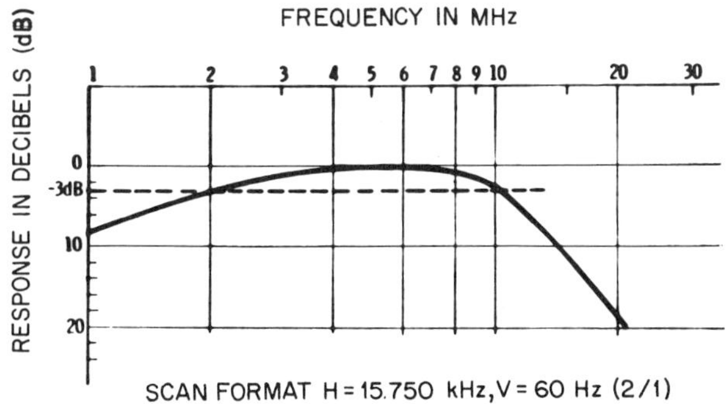

FIG. 5. MTF of fluid-film light valve.[21]

folded spectrum are included in Section III,A,3 and theoretical analyses are available elsewhere.[22–24] The resultant frequency response of the light valve is shown in Fig. 5. The modulation is 50% (or 3 dB down) at 10 MHz or 25 lp/mm and 10% at 20 MHz or 59 lp/mm. The response is a bandpass as predicted, extending from 2 to 10 MHz. At 30 frames/sec, this corresponds to a time bandwidth of 2.5×10^5 or a resolution at 50% modulation of 500×500. The time bandwidth can probably be extended to 10^6 and resolution to 1000×1000 points with various device corrections and improvements.

The dynamic range of all deformable devices, and most others, is limited either by intermodulation distortion or by the optical noise level produced by scatter from the light valve. The dynamic range of this fluid-film light valve has been measured by recording a 120-Hz sine wave balance-modulated on a 4.2-MHz carrier (the peak of the device's response curve). The strength of the signal was increased to 43 dB above noise until the intermodulation terms equaled the noise level. The input signal range was shown to be 50 dB by decreasing the signal strength until the 120-Hz spatial frequency peak was just above the noise level.

This measurement scheme in which the signal was recorded at the full two-dimensional time bandwidth of the device with a processing gain of 10^5 is compatible with the intended use of this light valve as a wide band recorder and two-dimensional spectrum analyzer. However, since the full two-dimensional time bandwidth of the device is not available in many multichannel signal processing applications, these light valve parameters summarized in Table II may not be directly appropriate for all optical data processing applications.

[22] D. Casasent, Optical signal processing, *in* "Applications of Optical Data Processing" (D. Casasent and H. J. Caulfield, eds.). Springer-Verlag, Berlin and New York, 1977.

[23] C. E. Thomas, *Appl. Opt.* **5**, 1782 (1966).

[24] D. Casasent and R. Kessler, *Opt. Eng.* **16**, 402 (1977).

TABLE II

G.E. FLUID-FILM LIGHT VALVE SPECIFICATIONS[a]

3 dB bandwidth	= 10 MHz (20 MHz)
Target size	= 20 × 20 mm^2 (40 × 40 mm^2)
Resolution at 50% modulation	= 25 lp/mm (50 lp/mm)
	= 500 × 500 points (1000 × 1000 points)
Limiting resolution	= 50 lp/mm
Space–bandwidth product	= 2.5 × 10^5 (10^6)
Cycle time (integration time to $1/e$)	= 30 frames/sec
Frame-space–bandwidth product	= 7.8 × 10^6 (3 × 10^7)
Dynamic range (in folded spectrum)	= 50 dB
Lifetime	> 3000 hr = 3 × 10^8 cycles
Cosmetic quality	= good
Optical quality	= fair, 2λ/cm
Storage time	= 20–300 msec
Erasure	= by decay

[a] Projected values in parentheses.

Comparable device parameters to those listed in Table II are provided for all light valves. The device's limiting resolution at 5% modulation is useful only for projection display applications, not coherent optical processing. The device's space or time bandwidth (SBW) is the product of its resolution at 50% modulation and its area. The frame-space bandwidth is the product of the light valve's SBW and its frame rate (reciprocal of cycle time).

The G.E. fluid-film light valve is the most developed of the devices to be considered and thus its shortcomings are better known than those of many of the other devices. The cosmetic quality of the device (freedom from flaws, scratches, pinholes, etc.) is excellent. However, its optical quality and flatness are poor by comparison to other coherent light valves with several optical fringes occurring over 1 cm. Improved oil application, damming, and adding a phase correction to the recorded data may improve this problem. Automatic optical sensing of the aberrations by interference techniques is required to determine the required phase correction. Such methods have previously been used to improve the recorded raster,[25] but convergence times of 100 sec for the routines used, and the dynamic and changing nature of the phase errors as new areas of the oil are recorded upon, are problems that remain to be overcome. In some reviews authors have confused raster and scan correction with correction for nonoptical flatness.

[25] T. Turpin, *Dig. Int. Opt. Comput. Conf.* p. 34. IEEE Cat. No. 74CH0862-3C (1974).

The most serious limitations of this light valve, besides its optical quality, are its lack of storage and its reliance on erasure by decay of the charge. This restricts its viable applications to cases where rapid and complete replacement of the input data is not required.

3. *Applications*[21,22,26]

As previously noted, the primary applications for this light valve have been as a wide band recorder and two-dimensional spectrum analyzer. Thomas[23] first described this method of optical spectrum analysis. Surveys of various applications of the folded spectrum are also available.[22,24]

In conventional optical spectrum analysis, the time bandwidth of the data is restricted to what can be recorded along a single line of the optical transducer. This does not utilize the two-dimensional nature and full two-dimensional time or space bandwidth of an optical processor. However, if the input signal is raster recorded on all N lines of the light valve with M points per line, the two-dimensional optical Fourier transform pattern (obtained at P_1 of Fig. 1 with the light valve placed at the input plane P_0) is a folded spectrum. For a signal of time length T_i recorded at a horizontal scan rate f_h and a vertical frame rate $f_v = 1/T_i$, the horizontal axis in the folded spectrum output will correspond to a coarse frequency axis, and the vertical axis to a fine frequency axis. With an input time bandwidth of 10^5, the entire spectrum of a 10-MHz bandwidth input signal can be compactly displayed in two dimensions with a frequency resolution of 100 Hz.

A typical folded spectrum output pattern[22] of a 12-MHz bandwidth of the rf spectrum is shown in Fig. 6a with the locations of several radio, TV, and FSK stations labeled. The fine-frequency axis is horizontal and the coarse-frequency axis vertical in this figure. The input pattern used was about 500 × 500 and the fine-frequency resolution obtainable was about 40 Hz. An isometric display of this data is shown in Fig. 6a. Real-time demodulation of the data on any FSK, AM, or FM station has been demonstrated by appropriate spatial filtering in the folded spectrum display.[22,24,26]

In one of the more recent and different applications of this light valve,[27] a radar chirp signal was recorded on successive lines of the light valve with the device placed in the input plane of an optical processor. The one-dimensional vertical Fourier transform of this input data (achieved with a cylindrical/spherical lens system) is shown in Fig. 7. The vertical coordinate of the output display is proportional to the target's Doppler and the horizontal coordinate to the target's range. The output patterns for targets at two different fine ranges are shown in Figs. 7a and 7b and for targets at two different fine Dopplers

[26] R. Markevitch, *Proc. Ann. Wideband Rec. Symp., 3rd RADC* (April 1969).

[27] Ampex Rept. RR 73-14 (March 1974).

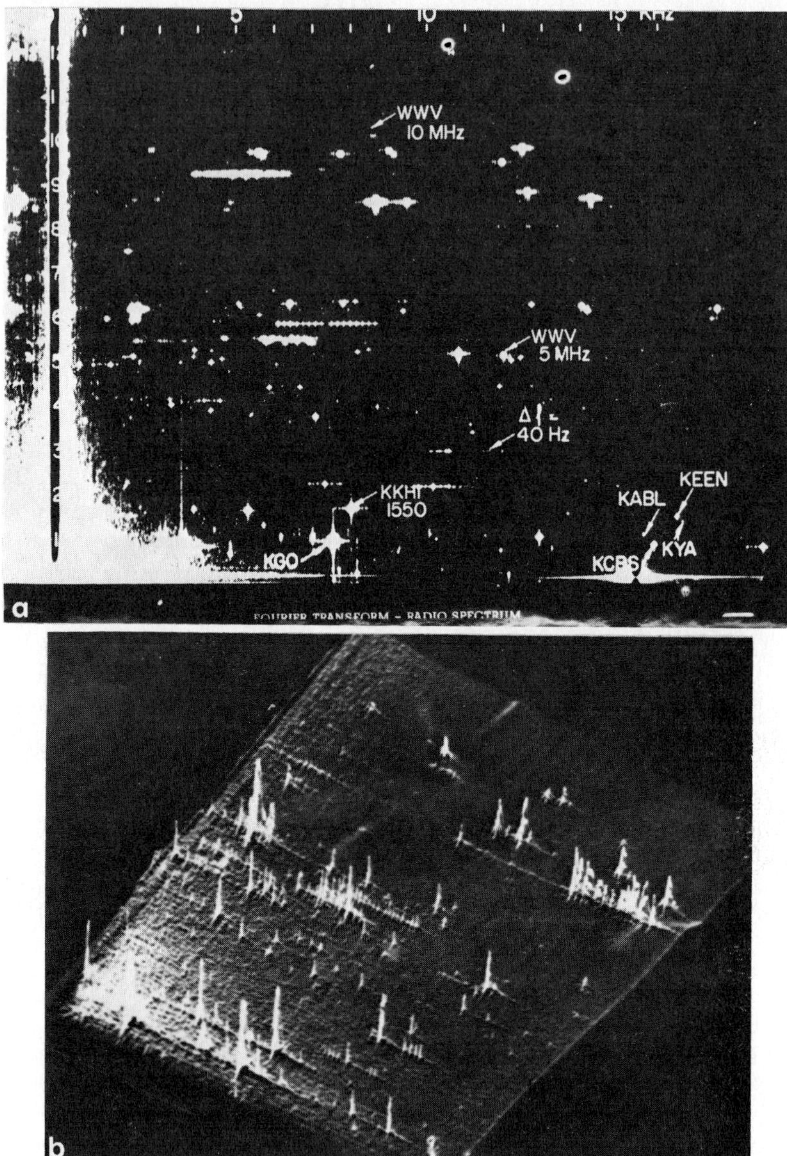

FIG. 6. Real-time folded spectrum optical output (a), and isometric display (b), of a 12-MHz wide band rf spectrum using the fluid-film light valve.[22] (Courtesy Ampex Corporation.)

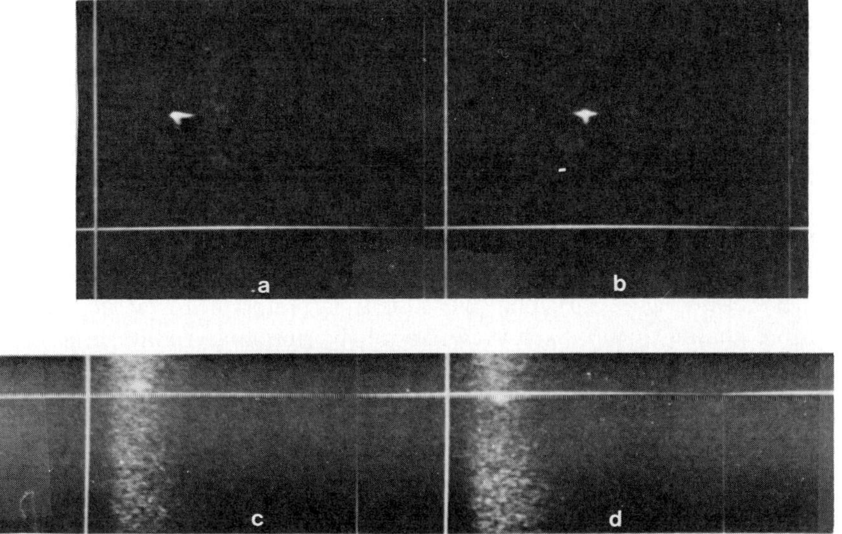

FIG. 7. Real-time linear FM radar processing, using the fluid-film light valve (Courtesy Ampex Corporation). (a) and (b) show different target ranges; (c) and (d) show different target Dopplers.

in Figs. 7c and 7d. These output patterns in Fig. 7 thus correspond to the locations of the peak of the ambiguity function for various simulated targets.

B. THERMOPLASTIC OPTICAL PHASE RECORDER (TOPR)[28, 29]

1. *Structure and Operation*

We now turn to a somewhat different electron-beam-addressed thermoplastic light valve (the TOPR). This light valve is again similar to a conventional CRT with a special faceplate consisting of a 5–10-μm thick thermoplastic (deposited by a gravity settling coating method) with a gold conductive layer between the thermoplastic and a quartz substrate. The primary applications of this light valve are as the input transducer for an optical processor for spectrum analysis and for synthetic-aperture radar image reconstruction. ERIM has developed two types of TOPR light valves for these purposes: a fixed plate device on which a 2 in. × 2 in. raster is recorded, and a 15 in. rotating disk target. Both a raster and a polar format pattern can be recorded. The polar format is useful in a specific synthetic-aperture radar system. In this scheme, the

---

[28] I. Cindrich, G. D. Currie, and C. Leonard, *Proc. Elec. Opt. Syst. Design Conf.* p. 301 (1975).

[29] G. D. Currie, I. Cindrich, and C. Leonard, *Proc. SPIE* **83**, 8 (1976).

electron gun always records data at the same location and the disk rotates so that data are recorded on radial lines between 2 in. and 6.5 in. radii. An antireflection coating is applied between the thermoplastic and conductive layers.

As in the G.E. liquid-film light valve, a charge pattern is scanned onto the thermoplastic target by a modulated electron beam in a fast or slow mode depending on the application. The TOPR uses an intensity-modulated electron beam, whereas the electron beam in the G.E. liquid-film light valve is velocity modulated. In the TOPR, there is more concern for higher bandwidth data recording and for more permanent storage of the recorded data.

The thermoplastic target in the TOPR is temperature cycled by direct electronic heating in the fixed plate device (by applying a voltage across the conductive layer) or by localized radiant heating in the rotating disk system. As the temperature of the thermoplastic target is increased from 50 to 70 °C, the viscosity of the thermoplastic decreases. As in the G.E. device, the electrostatic forces due to the charge deposited on the thermoplastic cause it to deform. The deformations grow until the electrostatic forces of deformation and the surface tension forces (that tend to restore the thermoplastic to its undeformed and unmodulated state) become equal. At high temperatures, the resistance of the thermoplastic decreases and the charge deposited on its surface conducts away to the transparent conducting layer.

Several temperature control modes are used to heat the thermoplastic target to its development temperature. In the first mode of operation, the plastic is heated to its development temperature ($\simeq$100 °C) by applying a short duration (200 msec) current pulse to the conductive layer. The total energy delivered to the plastic is small and the substrate is small, thus the thermoplastic cools rapidly and the deformations are frozen in. In this operating mode, patterns have been stored at room temperature for over three years. The stored pattern can be erased by applying a second heating pulse to reheat the plastic.

In the second mode of operation, the thermoplastic is kept at a constant temperature (50 to 60 °C). The deformations now develop within a few seconds and persist for several minutes. Development times of 0.5 sec and erasure by decay in 5 sec are simultaneously obtainable depending on the thermoplastic's temperature and the charge deposited (development is faster with more deposited charge, whereas erasure is slower with a larger deposited charge). Of course, the spatial-frequency response decreases as the plastic is heated. In this case, a pulsed current applied to the conductive layer has been found to result in a more uniform temperature distribution across the thermoplastic and to speed development and erase time. At present, a steady state heating current yields more reproducible cycle times and performance. Eventually (when perfected) pulsed heating is expected to be the best method.

However, in no instance can the cycle, development, or erase times be reduced appreciably below 1 sec. (A 10-Hz frame-rate system is presently under

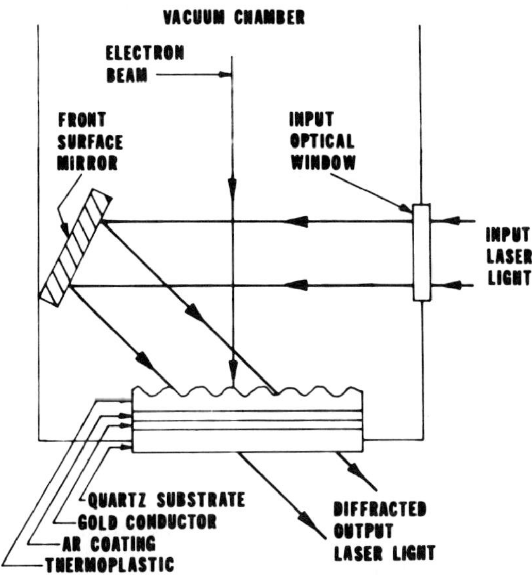

FIG. 8. Schematic diagram of the thermoplastic optical phase recorder (TOPR).

consideration.) This slow frame rate and the questions of the target material's lifetime are items that limit the device's use at present. Its main advantages over the G.E. fluid-film light valve include storage, larger bandwidth, superior optical quality and scanning electronics, and a resultant large transform plane space bandwidth.

A schematic of the transmission mode version of the fixed plate TOPR system is shown in Fig. 8. The electron beam is on axis, whereas the input laser beam to be spatially modulated passes through an input optical window, reflects from a front surface mirror mounted inside the vacuum chamber, and exits through the thermoplastic target at an angle. The use of a non-normal incident read light beam appears to cause no resolution loss at present. In a scaling correlator (input behind the Fourier transform lens), a resolution degradation would result. In practice, two vacuum pumps are used to decrease the effects of outgassing. The two chambers are separated by a small aperture in the neck of the gun through which the electron beam passes. A 7 kV accelerating potential with the anode at 7 kV is used. The effects and extent of outgassing appear to be minimal when the target is heated and a single vacuum chamber architecture appears to be possible if a demountable system is not required. A 16 μm (0.66 mil) spot size is achieved with about 0.5 μA of beam current. The entire assembly measures 27 in. × 30 in. × 40 in. Considerable care has been taken to correct the geometrical scan errors to 0.2% over the 20° deflection angle used and over a 5-MHz bandwidth using magnetic deflection and an analog scan correction module. Stable sweep circuitry insures reliable line-to-line spacing and reduces scan start jitter to a fraction of a resolution element.

2. *Specifications*

As with all light valves, no one set of specifications is adequate to describe the performance of the TOPR. Only the fixed-plate system is discussed here, since it is of the most general use. The TOPRs amplitude MTF for several line scan rates is shown in Fig. 9. The amplitude of the peak value of the first-order diffracted spot is plotted versus the frequency of the input signal. A peak beam current of 0.5 μA was used in obtaining curve *C*. Curve *D* is a theoretical estimate of the TOPRs performance with a peak beam current of 1 μA. At higher sweep speeds, the MTF shifts to higher frequencies within the limits of available beam current for exposure. A 20 μsec/in. scan and 40-MHz response at 50% modulation corresponds to a 32 lp/mm resolution over 50 mm for a 1600 $\times$ 1600 point resolution at 50% modulation and a time bandwidth of 2.5×10^6. Limiting resolution for curve *C* is 55 lp/mm. Projected performance such as that estimated from the data in curve *D* is far superior. The present TOPR specifications are listed in Table III with projected values in parentheses.

The optical quality of the TOPR is excellent with less than $\lambda/6$ peak-to-peak error over the 2 in. $\times$ 2 in. target area. The optical quality of the TOPR appears to be adequate after repeated cycles. However, because the target deforms to produce modulation as does the G.E. fluid-film light valve, the degree to which the surface returns to this original $\lambda/6$ optical flatness after each or many cycles of operation and the time required for this to occur remain to be further investigated.

The lifetime of prior thermoplastic light valves has been very limited. The TOPR device has exhibited a lifetime of over 200,000 cycles but exhaustive lifetime tests have not been conducted. This remains as another item requiring further consideration. The TOPRs dynamic range from the peak output light

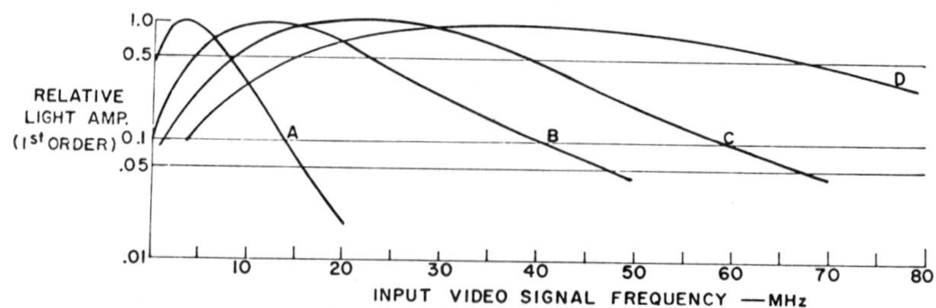

FIG. 9. MTF of the TOPR for different scan rates.[28] Curve A, 100 μsec/in.; curve B, 40 μsec/in.; curve C, 20 μsec/in.; curve D, 10 μsec/in.

TABLE III

TOPR FIXED PLATE LIGHT VALVE SPECIFICATIONS
(projected values in parentheses)

3 dB bandwidth	= 40 MHz (200 MHz)
Target size	= 50 × 50 mm (100 × 100 mm)
Resolution at 50% modulation	= 32 lp/mm (100 lp/mm)
	= 1600 × 1600 points (10,000 × 10,000 points)
Limiting resolution	= 50 lp/mm (150 lp/mm)
Space–bandwidth product	= 2.5×10^6 (10^8)
Write, erase, cycle time	= 0.5–1 sec (temperature control) (10 Hz)
Frame-space–bandwidth product	= 2.5×10^6 (10^8)
Dynamic range	= 400:1
Lifetime	= 10^5 cycles (limited, may be higher)
Optical quality	= $\lambda/8$ peak to peak
Storage time	= 1 yr
Erasure	= 1 sec (temperature control) (10 Hz)

level to the rms output noise is 400:1. The ratio of peak first-order output light to second-order diffracted light is only about 10:1 with a 15-V peak-to-peak 20-MHz input sine wave.

3. *Applications*

The main application for which the TOPR is being developed is as a real-time and reusable optical processor for synthetic aperture radar (SAR) data. The basic principles of SAR processing are reviewed elsewere[30–32] and are thus only briefly noted here. The radar in SAR is airborne and emits short pulses which are reflected from the ground. The return signals are mixed with the transmitted pulse pattern and summed so that as the small on-board antenna moves, the signals appear to come from a much larger antenna. The synthetic aperture that results is the product of the aircraft's speed and the time of signal integration.

A schematic of the ground-based SAR system used to produce data for the TOPR is shown in Fig. 10a. A fixed radar antenna is mounted on a tower with reflecting targets placed on a rotating 20-ft diameter platform. An SAR image of an object on the platform can be produced as follows. A point reflector at radius r from the center of the platform is located in range by a measurement of the time of return of short transmitted radar pulses. The point's azimuth is

[30] L. J. Cutrona, E. N. Leith, L. J. Porcello, and W. E. Vivian, *Proc. IEEE* **54**, 1026 (1966).

[31] E. N. Leith and A. L. Ingalls, *Appl. Opt.* **7**, 539 (1968).

[32] E. N. Leith, Synthetic aperture radar, *in* "Applications of Optical Data Processing" (D. Casasent, ed.) Springer-Verlag, Berlin and New York, 1978.

High Radar Antenna
Low Radar Antenna
41.5 m
26.3°
4.5°
37 m
Rotary Platform
(a)

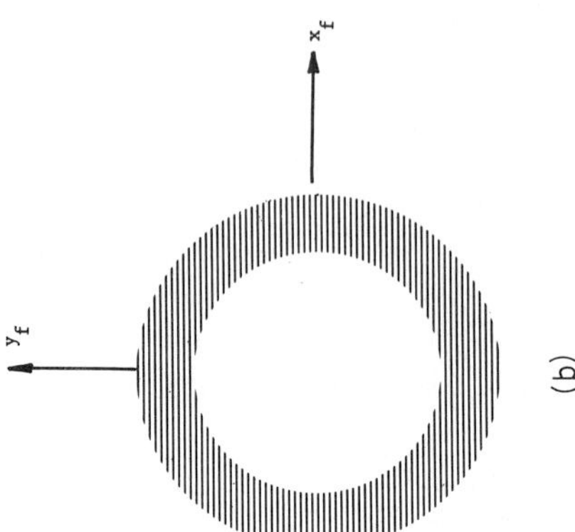

(b)

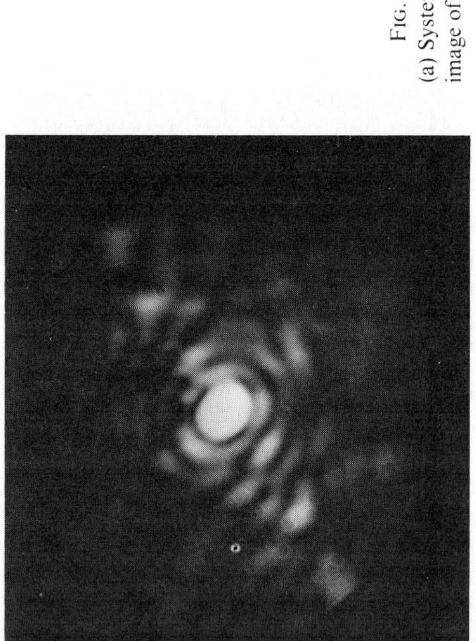
(c)

FIG. 10. Synthetic aperture radar processing using the TOPR.[29] (a) System schematic, (b) input record for point target, (c) output image of point target.

found from a measurement of the Doppler shift of a reflected pulse as the point moves. Point reflectors at the same radius *r* can be distinguished from a record of these range positions as a function of the table's rotation angle. The phase history of a point object is shown schematically in Fig. 10b. With such an input pattern recorded on the TOPR placed at the input plane P_0 of Fig. 1, the two-dimensional Fourier transform produced at P_1 by spherical lens L_1 will be an image of this point object. A greatly magnified output image of a point target so obtained in real-time on the TOPR is shown in Fig. 10c.

Images of more complex objects have been produced. In such cases, the returns from all of the separate point sources in the object are superimposed on the TOPR in formats similar to that shown in Fig. 10b. To separate the two overlapping conjugate images and zero-order dc term, an electronic offset carrier is added to the phase history before recording, and the zero-order and one first-order image blocked during processing by a suitable aperture. The highest image quality results when the data are recorded in a polar format and thus the rotating disk TOPR was devised. In this system the data records are recorded radially on the thermoplastic between radii of 2 and 6.5 in. For coherent recording, extremely accurate scan registration is essential; hence the system's high 0.02% geometrical linearity specification. Resultant real-time imagery produced on the TOPR is comparable to the best non-real-time photographically recorded and optically processed (off-line) data.

IV. DKDP LIGHT VALVES

The final electron-beam-addressed coherent light valve to be described achieves spatial modulation by electronically controlled variations in the index of refraction of the target material, which cause a spatial modulation of a collimated reading laser beam by the electro-optic effect. The target crystal used is a thin basal plate of DKDP (potassium dideuterium phosphate). Carnegie–Mellon University (CMU) in the United States and the Laboratoire d'Electronique Physica Appliqué (LEP) in France are still actively pursuing work on and applications of electron-beam-addressed DKDP (*e*-beam DKDP). CMU has emphasized the coherent optical data processing applications of this light valve and LEP its use as a large-screen color projection and display device. A detailed description of the device and selected applications of its use follow in Section IV,A.

The first optically addressed light valve to be described (photo-DKDP) is discussed in Section IV,B. It was developed at LEP where it has been used mainly for noncoherent optical processing with coherent optical processing applications of the device and its integration into systems the subject of current research at CMU.

A. e-BEAM DKDP LIGHT VALVE

1. *Structure and Operation*

The schematic of the CMU coherent e-beam DKDP light valve is shown in Fig. 11. A thin 250-μm thick, 0°, Z-cut basal plate of DKDP crystal is used as the target. A thin transparent grounded layer of CdO vacuum deposited on the crystal serves as the anode for the depressed cathode electron gun. The CaF substrate provides rigidity for the target assembly. The input signal is applied to the grid of the write gun and its beam current modulated in synchronism with the deflection of the electron beam. The spatial charge pattern deposited on the target crystal is proportional to the input data. The recording format chosen depends on the application, with a wide variety of formats employed in signal processing applications,[22] but a raster scan is the most used.

The spatial voltage distribution across the crystal and the associated electric field cause a point-by-point spatial variation in the crystal's index of refraction described by Eq. (2.11) and a corresponding spatial variation in the crystal's transmittance given by Eq. (2.12) where $V = V(x, y)$ is a spatial function. Two versions of the e-beam DKDP light valve are in present use. The CMU system[33] (shown in Fig. 11) uses as 6 kV accelerating potential write gun with a 50-μA beam current in a 25-μm spot. During writing, the front surface of the crystal is charged negatively toward the crystal's half-wave voltage. During erasure, a lower accelerating potential (600 V) gun is used. At this low accelerating potential, the secondary emission ratio of the DKDP is greater than one, the front surface of the crystal is charged positively toward a zero-volt level thereby erasing the previously recorded pattern. A defocused erase electron beam with a total current of 500 μA achieves erasure in 0.5 msec by flooding the target crystal (faster erase times are possible with higher current guns). The focusing and deflection system for the write gun are linear (1%) and well corrected for the off-axis geometry used.[34]

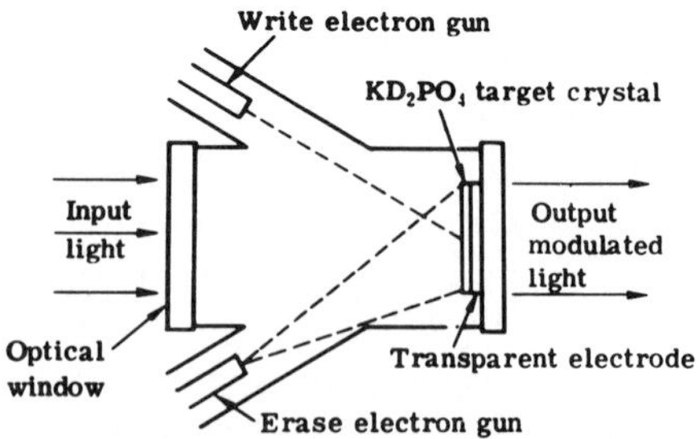

FIG. 11. Schematic of e-beam DKDP.

[33] D. Casasent and W. Keicher, *Proc. Elec. Opt. Syst. Design Conf.* p. 99 (1971).

[34] D. Casasent and F. Caimi, *J. Vac. Sci. Technol.* **10**, 1102 (1973).

In the CMU system, the data are recorded on the DKDP by the write gun in the normal 32-msec scan time of a conventional TV frame. During the first half of the vertical retrace time (0.5 msec) the laser is pulsed on, passes through the target crystal, and is spatially modulated by the recorded data. All processing actually occurs during this read or process time. The erase gun is then pulsed on during the second half of the vertical retrace time returning the target to its original unmodulated state by the start of the next frame. In coherent optical processing, these write, read, and erase cycles must not overlap since the Fourier transform of the input data is not valid until an entire frame is recorded. In certain instances (e.g., wide band recording), no major problems will arise if the transform of half of one frame and half of the next is formed. However, in most instances, incomplete erasure represents a major problem.

The LEP system[35,36] is primarily intended for real-time large-screen TV projection and display applications. The recording scheme used for such cases must not exhibit flicker (i.e., separate write, read, and erase cycles as used in Fig. 11 are not desirable). A scan line pattern in the output display is also objectionable in projection applications. To achieve these features, a fine wire grid (40–50 μm spacing) is mounted in front of the DKDP crystal. The input signal is applied to this grid and a low accelerating potential (500–1000 V) write gun with constant beam current is employed. As the electron beam is scanned across the target, it acts as a flying short circuit between the grid and the corresponding point being addressed on the crystal. In this recording scheme, the point on the crystal to which the electron beam is deflected is charged positively to the voltage on the grid at that time (the same voltage is applied uniformly to the grid). Writing is thus by secondary emission. A second grid at positive potential is used to collect the secondary electrons. The spacing of the grid from the crystal (about 40 μm) must be comparable to the crystal resolution desired. Since the charge at all points on the crystal's surface is constant between scans (or until readdressed), a flicker-free display results.

Secondary-emission writing, similar to that employed in the LEP systems, is used in the CMU system to measure the crystal's half-wave voltage, Curie and transition temperatures, secondary emission curve, and other parameters of DKDP needed to optimize operation of this light valve system.[37–39]

In this LEP device, the electron beam is on axis, a dielectric mirror is deposited on the electron beam side of the DKDP, and the recorded data are read out in reflection (otherwise the target structure is comparable to that of

---

[35] G. Marie and J. Donjon, *Proc. IEEE* **61**, 942 (1973).

[36] G. Marie, J. Donjon, and J. Hazan, Pockels-effect imaging devices and their applications, *in* "Advances in Image Pickup and Display" (B. Karzan, ed.), Vol. 1, p. 225. Academic Press, New York, 1974.

[37] D. Casasent and W. Keicher, *Dig. Int. Elec. Dev. Meeting, IEEE-ED Group*, p. 426 (1973).

[38] D. Casasent, *IEEE Trans. Electron. Dev.* **ED-20**, 1109 (1973).

[39] D. Casasent and W. Keicher, *J. Opt. Soc. Am.* **64**, 1575 (1974).

the CMU system). The target crystal used in the CMU system is 50 × 50 mm, whereas that used in the LEP system is about 25 × 25 mm. For coherent optical processing applications, the target assembly is wedged to reduce the effect of reflections.

Both *e*-beam DKDP light valves and the photo-DKDP device (Section IV,B) are operated at the target crystal's transition temperature $T_0 = -50°C$ (between its paraelectric and ferroelectric phase). At this temperature, the ratio of the crystal's longitudinal and transverse dielectric constants $\varepsilon_{33}/\varepsilon_{11}$ increases dramatically. The effective thickness b' of the crystal is $b/(\varepsilon_{33}/\varepsilon_{11})^{1/2}$, where b is the actual thickness of the DKDP. At room temperature $b' \simeq b$, whereas at T_0 we find $b' \simeq 0.07b$.

Thus, at T_0 the crystal's lines of force are concentrated along the optical axis, the deposited charge thus does not spread as it normally would, and the crystal's effective thickness is reduced by a factor of over 14. If we view a single element of the crystal as a parallel plate capacitor with the conducting electrode as one plate and the uniform electron-beam spot of diameter d at a voltage V as the other plate, one can show that the diameter to the $V/2$ point is $d + 0.44b'$. Thus, at room temperature a resolution element is approximately half the crystal's thickness or 125 μm, whereas at T_0 a resolution element is effectively the electron-beam spot size d or about 25 μm.

Other advantages besides resolution result from operation of the light valve at T_0. The time constant of decay for DKDP increases greatly at T_0 and a useful storage of over 1 hr results. The target's uniformity of response also improves dramatically at T_0, The half-wave voltage of DKDP is over 3000 V at room temperature but decreases to about 250 V at T_0. This also allows the use of lower electron gun accelerating potentials and makes complete erasure possible. Further details on these and other aspects of Curie point operation of DKDP are presented elsewhere.[37–39] The $T_0 = -50$ °C operating temperature of the *e* beam and photo-DKDP light valves is easily achieved by Peltier cooling cells around the target assembly.

2. *Specifications*

The specifications for the *e*-beam DKDP light valve are listed in Table IV. Existing systems operate at TV rates (30 frames/sec), a line scan rate of 53 μsec, and a complete write, read, and erase cycle every 33 msec. Operation at other rates is possible with different deflection coils and electron guns. The MTF for an *e*-beam DKDP light valve at 25 frames/sec with a 30-μm electron-beam spot and a 180-μm thick DKDP crystal is shown[40] in Fig. 12. The resolution at 50% MTF is seen to be 15 lp/mm with a limiting resolution of 30 lp/mm. This corresponds to a 750 × 750 point input at 50% modulation.

---

[40] J. Donjon *et al.*, *Acta Electron.* **18**, 187 (1975).

TABLE IV

e-BEAM DKDP LIGHT VALVE SPECIFICATIONS

Bandwidth	= 10 MHz
Target size	= 50 × 50 mm
Resolution at 50% modulation	= 15 lp/mm
	= 750 × 750 points
Limiting resolution	= 30 lp/mm
Space–bandwidth product	= 5×10^5
Write, read, erase cycle	= 33 msec
Frame-space–bandwidth product	= 1.5×10^7
Dynamic range	= 40–60 dB
Lifetime	= 3 years
Optical quality	= $\lambda/4$
Erasure	= by secondary emission in 0.5 msec or less
Storage	= over 1 hr

The device's contrast ratio depends on the degree of collimation of the input light with 100:1 values typical for 9° input light angles. For coherent light applications, well-collimated light is used and contrast ratios of 1000:1 or 60 dB have been measured. This contrast ratio also can be realized by the use of a quartz compensating plate.

DKDP crystals can be fabricated to high surface flatnesses with a $\lambda/4$ error over the surface of the device obtainable at present. Sealed DKDP light valves have been fabricated (by baking out the system) and operated successfully for over three years.

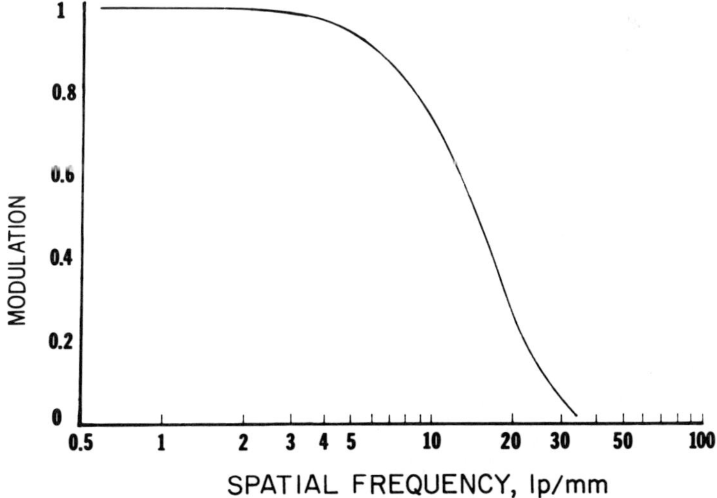

FIG. 12. Measured MTF of *e*-beam DKDP.

This represents the one electron-beam-addressed light valve with no foreseeable optical quality defects ($\lambda/4$ or better), high resolution (1000×1000 points), storage (1 hr), a fast and complete erase cycle, and real-time operation at 30 frames/sec with a complete write/read/erase cycle every 33 msec.

3. *Applications*

A complete description of the numerous optical data processing applications that have been demonstrated on *e*-beam DKDP would fill several chapters. A diverse selection of applications is thus included here with emphasis on the optical pattern recognition and correlation applications of these light valves. References for each application are included from which the interested reader may obtain more complete details.

In amplitude modulation of a Pockels-effect device, the input light is linearly polarized at 45° to the induced crystallographic axes. In this mode, the polarization of the zero order and all even-order diffracted terms are unchanged, whereas the first-order image light is rotated 90°. Thus the dc and even orders are automatically blocked by the output analyzer. This method is most effective when the average charge on the crystal is near zero. This condition is achieved by proper choice of the potential on the grid during erasure, i.e., the final uniform no-modulation state. This condition also reduces the crystal's ionic conduction and thus improves its performance.

In the *e*-beam DKDP light valve, one can block the dc component of the input signal and thus effectively subtract the dc term in the Fourier transform. This is known as zero-order suppression. It allows one to avoid having to record the normally overexposed zero-order term in the Fourier transform, thus reducing the noise introduced by scattered light and the dynamic range requirements of the recording media. To demonstrate this, an image of the letter "m" was recorded on the DKDP with the dc term in the video blocked. A Fourier-transform hologram of this image was then formed, the reconstructed real image is shown[41] in Fig. 13. Various constant signal levels can be similarly subtracted and high modulation levels used with only 1.6% distortion of the first order for full modulation with a sinusoidal grating input. Other special device features such as these are possible on the remaining devices. Several examples are included in Section IV,B,3.

The use and advantages of computer-generated binary holograms in optical data processing are well known.[42] A camera was focused on such a hologram of the letters "ICO" and a charge pattern proportional to the input data recorded on the DKDP. The real-time reconstruction of this computer-generated hologram's image is shown[43] in Fig. 14. The resolution and geometrical

---

[41] G. Groh and G. Marie, *Opt. Commun.* **2**, 133 (1970).

[42] A. W. Lohmann and D. P. Paris, *Appl. Opt.* **6**, 1739 (1967).

[43] G. Goetz, *Appl. Phys. Lett.* **17**, 63 (1970).

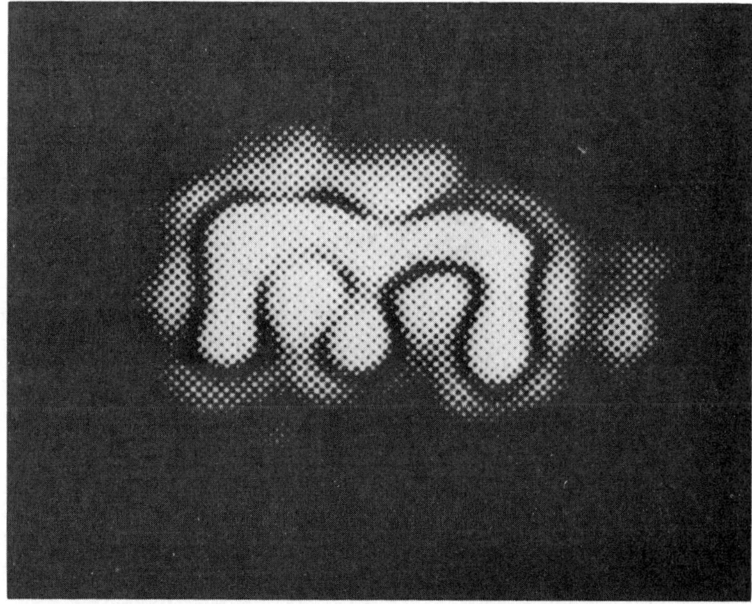

FIG. 13. Real-time baseline subtraction using *e*-beam DKDP.[41]

FIG. 14. Real-time reconstructed computer generated hologram using *e*-beam DKDP.[43]

FIG. 15. Real-time reconstructed acoustical hologram using *e*-beam DKDP.[45]

fidelity of the camera were the principal limitations in the quality of the reconstructed image.

Holographic imaging techniques can also be used in underwater imaging. When an ultrasonic signal reflected from an underwater object is received, the resultant acoustic hologram can be reconstructed to yield an image of the object.[44] Synthetic acoustic holograms (formed by oscillating a static holo-

[44] H. R. Farrah, E. Marom, and R. K. Mueller, An underwater viewing system using sound holography, *Acoust. Hologr.* **2**, 173 (1970).

gram through 45° and viewing it with a TV) have been recorded on *e*-beam DKDP and subsequently reconstructed (Fig. 15) in real time.[45]

Many noncoherent optical correlation applications such as speech spectrogram pattern recognition[46] have been demonstrated on *e*-beam DKDP. A wealth of extraordinary quality large-screen images in black and white and color (using three *e*-beam DKDP light valves) have also been produced.[47] However, we will continue to emphasize the coherent optical processing applications of all devices in this review.

A final example of coherent image processing using *e*-beam DKDP is coded aperture x-ray imaging.[48] In this scheme an object consisting of layers s_i (for the example discussed, four lead digits separated by 2 cm are used as the object) are multiprojected by an array p of x-ray tubes onto an image intensifier from which the superimposed pattern is recorded on the DKDP as

$$s' = \sum s_i * P_i \tag{4.1}$$

where P_i is the point spread function of layer s_i. These P_i differ only by a scaling factor.

In the example shown, p is projected with a pinhole camera at layer s_0. This pattern is recorded on the DKDP and its conjugate transform $P_0{}^*$ formed and recorded in the transform plane. With s' recorded on the DKDP at the input plane and a filter described by $P_0{}^*$ placed in the transform plane, the output pattern in the correlation plane is

$$s'' = s_0 * p_0 \circledast p_0 + \sum_{i \neq 0} s_i * p_i \circledast p_0 = s_0 \tag{4.2}$$

or the desired image of layer s_0. As the position of the filter hologram is shifted along the optical axis, the deconvolution of other layers occurs, and their images appear in the output plane. The images of the four layers in the example used are shown[49] in Fig. 16 for a Golay nonredundant ten-point source distribution. They were all obtained in real time on the *e*-beam DKDP and are comparable to the imagery obtained by all other reconstruction methods.

Numerous coherent pattern recognition or correlation applications have been reported using both the FPC or JTC system with the *e*-beam DKDP used in the input plane. Two examples are shown in Figs. 17 and 18. The runway section (Fig. 17b) of the entire input image (Fig. 17a) was recorded on the DKDP and a matched filter of it formed. With the entire input image (Fig. 17a) recorded on the DKDP, the correlation plane pattern of Fig. 17c occurs at the

---

[45] G. Goetz, R. Koppelmann, and R. K. Mueller, *Proc. Elec. Opt. Syst. Design Conf.* p. 202 (1972).

[46] H. Weiss, *Proc. Int. Joint Conf. Patt. Recog.* IEEE Cat. No. 73CH0821-9C (1973).

[47] G. Marie, *Ferroelectrics* **10**, 9 (1976).

[48] H. H. Barrett, *Proc. IEEE* **65**, 89 (1977).

[49] H. Weiss, *Dig. Int. Opt. Comput. Conf.* p. 41. IEEE Cat. No. 74CH0862-3C (1974).

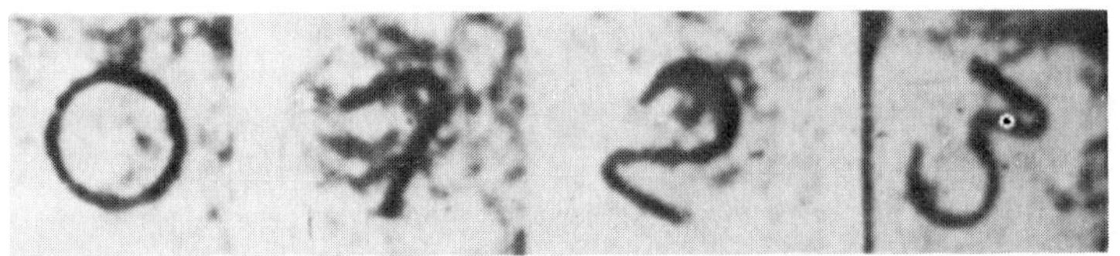

FIG. 16. Real-time reconstructed coded aperture x-ray images using e-beam DKDP.[49]

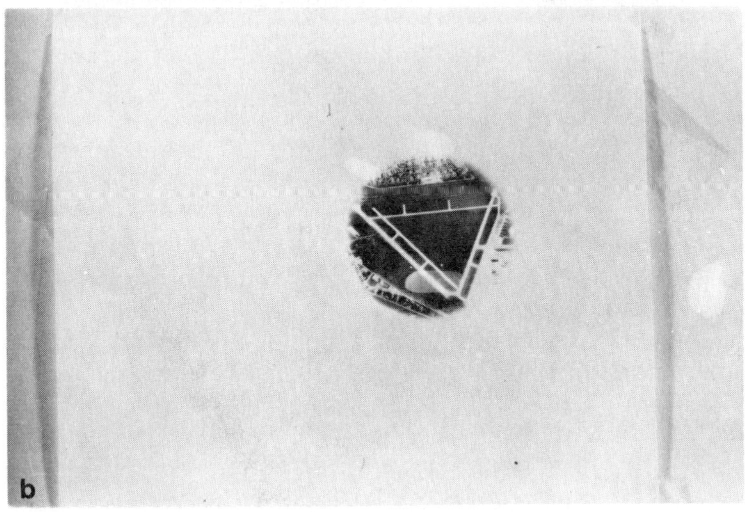

FIG. 17a,b. Real-time image pattern recognition using e-beam DKDP.[50] (a) Input pattern, (b) searched-for image.

FIG. 17c. Real-time image pattern recognition using *e*-beam DKDP.[50] Output correlation peak.

output plane. It consists of a single peak of light whose position is proportional to the location of the runway in the input pattern.[50] In the second example shown (Fig. 18), the input (Fig. 18a) was a paragraph of text with seven occurrences of the word "God." As before a matched filter of the word "God" was formed from the data on the DKDP. With the paragraph of Fig. 18a recorded on the DKDP at the input plane and the matched filter of the word "God" at the transform plane, the output correlation plane pattern shown in Fig. 18b appears. It consists of seven peaks of light whose positions correspond to the locations of the seven occurrences of the word "God" in the input pattern.[51]

Numerous signal-processing applications of *e*-beam DKDP have also been demonstrated including wide band folded spectrum analysis,[24] cw Doppler radar processing,[24] linear and planar phased array radar processing,[52] pulsed Doppler, and FM step radar processing.[22,53,54] Only two examples are included here for brevity.

In the optical processing (not beam forming) of linear phased array radar data,[52] the time history of the heterodyned signals from N elements of the

---

[50] D. Casasent and W. M. Sterling, *IEEE Trans. Comput.* **C-24**, 348 (1975).

[51] D. Casasent, *Opt. Eng.* **13**, 228 (1974).

[52] D. Casasent and F. Casasayas, *IEEE Trans. Aerosp. Electron. Syst.* **AES-11**, 65 (1975).

[53] D. Casasent and F. Casasayas, *Appl. Opt.* **14**, 1364 (1975).

[54] D. Casasent and E. Klimas, *Appl. Opt.* **17**, 2058 (1978).

God performed many miracles in Indonesia because we had many needs. Americans do not have the same needs.

One day I was talking at a school in America, and one of the boys said, "That's what we need in America. We need miracles like that. We need to have water turned into wine in our church." And I said to that brother, "Well, there's no sense in God turning water into wine here in America. You have wine here. But grapes do not grow in Indonesia. Therefore, we have no wine. And we have no bread. So many times God needed to perform miracles. God always does miracles for a purpose. In America you have other needs. You need the power of God to reach souls for Jesus Christ. You can expect God to give you that. But if you have grapes, it's silly to ask God for grape juice."

a

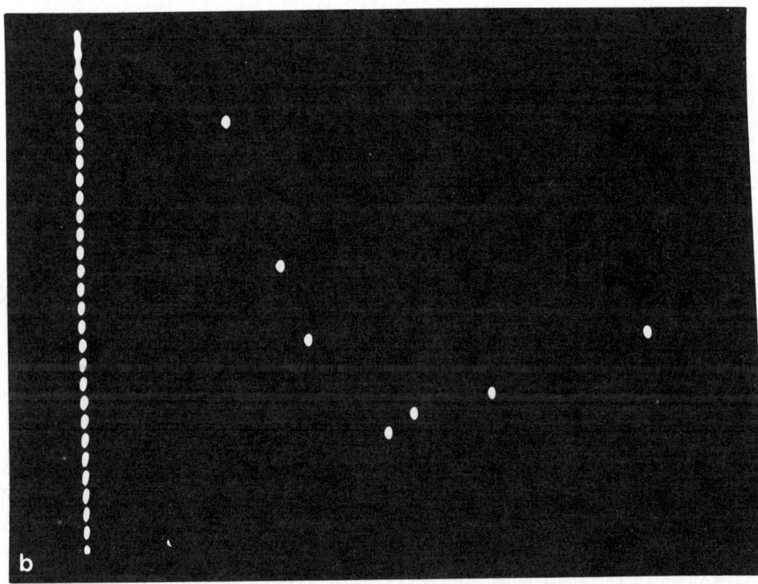

FIG. 18. Real-time text recognition using *e*-beam DKDP.[51] (a) Input pattern, (b) output correlation plane pattern.

array are recorded on successive lines on the *e*-beam DKDP light valve. The vertical coordinate of the first-order off-axis peaks of light in the Fourier transform of this data (Fig. 19) is proportional to the target's azimuth angle. For planer phased array data, the coordinates of the output first-order peaks of light are proportional to the target's azimuth and elevation angles.

In the optical processing of pulsed Doppler and FM stepped data,[53] successive received radar echoes are recorded on successive lines on the DKDP. The coordinates of the first-order off-axis terms in the transform of this data are now proportional to the fine Doppler or fine range of the target. The output pattern in Fig. 20 (optically processed in real time on the DKDP using real radar data) shows a pair of off-axis peaks of light, whose locations correspond to the fine range of a real radar target obtained from real linear FM stepped radar data. Extensions of these optical processing concepts using a multichannel correlator and coded wave forms have also been performed.[54] These systems yield higher SNR outputs in which both coordinates of the output plane contain target data (such as azimuth/Doppler and range/Doppler).

B. Photo-DKDP Light Valve

1. *Structure and Operation*[55,56]

The first optically addressed light valve considered also uses DKDP as the target crystal. The structure of this photo-DKDP light valve is shown in Fig. 21 with the electrical connection required for write (W), read (R), erase (E), and storage (S) of data on the device.

The approximate divisions of the applied voltage across the various layers in the device during its various cycles of operation are shown in Fig. 22. A uniform voltage $V_0 = 100$ V is initially applied between the electrodes with the Se photoconductor side positive. Since the dielectric constant for DKDP is far larger than for Se (600 versus 6) almost all of this V_0 appears across the Se (curve a). With this V_0 present, the Se side of the light valve is exposed to a spatially modulated light beam (in parallel or scanned point by point) at λ_W (in the uv or visible, usually at 440 nm). This λ_W light produces photocarriers in the Se which drift to the DKDP interface causing the voltage across the 10-μm thick Se to drop and the voltage across the 150-μm thick DKDP crystal to increase by an amount proportional to the input light intensity (curve b).

During readout, the electrodes are shorted. This presents a reduction in contrast by capacitive coupling of the electric field through the Se. The voltage V_1 at the interface (curve c) and present across the DKDP is now positive and proportional to the input exposure.

---

[55] J. Donjon *et al*, *IEEE Trans. Electron. Dev.* **ED-20**, 1037 (1973).

[56] F. Dumont, J. P. Hazan, and D. Rossier, *Philips Tech. Rev.* **34**, 274 (1974).

FIG. 19. Real-time phased array radar signal processing using *e*-beam DKDP.[52]

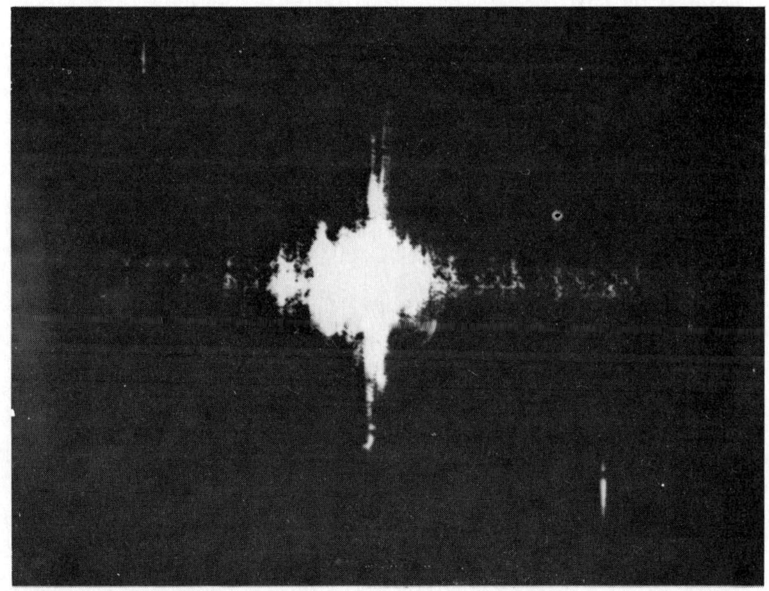

FIG. 20. Real-time FM step radar signal processing using *e*-beam DKDP.[53]

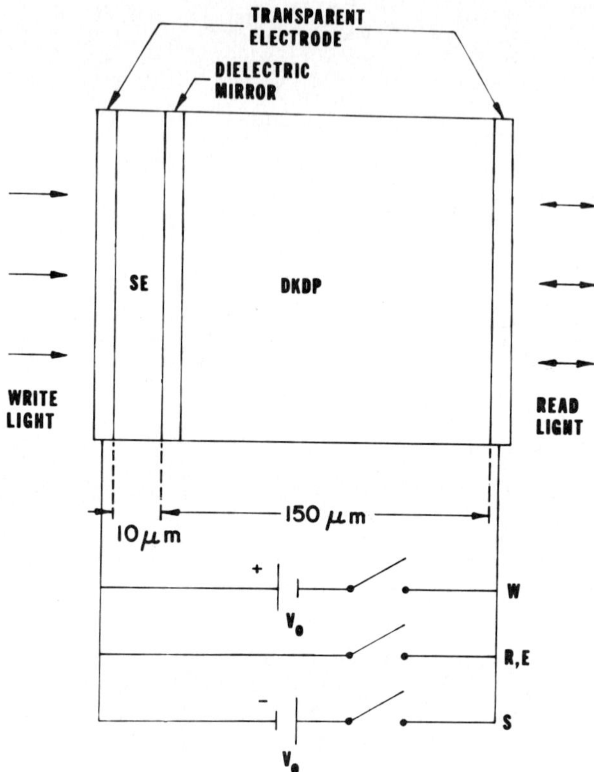

FIG. 21. Schematic of the photo-DKDP light valve.

With a spatial variation in λ_E a spatial variation $V(x, y)$ in V_1 occurs and by the Pockels effect, Eqs. (2.11) and (2.12) the phase or amplitude of the output light at λ_R (usually 504 or 633 nm) will be spatially modulated. To erase the recorded data, the Se side of the device is flooded with broadband light in the λ_W spectral region (curve d).

In the operating mode just described, writing requires the movement of electrons while erasure is achieved by the flow of holes. Operation with the reverse polarity for V_0 (Se side negative during writing) is also possible. In one case, writing is slow (10 μsec) and erasure fast (1 μsec), whereas opposite speeds result for the opposite polarity of V_0. The target crystal in the photo-DKDP light valve is operated at the crystal's transition temperature ($t_0 \simeq -50$ °C) as in *e*-beam DKDP light valves. The useful storage time for both devices is about a half hour. For long-term storage of the data in photo-DKDP the applied voltage can be reversed to $-V_0$. The voltage at the interface is now negative and charge flow greatly reduced.

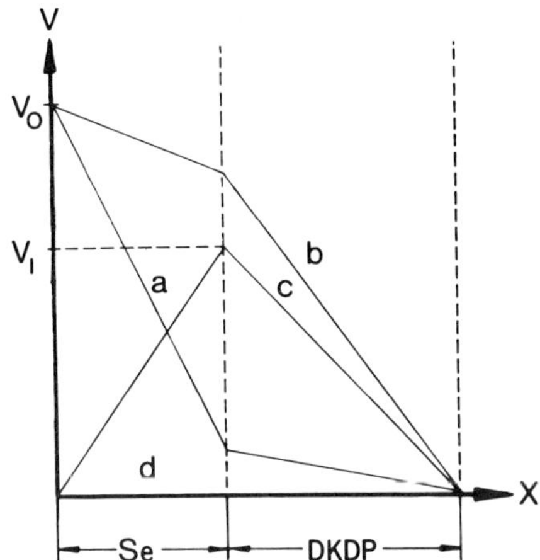

FIG. 22. Voltage distribution across the photo-DKDP light valve at various stages of its operation.

One can subtract various constant levels from a latent image stored on the photo-DKDP light valve by applying a second uniform exposure with V_0 reversed. This is commonly referred to as baseline subtraction or level slicing and is one of the special device features referred to in Table I. By subtracting more charge than the average present in an image, a complete reversal of the stored pattern results or a positive-to-negative image conversion. These and several other examples of baseline subtraction (including addition and subtraction of images) will be shown in Section IV,B,3.

Se was chosen as the photoconductor because it can be vacuum deposited on DKDP and because of its good sensitivity and low dark current at the device's $T_0 = -50$ °C operating temperature. Se also exhibits better charge storage than CdS and other more conventional photoconductors. At $T_0 = -50$ °C, the product of the mobility and lifetime of holes and electrons in Se are approximately equal. Thus, this light valve is electrically symmetrical and subtraction of images is far easier than in electrically nonsymmetrical light valves such as the PROM (Section V).

2. *Specifications*

The optically addressed photo-DKDP and PROM light valves are often compared[56,57] since they both utilize the Pockels effect and exhibit storage and comparable resolution. The limiting measured resolution for both devices

[57] S. Iwasa and J. Feinleib, *Opt. Eng.* **13**, 235 (1974).

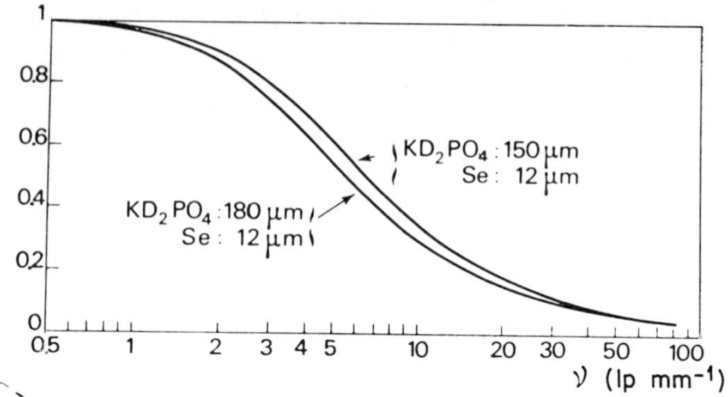

FIG. 23. MTF for two thicknesses of photo-DKDP light valve.[40]

is about 80 lp/mm. The theoretical MTF curves for two thicknesses of photo-DKDP and PROM crystal are shown[40] in Figs. 23 and 31 (Section V), respectively. These curves were obtained by solving Laplace's equation in each light valve. As shown, the MTF curves for both light valves are comparable. In this review, we will use 10 lp/mm as the resolution of both devices at 50% modulation. These valves differ somewhat (10 lp/mm versus 20 lp/mm),[58] from other measured PROM resolutions. However, these higher values were obtained interferometrically and this may be one possible explanation for the differences.

High quality images have been formed on photo-DKDP with light flashes as short as 10μsec. However, for both the PROM and photo-DKDP light valves, a maximum cycle rate of 100 frames/sec is used in the tabulated comparison data since this seems to be the maximum realistic rate at which new input data can be prepared and imaged or scanned onto the device. We also note the response and erase times for both devices in the tabulated device specifications.

The sensitivity (the exposure in $\mu J/cm^2$ required to change the readout light by 33%) and maximum readout transfer ratio (MRTR) (percent of the input light that leaves the device) of all optically addressed light valves will also be listed. The MRTR for photo-DKDP (50%) greatly exceeds that of the PROM (8%) because of the higher transmittance of DKDP and its electrodes, the higher electro-optic sensitivity of DKDP, and higher transfer efficiency of DKDP.[56]

3. *Applications*[47,55]

Photo-DKDP, like *e*-beam DKDP, has been used in numerous optical data processing applications. Matched spatial filters have been recorded on it and

[58] R. A. Sprague and P. Nisenson, *Proc. SPIE* **83**, 51 (1976).

successful correlations performed. Such correlation applications were noted in Section IV,A in which the *e*-beam DKDP light valve was used at the input plane. In this section, we emphasize the use of photo-DKDP in many of the special-device features noted in Table I. Many of these operations rely on the ability to add and subtract images or constant levels from a data pattern recorded on the device (as discussed in Section IV,B,1).

Analysis of data such as the reconstruction shown in Fig. 24a of an image recorded on photo-DKDP can often be enhanced by subtracting a constant light level from the image. This is done by exposing the DKDP with the image of Fig. 24a stored on it to a second uniform exposure with $-V_0$ across the device. As shown in the new reconstruction in Fig. 24b, contours of a particular gray level now appear black. This is the basis of contrast enhancement, positive to negative image conversion, equiluminance contour generation, baseline subtraction, etc. If the second exposure with $-V_0$ applied is made with a second image (the original image slightly displaced) rather than a uniform exposure, the resultant reconstructed image will appear differentiated in one dimension (first-order differentiation). If the second exposure with $-V_0$ applied is made with a defocused version of the first image, the reconstruction will be a second-order differentiation of the original image (Fig. 25).

The use of such subtraction features for change detection is also obvious. To demonstrate this, a photograph of a man was imaged onto and recorded on the photo-DKDP light valve. A second image of the man rotated 180° was then recorded with $+V_0$ applied across the device in both cases. The new reconstructed image (Fig. 26a) shows the addition of two scenes. The original image was then recorded again with $-V_0$ present. The output (Fig. 26b) shows the successful subtraction of the images.

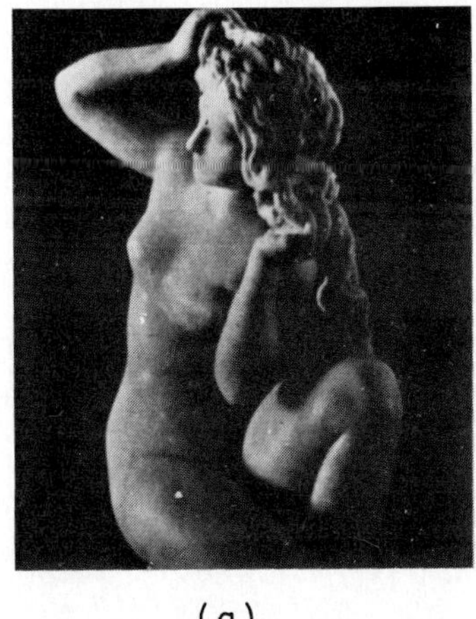

(a)

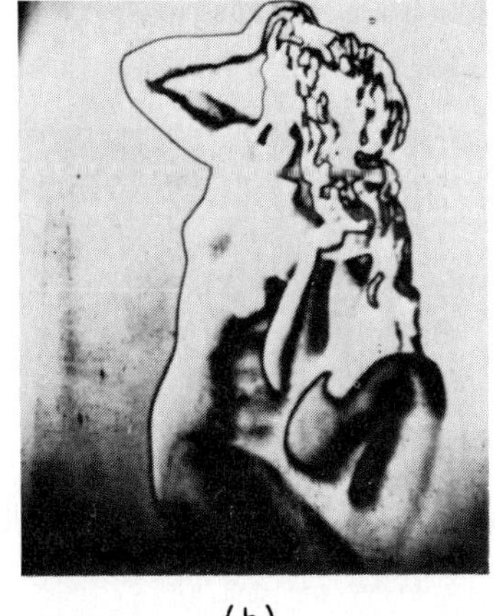

(b)

FIG. 24. Real-time equiluminance contour generation or level slicing using the photo-DKDP light valve.[56] (a) Original reconstruction, (b) level-sliced reconstruction.

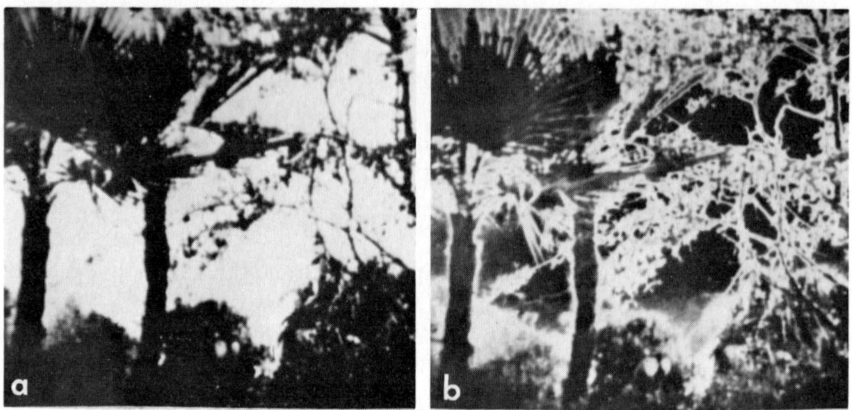

FIG. 25. Real-time spatial differentiation using photo-DKDP.[47] (a) Original reconstruction, (b) differentiated reconstruction.

Whereas spatial integration is useful in enhancing fixed aspects and fine details of an image, temporal integration is useful in increasing the SNR of an image affected by uncorrelated noise. The results of the temporal integration on photo-DKDP are shown in Fig. 27. A reconstructed image (Fig. 27a) was obtained by projecting an image onto the DKDP with a diffuser in front of the input image. When the diffuser was rotated during the exposure, the desired image reconstruction shown in Fig. 27b results.

As a final different application of an OTTO, we refer to Fig. 28 which is a visible reconstruction of an image projected onto photo-DKDP with x-rays. The x-ray sensitivity of the Se photoconductor makes this possible and demonstrates another wide area of application for such devices as image converters.

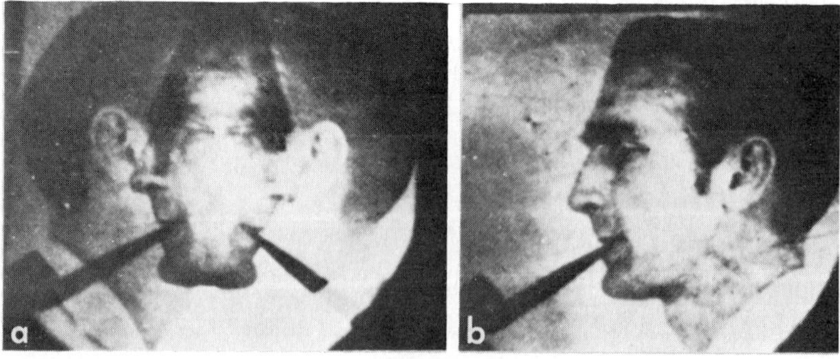

FIG. 26. Real-time addition (a), and subtraction (b), of imagery using photo-DKDP.[55]

FIG. 27. Real-time temporal integration on photo-DKDP.[56] (a) Input corrupted with uncorrelated noise, (b) temporally integrated reconstruction.

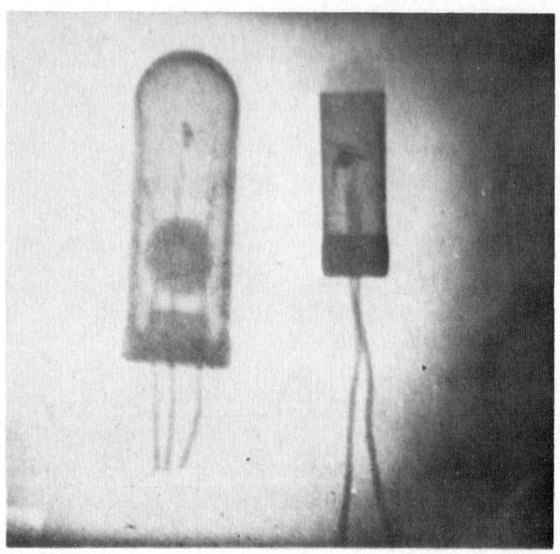

FIG. 28. Real-time reconstruction using photo-DKDP as an x-ray-to-visible converter.[56]

TABLE V

PHOTO-DKDP LIGHT VALVE SPECIFICATIONS

Target size	= 30 × 40 mm^2
Resolution at 50% modulation	= 10 lp/mm, 300 × 400 points
Limiting resolution	= 50 lp/mm
Space–bandwidth product	= 1.2×10^5
Cycle time	= 100 frames/sec (or faster)
Write, read time	= 10–100 μsec
Frame-space–bandwidth product	= 1.2×10^7
Dynamic range	= 100–1000:1
Lifetime	= long, 3 years
Cosmetic quality	= excellent
Optical quality	= $\lambda/2$
Storage time	= 1 hr
Sensitivity	= 1 $\mu J/cm^2$
MRTR	= 50%
Required voltage	= 100 V

V. PROM LIGHT VALVE[57–59]

1. *Structure and Operation*

The PROM (Pockels Readout Optical Modulator) has been under research development at Itek Corporation since 1966 for many diverse applications. The PROM (like photo-DKDP) is optically addressed; the index of refraction of the target crystal ($Bi_{12}SiO_{20}$) is spatially varied proportional to the input write light ($\lambda_W = 440$ nm) distribution, and spatial amplitude or phase modulation of the readout light ($\lambda_R = 633$ nm) is produced by the Pockels effect (Section II). However, since $Bi_{12}SiO_{20}$ is both electro-optic and photoconductive in blue write light, the PROM structure (Fig. 29) differs from that of the photo-DKDP light valve. The device is written on from one side and read in reflection from the other. A voltage division across the $Bi_{12}SiO_{20}$ is required as in photo-DKDP. To achieve this, a thin 6-μm insulating layer of parylene is coated onto the typically 800-μm thick $Bi_{12}SiO_{20}$ crystal. The control voltage across the device (1500–2000 V versus 100 V for photo-DKDP) is applied between the transparent outer electrodes of indium tin oxide. Polarized light is required as in all Pockels-effect light valves.

The voltage values and polarities present during writing (W), reading (R), and erasure (E) are indicated on Fig. 29. The operation of the device is best described with reference to Fig. 30 which shows the voltages across the various layers of the device at various portions of its cycle. A voltage $V_0 \simeq 1500$ V ($0.25\ V_{1/2}$, where $V_{1/2}$ is the crystal's half-wave voltage) is initially applied between electrodes (left or right side of the device positive) and capacitively divides between the layers (curve a). The right (or negative) side of the device is then flooded with erase light (uv or blue λ_W light) and mobile carriers are generated within the $Bi_{12}SiO_{20}$. The electrons produced drift through the crystal to the left parylene/crystal interface until the net field in the crystal is zero. (Holes are essentially immobile in this crystal and photoconductivity is due only to electrons.) This constitutes erasure (curve b). About 750 V are now across each 6-μm parylene layer.

The erase light is now turned off and the applied voltage V_0 reversed (left side negative now). After capacitive division of the voltage, about $0.5\ V_0$ is across each parylene layer and about $2V_0 = 0.5V_{1/2}$ across the $Bi_{12}SiO_2$ (curve *c*). The crystal is now exposed to spatially modulated λ_W light incident from the left on the negative side of the crystal. As before, photocarriers are generated in the $Bi_{12}SiO_{20}$. If blue write light is used, it will be strongly absorbed in a thin layer and electrons will be generated close to the dichroic mirror. These electrons now drift through the crystal to the right or positive side, decreasing the voltage across the crystal in those bright, illuminated areas (curve d).

---

[59] P. Nisenson and R. A. Sprague, *Appl. Opt.* **14**, 2602 (1975).

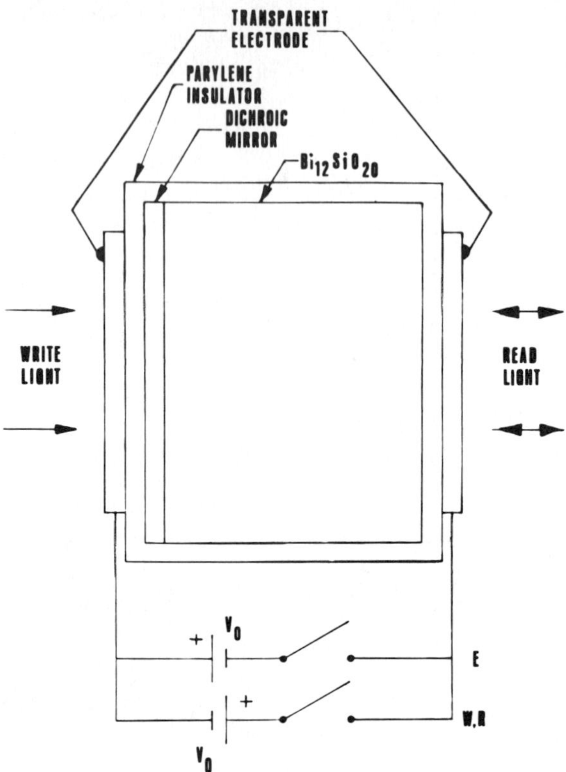

FIG. 29. Schematic of the PROM light valve.

Thus, the voltage division for dark areas is described by curve c and for bright areas by curve d in Fig. 30. For a fully exposed area there will be no voltage across the $Bi_{12}SiO_{20}$. The write light is now turned off and the crystal illuminated from the right with read light. This light passes through the crystal twice, emerges fully retarded by $\pi/2$ in those dark areas (where $2V_0$ is across the crystal), and unchanged in fully exposed areas (where there is no voltage across the crystal). With polarizers in the entering and reflected read beams, the output image will be a contrast-reversed version of the input light distribution.

If a positive voltage (left side positive) is applied during readout, a positive output image results, and thus contrast reversal of a recorded image can be achieved, as in photo-DKDP. With intermediate voltages applied during readout, baseline subtraction and other similar operations can be performed (see Section IV,B).

2. *Specifications*

The PROM has been rather well analyzed and, as for the other light valves discussed, it is difficult to list one set of device specifications without various

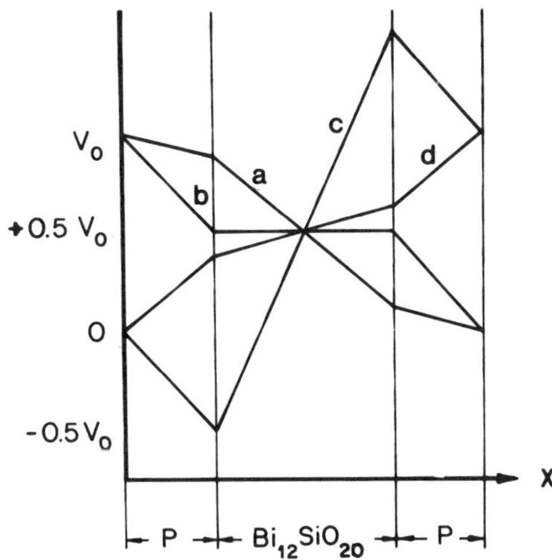

FIG. 30. Voltage distribution across PROM light valve at various stages of its operation.

qualifying remarks. The material's sensitivity is optimum near 430 nm, thus either a blue laser at 442 nm is used for writing or the image is noncoherently focused onto the device. About 10 $\mu J/cm^2$ is required to reduce the voltage across the crystal to the $1/e$ point. When readout is at 633 nm, the material is less sensitive and an exposure of 100 mJ/cm^2 can be used before the recorded image has decayed to the $1/e$ point. However, readout with a 50-mW laser (2 mm^2 beam assumed) can only last for 50 sec before the recorded data is erased. We thus list storage time (defined as during readout) as 1 min in the PROM specifications noted in Table VI.

TABLE VI

PROM LIGHT VALVE SPECIFICATIONS

(projected values in parentheses)

Target size	= 25 × 25 mm^2 (40 × 40 mm^2)
Resolution at 50% modulation	= 10 lp/mm (20 lp/mm)
	= 250 × 250 points (800 × 800 points)
Limiting resolution	= 80 lp/mm
Space–bandwidth product	= 6.2×10^4 (6.4×10^5)
Cycle time	= 100 frame/sec (or faster)
Frame-space–bandwidth product	= 6.2×10^6 (6.4×10^7)
Dynamic range	= 5000:1
Lifetime	= long
Cosmetic quality	= excellent
Optical quality	= $\lambda/4$
Storage time	= 1 min (under readout)
Sensitivity (to $1/e$)	= 10 $\mu J/cm^2$
MRTR	= 8%
Required voltage	= 1500–2000 V

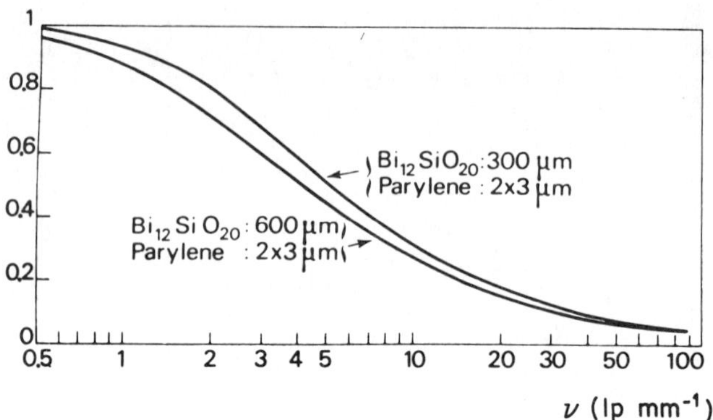

FIG. 31. MTF for two thicknesses of the PROM light valve.[40]

The theoretical MTF curves for two quite thin PROMs are shown[40] in Fig. 31. Other quoted resolutions[58] obtained with narrow cone angles by interferometric methods from measured diffracted light intensities are not typical of most applications envisioned. Thus for reasons advanced earlier, we use 10 lp/mm as the resolution at 50% modulation and 80 lp/mm as the limiting resolution. Below the 50% modulation point, lower exposures (e.g., 10 $\mu J/cm^2$) yield better resolutions, with higher exposures (6000 $\mu J/cm^2$) producing better results at higher spatial frequencies. PROM MTF curves have also been obtained using baseline subtraction, but such data not directly usable, since the MTF can only be optimized at one spatial frequency. Thus the use of such MTF improvements is questionable for imagery with a continuous range of spatial frequencies. PROM MTF curves have also been obtained using a superprime mode of operation. These data cannot always be used either since in this mode the voltage across the PROM must be reversed during exposure and this is not always possible (e.g., when the PROM is used as a SOALM).

The sensitivity of the PROM is also quite dependent on many other device parameters. Thinner PROMs generally have better sensitivity at shorter wavelengths, whereas thicker PROMs are superior at longer wavelengths due to the absorption and penetration depths of the light. Sensitivity S also varies with the thickness b of the PROM, but the dependence is far less severe than the device's wavelength dependence. At 442 nm, S increases with b; at 365 nm it decreases; for 404 nm no clear relationship exists. These PROM specifications are summarized in Table VI.

3. *Applications*

Use of the PROM in various baseline subtraction applications, such as those noted in Section IV,B,3, has been demonstrated. These included: contrast reversal of imagery (inversion from positive to negative), level slicing (generation

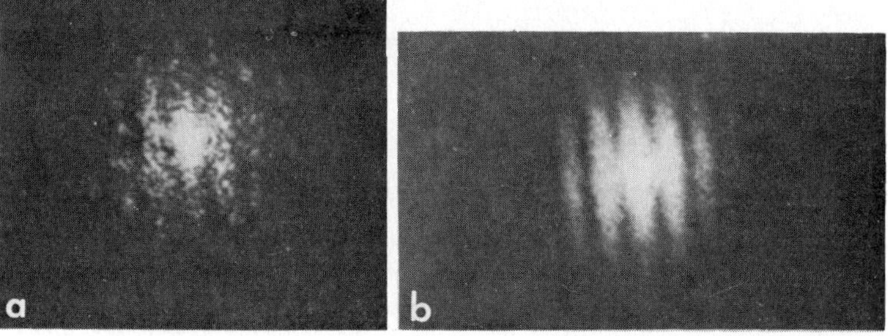

FIG. 32. Real-time speckle interferometric integration using the PROM light valve.[58] (a) Output after one frame, (b) output after 50 frames.

of constant intensity contours), subtraction or reduction of the zero-order term in the power spectrum of an image, etc.[57,58] The applications selected for inclusion here are chosen to show two additional new applications of real-time devices in optical computing.

A real-time astronomical speckle camera using the PROM is currently under development at Itek. In this application,[60] images of astronomical objects are recorded through a telescope and image intensifier with a fiber-optic input to the PROM with short enough exposure times (20 msec) to freeze atmospheric turbulence. The two-dimensional optical Fourier transform of each frame is formed on film, and transforms of successive frames integrated to allow recovery of object data out to the diffraction limit. The output reconstructions of double-star speckle input images after one and 50 frames of integrating using the PROM optical speckle interferometer, are shown in Fig. 32. Baseline subtraction of the zero-order term was used to enhance the SNR of these outputs for this application.

Acoustical holograms have also been reconstructed using the PROM[58] and the device has been used in the filter plane of the systems in Figs. 1 and 2 for pattern recognition and correlation of simple single-letter input objects.[59]

In the final example to be discussed, the PROM is used in the filter plane of a system similar to that of Fig. 1a. However, the pattern on the PROM is not formed from the input data at P_0 but rather is formed point by point by a separate krypton laser ($\lambda = 476$ nm), sequential modulator, and two-dimensional deflector.[58] A bar target input was used and the positions on the filter-plane PROM, between crossed polarizers, corresponding to various orders in the transform of the input, were addressed by the krypton laser. This allows selected orders in the transforms of the input to be passed. With the zero and +1 order spots passed, the image of Fig. 33a results; with the zero and first four orders

[60] P. Nisenson and R. Stachnik, *Tech. Dig. AAS Top. Meeting Imaging in Astron. Cambridge, Massachusetts* (1975).

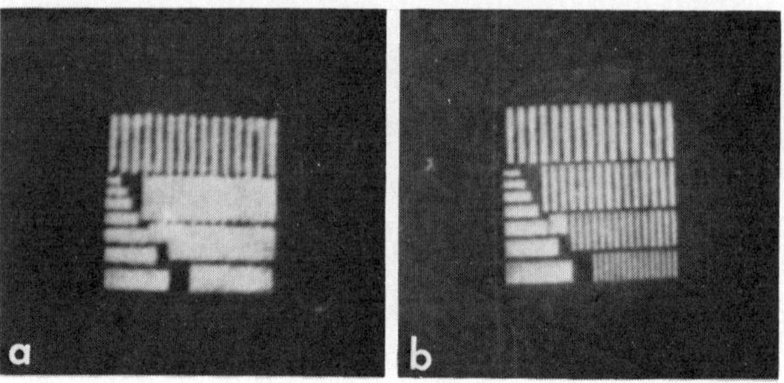

FIG. 33. Use of the PROM light valve as a computer-controlled spatial filter to pass (a) only the zero and first orders, and (b) the zero through fourth orders of the transform of the test pattern.[58]

passed, the image of Fig. 33b results. This use of a computer-controlled light valve filter represents yet another use of this and other coherent light valves.

VI. HYBRID LIQUID-CRYSTAL LIGHT VALVE

Over 4000 liquid-crystal references exist and over 15 electro-optic effects have been identified in these materials. However, as a coherent light valve, only one liquid-crystal device is considered; the hybrid field effect liquid-crystal light valve developed by Hughes Research Laboratories.[61–63] We will thus restrict our discussion to this one device and to the liquid-crystal background required to understand its operation.

The scattering mode effect (so often used) by its very nature destroys the coherence of the input light and is not appropriate for use in a coherent light valve. Similarly, dc liquid-crystal devices have limited lifetimes and thus only ac light valves should be considered. Devices using field rather than current effects promise both speed and lifetime and are thus clearly preferable. The hybrid liquid-crystal device reflects these and other considerations.

Electrodes and a photoconductor are placed on all liquid-crystal light valves and a voltage applied between the electrodes as in other optically addressed light valves. However, the transmission of the liquid crystal depends on the orientation of the liquid-crystal molecules. Liquid-crystal molecules align parallel (or perpendicular) to an applied field depending on whether the material's dielectric anisotropy is positive or negative. For a material with positive dielectric anisotropy, $\varepsilon_{\parallel} > \varepsilon_{\perp}$. The terms homogeneous (or homeotropic) alignment are also used in describing liquid-crystal devices. These terms

[61] J. Grinberg *et al.*, *Opt. Eng.* **14**, 217 (1975).

[62] P. G. Reif, A. D. Jacobson, W. P. Blehn, and J. Grinberg, *Proc. SPIE* **83**, 34 (1976).

[63] J. Grinberg and A. D. Jacobson, *J. Opt. Soc. Am.* **66**, 1003 (1976).

refer to whether the long axes of the liquid-crystal molecules are aligned parallel (or perpendicular), respectively, to the electrode surfaces. This initial molecular alignment is determined by the material's anisotropy and the preparation of the electrode surface. Ion-beam etching, chemical treatment, thin film deposition, and proper rubbing of the surface have been used to control the alignment of the liquid-crystal molecules at the electrode surfaces. In the presence of a field between the electrodes, the elastic forces tending to realign the molecules to their original orientation compete with the dielectric forces tending to orient the molecules in the electrical field.

In the hybrid field effect liquid-crystal light valve, the off-state of a resolution element is determined by the twisted nematic effect and the on-state by the optical birefringence effect. Neither electro-optic effect is by itself suitable for use in a coherent light valve; however, the hybrid combination of the two produces excellent results. Since the liquid crystal used in the hybrid field effect device has a positive dielectric anisotropy, the twisted nematic and optical birefringence effects are now described for such a case.

The liquid-crystal molecules are aligned parallel to each electrode in the twisted nematic effect in the absence of a field. However, there is a 45° difference between the angular alignment of the molecules at the top and bottom electrodes. The liquid-crystal molecules in successive layers are thus rotated slightly while remaining parallel to the electrodes with a total twist angle of 45° between the top and bottom layers. Consider the case of a dielectric reflecting layer placed on one surface of the device and linearly polarized light incident on the opposite surface with the direction of polarization parallel to the direction of alignment of the liquid-crystal molecules at the entrance electrode surface. The plane of polarization of the input light will then be rotated through the 45° twist angle of the liquid-crystal molecules as it passes through the device. Upon reflection from the mirror, the plane of polarization of the light is rotated back through the same 45° angle and thus emerges polarized in the same direction as the entering light. With crossed polarizers in the entering and exiting beams (a polarizing beam splitter is used in practice), the output light is blocked by the last polarizer. No output light results and thus the device's off-state (no voltage between the electrodes) is determined by the twisted nematic effect.

In the presence of a field (a voltage between the electrodes), the liquid-crystal molecules tilt from their homogeneous alignment (parallel to the electrodes) state toward the homeotropic alignment (perpendicular to the electrodes) state. In the full homeotropic alignment state, the molecules will not affect the polarization of the input light, and a dark on-state (no output light through the final polarizer) will also result. However, between full-off and full-on, the molecules are tilted between 0 and 90°. Their optical birefringence will now affect the polarization of the input light and the exiting light will not be linearly polarized. The amplitude of the exiting light transmitted through the final polarizer will then be proportional to the voltage applied between the electrodes and a voltage-controlled on-state with gray-scale results.

The twisted nematic effect was chosen for its fast response times. Since decay time is shorter for a homogeneous than for a homeotropic alignment, a positive dielectric anisotropy crystal with molecules aligned parallel to the electrodes (in the off-state) was selected. The response time of a liquid crystal is also proportional to the square of the thickness of the liquid crystal and thus thin (2-μm thick) crystals are used. However, for such a thin crystal, the angle by which the polarization of the input light is rotated as it passes through the crystal is no longer independent of the thickness of the liquid crystal or the wavelength of the input light, as it is for thick twisted nematic crystals. Thus, to achieve both high contrast and speed in this hybrid field effect device, the thickness of the liquid-crystal cell must be optimized for the wavelength of the read light used.[63]

1. *Structure and Operation*[61,62]

The schematic of the hybrid field effect liquid-crystal light valve is shown in Fig. 34. It is topologically similar to that of photo-DKDP with an active medium (the liquid crystal) and photoconductor separated by a dielectric mirror with electrodes on both outer edges. The write light (510–550 nm) is incident from the right on the CdS photoconductor (usually a 12-μm thick sputter-deposited layer). The 2-μm thick CdTe light blocking layer (sheet resistance of 10^{11} Ω/square and an optical absorption of over 10^5) and dielectric mirror are included to provide isolation of the CdS from the read light λ_R (usually 633 nm) incident from the left. The dielectric mirror consists of alternating $\lambda/4$ films of

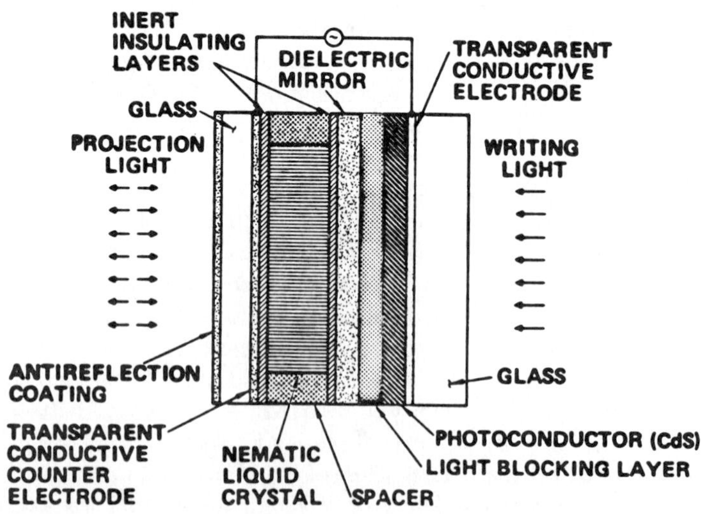

FIG. 34. Schematic of the hybrid field effect liquid-crystal light valve.[61]

TiO_2/SiO_2 of high (2.5) and low (1.45) index of refraction. The dielectric mirror can be designed for a reflectivity above 90% over the visible spectrum or tuned to a narrow spectral region, depending on the application. A 2-μm thick cyanobiphenyl nematic liquid crystal mixture (British Drug House mixture E-7) with high anisotropy ($\Delta\varepsilon = 11$) and birefringence ($\Delta n = 2.7$) is used. The transparent electrodes are thin *e*-beam films of indium–tin oxide with low 30 Ω/square sheet resistance. The entire assembly is sandwiched between two $\lambda/10$ flat 12-mm thick glass substrates.

Thin films of SiO_2 are sputter-deposited on the electrodes. Shallow angle ion-beam etching of these layers at an angle less than 20° to the surface has been found to provide highly uniform homogeneous (parallel) alignment of the liquid-crystal molecules. The outer surface of the counter electrode and the front surfaces of the glass substrate are antireflection coated to reduce the contrast of the multiple reflection interference fringes. The thickness of the counter electrode is $\lambda_R/2$ to provide an optical match between the glass and liquid crystal.

Operation of the device requires only a low 6-V rms ac signal at 10 kHz between the electrodes. As in photo-DKDP, most of this voltage initially appears across the CdS. When illuminated with a spatially modulated λ_W beam, the resistivity of the CdS is spatially reduced and a spatial voltage distribution proportional to the input light distribution appears across the liquid crystal. The amplitude of the reflected λ_R beam after passing through a crossed polarizing beam splitter is thus spatially modulated proportional to the input light distribution.

2. *Specifications*[62]

The specifications for the hybrid field effect light valve are listed in Table VII. Its MTF curve shown in Fig. 35 is flat out to 20 lp/mm with 30 lp/mm resolution at 50% modulation making it the best of all the available optically addressed light valves. Its only disadvantage is a lack of storage with erasure by decay in 15 msec (to 10%). The device is thus most useful as an OTTO or as a filter plane material at P_1 in the JTC system of Fig. 2 where storage is not essential. The ability to simultaneously write and read the device is also of considerable use in many applications. The device's rise time is 100 msec (to 90%) but the device can and has been addressed with laser pulses as short as 10 nsec with no reciprocity failure.

The device's listed sensitivity is the product of the 10 msec excitation time and its threshold (3.3 $\mu W/cm^2$) or full contrast (160 $\mu W/cm^2$) sensitivity. Its 1.6 $\mu J/cm^2$ sensitivity at full contrast is also the best of all optically addressed light valves. An optical flatness of $\lambda/4$ can apparently now be achieved since problems of bending the flats during deposition of the many thin film layers required have been overcome.

TABLE VII

HYBRID FIELD EFFECT LIQUID-CRYSTAL LIGHT VALVE SPECIFICATIONS
(projected values in parentheses)

Target size	$= 25 \times 25\ mm^2$ ($50 \times 50\ mm^2$)
Resolution at 50% modulation	= 30 lp/mm = 750 × 750 points
Limiting resolution	= 70 lp/mm
Space–bandwidth product	$= 5.6 \times 10^5$ (2.2×10^6)
Write time (to 90%)	= 10 msec (response time)
Erase time (to 10%)	= 15 msec (by decay)
Frame-space–bandwidth product	$= 2.2 \times 10^7$ (8.8×10^7)
Dynamic range	= 100:1 or larger
Lifetime	= long
Cosmetic quality	= good
Optical quality	$= \lambda/4$
Storage time (to 10%)	= 15 msec (erasure by decay)
Sensitivity (full contrast)	$= 1.6\ \mu J/cm^2$
MRTR ($100\ mW/cm^2$ max)	= 70%
Required voltage	= 6-V rms at 10 kHz

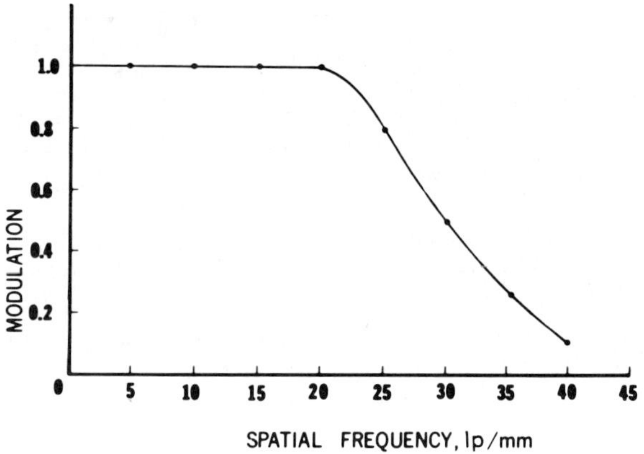

FIG. 35. MTF for the liquid-crystal light valve.[62]

This light valve is also of particular interest since the many thin film layers required can apparently now be deposited with adequate quality and quite good reproducibility.

3. *Applications*[62,64]

The major applications for this light valve have been as a noncoherent to coherent converter. It has also recently been interfaced to a channel intensifier

[64] A. D. Gara, *Appl. Opt.* **16**, 149 (1977).

with a fiber-optic output window. Resolution at 50% modulation was 12 lp/mm with a limiting resolution of 20 1p/mm for this breadboard system. This new innovation greatly extends the device's spectral respnse beyond the 510–550 nm range of the CdS. The image intensifier can convert broadband spectral light to a narrow spectral region in which the light valve is sensitive, thus enhancing both the spectral response and sensitivity of the device as well as its storage ability. A research effort is also currently in progress to interface this liquid crystal to a CCD array thus providing a compact input EALM,

The one application of this device that we discuss here is its use as the input transducer in an optical correlator of three-dimensional objects. The basic

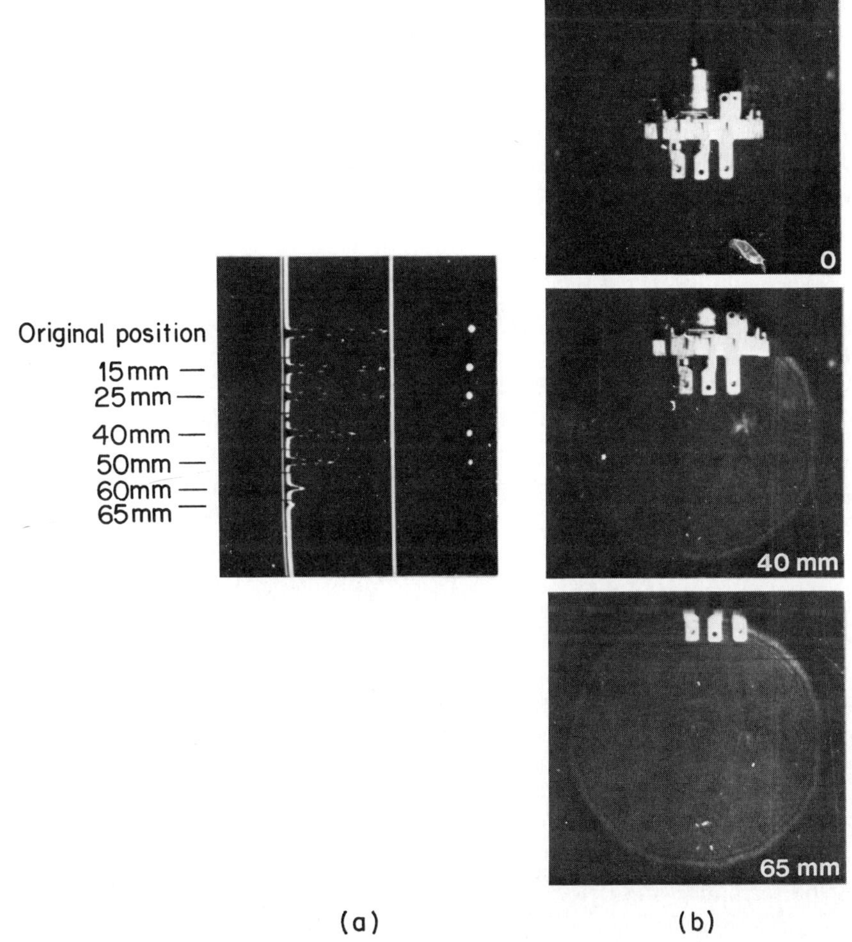

FIG. 36. Real-time correlation of three-dimensional object using the liquid-crystal light valve.[64] (a) Cross section and correlation peaks for (b) different positions of the input object.

FPC system of Fig. 1 is modified for use with a reflex mode input light valve with the read light incident through a polarizing beamsplitter. An electromechanical relay was used as the three-dimensional input object. It was imaged onto the liquid-crystal light valve with a tungsten–halogen lamp. A linearly polarized and collimated He–Ne laser beam ($\lambda_R = 633$ nm) was passed through a polarizing beamsplitter which reflected the vertical polarization component of the beam onto the liquid crystal and transmitted the horizontal component. Upon reflection from the liquid crystal, the vertical component of the input beam emerges with a horizontal component whose amplitude is proportional to the noncoherent intensity of the input image. This horizontal reflected component passes through the polarizing beamsplitter and a Fourier transform lens onto a film plate where it is interfered with the component of the input light originally transmitted by the beamsplitter. A matched spatial filter of the input object is thus formed on the film plate.

As the input object moves across the input field of view, an image of it in various positions is continuously recorded on the liquid crystal and read in reflection (the reference beam used in filter synthesis is now blocked). The Fourier transform of each new three-dimensional input object is formed at the matched spatial filter plate and the correlation of the two objects appears in the back plane of a second transform lens behind the filter plane as in the FPC of Fig. 1.

The sensitivity of the correlation to angular and translational object displacements and the sensitivity of the filter to cross correlations have been analyzed.[64] In Fig. 36, we show the resultant correlation peaks (Fig. 36a) for various displacements of the object (Fig. 36b). The cross-sectional amplitude of the correlations is seen to decrease as the object is translated across the input field of view. From these and similar data the position and angular orientation of a complex three-dimensional object can be determined in real time.

VII. SUMMARY

The six leading coherent light valves have been discussed. The structure and operation of each, together with tabulated device specifications and representative application examples of each have been presented. A composite summary table of all devices and their specifications is provided in Table VIII.

The fluid-film light valve is the most developed but has poor optical quality, very limited storage, and no erase mechanism other than decay. The TOPR has excellent storage and optical quality, but a maximum cycle time of 1 sec and questionable lifetime. The *e*-beam DKDP light valve offers the potential of the best performance of all electronically addressed light valves with good optical quality, lifetime, storage, and an erase mechanism.

The liquid crystal actually has the best resolution of all optically addressed light valves, and if its optical quality and lifetime prove adequate, should be the

TABLE VIII Composite Specifications of all Major Coherent Spatial Light Modulators[a]

Parameter[a]	Device					
	G.E. light valve	TOPR thermoplastic	Photo-DKDP light valve	PROM	*e*-Beam DKDP	Liquid crystal
3 dB bandwidth	10 MHz (20 MHz)	40 MHz (200 MHz)	—	—	10 MHz	—
Target size	20×20 mm^2 (40×40 mm^2)	50×50 mm^2 (100×100 mm^2)	30×40 mm^2	25×25 mm^2 (40×40 mm^2)	50×50 mm^2	25×25 mm^2 (50×50 mm^2)
Resolution at 50% modulation	25 lp/mm at 500×500 pts (50 lp/mm at 1000×1000 pts)	32 lp/mm at 1600×1600 pts (100 lp/mm at $10{,}000 \times 10{,}000$ pts)	10 lp/mm at 300×400 pts	10 lp/mm at 250×250 pts (20 lp/mm at 800×800 pts)	15 lp/mm at 750×750 pts	30 lp/mm at 750×750 pts
Limiting resolution	50 lp/mm	50 lp/mm (150 lp/mm)	50 lp/mm	80 lp/mm	30 lp/mm	70 lp/mm
Space–bandwidth product	2.5×10^5 (10^6)	2.5×10^6 (10^8)	1.2×10^5	6.2×10^4 (6.4×10^5)	5×10^5	5.6×10^5 (2.2×10^6)
Cycle time	30 frames/sec	0.5–1 sec (10 Hz) (temperature controlled)	100 frames/sec	100 frames/sec	30 frames/sec	30 frames/sec
Frame-space–bandwidth	7.8×10^6 (3×10^7)	2.5×10^6 (10^8)	1.2×10^7	6.2×10^6 (6.4×10^7)	1.5×10^7	2.2×10^7 (8.8×10^7)
Dynamic range	100–1000:1	400:1	100–1000:1	500:1	100–1000:1	100:1
Lifetime	~3000 hr = 3×10^8 cycles	10^5 cycles (limited, maybe larger)	3 yr	Long	3 yr	Long
Cosmetic quality	Good	Excellent	Excellent	Excellent	Excellent	Good
Optical quality	Fair, 2 λ/cm	$\lambda/4$–$\lambda/8$	$\lambda/2$	$\lambda/4$	$\lambda/4$	$\lambda/2$
Storage time	20–300 msec	Months	1 hr	1 min	1 hr	15 msec
Erasure	20–300 msec by decay only	1 sec by temperature control	0.5 msec by light flash	1 msec by light flash	0.5 msec or less by secondary emission	15 msec by decay only
Write, read time	33 msec/frame	33 msec/frame	50 nsec	10 nsec	33 msec/frame	10 nsec
Sensitivity	5 μA/spot	1 μA/spot	1 μJ/cm^2	1000 μJ/cm^2	50 μA/spot	1.6 μJ/cm^2
MRTR	—	—	50%	8%	70%	70%
Required voltage	5000 V	7000 V	100 V	1500–2000 V	2500 V	6 V at 10 KHz

[a] Projected values in parentheses.

best device in applications where storage is not required. When storage is needed, photo-DKDP or the PROM must be used. The PROM appears to offer the best eventual performance; however, research still remains to be done on all light valves to completely answer these questions.

From these pages, one can see that major advances have recently been made in the development of real-time and reusable spatial light modulators. These devices are rapidly becoming more and more available, their specifications continue to improve and with this new avenues and applications for real-time coherent optical processing are emerging.

ACKNOWLEDGMENT

The financial support of the Office of Naval Research on Contracts NR 048-600 and NR 350-011, and the Air Force Office of Scientific Research on Grant AFOSR-75-2851 for the author's work included in this review is gratefully acknowledged. The cooperation of many companies and individuals in providing information and the imagery presented is also gratefully acknowledged: Bob Markevitch, Ampex Corp.; Dan Currie, ERIM; G. Marie, LEP; Bob Sprague, Itek Corp; and Alex Jacobson, Hughes Research Laboratories.

CHAPTER 6

Scanning Devices and Systems

GERALD F. MARSHALL

Department of Optics, Energy Conversion Devices, Incorporated, Troy, Michigan

I. INTRODUCTION

A flashlight or a searchlight, of which we are all aware, is the most elementary of scanning devices. When these devices are handled to seek out and identify an object, we call the combination of the device and the control mechanism a scanning system.

Today, scanning devices and scanning systems are far more complex, and involve precision mechanical and optical engineering with intricate electronics.

The purpose of this chapter is to look into the fundamental considerations of the principles of scanning devices and systems; this includes the light beam,

ISBN 0-12-408606-3

the scanning motions and scanning patterns, and optical, mechanical, and electronic engineering considerations.

The theme that has been adopted is one in which the subject is presented in a practical engineering manner, devoid of any mathematical development but expressing formulas where necessary. Although topics covering the state-of-the-art techniques and beyond are of important interest to the specialist, they will only be mentioned in passing and not discussed. It is my intention to provide engineers, who have a knowledge of science and engineering in other disciplines, with a stepping stone by which they may more readily cross over into the field of scanning devices and systems with a basic understanding of them for consideration as an application to solving problems.

Because this chapter relates to applied optics and optical engineering, care has been taken to avoid being overly pedantic in order to communicate more freely, as is experienced in engineering practice. By way of example: light beams are depicted by single lines in diagrams; in the text the term light beam is used frequently, where correctly the term light ray, which in reality does not exist other than as a geometrical concept, may be more precise. In practice, however, we deal and talk in terms of rays and beams of light interchangeably.

A. Classification of Systems and Devices

1. *Systems*

The purpose of almost all scanning systems is either to interrogate an object scene and collect data, or reproduce information from processed data. Any one scanning system may be designed to perform either function separately or in combination. Hence, scanning systems fall into two corresponding basic groups; namely,

(a) reading systems and
(b) reproducing systems.

Essentially, reading systems receive data as a modulated light signal, while reproducing systems transmit data as a modulated light signal (Fig. 1).

a. Reading Systems. Reading Systems fall into two subgroups:

(c) passive and
(d) active.

The terms passive and active relate to the techniques of gathering data. The *passive* reading system is fundamentally an optical directional receiver that collects only an orderly series of light beams from a target scene; and processes the signal data and stores it, records, and/or displays the information. The *active* reading system is an optical directional transmitter-cum-receiver that

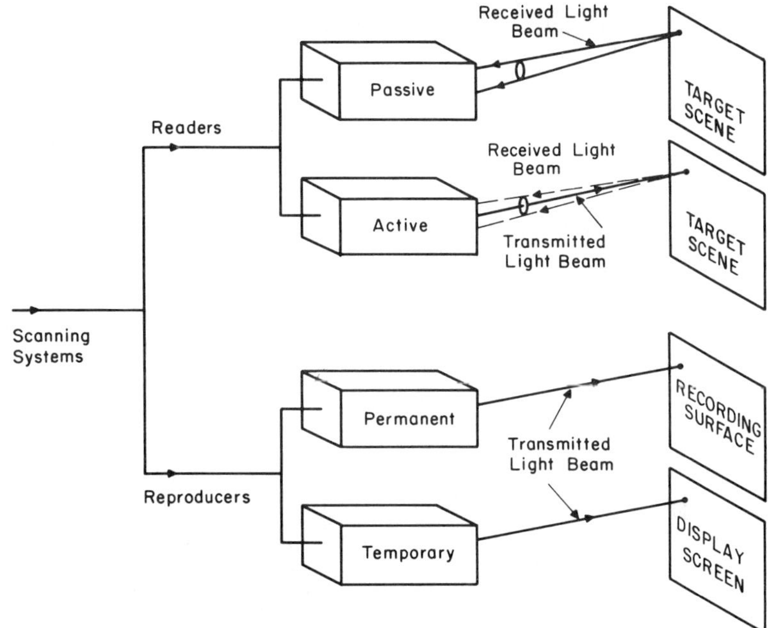

FIG. 1. A classification of scanning systems.

radiates a scanning light beam; collects the modulated reflected light from a target scene; and processes the signal data, stores, records, and/or displays the information. The key difference between a passive and active system is that the passive system relies on external light emitted from or reflected by the object target, while an active system incorporates its own light source, which illuminates the target as it reads.

b. Reproducing Systems. Reproducing systems also fall into two subgroups:

(e) permanent and
(f) temporary.

The terms permanent and temporary express the end product form of the information reproduction.

The reproducing system that provides a permanent copy output is an optical directional projector that records information on a photosensitive material. These systems are more commonly referred to as recorders. They include contact printing, photo typesetting, and facsimile reproduction. The reproduction system that provides a temporary output is an optical directional projector that displays the information on a nonphotosensitive screen for real-time visual observation.

Common to both reading and reproducing systems is a scanning mechanism which is responsible for the direction of the outgoing light beam or the directional location of the incoming light beam. Such scanning mechanisms, which are referred to as scanning devices, are crucial to any scanning system and a prime topic of this chapter (Sections III, IV, and V).

2. *Devices*

Fundamentally, scanning devices are light beam deflectors. Deflectors can be categorized into types based upon the principal optical property used to produce the light beam deflection; these are:

(a) reflective,
(b) refractive, and
(c) diffractive.

The inclusion of a device such as a prism, which is seemingly a refracting component, into the reflecting group may, at first, appear to be paradoxical. However, its inclusion takes meaning when it is understood that it is the reflecting properties of the facets within the prism that are being used to effect the light beam scanning sweep.

The term scanning has become synonymous with light beam deflecting. The path traced out by a scanning light beam is called a scan trace or scan pattern, while scanning devices are simply referred to as scanners.

There are other scanning devices, such as holographic scanners[1,2] which, because of their being a state-of-the-art development, are only referenced.

II. FUNDAMENTAL CONSIDERATIONS

Scanning systems and devices inherently involve optical components and almost always involve mechanical structures and electrical/electronic circuitry. It is appropriate to comment briefly on some of the more important aspects associated with scanning, light radiation, the mechanical properties of materials, and the electrics that drive and control the devices.

A. Scanning

Scan patterns and scanning motions are inextricably interrelated involving both beam displacement and beam-displacement rate (velocity), and beam deflection and beam-deflection rate.

---

[1] L. Beiser, Holofacet laser scanning, *Proc. EOSD Conf.* pp. 75–81 (1973).

[2] L. Beiser, Advances in holofacet laser scanning, *Proc. EOSD Conf.* pp. 333–340 (1975).

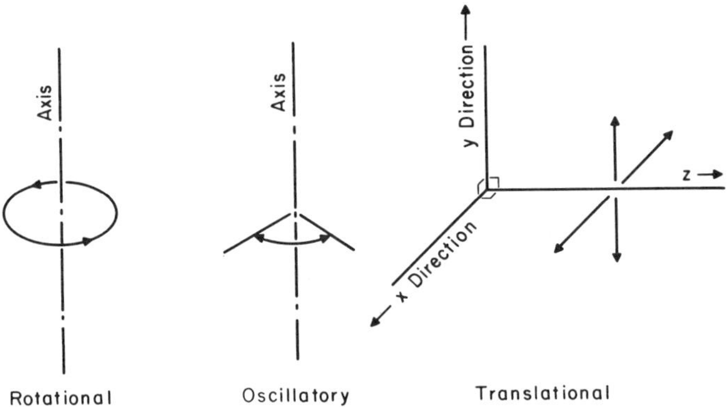

FIG. 2. Three primary scanning motions.

1. *Scanning Motions*

Scanning motions essentially fall into three basic movements (Fig. 2):

(a) rotational,
(b) oscillatory, and
(c) translational.

a. Rotational Motion. A rotational motion is a natural uncomplicated movement associated with mechanical scanners. Although the angular speed for any one scanning device is usually at a set rate, the range of speeds of rotational devices is wide. The limits are determined by the bearing design and performance, and by the dynamic internal stresses of the moving parts.

b. Oscillatory Motion. An oscillatory motion may be one of the most natural physical movements but is probably the most complicated of all motions when associated with mechanical scanners. The complexity can, in part, be attributed to the continually changing direction, speed, and acceleration, but it is compounded when the natural dynamic characteristics of the moving parts are constrained to conform to application requirements. For any given fundamental periodicity, the dynamic requirements can range from a simple harmonic motion for which the displacement varies sinusoidally with time, to a reciprocating motion where the displacement is proportional to time and has a triangular or sawtooth (ramp) displacement against time (Fig. 3).

Knowledge of Fourier analysis informs us that the triangular and sawtooth waveform, or any periodic waveform, are analytically made up of a series of cosine and sine waveforms as expressed mathematically by

$$x = \sum_{1}^{\infty} a_k \cos(kt) + \sum_{0}^{\infty} b_k \sin(kt) \tag{2.1}$$

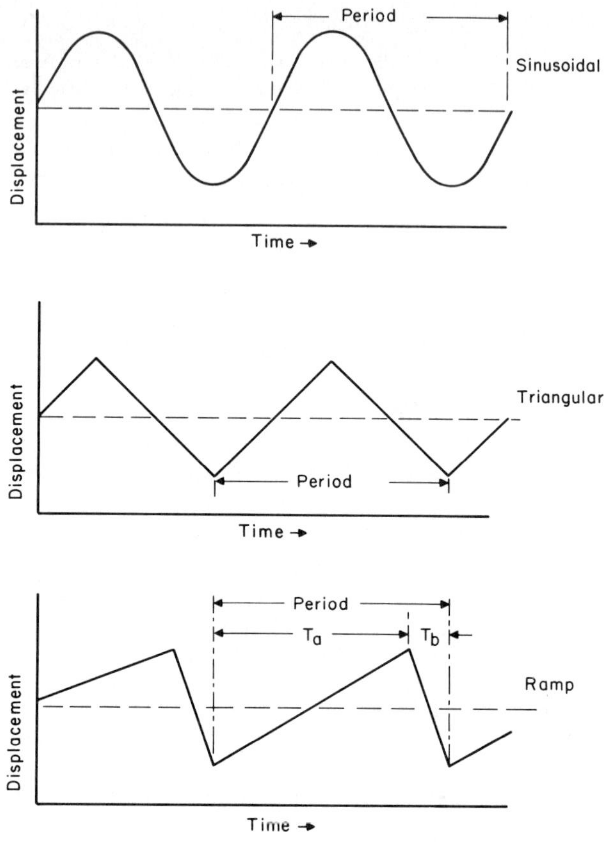

FIG. 3. Ideal oscillatory motions.

For the triangular and sawtooth oscillatory motions it is the instant in time where the discontinuity takes place to which attention must be given. Ideally, one might desire instantaneous change of direction, but this infers an infinitely high retardation and acceleration with accompanying infinitely high g stresses. Long before these stresses could be achieved such stresses would tear physical materials apart. In order that discontinuities lead to finite g stresses within the physical stress limitations of materials, all discontinuities must take place over a short but finite time interval, such that the apparent sharp waveform corners are, in fact, rounded.

By nature, physical components all have an inherent frequency response characteristic with at least one "natural" resonant frequency. The response characteristic is a function of the material elastic parameters and component geometry, together with the number and the location of the supports and bearings. Since a triangular or sawtooth waveform is, by Fourier analysis, effectively composed of a special series of simple harmonic motions, each having a frequency that is a multiple of the fundamental, there is an interaction between the input oscillatory power and the natural frequency response characteristic of the oscillatory elements. Consideration of this interaction is of great importance in controlling unwanted scan perturbations and in taking advantage of the natural response characteristics. In some designs one may need to stay away from the resonant frequency, while in another design one may wish to use the resonant frequency at which the driving power input for sustained forced oscillation is a minimum.

c. Translational Motion. A translational motion or transverse motion is a sideways movement at right angles without change of direction (Fig. 2). Conceptually, a transverse motion may be considered as an arcuate motion with an axis or center of rotation at an infinite distance away. With the collimating and focusing properties of lenses, an angularly scanning beam from a rotational or oscillatory device can be transformed into a transversing scanning beam, and vice versa (Fig. 4).

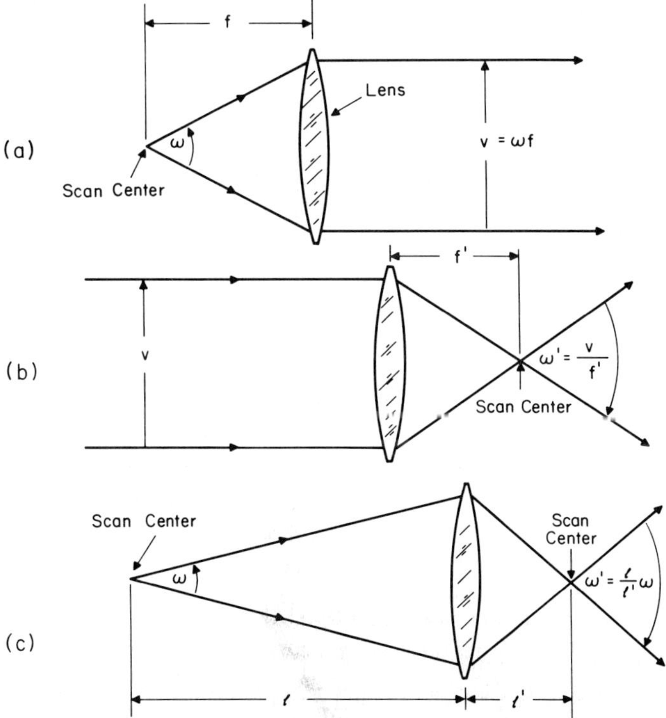

FIG. 4. Transformation of (a) an angular beam scan into a traversing beam scan, (b) traversing scan into an angular beam scan, (c) magnification of deflection angle and scan angular velocity. ω is the angular velocity; v is the linear velocity.

2. *Scan Patterns*

Scan patterns, the patterns traced out by the light beam of a reproduction system or the search path followed by a reading system, are numerous. A few of the basic patterns will be covered. The reading and writing system scanning patterns are determined initially by the desired reading and writing design requirements, the performance and limitations of scanning devices, and finally by the production cost effectiveness.

a. Straight Line Pattern. The simplest of all patterns is the straight line scan. Conceptually, the most desirable straight line scan is one in which the beam sweeps at a constant speed and the trace is repetitive (Fig. 5a). Close approximations to such scan traces are achievable with continuous and oscillatory mechanical devices as well as solid state acousto-optical deflectors. The most fundamental of all straight line scans is one in which the beam scans in simple harmonic motion for which the rate of scan varies sinusoidally such that the trace is reciprocating (Fig. 5b). Simple harmonic motion scan traces are inherently the simplest to achieve with oscillatory mechanical devices and electronically controlled solid state acousto-optical deflectors. Such a trace is also possible with devices employing continuous motion techniques (Section III, C and Section IV, A).

b. Raster Pattern. The combination of two constant-speed scan motions in mutually perpendicular directions x and y, produces a TV-type raster pattern (Fig. 5c). The lines per frame correspond directly to the ratio of the periodicities (frequencies) of the x and y scan motions. The lines will appear essentially parallel to the x axis when the x scan frequency is many times that of the y scan and vice versa. As in a television picture, the pattern will appear to roll if the two scan frequencies are not synchronized.

An actual TV raster is slightly more complicated. In order to double the number of raster lines and provide a finer picture structure without either doubling the x scan frequency, which would be more demanding electronically, or halving the y scan frequency, which would result in a visual flicker, successive frames are interlaced. Interlacing is simply the staggering of alternate frames such that the lines of one frame raster lie between the lines of the next frame raster. Thus, for an interlaced TV-type picture raster, the x and the y scan frequencies will be related by

$$2f_x/f_y = \text{odd integer} \tag{2.2}$$

The odd integer represents the number of lines per picture.

A raster pattern may or may not have a flyback scan. A flyback scan exists when the scan is generated by an oscillatory scanning device but will be absent when the trace is generated by a continuous motion scanning device such as rotating polygons (Sections III,A, III,B, and IV,C). However, the absence of a flyback trace does not preclude the presence of dead time (Section III,A,5,*b*).

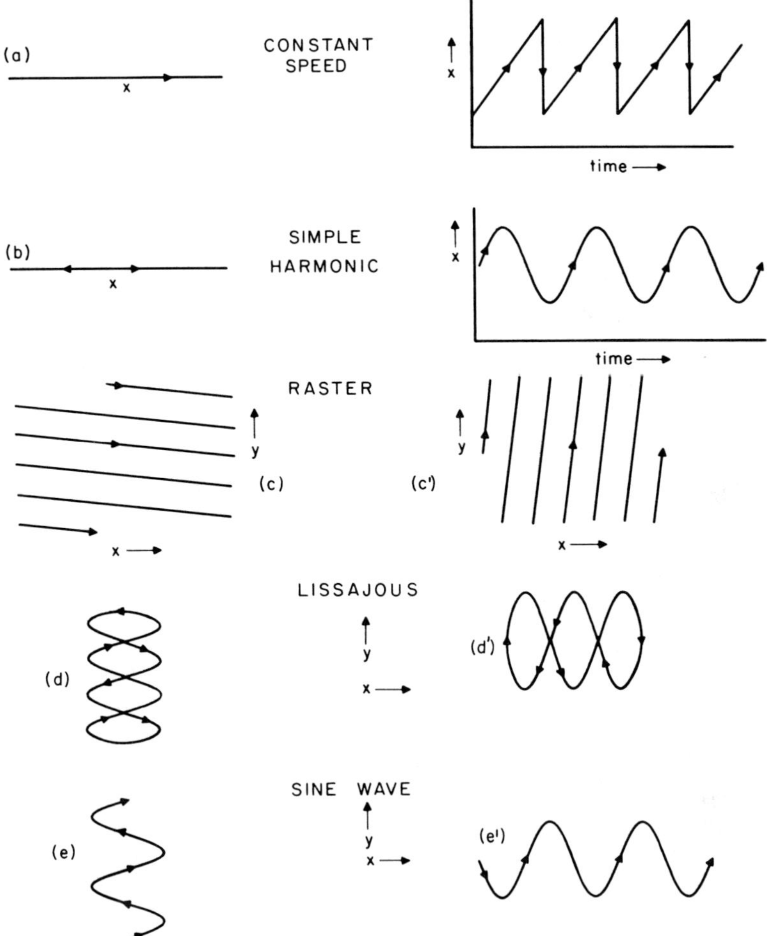

FIG. 5. Basic scanning motions and patterns. (a) And (b): two fundamental linear scan motions; (c), (d), and (e): combination of two orthogonal linear scan motions. (c) Both x and y scan motions constant speed $f_x \gg f_y$. (c′) $f_x \ll f_y$. (d) Both x and y scan motions sinusoidal $f_x > f_y$. (d′) $f_x < f_y$. (e) y Scan motion constant speed, x scan motion sinusoidal $f_x > f_y$. (e′) x Scan motion constant speed, y scan motion sinusoidal $f_x < f_y$.

c. Lissajous Pattern. The combination of two simple harmonic scan motions in mutually perpendicular directions x and y, will produce the classical patterns credited to Lissajous and known as Lissajous figures. The pattern is characterized by its continuity of trace (Fig. 5d). The pattern configuration depends on three factors associated with the two harmonic motions:

(a) frequency ratio,
(b) phase differences, and
(c) amplitudes.

The ratios of frequencies are usually simple multiples or fractions, such as $\frac{1}{1}, \frac{1}{2}, \frac{1}{3}, \frac{2}{3}$, or $\frac{3}{4}$. Typical Lissajous patterns are illustrated in textbook publications.[3] Sometimes we require a stationary pattern, in which case the frequencies need to be synchronized, and their ratio is fixed. Other times we require a visibly moving pattern, one where the whole pattern appears as if it is rolling about the x direction or turning around the y direction. The pattern seems to move in the x direction when the y scan frequency f_y is greater than the x scan frequency f_x, and the scan ratio (f_y/f_x) does not reduce to a whole number. For example, if $f_y/f_x = 3.001$, the integral part 3 determines the general pattern configuration, while the decimal portion 0.001 is a measure of the migration rate. When migrating, the movement gives the impression that one is watching a three-dimensional display of a revolving carousel turning about an axis parallel to the y direction (Fig. 5d′).

The pattern exhibits a slow movement in the y direction when the x scan frequency is greater than the y scan frequency and the ratio f_x/f_y does not reduce to an integer. In the same manner as illustrated above, the decimal portion of the ratio is a measure of the migration rate but the movement gives the impression that one is watching a three-dimensional display of a carousel on its side rolling about an axis parallel to the x direction (Fig. 5d).

For a pattern to migrate at a uniform rate, it is important that the frequency ratio be synchronized, otherwise the pattern will drift, reverse, and speed up at random.

Lissajous scan patterns and other scan patterns which are being considered for an application can frequently be generated on an oscilloscope display. In this way and for many applications, a feasibility study for the scan pattern effectiveness can be carried out before developing and investing in optical and electro-optical and electromechanical hardware.

d. Sinusoidal Pattern. The combination of one constant-speed scanning motion and a simple harmonic scanning motion in mutually perpendicular directions x and y, produces a sinusoidal trace (Fig. 5e). The number of sinusoidal cycles (spatial wavelengths) is directly related to the ratio of the periodicities of the two motions. If the ratio is an odd multiple it will interlace and have the appearance of a Lissajous pattern described above (Section II,A,2,*c*). The same general comments concerning migration of the pattern given above apply, assuming that the harmonic scan frequency is always greater than the constant-speed scan frequency.

B. Optical Aspects

Scanning devices are far from being restricted to using radiation in the visible region. They are used for radiation in the invisible regions of the spectrum

[3] B. L. Worsnop and H. T. Flint, "Advanced Practical Physics," p. 44. Methuen, London, 1959.

just before and just beyond the visible region where the radiation wavelengths are less than 360 nm and greater than 720 nm. When using the general term "light" throughout the text it will be assumed that we are referring to electromagnetic radiation that embraces all three spectral regions, ultraviolet, visible, and infrared.

Scanning devices and systems are used for monochromatic light; a single wavelength or a series of single wavelengths simultaneously. They are also used for a band or a series of bands of wavelengths. Because many scanning systems use lasers as monochromatic light sources, it is often mistakenly thought that scanners are used exclusively with lasers. Where and when appropriate, scanning devices use white light which contains a broad band of wavelengths that may extend through the visible and beyond into both the uv and ir regions.

1. *Focused Beam Spot Diameter*

By geometrical optics rules, a perfectly designed focusing lens focuses a set of light rays from a point to a point. In practice, lenses always focus light to a spot rather than to a point. The spot size may partly be attributed to the lack of perfection in the manufacture of a lens, but more importantly, even an ideal lens design manufactured to perfection produces a spot of a finite size that is attributed to the wave nature of light itself.

By physical optics rules, which take into account the wave nature of light and its interaction with the finite dimensions of optical components, a perfectly designed and manufactured focusing lens focuses light to a spot surrounded by faint rings, see Fig. 6.

The predicted spot diameter D is given by

$$D_a = 2.44\lambda(l/A) \tag{2.3}$$

where A represents the beam diameter or the lens aperture, whichever is the smaller, and where l represents the spot distance from the lens which, for a collimated incident beam, is equal to f, the focal length of the lens.

The intensity distribution of the spot is depicted in Fig. 7 and is accompanied by a series of light and dark rings. Such an image and distribution is called an Airy pattern, and such a precisely made lens that can produce this pattern is said to be diffraction limited. The central spot is called the Airy disk.

The intensity distribution of the Airy pattern assumes that the incident light is monochromatic and has a uniform intensity distribution across the lens aperture. It is important to consider the spot diameters and their intensity distributions when the incident monochromatic beam has a Gaussian intensity distribution across the lens aperture and when the aperture truncates the Gaussian distribution.

2. *Focused Gaussian Beam Spot Diameter*

The effective diameter D of a beam or spot that has a Gaussian intensity distribution is considered to be that diameter corresponding to 4σ, where σ

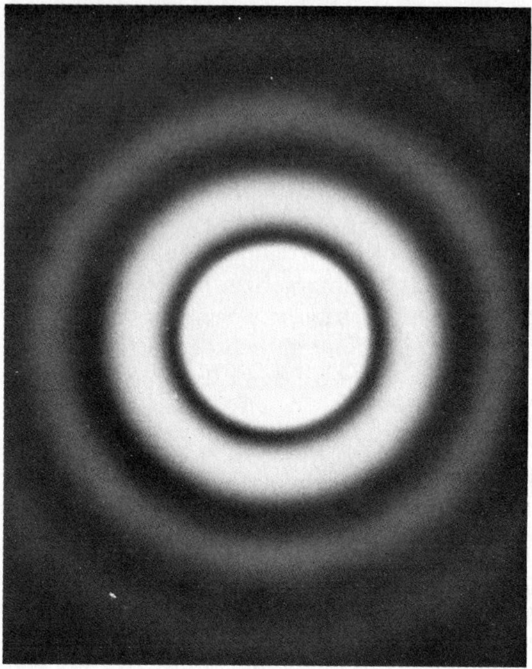

FIG. 6. Airy pattern. (Photograph courtesy University of Rochester, N.Y.)

is the standard deviation of a Gaussian beam intensity distribution I represented by

$$I = I_0 e^{-r^2/2\sigma^2} \tag{2.4}$$

where I is the intensity at a radius r from the axis of the beam at which the peak intensity is I_0 (Fig. 8).

The effective spot diameter D_s is related to the effective beam diameter D_b as follows[4]

$$D_s = \frac{4}{\pi} \lambda \frac{l}{D_b} \tag{2.5}$$

where

$$D_s = 4\sigma_s \tag{2.6}$$

and

$$D_b = 4\sigma_b \tag{2.7}$$

(l/D_b) represents the effective relative aperture of the focused beam.

---

[4] H. M. Haskel and A. N. Rosen, Power and focusing considerations for recording with a laser beam in the TEM mode, *Appl. Opt.* **10**, 1354 (1971).

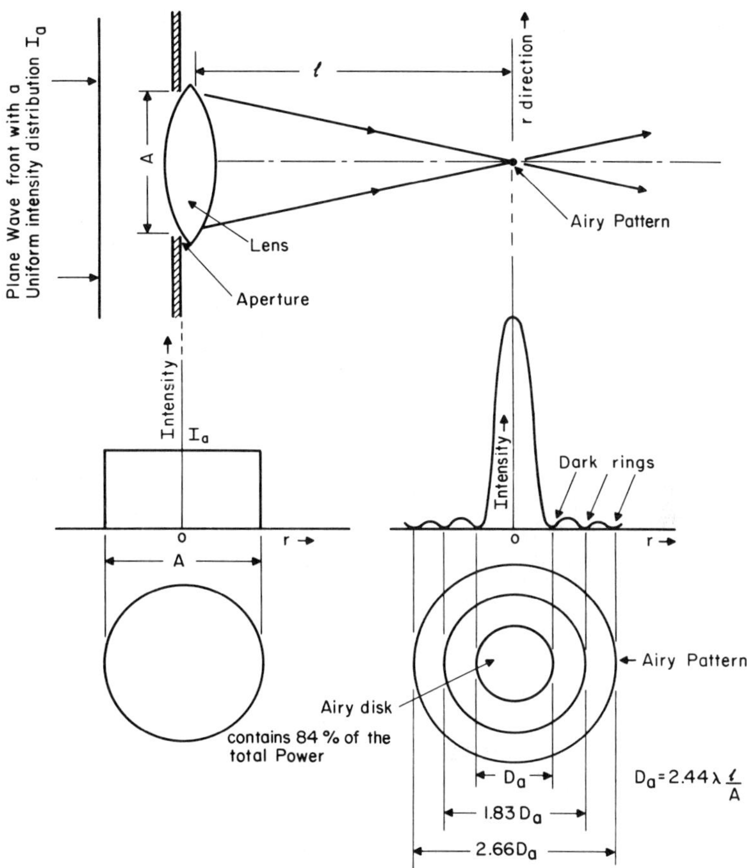

FIG. 7. Aperture and image plane intensity distribution. Scale factors of axes are arbitrary for clarity.

3. *Beam/Spot Power*

If the total power, that is, the integrated intensity, within the entire beam or spot is represented by W, then the fraction of power p enclosed in concentric circles of diameter d within the beam or spot is given by

$$p = 1 - e^{-(d^2/8\sigma^2)} \tag{2.8}$$

Equation (2.8) is expressed graphically in Fig. 9 and both Eqs. (2.4) and (2.8) are tabulated in Table I.

It should be mentioned that although the two columns 3 and 4 coincidently and conveniently sum to a constant, column 3 relates to an equation with two variables $I(r)$, and column 4 relates to an integral with three variables $\iint I(r, \theta)\, dr\, d\theta$ which corresponds to the "volume" under a three-dimensional

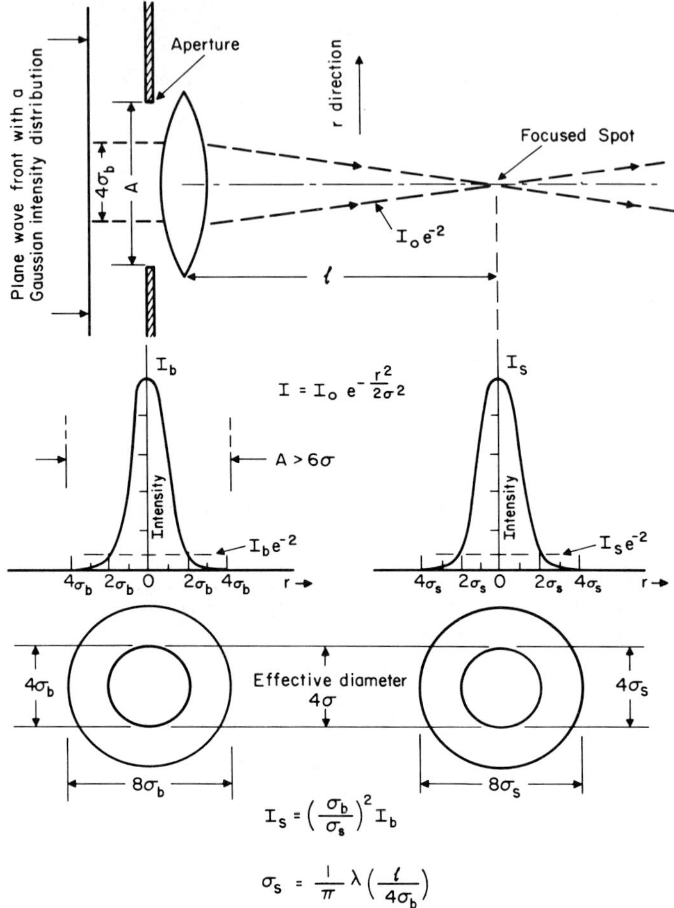

FIG. 8. Wave front with a full Gaussian intensity distribution focuses to a spot which also has a Gaussian intensity distribution.

bell shape of a Gaussian curve. This constant-sum feature can readily be attributed to the special mathematical properties of a Gaussian distribution itself. Other distributions do not necessarily exhibit this simple relationship, as one can ascertain by examining the case of a uniform intensity distribution.

From Table I we see that at the effective beam/spot diameter of 4σ, the intensity level drops to 14% of its peak, but contained within the central circle of that diameter is 86% of the total beam/spot power. It follows that 14% of the total beam/spot power also lies outside the effective diameter.

4. *Truncated Beams*

Truncation of a beam diameter, sometimes called aperturing, is important because the mirror facets of scanning devices are themselves apertures which,

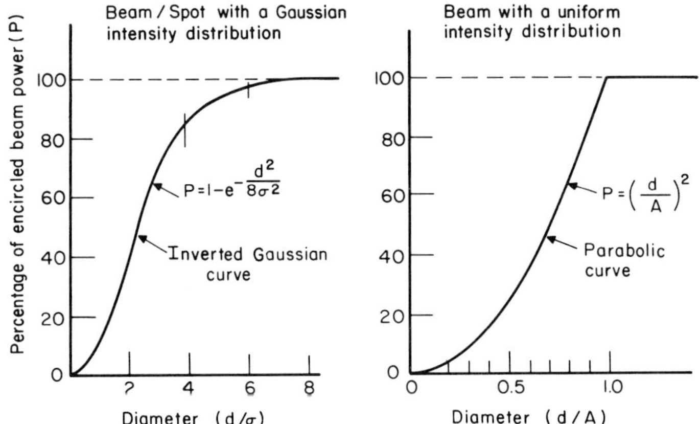

FIG. 9. Relative beam/spot power encircled versus diameter.

when viewed obliquely, project various shapes; elliptical, rectangular, trapezoidal, sectors, etc. These mirror apertures frequently move across light beams and truncate the intensity distributions asymmetrically with a consequent variation in the effective spot diameters, shapes, and intensity distributions.

When a beam with a Gaussian distribution is passed through an aperture it will be truncated more sharply as the effective incident beam diameter $4\sigma_b$, which we shall assume is variable, becomes equal to and greater than the aperture diameter A (Fig. 10).

The emerging beam intensity distribution from the aperture is neither full Gaussian nor uniform, so that when focused to a spot the intensity distribution across the spot is no longer Gaussian. For a given aperture, diameter A, as the beam diameter is expanded the truncation of the distribution is more

TABLE I

RELATIVE INTENSITY AND POWER VERSUS RADIUS AND DIAMETER IN A GAUSSIAN DISTRIBUTION

Beam or spot: Radius r/σ	Beam or spot: Diameter d/σ	Intensity level (%) I/I_0	Fraction of power within beam spot diameter (%) $p = 1 - e^{-(d^2/8\sigma^2)}$
0.50	1.00	94	6
0.76	1.52	75	25
1.00	2.00	61	39
1.175	2.35	50	50
1.665	3.33	25	75
2.00	4.00	14	86
3.00	6.00	1.1	98.9
4.00	8.00	0.034	99.966

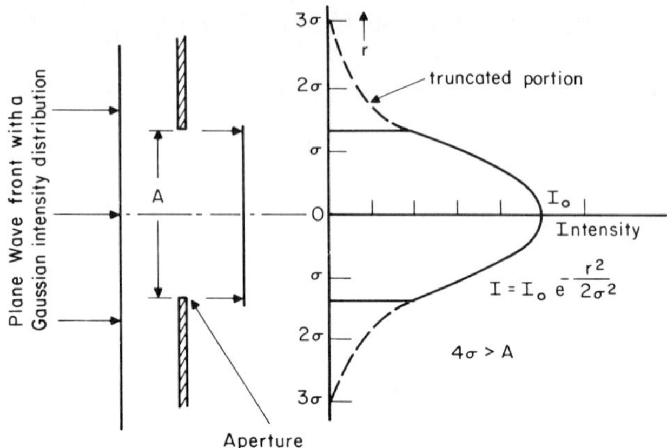

FIG. 10. Aperture truncates a beam with a Gaussian intensity distribution.

severe but the intensity variation across the aperture becomes more uniform. When $\sigma = A$ (see Table I) $I/I_0 = 94\%$. This represents a variation of 6% from a uniform intensity distribution. Thus, as a Gaussian beam expands and is more and more truncated, the focused spot evolves into an Airy pattern, and the central spot becomes the Airy disk.

In general,[5] for beams truncated by apertures of different shapes, circular, square, triangular, etc., the effective spot diameter is given by

$$D_s = k\lambda(l/A) \qquad 1 < k < 3 \tag{2.9}$$

where k is a constant related to the aperture shape and A is the effective aperture dimension.

C. Mechanical Aspects

All scanning devices are made in one form or another from physical materials that have elastic properties which are of fundamental importance to the design and performance of a scanner. These properties in particular include stability, stiffness, and stress limitations. The dimensional stability of the component materials with time and temperature changes has to be considered in relationship to design and manufacturing tolerances, while these tolerances in turn have to be directly related to the wavelength of light employed and design requirements.

By way of example, a component operating in the infrared at a wavelength of, say, 10 μm (400 μ in.) need be 10 times less precise than one at 1 μm (40

[5] L. Beiser, Laser scanning systems, *in* "Laser Applications" (M. Ross, ed.), Vol. 2, p. 57. Academic Press, New York, 1974.

μ in.) and 20 times less precise than one at 500 nm (20 μ in.) which corresponds to the green region of the visible spectrum.

Stiffness properties of materials determine the degree of dimensional changes and bending when subject to the gravitational force of its own weight and the inertial forces of acceleration in motion. Plane mirrors flex especially under an oscillatory motion, a motion in which the acceleration is constantly varying.

1. *Mirror Flexing*

The amount a vibrating mirror flexes depends on the stiffness of the substrate material and the geometrical form, the inertial stresses, and the driving oscillatory frequency in relation to the natural frequency response characteristics of interrelated structural components.

A formula given by Beiser[6] shows that the maximum deformation of a round mirror oscillating about a diameter D is proportional to

$$D^5 \rho / h^2 E \tag{2.10}$$

where ρ represents the material density, E Young's modulus of elasticity, and h the thickness.

2. *Stress Limits*

Elastic stress limits of a material limit the safe speed or frequency to which a component can be driven. High tangential and radial stresses are produced by acceleration when a component is subjected to rotational or vibrational movements.

It is not surprising that the elastic stress limits differ from material to material, but the applied limits for a particular material also vary according to the factor of safety a designer chooses to select for stability and repeatability. There are two stress limits of special interest; namely, the precision elastic limit and the yield point.

The precision elastic limit is that stress below which there is an absence of hysteresis and a material is considered to fully recover when the stress is released. The yield point is that stress beyond which a material fails to recover and is approaching a catastrophic failure.

Consider a uniform thin rod of length L and density ρ allowed to rotate about an axis through one end and normal to its length. In the static state the maximum stress at the support end for the hanging rod expressed in engineering units will be

$$S_s = \rho L \tag{2.11}$$

[6] L. Beiser, Laser scanning systems, *in* "Laser Applications" (M. Ross, ed.), Vol. 2, p. 90. Academic Press, New York, 1974.

In the dynamic state when the rod rotates at a uniform angular speed ω, the radial stress in the rod due to motion will be a maximum at the support end and equal to

$$S_{\mathrm{d}} = \rho L[(\omega^2 L/2)/g] \tag{2.12}$$

where g is the acceleration due to the earth's gravitation. This maximum dynamic stress expressed in g's is given by dividing Eqs. (2.10) and (2.11)

$$S_{\mathrm{d}}/S_{\mathrm{s}} = (\omega^2 L/2)/g \tag{2.13}$$

Example: Let $L = 2$ in., $\omega = 2\pi f$, $f = 600$ rpm $\equiv 10$ rps, and $g = 981$ cm sec^{-2} $\equiv 386.4$ in. sec^{-2}. This leads to

$$S_{\mathrm{d}}/S_{\mathrm{s}} = 10.65 \tag{2.14}$$

Inspection of Eq. (2.12) shows that the number of g's increase rapidly and according to the square of the rotational speed. Thus, a two-fold increase in speed quadruples the g's and, therefore, the inertial stresses.

The tip velocity V at the unsupported end of the rod is given by

$$V = \omega L \tag{2.15}$$

Substituting in Eq. (2.13) leads to:

$$V_{\mathrm{lim}} = \left(\frac{2gS_{\mathrm{lim}}}{\rho}\right)^{1/2} \tag{2.16}$$

Inspection of Eq. (2.14) shows that the limiting tip velocity, V_{lim}, is proportional to the square root of the material stress limit. A twofold improvement in the stress limits between materials will, therefore, produce approximately only a 41% advantage in the tip velocity limit. Furthermore, the tip velocity limit is independent of the length of the rod.

Stress characteristics within a rotating disk are important because they relate closely to the rotor of a rotating scanner.

There are two special cases that shall be considered; a disk

(a) without a central hole, or
(b) with a central hole.

For the simple rod described above only radial stresses existed; for a disk both radial and tangential stresses exist.

3. *Disk without a Hole*

Suppose a solid cylinder of radius R and density ρ rotates about its own axis and at an angular speed ω then, according to Roark,[7] at any point distant r from the center, the radial tensile inertia stress is

$$S_{\mathrm{r}} = \tfrac{1}{8}(\rho\omega^2/g)[(3 + \gamma)(R^2 - r)] \tag{2.17}$$

[7] R. J. Roark, "Formulae for Stress and Strain," p. 360. McGraw-Hill, New York, 1965.

and the tangential stress is

$$S_t = \tfrac{1}{8}(\rho\omega^2/g)[(3 + \gamma)R^2 - (1 + 3\gamma)r^2] \tag{2.18}$$

The radial and tangential stresses are at a maximum and equal at the center and are

$$S_r(\max) = S_t(\max) = \tfrac{1}{8}(\rho\omega^2/g)(3 + \gamma)R^2 \tag{2.19}$$

where γ represents Poisson's ratio.

4. *Disk with a Hole*

A solid cylinder with central hole of radius R_0 has a set of modified formulas which, according to Roark,[8] has a radial tensile inertia stress

$$S_r = \frac{1}{8}\frac{\rho\omega^2}{g}(3 + \gamma)\left[(R^2 - r^2) + R_0{}^2\left(1 - \frac{R^2}{r^2}\right)\right] \tag{2.20}$$

and a tangential tensile inertia stress

$$S_t = \frac{1}{8}\frac{\rho\omega^2}{g}\left[(3 + \gamma)R^2\left(1 + \frac{R_0{}^2}{r^2} + \frac{R_0{}^2}{R^2}\right) - (1 + 3\gamma)r^2\right] \tag{2.21}$$

The radial stress is a maximum at $r = (RR_0)^{1/2}$ and is

$$S_r(\max) = \frac{1}{8}\frac{\rho\omega^2}{g}(3 + \gamma)(R - R_0)^2 \tag{2.22}$$

The tangential stress is a maximum at the perimeter of the hole, $r = R_0$, and is

$$S_t(\max) = 2\left[\frac{1}{8}\frac{\rho\omega^2}{g}\{(3 + \gamma)R^2 + (1 + \gamma)R_0{}^2\}\right] \tag{2.23}$$

As has been pointed out by Hayosh,[9] it should be noted from Eqs. (2.19) and (2.23) that the maximum tangential stress of a disk with a small hole at the center is twice that of a solid disk no matter how small the hole is made. It is for this reason that high-performance rotary scanner rotors are machined from a solid block of material. These formulas are best understood by referring to Figs. 11 and 12.

5. *Distortion*

Mirror surfaces distort under stresses. The degree of distortion depends on the substrate material density ρ, its Poisson's ratio γ, and its Young's modulus of elasticity E. A dynamic distortion factor J may be represented by

$$J = \gamma\rho/E \tag{2.24}$$

---

[8] R. J. Roark, "Formulae for Stress and Strain," p. 361. McGraw-Hill, New York, 1965.

[9] T. Hayosh, Technical considerations of rotating mirror deflectors, simple and sophisticated, *Proc. SPIE* **84**, 128 (1976).

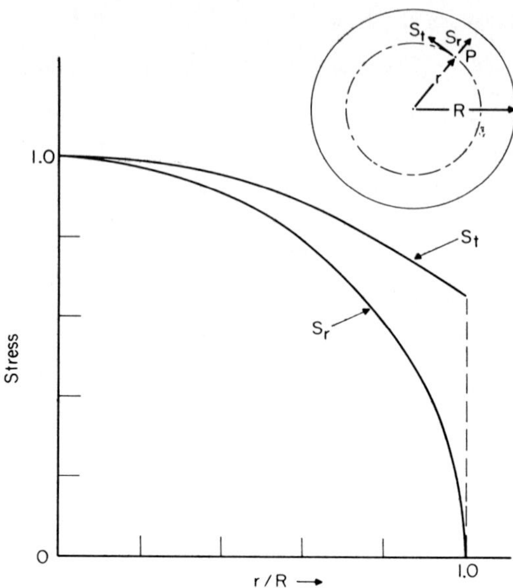

FIG. 11. Radial and tangential tensile stress distributions within a rotating solid cylinder/disk.

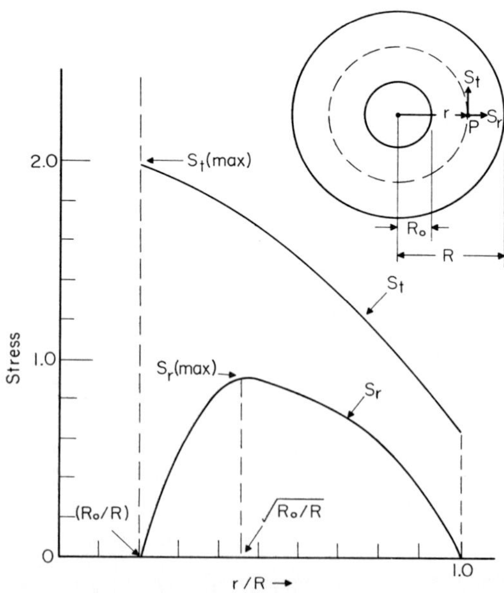

FIG. 12. Radial and tangential tensile stress distributions for a rotating cylinder with a central hole. Stress units are relative to the maximum stress value for a solid cylinder in Fig. 11.

TABLE II
COMPARISON OF SCANNING MATERIAL SUBSTRATES

Material	Density kg m^{-3} (lb in^{-3})	Young's modulus E Newtons m^{-2} (lbwt in.$^{-2}$)	Poisson's ratio	Relative distortion factor[a]	Limiting[b] peripheral velocity m sec^{-1} (in. sec^{-1})
Beryllium	1.85×10^3 (0.067)	3.08×10^{11} (4.47×10^7)	0.03	2	445 (17500)
Aluminum	2.76×10^3 (0.1)	0.69×10^{11} (1×10^7)	0.33	132	220 (8700)
Stainless steel	7.7×10^3 (0.28)	2.07×10^{11} (3×10^7)	0.3	112	570 (22400)
Quartz[c]	2.2×10^3 (0.08)	0.70×10^{11} (1.02×10^7)	0.17	53	200 (7850)

[a] Modified data. (*SPIE Proc.* **84**, 127, Table I (1976).)

[b] Limiting velocity depends on the margin of safety embodied in the limiting stress value chosen.

[c] Fused quartz–fused silica properties. (Courtesy of Heraeus-Amersil Inc., Sayreville, New Jersey.)

When E is expressed in engineering units, grams weight per square centimeter or pounds weight per square inch, J has dimensional units identical to those of curvature.

The derived parameters for distortion and limiting peripheral velocity for selected materials are given in Table II. It will be noted that stainless steel can achieve a high limiting peripheral velocity in excess of 1000 mph, while beryllium has a distortion factor significantly lower than any other known material.

6. *Bearings*

The performance and life of a scanning device in a very real sense hinges on the type of bearing used, be it rotating or oscillatory.

Essentially, for rotating devices the types of bearings used are (a) ball, (b) liquid, and (c) gas. Ball bearings are used successfully at lower speeds, while gas bearings are essential at high speeds. Liquid bearings are deprecated because of outgassing that produces an unwanted film deposit on the optical surfaces.

Gas bearings are of two types—hydrodynamic and hydrostatic. Hydrodynamic gas bearings are self-activating, which by design automatically pump air or the ambient gas between the bearing surfaces. Hydrostatic gas bearings are not self-activating but have a gas, nitrogen or helium, forced in between the bearing surfaces. Such bearings permit the construction of high-performance scanners operating above 150,000 rpm. A comparison of these bearing types is given in Table III.

Oscillatory devices have the advantage of an additional type of bearing support; a mechanical coupling which either undergoes a torsional or flexing motion. Such coupling supports have many advantages, but material fatigue is an important aspect that has to be considered.

TABLE III

PERFORMANCE COMPARISON OF BEARING PARAMETERS[a]

Parameter	Ball bearings	Liquid bearings	Air bearings
Life	Poor	Excellent	Excellent
Radial stiffness	Poor	Excellent	Excellent
Support equipment	None	Pump and condenser	Compressor[b]
Contamination	Poor	Poor	Excellent
Torque	Excellent	Good	Excellent
Axial stiffness	Poor	Excellent	Excellent
Thermal conduct	Poor	Excellent	Adequate
Noise	Poor	Excellent	Excellent
Res. shock	Poor static	Excellent	Excellent
Res. vibration	Good running	Excellent	Good

[a] T. Holland, R. Sherman, and B. J. Thompson, "Laser Scanning Techniques." WESCON, 1968.

[b] Compressor required for film gate, so air bearing does not add to support equipment requirements.

D. ELECTRICAL ASPECTS

There are two states in the motion of scanning devices; these are (a) transient state, and (b) steady state. The transient state is that period of time from start-up to the device reaching the operational steady state. This applies to both rotating and oscillatory scanners. The transient state is also that period of time in changing from one operational steady state to another. For rotating devices this corresponds to a change in speed; for an oscillatory device it represents a change in frequency, amplitude, or waveform. This transient time is a factor to be considered since it may represent a significant and unexpected time duration. Although the transient time can inherently never be eliminated, it can often be controlled and minimized to an acceptable level by the design of the electrical drive and control system.

The steady state itself has to be controlled to ensure that when a device has reached its dynamic operational characteristics it maintains these characteristics with the minimum of variation. Steady state performance characteristics are set by electrically driving the device either in an open loop control system or in a closed loop control system.

The open loop control system relies on the electromechanical parameters and characteristics of the device and regulated supply inputs. The closed loop control system, which is a servocontrol system, continuously monitors the instantaneous speed and position of the scanner and generates the necessary feedback signal to apply an appropriate compensating electrical drive torque.

The instantaneous monitoring of position usually employs a capacitive, magnetic, or optical encoding technique. An optical encoder (Fig. 13) for rotating scanning devices takes the form of a disk with equiangular sectors, black and white for sensing by reflection, or sectors opaque and transparent for sensing by transmission. The encoder is mounted on the rotating scanner shaft. In some cases, the sensing can be achieved by using the angular placement and reflecting properties of the scanner facets themselves and, thereby, perform the duty of an encoder.

Table IV presents a qualitative comparison of motor parameters with motor types for rotating scanning devices.

The power to maintain the steady state speed against drag from the surrounding air (windage) is higher than one might at first expect. Because the rotating member acts like a fan with tip velocities that may be approaching the speed of sound, the motor power requirement should be analyzed and calculated for coping with not only the steady state but also the transient time to reach the steady state.

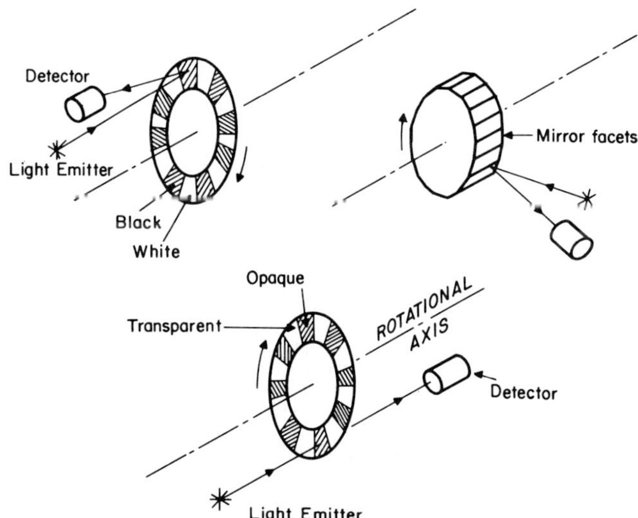

FIG. 13. Optical encoder. Closed loop sensing techniques to monitor instantaneous position and speed of a rotating scanner.

TABLE IV

MOTOR PARAMETERS VERSUS MOTOR TYPE[a]

Factor	Motor Type						
	dc motor	Induction		Hysteresis sync.	Reluctance		Brushless dc motor
		Squirrel cage	Solid		Squirrel cage	Hysteresis	
Highest speed	5000 rpm	30,000 rpm	150,000 rpm	150,000 rpm	30,000 rpm	150,000 rpm	30,000 rpm
Output torque	High	High	High	Medium	Medium	Low	Medium
Rotor heating	$I^2 r_m$	Load dependent	Load dependent	Minor loop	Load dependent	High minor loop	None
Complexity for simple speed control	Easy Back emf	Hard Requires sensor	Medium Requires sensor	Self-regulating	Hard Requires sensor	Self-regulating	Easy Back emf
Complexity for phase lock performance	Medium Requires encoder	Hard linearization required Requires encoder	Medium Requires encoder	Hard linearization required Requires encoder	Hard linearization required Requires encoder	Hard linearization required Requires encoder	Medium Encoder already used
Motor efficiency	High	Medium	Medium	Medium	Medium	Low	High
System power factor	High	Medium	Low	Low	Medium	Low	High
Relative cost	Low	Low	Medium	Medium	Medium	Medium	High

[a] T. D. Hayosh, Motors and control systems for rotating mirror deflectors, *Proc. SPIE* **84**, 104 (1976).

III. REFLECTING SCANNERS

A plane mirror is the basic optical element that forms the simplest of all scanning devices. It becomes a device when the mirror is mounted and capable of moving about an axis of rotation.

The direction and position of the rotational axis with respect to the plane of the mirror and with respect to the direction of the incident light beam, are key factors in the operation of mirror scanning devices. We shall look at a number of particular arrangements that most frequently exist in practice.

The general characteristics described in Sections III,A and III,B are important because the departures from the special case conditions provide an insight into both the effects of manufacturing dimensional and angular tolerances, and the effect of the misalignment of light beams in relation to mechanical axes. Furthermore, these characteristics form the basis for the understanding of the action of mirror polygon scanning devices.

A. Plane Mirror Parallel to Rotational Axis

1. *General Characteristics*

Consider a plane mirror (Fig. 14) capable of rotating about an axis parallel to the surface of the mirror. Let us further assume a more special case, where

(a) the incident beam is normal to the axis of rotation,
(b) the incident beam is directed through the axis of rotation, and
(c) the axis of rotation lies in the surface of the mirror.

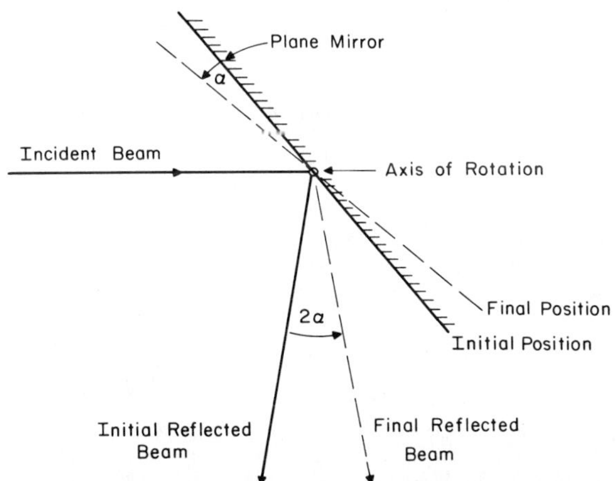

Fig. 14. Illustration of how the reflected beam sweeps through twice the angular motion of a plane mirror in which the axis of rotation lies in the reflecting surface of the mirror.

Suppose the mirror rotates in a counterclockwise direction through an angle α. As a consequence of the laws of reflection (angle of reflection equals the angle of incidence), the reflected beam will sweep through an angle 2α. It follows that if the mirror rotates at an angular speed ω, then the reflected beam sweeps (scans) with an angular speed of 2ω. We also note that the reflected beam sweeps out a plane which contains the incident beam.

We have dealt with a special case. Let us consider the effect of changing conditions (a), (b), and (c) above separately.

a. Beam Inclined to Rotational Axis. Suppose the incident beam makes an acute angle β with the axis of rotation (Fig. 15), then the reflected beam sweeps out a cone whose half angle is β and whose axis coincides with the axis of rotation. This is a more general case of that described above, but when we allow $\beta = 90°$, that is, the incident beam is normal to the axis of rotation, the cone becomes a plane, as we should expect and have already shown in the special case. We should note that in both instances the reflected beam throughout the sweep radiates from a point; namely, the point of incidence.

b. Beam Displaced from Rotational Axis. Consider the second general case in which the incident beam is normal to the axis of rotation but that its line of direction passes the rotational axis at a distance s (Fig. 16). The reflected beam still sweeps at twice the angular rate of the mirror but the apparent center of scan of the reflected beam is no longer a stationary point; it is a locus.

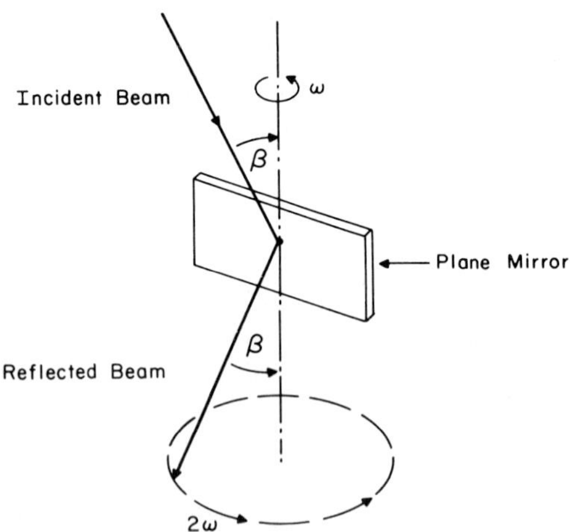

FIG. 15. Illustration of a conical scan produced when the incident beam is inclined to a plane mirror in which the axis of rotation lies in the reflecting surface of the mirror. The angular sweep rate is twice the rotational rate of the mirror.

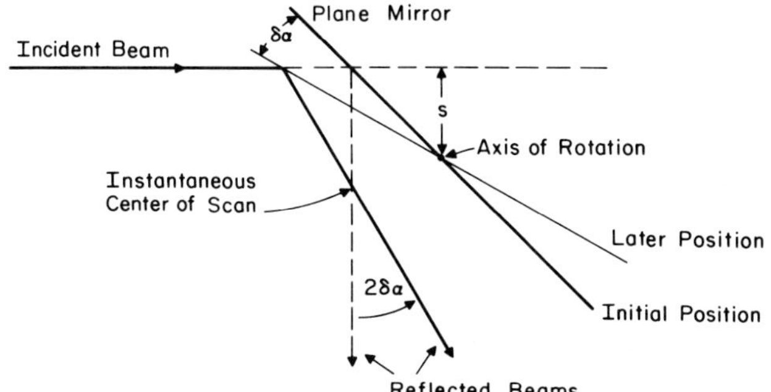

FIG. 16. Incident beam directed past the axis of rotation at a finite distance results in a reflected beam whose center of scan has a locus.

This locus reduces to a single point when $s = 0$ and the configuration corresponds to the special case shown in Fig. 14.

c. Mirror Displaced from Rotational Axis. Similarly, consider the third general case in which the axis of rotation is displaced by a distance p from the plane of the mirror but remains parallel to surface (Fig. 17). The reflected beam still sweeps at twice the angular rate of the mirror but the apparent center of scan of the reflected beam is no longer a stationary point; it is a locus. The locus becomes a single point only when the axis of rotation passes through the point of incidence at the mirror ($p = 0$) as in Fig. 14.

In reality, because of manufacturing, assembly, and optical alignment tolerances, a general case with compound effects will always exist to a greater or lesser extent depending on the tolerance limits.

For the parallel faceted polygons described later in this section, departures from conditions (a), (b), and (c) are by design.

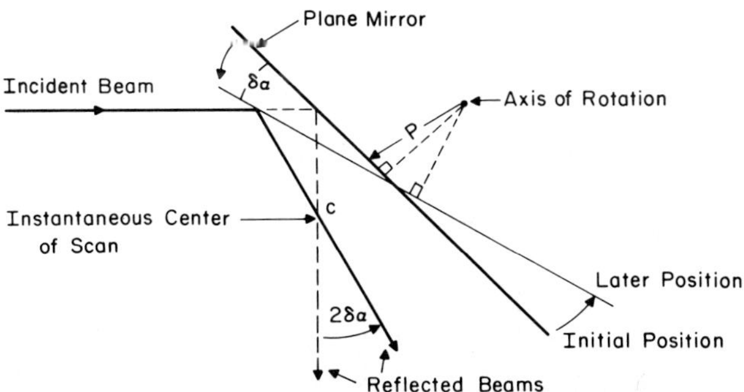

FIG. 17. Parallel displacement of the axis of rotation from the mirror surface results in a reflected beam whose center of scan has a locus.

2. *Galvanometer*

A galvanometer scanner is essentially a simple motor in which the moving-iron rotor is limited to angular displacements of about $\pm 15°$ from the neutral position. The rotor is restored to the neutral position by a torsional spring.

The torsional spring may be coiled or straight according to the stiffness required. Piano wire is often used for this spring action. The mirror light deflector is rigidly attached to the rotor mass (Fig. 18). In Fig. 18 the interconnected elements of the rotor are depicted but the electromagnetically excited stator has been omitted for clarity. For a given angular acceleration the driving torque is, by Newton's second law of motion, proportional to the moment of inertia of the rotor assembly of elements. The moment of inertia of a cylinder rotating about its axis is proportional to the fourth power of its radius. Thus, the driving torque is proportional to the fourth power of the rotor radius. For this reason galvanometer rotors are slender to optimize the dynamic frequency response characteristics.

The simple galvanometer operates most efficiently when driven at its natural frequency f_n but as will be seen in Fig. 19, slight changes in frequency from the natural frequency produce a sharp fall-off in amplitude. This sensitivity to frequency drift requires consideration so that it can be regulated by feedback in either the design of the system or in the galvanometer itself.

In the more advanced design of a galvanometer the instantaneous position or velocity of the rotor is monitored within the galvanometer assembly by capacitive or magnetic probes. This permits the galvanometer to more closely follow the applied drive waveforms ranging in frequency from dc to 1.0 kHz

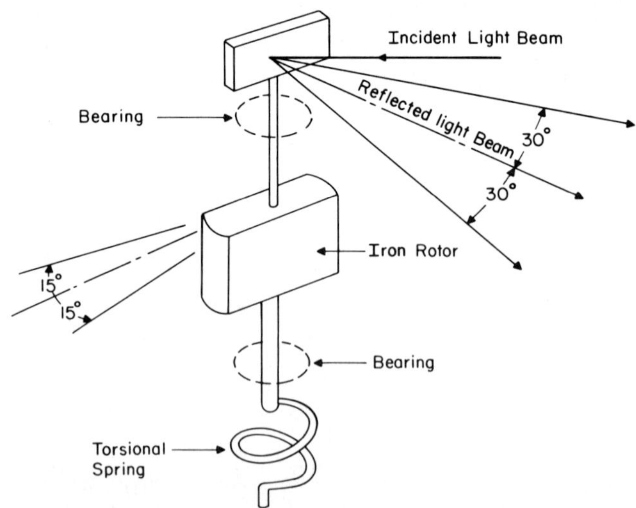

FIG. 18. Basic components of a galvanometer rotor.

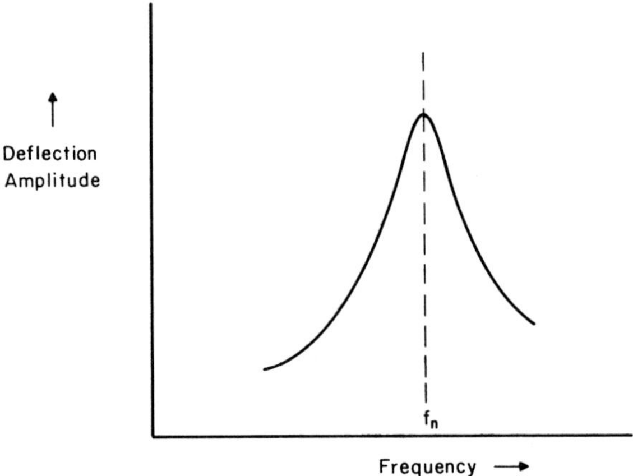

FIG. 19. Simple galvanometer response characteristic.

and including sine, square, triangular, and sawtooth waveforms. The simple open loop galvanometer operating frequencies typically range from dc to 200 Hz.

a. Resonant Galvanometer. It was pointed out that when the simple galvanometer is operated at or close to its natural resonant frequency, its amplitude varies sharply with small changes in frequency. Even though one may wish to operate the galvanometer only at the resonant frequency, it requires strict frequency control for consistent performance. This objectionable characteristic has been overcome by the design of a rotor with two natural frequencies based upon the principle of the dynamic vibration absorber,[10] invented by Frahm in 1909.[11]

Essentially, the rotor consists of two inertia masses connected by a torsion spring such that the rotor has two degrees of freedom. The galvanometer frequency response characteristic is shown in Fig. 20.

The galvanometer is designed to operate at a frequency f_0, a frequency between the two natural resonant frequencies of the rotor assembly. In contrast to the sharp amplitude versus frequency sensitivity at the natural frequencies f_1 and f_2, the amplitude response is insensitive to small changes in frequencies. We should note that such devices can only produce a sine waveform output but can be driven by any periodic input waveform.

---

[10] E. F. Burke and J. E. Muckenhaupt, A stable fixed frequency scanning system, *Proc. SPIE* **84**, 57–61 (1976).

[11] J. P. den Hartog, "Mechanical Vibrations," 4th ed., pp. 87–91. McGraw-Hill, New York, 1956.

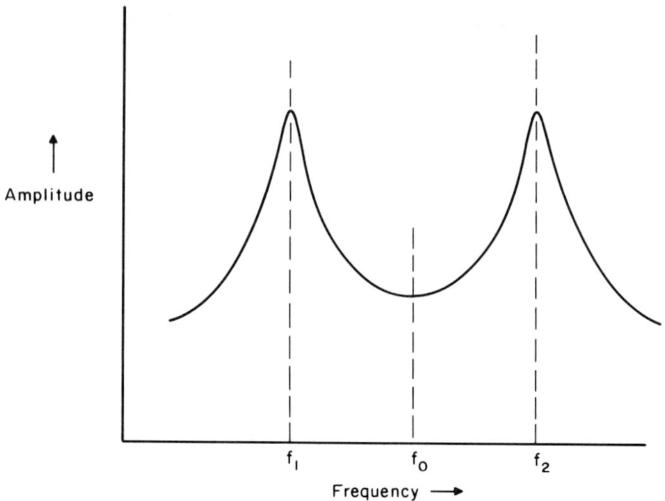

FIG. 20. Resonant galvanometer response characteristic.

It is of interest to mention a paradox that exhibits itself in such a device when operating at the tuned resonant frequency f_0; namely, the main body of the rotor remains stationary while the mirror mass, the dynamic absorber, oscillates.

b. Tangent Correction. To a first order, the rotation angle θ of the rotor of a simple galvanometer is proportional to driving current i in the electromagnetic stator. At the same time the deflection of the light beam spot across a plane surface (Fig. 21) is proportional to the tangent of 2θ, thus:

$$\theta = ki \tag{3.1}$$

$$y = R \tan(2ki) \tag{3.2}$$

By tapering the rotor iron mass, the trigonometrical tangent function dependence of Eq. (3.2) can be eliminated such that the light spot deflection across a flat screen is directly proportional to the driving current i.[12]

Hence

$$y = Ki \tag{3.3}$$

where K is a constant. Compensation by shaping the rotor mass is so effective that one can overcorrect and thereby design a galvanometer to effect a spot deflection onto a concave, plane, or convex surface that is proportional to the driving current.

[12] U.S. Patent 3,177,385 (1965). Compensation Achieved by Shaping Rotor.

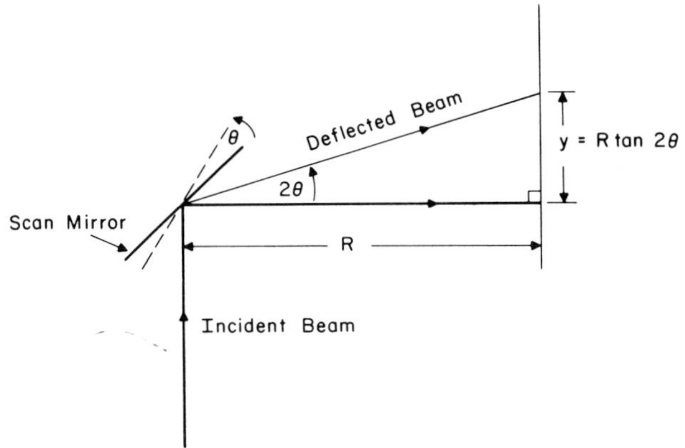

FIG. 21. Trigonometrical relationship of scan deflection y versus deflection angle 2θ.

3. *Nodding Mirror*

A nodding mirror is a mirror oscillating back and forth to produce a triangular waveform motion. Scanning mirrors of this type are usually large and have a high inertia compared to that mounted on a galvanometer. Nodding mirrors are so special that they need to be custom designed to achieve the required optical and dynamic mechanical characteristics.

4. *Steering Mirror*

A steering mirror is a form of scanning mirror that is gimballed and driven to compensate for the pitch,[13,14] yaw, and roll of a carrying vehicle in order that the receiver or transmitter can follow the target and thereby stabilize the resultant imagery with respect to a system (Fig. 22).

Because of the laws of reflection, a steering mirror is driven back through half the angular displacement of the vehicle. The mirror always assumes a mid position between the nominal position and the position that the mirror would take up if it were fixed and moved with the vehicle. The same remark concerning custom design mentioned about nodding mirrors Section III,A,3 equally applies.

5. *Parallel Polygon*

A parallel polygon scanner is basically a motor assembly with a rotor or armature which carries a regular polygon of mirrors each parallel to the axis of

---

[13] G. F. Marshall, System and production engineering of a beryllium optic, *Opt. Eng.* **13**, 354–361 (1974).

[14] G. F. Marshall, System and production engineering of a beryllium optic. *Proc. SPIE* **54**, 41–49 (1974).

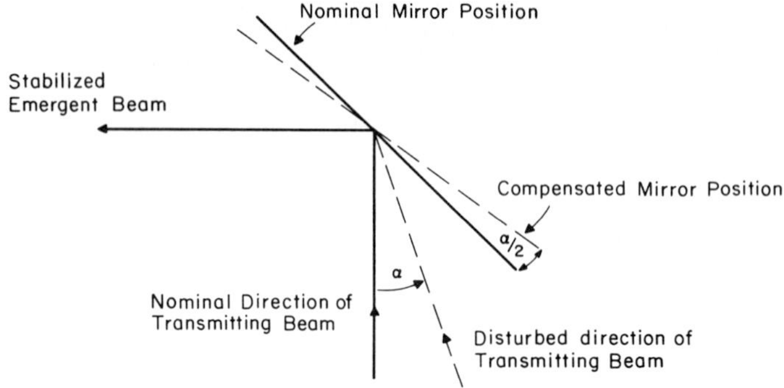

FIG. 22. Steering scanning mirror stabilizes the direction of light beams.

rotation, Fig. 23. Inherently the mirror facets are displaced from the axis of rotation so that the center of scan is always a locus (Section III,A,1,*c*, Fig. 17) as opposed to being a single point. The maximum angle ϕ in degrees through which the beam can sweep is given by

$$\phi = 720/N \tag{3.4}$$

where N is the number of facets. If $N = 20$ then $\phi = 36°$.

a. Important Parameters. There are four essential design parameter tolerances necessary to satisfactorily specify a polygon scanner. These tolerances relate to the

(a) height of each facet from the axis of rotation (Fig. 24),
(b) angle between adjacent facets (Fig. 25),
(c) angle of each facet to the axis of rotation (Fig. 24), and
(d) flatness of each facet.

These tolerances directly relate to the consistency of the beam scan start, overlay, and focus of successive scan traces from consecutive facets.

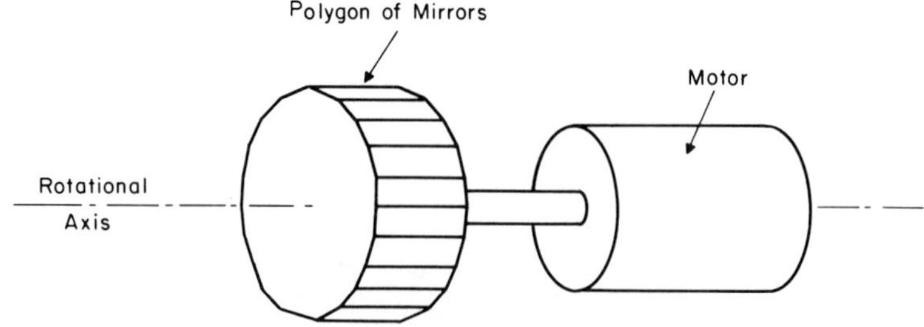

FIG. 23. Schematic of a parallel polygon scanner.

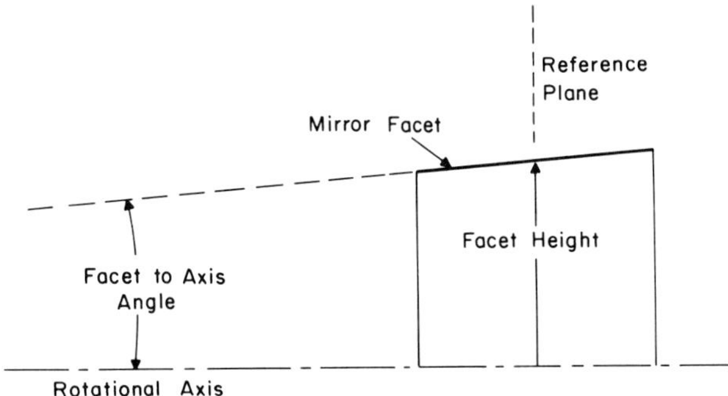

FIG. 24. Two key parameters affected in mounting a polygon.

For sophisticated polygon scanners the polygon is manufactured as an integral part of the rotor to achieve close tolerances on these parameters. For simple polygon scanners a separated mirror polygon is mounted on the shaft of the rotor. Unless the mirror polygon axis is mounted so that it is collinear with the rotational axis, the essential design parameters, height of each facet from the rotational axis, and angle of each facet to the axis of rotation are destroyed. This will occur even if the mirror polygon is manufactured to perfection. Thus, the alignment of mounting a separate polygon onto a rotor shaft is critical. It is for this reason that polygon scanner manufacturers prefer to undertake the design and manufacture of the entire subassembly.

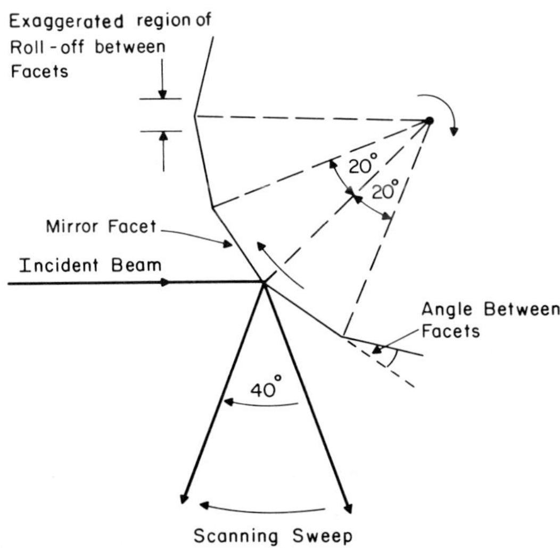

FIG. 25. Deflection sweep of a regular polygon.

TABLE V

TYPICAL DYNAMIC STATE TOLERANCES OF PARALLEL AND PYRAMIDAL SCANNERS

Parameter	Production	Custom
Facet height above axis	$\pm 25\ \mu m$ (± 0.001 in.)	$\pm\ 5\mu m$ (± 0.0002 in.)
Facet to facet	± 30 arc sec	± 5 arc sec
Facet to axis	± 30 arc sec	± 5 arc sec
Flatness	$\lambda/2$	$\lambda/8$
rpm	3600	>0
Features	Separate polygon and rotor	Integral polygon and rotor
Bearings	Ball	Gas
Facet roll-off	$625 \pm 125\ \mu m$	$250 \pm {75 \atop 50}\ \mu m$
	$(0.025 \pm 0.005$ in.)	$\left(0.010 \pm {0.003 \atop 0.002} \text{ in.}\right)$

Table V outlines typical tolerances of parallel and pyramidal scanners obtainable under dynamic conditions. Custom-made high-performance scanners with dynamic angular tolerances equal to or less than ± 0.5 arc sec can be achieved by patience and painstaking effort of opticians and the engineering of the entire subassembly with special attention to the dynamic balance and the bearing design. Table V is in no way intended to be absolute but rather a guideline for design consideration.

b. Facet Roll Off. The specification for flatness of a mirror or polygon facet can never reach to the very edge of the surface. There is always a finite region close to the edges (Fig. 25) where the surface begins to "roll off." This roll off may typically be in the order of 0.5 mm (0.020 in.) and depends upon polishing techniques.

For a mirror polygon this roll off between two adjacent facets may amount to 1 mm (0.040 in.). One millimeter, according to the optical design, often represents a significant portion of the light beam diameter; its effects should be investigated and the dimension specified.

B. PLANE MIRROR 45° TO ROTATIONAL AXIS

1. *General Characteristics*

Consider a plane mirror (Fig. 26) capable of rotating about an axis inclined at a substantial angle to the mirror normal. Let us assume the special case in which

(a) the mirror plane is inclined at an angle of 45° to the axis of rotation, and
(b) the incident beam is collinear with the axis of rotation.

When the mirror revolves about the rotational axis through an angle α, the reflected beam will also sweep through the same angle α. It follows that if the mirror rotates at an angular speed of ω then the reflected beam also scans at the same angular speed ω. We further note that the reflected beam sweeps out a plane which is normal to the incident beam and, therefore, normal to the axis of rotation.

We have dealt with the special case. Let us consider the effect of changing conditions (a) and (b) separately.

a. Mirror-to-Rotational-Axis Not Exactly 45°. Suppose the mirror plane is inclined at an angle slightly less than 45° by a small angle ε such that the angle of incidence is $(45° - \varepsilon)$, then the reflected beam sweeps out a cone with a half angle given by $(90° - 2\varepsilon)$, and whose axis coincides with the axis of rotation (Fig. 27). This is a general case of that described above, but when we allow $\varepsilon = 0°$, the reflected beam becomes normal to the axis of rotation, and the cone becomes a plane, as we should expect and have already shown for the special case (Fig. 26).

We should again note that in both instances the reflected beam throughout the sweep radiates from a point; the point of incidence.

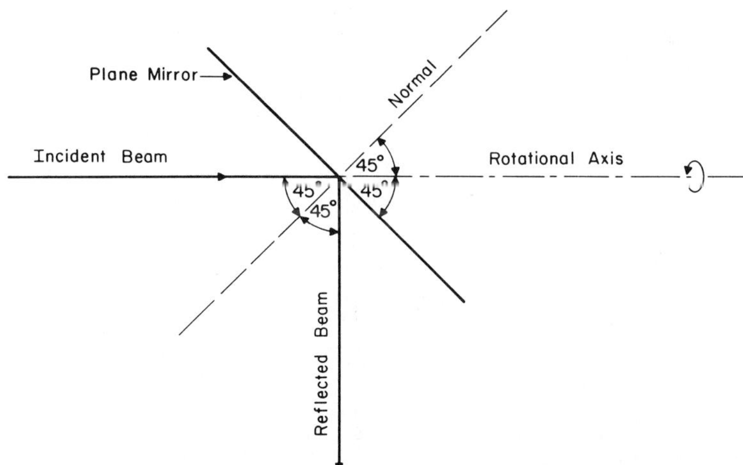

FIG. 26. Incident beam is collinear with mirror rotational axis, while the plane mirror is inclined to the axis at an angle of 45°. Reflected beam sweeps out a plane normal to the axis of rotation at the same angular rate.

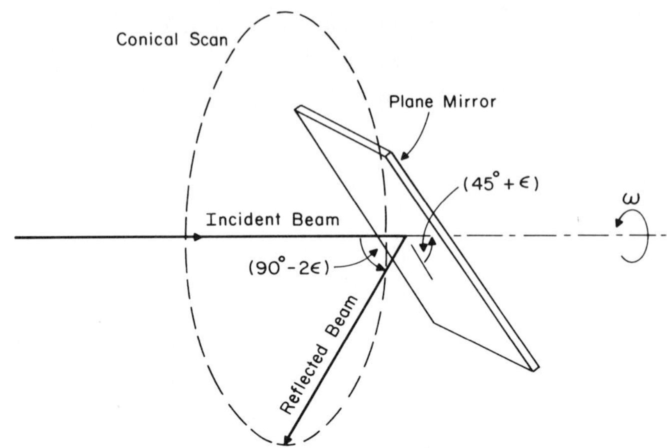

FIG. 27. Incident beam collinear with the mirror rotational axis, while the mirror plane is inclined to the axis at an angle ($45^\circ + \varepsilon$). The reflected beam sweeps out a cone whose half angle is ($90^\circ - 2\varepsilon$) to the axis of rotation. The beam scan rate corresponds to the rotational rate of the mirror.

b. Beam Displaced from the Rotational Axis. Consider the second more general case in which the incident beam is displaced from the axis of rotation but remains parallel to it (Fig. 28). The reflected beam sweeps at the same angular rate as that of the mirror and at every instant is normal to the axis of rotation. However, the apparent center of scan is no longer a point but a locus. This locus is the line produced by the reciprocating point of incidence which on the surface of the mirror traces out an ellipse.

When $\varepsilon = 0$, the reflected beam still radiates in a direction normal to the axis of rotation but the reflected beam no longer sweeps out a plane but, rather, a "plane" with a wave in it. Likewise, when $\varepsilon \neq 0$ the surface swept out is a "cone" with a wave in it.

2. *Single 45° Mirror*

A 45° mirror scanner (Fig. 29) is more complicated in design than it might appear. The complexity arises from the asymmetry of the mass distribution of the reflector along and about the axis of rotation. For this reason these scanners are best custom-designed and balanced for the dynamic state.

3. *Pyramidal Polygon*

A pyramidal polygon is basically a rotor assembly with a rotor armature which carries a regular polygon of mirrors inclined at an angle to the axis of rotation (Fig. 30).

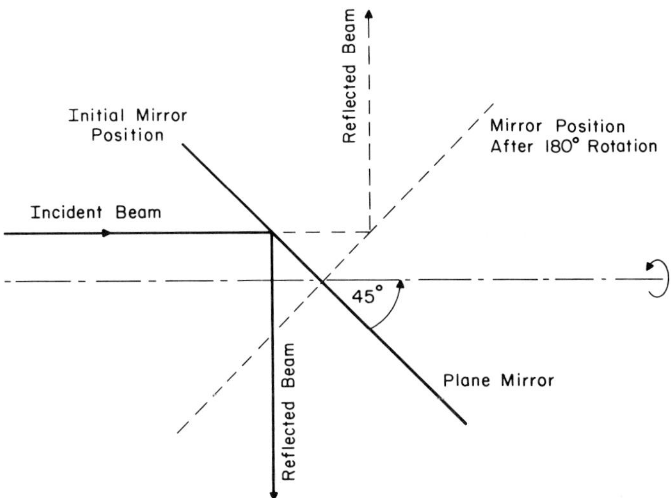

FIG. 28. Incident beam displaced from the axis of rotation. Reflected beam is normal to the rotational axis but shifts with rotational position.

The maximum angle ϕ in degrees through which the beam can sweep is given by

$$\phi = 360/N \tag{3.5}$$

where N is the number of facets. Equation (3.5) should be compared with Eq. (3.4). The same consideration concerning parameters as given in Section III,A,5 and Table V equally applies.

4. *Irregular Polygon*

With a regular polygon it is important that the facet-to-axis-of-rotation angle tolerance is close so that the overlay of scan traces is precise. A TV-type raster pattern (Fig. 5) is accomplished by synchronizing a second scanner. A synchronized raster scan may be achieved in one step by arranging each facet

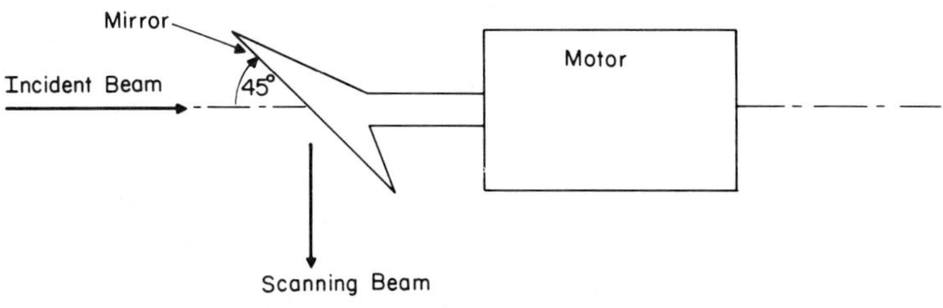

FIG. 29. Single 45° mirror scanner.

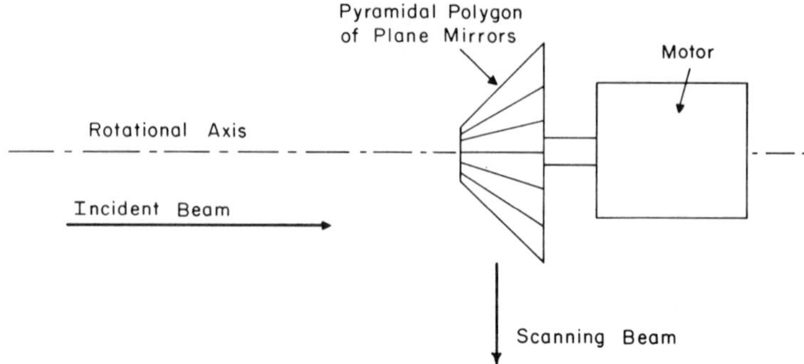

FIG. 30. Schematic of a pyramidal polygon scanner.

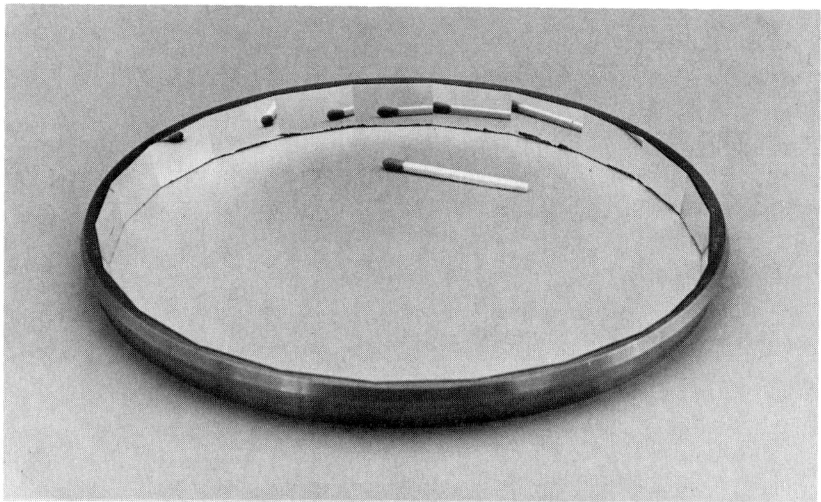

FIG. 31. Replicated irregular polygon scanner. (Photograph courtesy PTR Optics Inc.) There are twenty-three internal facets.

of a polygon to be tilted at a slightly different angle to the axis of rotation. This brings about a regular displacement of successive scan traces.

The manufacture of an irregular polygon is difficult but once the investment is made it can readily be replicated (Fig. 31).[15–18]

---

[15] T. T. Saito, Diamond turning of optics. *Opt. Eng.* **15**, 432 (1976).

[16] G. F. Marshall and F. Denton, Precision replication, *Design News* **31**, 72–76 (1976).

[17] H. M. Weissman, Epoxy replication of optics, *Opt. Eng.* **15**, 435–441 (1976).

[18] H. M. Weissman and F. Denton, Ten scanners for the effort of one, *Proc. SPIE* **84**, 35–40 (1976).

C. Plane Mirror Normal to Rotational Axis

Consider a plane mirror (Fig. 32a) capable of rotating about an axis normal to the surface of the mirror. Let us assume a special case where

(a) the incident beam is collinear with the axis of rotation and
(b) the mirror plane is at 90° (normal) to the axis of rotation.

The geometry of this idealized arrangement is such that when the mirror rotates it reflects the incident beam back on itself along a reciprocal and coincident path. This special case acts as a starting point for discussion and understanding of the more generalized variants.

1. *Inclined Beam*

Inspection of the diagram in Fig. 32b shows that when an incident beam is directed at any part of the rotating mirror, and at any angle, the reflected beam will remain stationary as if the mirror were motionless.

2. *Mirror-to-Rotational-Axis Angle not Exactly* 90°

Suppose the mirror normal is tilted at a small angle ψ to the axis of rotation to which the incident beam is collinear. When the mirror rotates, the reflected beam sweeps out a cone whose half angle is 2ψ, and whose axis coincides with the incident beam (Fig. 33). A device incorporating such a rotating mirror is referred to as a nutating scanner.

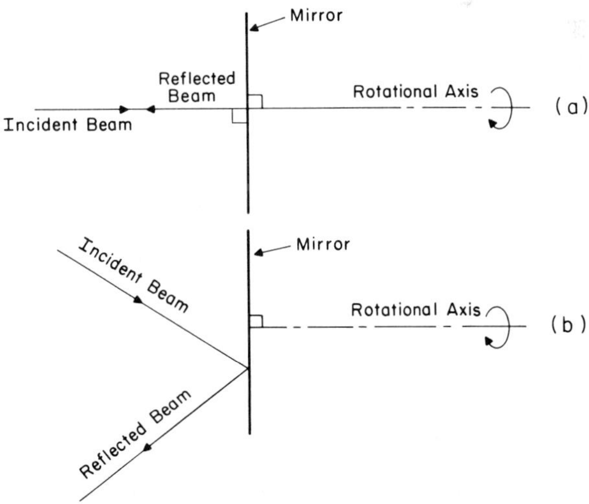

FIG. 32. A plane mirror rotating about a normal to the mirror. Reflected beam remains stationary for all angles of incidence.

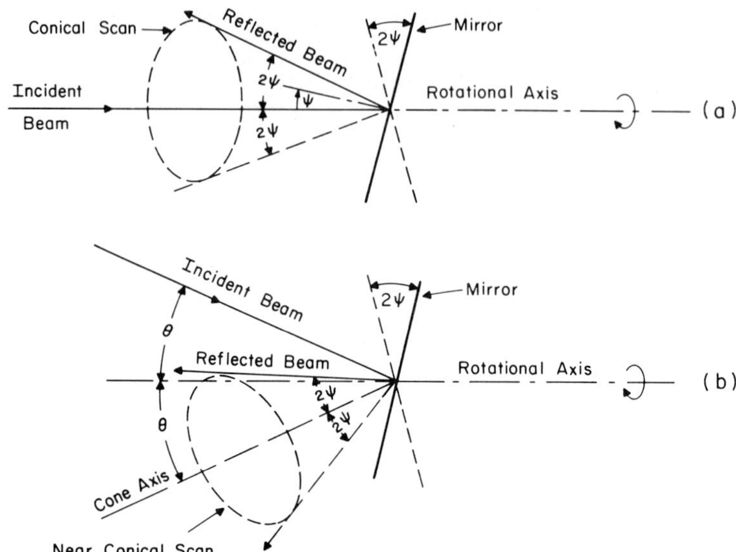

FIG. 33. A conical scan and a near conical scan produced by a plane mirror rotating about an axis inclined to the normal. (a) Incident beam is collinear with the rotational axis, (b) incident beam is inclined to the rotational axis.

3. *Inclined Incident Beam and Tilted Mirror to Axis of Rotation*

Suppose the incident beam makes small angle θ with the axis of rotation: the axis of the near conical beam sweep is tilted by virtue of the laws of reflection to the opposite side of the rotational axis through an equal angle (Fig. 33b). The half angle of the near cone remains at 2ψ.

For completeness, if we substitute $\psi = 45°$ we arrive at the condition described in Section III,B. When ψ is small the scan represents the reflective equivalent of the rotating prism wedge described in Section IV,A,2.

4. *Two Rotating Mirrors*

Figure 34 represents the reflector equivalent of two contrarotating wedge prisms in series given in Section IV. With the descriptions given, it is sufficient to mention that the elliptical conical sweep has a semimajor angular diameter of $2(\psi_1 + \psi_2)$ and a semiminor angular diameter of $2(\psi_1 - \psi_2)$.

D. PLANE MIRROR GENERAL CASE

In Sections III,A, B, and C, we dealt with rotating mirrors mounted parallel, 45°, and normal to the axis of rotation, respectively. For completeness, I now wish to develop and present the basic formulas that cover the general case for a

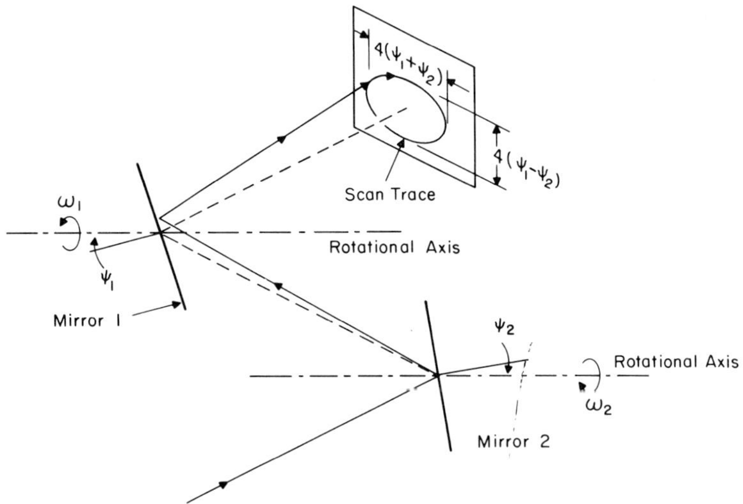

FIG. 34. Two contrarotating mirrors in series.

rotating mirror whose normal makes any angle with the axis of rotation, and in which the incident beam makes any angle with the axis of rotation (see Fig. 35).

1. *Derivation of Formulas*

Let the rotational axis be the z axis. Let the incident beam AP lie in the x–z plane and make an angle μ with the z axis, intersecting it at the point P. Let the mirror normal PQ make an angle β with the z axis, and intersect the x–y plane at the point Q. Let PB represent the reflected beam which intersects the x–y plane at B and forms an angle ϕ with the z axis.

Suppose θ is the angle of incidence at a time t, and the corresponding angle of reflection is θ'. Let α and β be the projected angles of θ and θ', respectively, onto the plane x–y. Angle α also represents the angular position of the rotor at a time t, while ξ represents the scan angle with respect to α also at a time t.

Consider the plane of incidence APQ, and the plane of reflection QPB. From Fig. 35 and three-dimensional analytical geometry, the angle of incidence θ is expressed by the relation:

$$\cos\theta = \cos\mu\cos\beta + \sin\mu\sin\beta\cos\alpha \tag{3.6}$$

Likewise, the angle of reflection θ' is expressed by the relation

$$\cos\theta' = \cos\phi\cos\beta + \sin\phi\sin\beta\cos\xi \tag{3.7}$$

By the first law of reflection—namely, the angle of reflection equals the angle of incidence—we have

$$\theta' = \theta \tag{3.8}$$

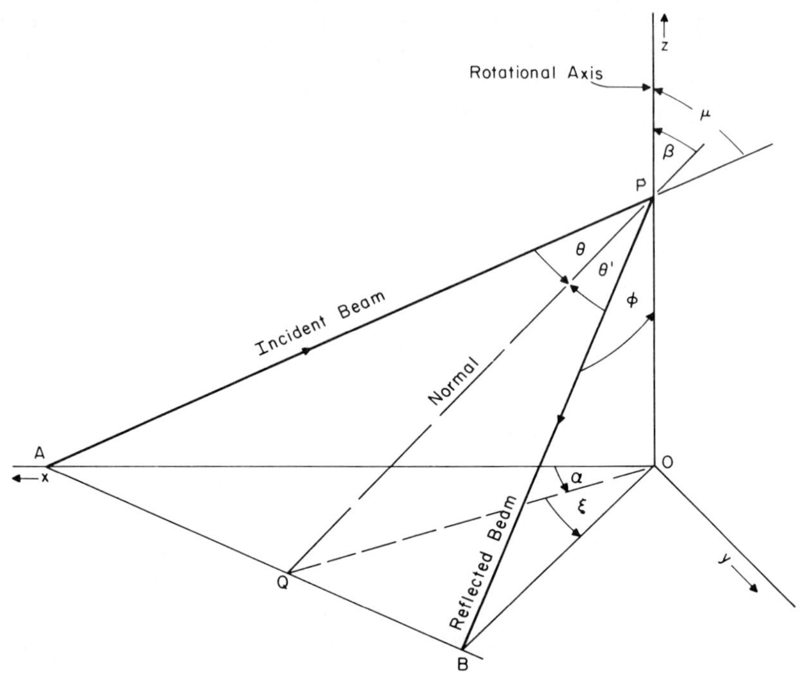

FIG. 35. Diagram depicting the key parameters relating to a rotating plane mirror scanner.

Hence, equating Eqs. (3.6) and (3.7) leads to:

$$\frac{\cos\mu - \cos\phi}{\sin\phi\cos\xi - \sin\mu\cos\alpha} = \tan\beta \tag{3.9}$$

By the second law of reflection—namely, the incident beam and the reflected beam lie on opposite sides of the normal and all three in a common plane—we have

$$\cos(\theta + \theta') = \cos\mu\cos\phi + \sin\mu\sin\phi\cos(\alpha + \xi) \tag{3.10}$$

Since in Eq. (3.8) $\theta' = \theta$, Eq. (3.10) becomes:

$$2\cos^2\theta - 1 = \cos\mu\cos\phi + \sin\mu\sin\phi\cos(\alpha + \xi) \tag{3.11}$$

Substituting for $\cos\theta$ from Eq. (3.6) into Eq. (3.11) gives:

$$\cos\mu\cos\phi + \sin\mu\sin\phi\cos(\alpha + \xi) = 2[\cos\mu\cos\beta + \sin\mu\sin\beta\cos\alpha]^2 - 1 \tag{3.12}$$

Equations (3.9) and (3.12) are two independent, simultaneous equations which are necessary and sufficient to solve the two unknown angles ξ and ϕ. These two angles define the direction of the reflected beam at any instant in time t.

The two general equations for ξ and ϕ in terms of μ, β, and α are explicit but cumbersome trigonometrical functions. To simplify computations and to minimize errors, one is well advised to consider Eqs. (3.9) and (3.12) as being the basic general equations. Thus, into these two equations one first substitutes the fixed values of parameters for any particular case under investigation before solving the equations.

2. *Scanning Velocity*

To determine the beam scan velocity, one first differentiates the two basic equations, (3.9) and (3.12), with respect to t, and then one substitutes the parameters of the chosen particular case before solving for $d\xi/dt$ and $d\phi/dt$.

3. *Scanning Sweeps*

Only in two special cases does the beam sweep out a regular cone. These are when:

(a) The mirror is coplanar with the axis of rotation, $\beta = 90°$ (Section III,A). Thence $\xi = \alpha$, $d\xi/dt = d\alpha/dt$, and $d\phi/dt = 0$. Since ξ is defined relative to α (see Fig. 35), the meaning of these values is that the scan advances at twice the rotational sweep of the mirror.

(b) The incident beam is collinear with the axis of rotation, $\mu = 0$ (Section III,B and C). Thence $\xi = 0$, $d\xi/dt = 0$, and $d\phi/dt = 0$. Again, since ξ is defined relative to α (see Fig. 35), the meaning of these values is that the scan sweep is in phase with the rotational sweep of the mirror.

In both cases the significance of $d\phi/dt = 0$ is that the axis of the conical scan coincides with the rotational axis.

In all other cases, for instance, as the incident beam departs from being collinear with the rotational axis, $\mu \neq 0$, or as the mirror departs from being coplanar with the axis of rotation $\beta \neq 90°$, the scan cone departs from a circular cross section and becomes more pear shaped and beyond. At the same time, the scan sweep velocity, which has both a tangential ($d\xi/dt$) and a radial ($d\phi/dt$) component, varies, while the range between the maximum and minimum velocity increases with the eccentricity of the scan cross section.

A qualitative appreciation of eccentric scan cross sections and velocity variations can readily be gained by first mounting a piece of a plane mirror at approximately 45° to the rotational axis of a simple motor, for example, a toy motor. Then, by directing a helium–neon laser beam along ($\mu = 0$) the rotational axis to the mirror, one should observe the scan sweep and cross section as the direction of the laser beam is gradually swung through 90° ($\mu = 90°$). Out of interest, particular note should be made of effects at the discontinuity that occurs when the incident beam is approximately 45° to the axis of rotation; more precisely when the beam strikes the mirror at grazing incidence $\mu = 90° - \beta$.

E. Prism

1. *A 45° Right Prism*

A 45° prism whose refracting angle is 90° (Fig. 36) may be used to produce a lateral and parallel sweep scan of a light beam. The translational sweep of the beam effectively corresponds to a center of scan at an infinite distance away. To effect this sweep, the prism itself undergoes a corresponding lateral movement in the same direction, which is both normal to the refracting edge of the prism and the incident beam.

2. *Roof Prism*

According to requirements and prism configuration, the emergent scanning beam may be conveniently directed either in the opposite direction to the incident beam (Fig. 36a) or at right angles to it (Fig. 36b). In either case, the beam displacements and sweep rates are twice those of the prism element. Furthermore and more importantly, the geometrical distance and optical path

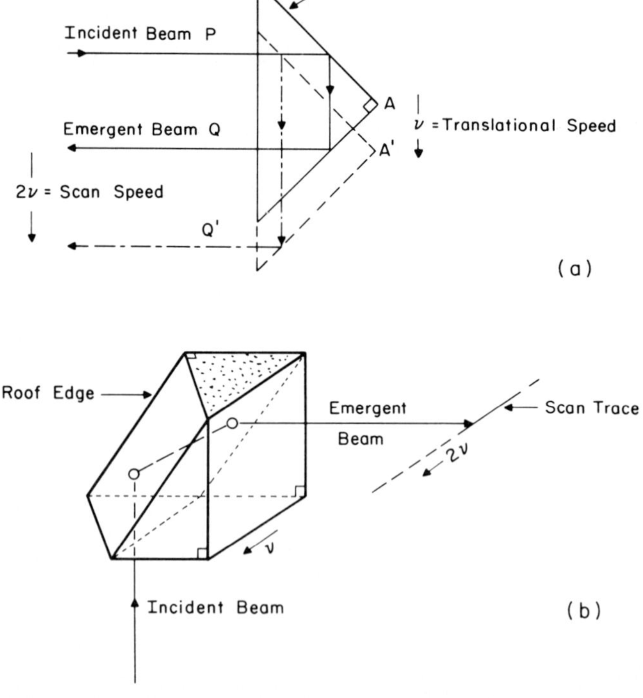

FIG. 36. Translational scan traces produced by (a) 45° right prism: incident beam and scan plane are coplanar; (b) roof prism: incident beam is normal to scan plane.

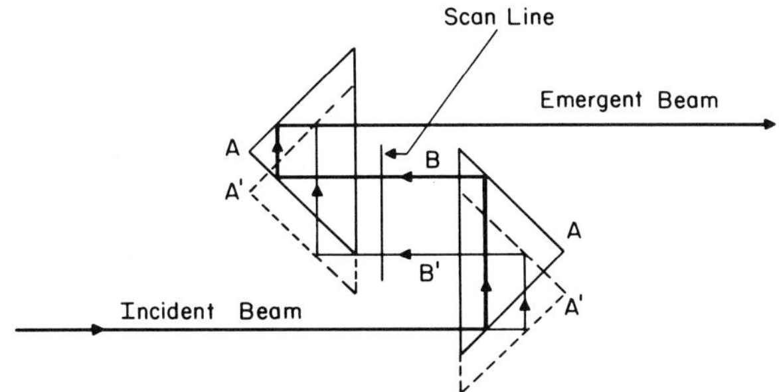

FIG. 37. Coupled right prisms scan and descan along a line.

difference from the point of incident to the point of emergence is a constant for any one particular arrangement. This property of constant distance is of key importance in an image-producing system.

3. *Coupled Prisms*

The combination of two mechanically coupled prisms, such as in Figs. 36 and 38, is used to scan a line and then to descan the scanning beam so the incident and the emergent beam to and from the combination remain fixed with respect to a system reference frame (Fig. 37). This device and technique have been

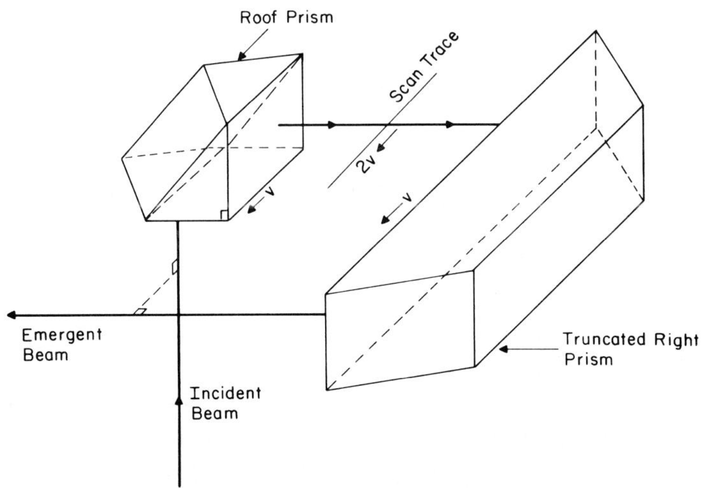

FIG. 38. Two types of prisms rigidly coupled and translated at a speed v in a direction normal to both a fixed incident beam and a fixed emergent beam produce a lateral beam scan between the two prisms with a scan speed of $2v$.

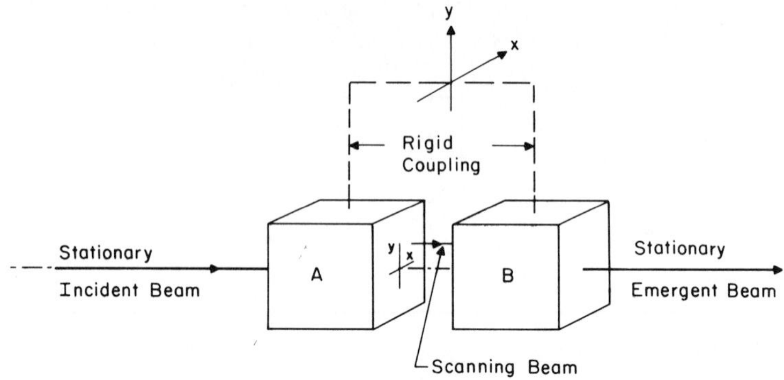

FIG. 39. Schematic of the general case of coupled components to provide a translational scan in two directions x and y.

described in detail.[19,20] The technique is not restricted to the use of prism elements; in principle the technique may use any pair of components (fiber optic, lens, etc.), which as individual components are so arranged or designed that when they are moved with respect to the incident beam they translate the emergent beam and maintain a constant optical path length. By suitable design a two-dimensional field scan may be achieved (Fig. 39).

IV. REFRACTING SCANNERS

A. WEDGE PRISM

1. *Single Wedge Prism*

Consider a light beam incident normally at the surface of a wedge-shaped prism. Let the wedge angle, technically called the refracting angle, measure A units. If A is less than 10°, then the angle of deviation D of the emergent beam from the incident direction is approximately given by

$$D = (n - 1)A \tag{4.1}$$

where n represents the refractive index of the material (Fig. 40).

The direction of deviation is away from the refracting edge (the sharp edge) of the prism towards the thicker part, and for normal incidence the plane of deviation, formed by the directions of the incident and emergent beams, is always normal to the refracting edge.

---

[19] G. F. Marshall, British Patent 1,192,465, Improvements Relating to Optical Systems.
[20] G. F. Marshall, Canadian Patent 855,454, Optical Systems.

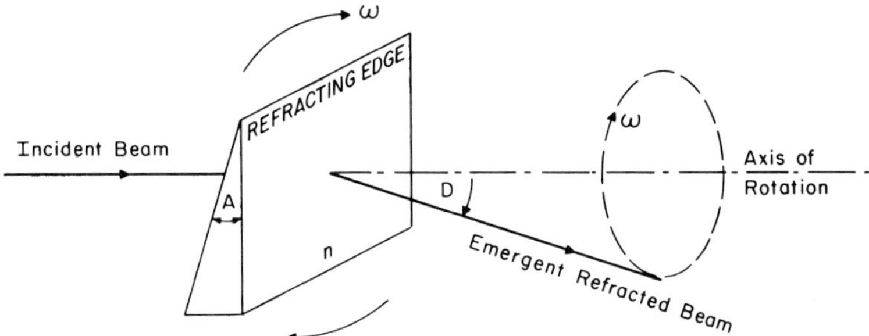

FIG. 40. The emergent refracted beam from a rotating prism wedge sweeps out a cone whose full angle approximately equals the wedge angle A of the prism.

When the prism is rotated about an axis collinear with the incident beam the emergent beam sweeps out a cone whose half angle is D. Typically, the value of n is in the order of 1.5; using Eq. (4.1) this leads to

$$D = \tfrac{1}{2}A \tag{4.2}$$

Thus, a rotating prism wedge with a refracting angle of 8° will sweep out a cone whose full angle measures approximately 8°.

2. *Two Wedge Prisms*

Suppose we add a second prism such that the emergent beam from the first prism passes on through the second prism (Fig. 41). The total deviation $\mathbf{D}_\mathbf{R}$ is the vector sum of the separate beam deviations $\mathbf{D}_1$ and $\mathbf{D}_2$ from each prism.

$$\mathbf{D}_\mathbf{R} = \mathbf{D}_1 + \mathbf{D}_2 \tag{4.3}$$

If the first prism rotates at an angular rate ω and the second prism rotates in a counter direction at an equal rate, the resultant beam sweeps out an elliptical

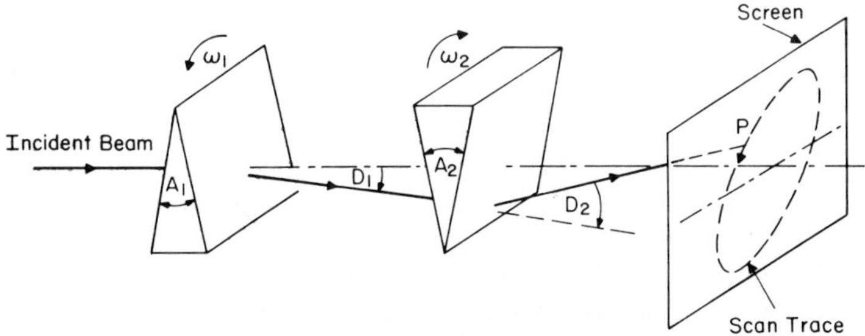

FIG. 41. Generation of an elliptical scan pattern by two contrarotating wedge-shaped prisms in tandem. The pattern is reduced to a line scan when deviations by prism elements are identical and $\omega_1 = \omega_2$. The scan pattern precesses when rotational speeds are unsynchronized; i.e., $\Delta\omega \neq 0$.

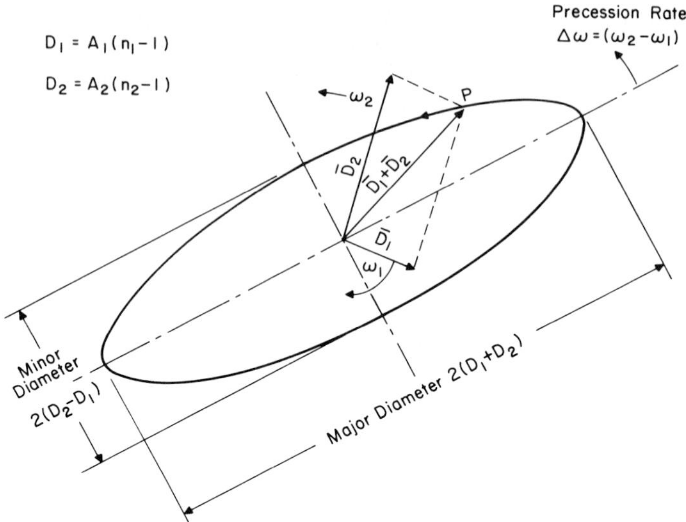

FIG. 42. Elliptical scan pattern generated by two contrarotating wedge prisms. Overbars indicate vectors.

cone. The elliptical cross section of this cone has major and minor diameters directly proportional to the sum and difference of the angular deviations of each prism (Fig. 42). When the prisms produce the same deviations, which could be when they are identical, having the same refractive index and the same refracting angle, then the minor diameter of the elliptically shaped conical sweeping beam reduces to zero. Under these circumstances the beam is effectively sweeping out a straight-line scan in simple harmonic motion whose angular deviation D_{R} at time t is given by

$$D_{\mathrm{R}} = 2D_1 \sin(\omega t) \tag{4.4}$$

When one counter-rotating prism rotates at a rate which is not exactly equal to that of the other prism, in other words, there is an angular speed differential $\Delta\omega = (\omega_1 - \omega_2)$, then the line scan precesses at the speed differential rate $\Delta\omega$.

B. LENS

A lens may be considered as being made up of a continuous series of truncated prism wedge elements whose refracting angles are small but which gradually and progressively change (Fig. 43).

The refracted beam deviation is proportional to the wedge angle of each element, as expressed in Eq. (4.1). If a lens (Fig. 44) is moved transversely to its

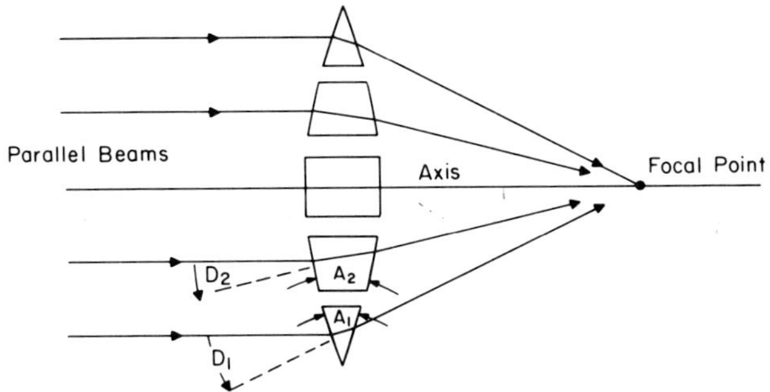

FIG. 43. Truncated prism wedge elements of a lens.

axis at a speed v to a fixed incident beam, the emergent refracted beam sweeps at an angular rate ω given by

$$\omega = v/f \tag{4.5}$$

where f represents the focal length of the lens. Since the lens itself has focusing properties, an auxiliary lens in the incident beam may be added to compensate or augment the power of the scanning lens.

A series of lenses mounted with their axes parallel at a fixed radius to form a circle on a disk, or on the surface of a cylinder with the lens axes radial, provides the basis for a scanning device equivalent to a reflecting polygon.

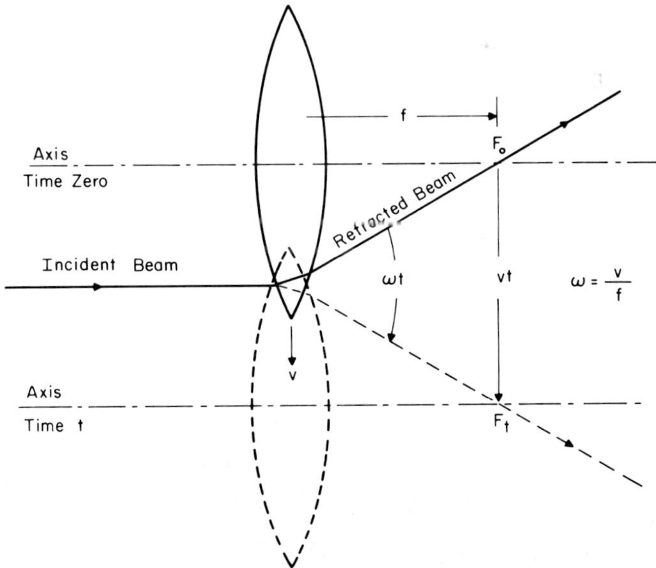

FIG. 44. Translating the lens normally to its axis produces an angular beam scan.

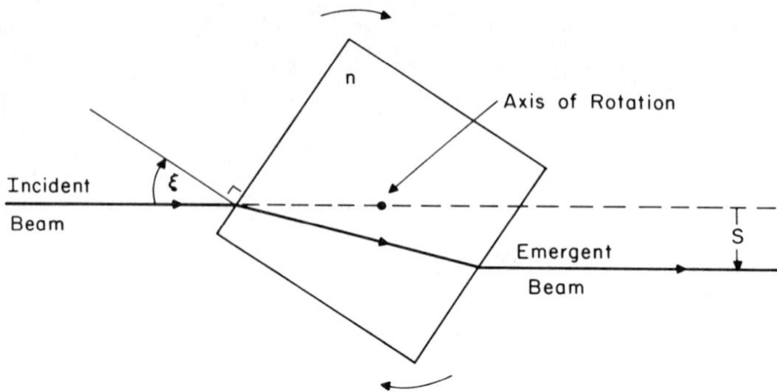

FIG. 45. Simplest multifaceted refracting polygon scanner.

Displacement of these lens elements to form a spiral array on a disk or a helix on a cylinder generates a self-synchronized raster.

C. REFRACTING POLYGON

A beam of light transmitted through a plane parallel window is undeviated but is translated through a distance s depending on the angle of incidence ξ and the thickness t of the window.

This beam displacement versus angle-of-incidence dependence is used to provide a periodic translational scan by directing a beam through a regular multifaceted polygon whose opposite facets are parallel (Fig. 45) and parallel to the axis of rotation. The displacement s is given by[21]

$$s = t \sin \xi \left[1 - \frac{\cos \xi}{(n^2 - \sin^2 \xi)^{1/2}} \right] \tag{4.6}$$

When ξ is small, Eq. (4.6) approximates to

$$s = \xi[(n - 1)/n]t \tag{4.7}$$

If $n = 1.5$, then $s = \frac{1}{3}\xi t$.

V. DIFFRACTING SCANNERS

A. GRATINGS

Scanning can be accomplished by using the light deflection principles of diffracted light in transmission or reflection. Diffraction of light has been covered

[21] L. Levi, "Applied Optics," Vol. 1, p. 355. Wiley, New York, 1968.

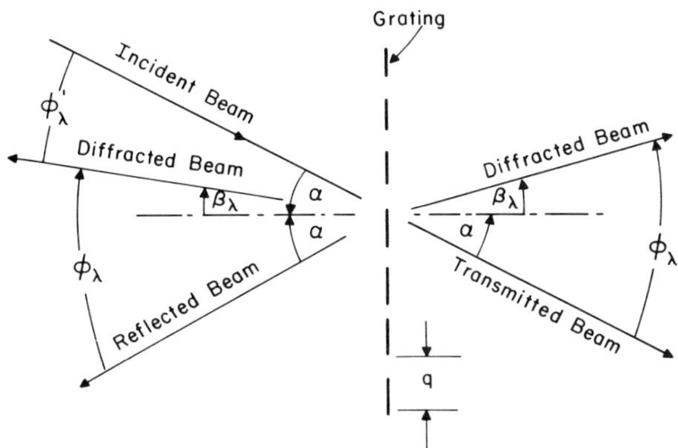

FIG. 46. Schematic of diffracted beams: reflection from and transmission through a grating. One order only is depicted for clarity.

by Walther (Volume I, Chapter 7 of this series), while diffraction gratings have been dealt with in detail by Richardson (Volume V, Chapter 2 of this series).

The diffraction angle β_λ from the normal (Fig. 46) is given by the grating equation:

$$m\lambda = q[\sin\alpha \pm \sin\beta_\lambda] \tag{5.1}$$

Inspection of Eq. (5.1) shows that β_λ is a function of the grating spacing q, angle of incidence α, the wavelength of light λ, and the diffraction order m. When α and β_λ are numerically small, Eq. (5.1) approximates to

$$m\lambda = q(\alpha \pm \beta_\lambda) \tag{5.2}$$

Thence

$$\phi_\lambda' = \alpha - \beta_\lambda = 2\alpha + m\lambda/q \tag{5.3}$$

$$\phi_\lambda = m\lambda/q \tag{5.4}$$

1. *Reflection Grating*

The diffraction grating in a reflection mode functions as a spectral scanner when the grating is rotated about an axis parallel to the rulings. Because of the wavelength dependence on the diffracted beam angle ϕ_λ, a simple rotation allows a succession of wavelengths to illuminate and emerge from the exit slit aperture of a system, Eq. (5.3).

2. *Transmission Grating*

In contrast to the reflected diffraction beam it can be seen from Eq. (5.4) that for all practical purposes, the direction of the transmitted diffracted beam is insensitive to a rotational movement of the grating about an axis parallel to the rulings. For any given wavelength and order, the most sensitive parameter is the grating spacing q. If the spacing can be varied, the direction of the diffracted beam will vary, making a scanning device. Such a device exists in the acousto-optical scanner (see Chapter 4 of this Volume).

B. Acousto-Optical Scanner

Essentially, an acousto-optical scanning device is a grating within a transparent block of material. Light directed through the cell block is diffracted by the grating formed by the variations of refractive index within the block. The regular variations are established by the passage of a travelling ultrasonic wave in a direction transverse to the light beam. The acoustical wave propagates a regular series of compressions through the material and these develop a corresponding regular series of increments and decrements in the refractive index of the medium through which they travel at the speed of sound (Fig. 47).

The scanning action is produced by frequency modulation of the ultrasonic wave which varies the effective spatial frequency of the grating. The grating spacing q is the wavelength of the ultrasonic wave within the optical material.

$$q = V/f \tag{5.5}$$

where V is the velocity of sound and f is the central deflection frequency. When $V = 617$ nm/μsec and $f = 75$ MHz, then $q = 8.2$ μm, which corresponds to a spatial grating frequency of 122 cycles/mm.

Using an ultrasonic carrier frequency of 75 MHz, the grating equation, Eq. (5.1), leads to a first diffraction order ($m = 1$) with a beam deflection angle of 61 mrad $\equiv 3.5°$. The second diffraction order ($m = 2$) beam deflection is 7°. The frequency modulation bandwidth that produces the beam scan is

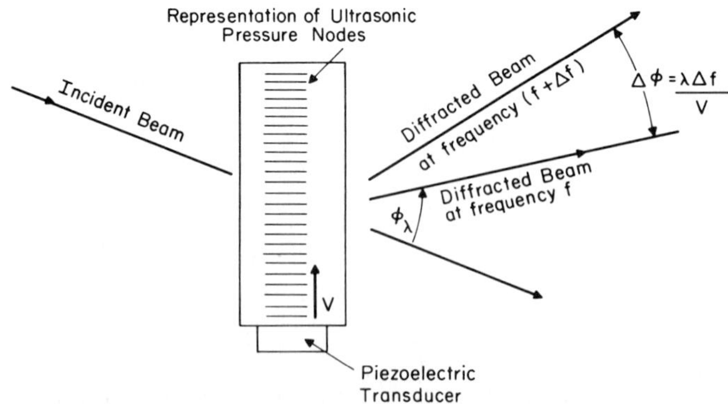

Fig. 47. Schematic of an acousto-optical deflector.

limited to one octave to avoid the overlapping of diffraction orders. Typically, the depth of modulation is restricted to one half of one octave. The scan angle is small but may be readily magnified by using a lens in a manner similar to that depicted in Fig. 4.

The response time to a change in the deflection angle is physically limited by the time it takes for the ultrasonic wave of a new frequency to traverse the light beam. If the light beam diameter is 6.17 mm, using an ultrasonic velocity of 617 nm/μsec gives a transit time response of 10 μsec. This transit response time is also responsible for limiting the spot resolution performance capability of the related acousto-optical modulator that controls the beam intensity. It is this response time which limits the performance of acousto-optical systems, but design techniques for these devices are employed to overcome this problem.[22]

The primary advantage of acousto-optical deflectors is their lack of inertia and, therefore, their high scan speeds. These devices are primarily used for monochromatic light from a laser beam and not white light, although conceivably they can readily be used for spectral scanning, and have been used for color image scanning.[23]

VI. READERS

A. Active

One of the earliest scanning systems that falls into the generic class of readers outlined in Section I,B was that suggested by Arago[24] and realized by Foucault and Fizeau independently in the midnineteenth century to measure the velocity of light in air. In essence, his system was a range finder in reverse. A beam of light was transmitted along a retroreflective path of about 20 m (Fig. 48). In the path was a plane mirror rotating about an axis parallel and through the surface of the mirror. Because of the finite time taken by the light in travelling a return path of 40 m and the new position taken up by the plane mirror in this time interval, the return image was deflected from that position of the image when the mirror was stationary. Hence, one of the earliest determinations of the velocity of light was made: 298,000 km/sec.

Foucault's system falls into the class of an active reader system. As a range finder this technique for measuring time intervals has been long superseded

[22] L. Beiser, Laser scanning systems, *in* "Laser Applications" (M. Ross, ed.), Vol. 2, p. 116. Academic Press, New York, 1974.

[23] P. Das, D. Schumer, and F. M. Mohammed Ayub, Color image scanning using acousto-optic interaction with surface waves, *Proc. SPIE* **84**, 91–96 (1976).

[24] F. A. Jenkins and H. E. White, "Fundamentals of Optics," 2nd ed., p. 383. McGraw-Hill, New York, 1951.

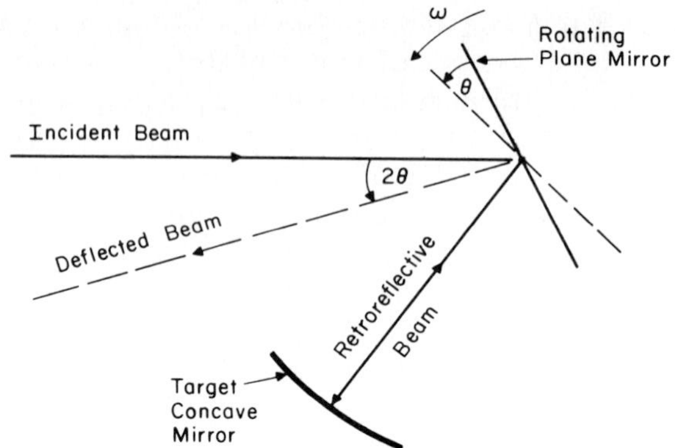

FIG. 48. Essential features of Foucault's scanning system to measure the velocity of light.

by pulse-modulation techniques and employing CRT time-base displays. Nevertheless, the system concept is still useful for applications other than measuring of small time intervals.

A number of scanning systems use rotating mirrors and galvanometer devices. These systems are, in principle, similar to that of Foucault, but where the objective is to collect the intensity modulations of the return signals produced by the spatial reflective characteristics of the target scene. The target can range from a bar code (Figs. 49 and 50) to the topography of a terrain.[25,26] They differ only in the details, such as the distances, dimensions, and signal strengths relating to its function.

B. PASSIVE

As mentioned in the Introduction, passive systems do not transmit light but scan and detect the intensity-modulated light reflected or emitted from a target scene (Fig. 51). Schematically such a system is shown in Fig. 52.

1. *Flying Spot Scanner*

In the earlier sections we related scanning devices to those mechanisms which deflect light beams. The scanning motion can also be achieved by moving the light source itself. This has been best accomplished by using the light emitted from a spot on the face of a cathode ray tube (CRT) and imaging it in the plane of the transparency target (Fig. 53). The modulated transmitted light is then collected by a sensor to produce a video signal for processing and display.

---

[25] G. F. Marshall, Lasers in the supermarket, *Phys. Technol.* **7**, 160–165 (1976).
[26] G. F. Marshall, Point of Sale, *Opt. News* **2**, 5–10 (1976).

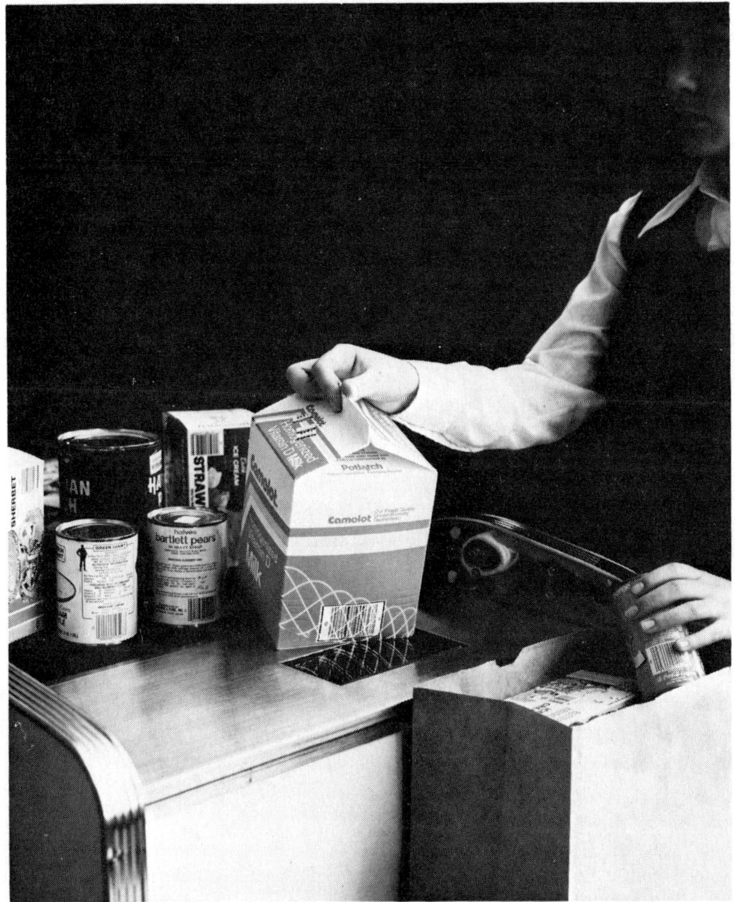

FIG. 49. A POS laser-beam scan is seen reading the UPC symbol on a supermarket product as the cashier bags the merchandise at the checkout counter. (Photograph courtesy of Schiller Industries Inc. Reprinted with permission of Cahners Publishing Co.)

This type of system is known as a flying-spot scanner, has the distinct advantage of having no inertia, and, therefore, a high flexibility to electronically switch from one scan pattern to another. However, in some applications such as point of sale, the spot image illumination at the target is too low to produce an adequate signal-to-noise ratio for the depth of field demanded. This may also be the limitation for other possible applications.

2. *Sensor Array Scanners*

Sensor array scanners, often referred to as solid state scanners, contrast to flying-spot scanners. Instead of moving the light source, an integrated-circuit

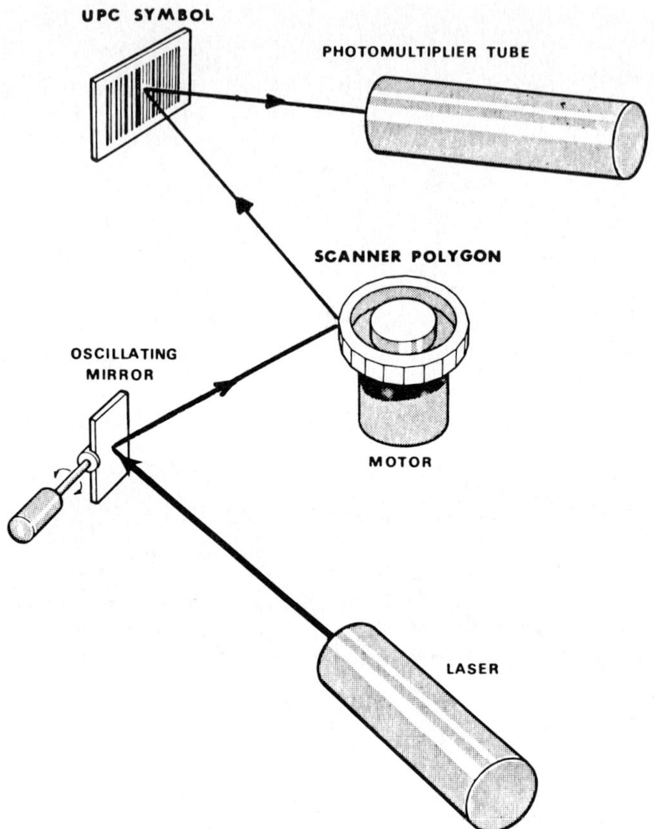

FIG. 50. Schematic diagram of an optical system of a laser-beam scanner UPC symbol reader. (Photograph courtesy of Schiller Industries Inc. Reprinted with permission from *Optics News* **2**, No. 1, 5–10 (1976).)

array of tiny sensors in the image plane is successively addressed and performs a scan of the image plane (Fig. 54). These tiny sensor elements, approximately at 12.5 μm spacing, define and limit the size of the picture elements. These sensor arrays in either one or two dimensions are of increasing importance, simple to use, and a powerful device for reader systems because of the precise dimensions between picture elements. In their simplest application they are used as an electro-optical equivalent of a shadowgraph.

In particular, they provide an elegant and rapid means for evaluating the MTF of lenses. Using a Ronchi grating in the object plane (Fig. 55) and a diode array in the image plane, the spatial frequency response characteristics of a lens across its field and along its axis may be displayed on an oscilloscope.

FIG. 51. Photographs of real-time display images from a passive infrared rotating mirror line-scan system. (b) Note that clothing in contact with the human body is warmer (lighter gray) than clothing not in contact. (c) City of Alexandria, Egypt. (Photographs courtesy Hawker Siddeley Dynamics Ltd., England. Reprinted with permission of Cahners Publishing Co.)

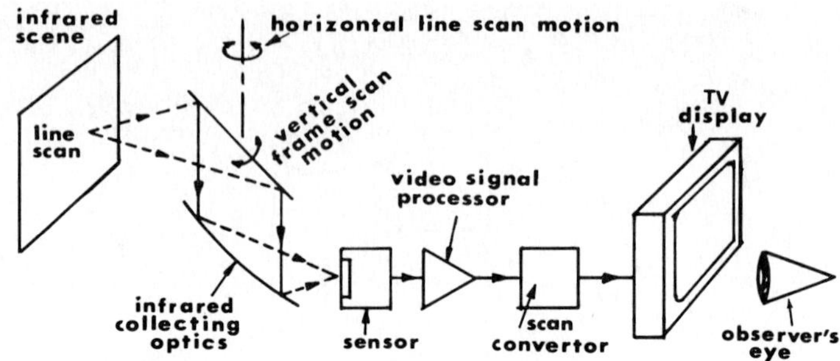

FIG. 52. Schematic diagram of a line scan reader system. Scene is mechanically scanned by a plane mirror. (Reprinted with permission of Cahners Publishing Co.)

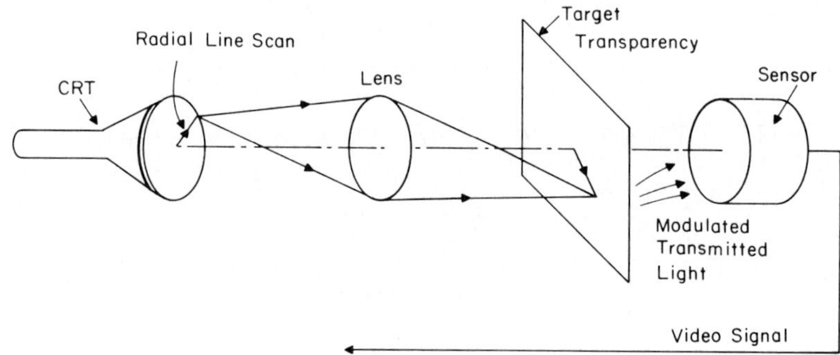

FIG. 53. Fundamentals of a flying-spot scanner reader system.

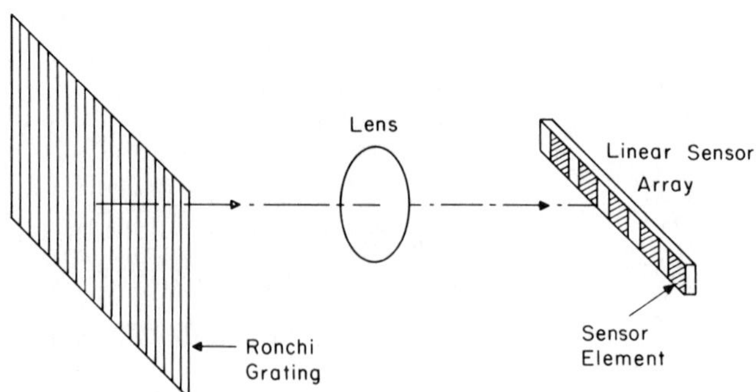

FIG. 54. Linear sensor array schematic for evaluating lenses.

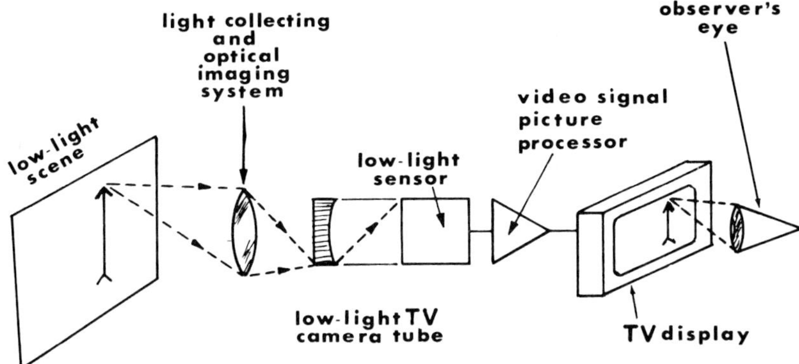

FIG. 55. Passive low-light-level electronic scanning system. (Reprinted with permission of Cahners Publishing Co.)

VII. REPRODUCERS

As outlined in Section I,B, reproducing scanning systems either write modulated data to provide a permanent record or project data to provide a real-time display.

A. RECORDERS

Probably the simplest of all recorders that brings out the underlying principles is that of contact printing (Fig. 56).[27]

The recording light beam which exposes the film is modulated by the contact negative as the beam scans and produces the corresponding exposure record; namely, the contact print. All other light-recording systems employ essentially the same fundamental scanning system concept but with differences in the scanning devices and the location of the signal intensity modulator.

Modulators are usually acousto-optical, and because of the comparatively large separation between the modulator and the recording surface, detailed consideration to the system design and the precision control of the beam spot position is of prime importance. It is of importance because of the precision registration required between successive scan lines, the regularity of each scan line, and the separation between adjacent picture elements of information being recorded. In the contact printing system the distance between the

[27] B. J. Thompson, Application of lasers to printing and recording. *Image Tech.* **2**, 16–24 (1969).

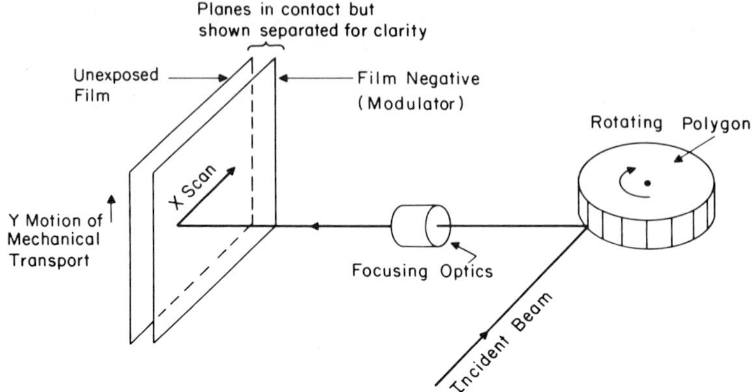

FIG. 56. Basic schematic of a recorder system with a contact printer.

recording surface medium and the modulator (negative film) is by design zero, which virtually overcomes the major difficulty of registration problems.[28]

B. DISPLAYS

The essential difference between reproducers that provide a permanent record and a reproducer that displays the information for instant viewing, is the surface on to which the data are projected. There is no fundamental difference between systems which provide a permanent record and those systems which display the information for instant viewing on a screen. Even with the high laser beam intensity it is difficult to match the convenience, versatility, and performance of a TV CRT display system.

For recorders one-directional scanning is sufficient, while the other direction is performed by a film transport system. For a display system both directions must be scanned and synchronized to produce a two-dimensional picture. In order to be acceptable, scanning display systems inherently need to meet the minimum visual requirements which TV CRT systems have routinely met over the years. However, the feasibility of optical scanning display systems has been demonstrated.

[28] G. T. Morgan and M. J. Fivel, A laser composer for the Graphics DLC 1000, *SPSE Annu. Fall Symp. Business Graphics, 16th* (1976).

CHAPTER 7

Coherent Optical Processing in Mapping

N. BALASUBRAMANIAN

Cupertino, California

and

ROBERT D. LEIGHTY

Center for Coherent Optics, USAETL, Fort Belvoir, Virginia

I. INTRODUCTION

The remarkable growth and success of digital computers and digital techniques has rendered today's data processing environment to be biased against optical processing techniques. The coherent optical processor expectations of about 15 years ago have not been realized for many assorted reasons; however, recent literature does indicate that the field of coherent optical

ISBN 0-12-408606-3

processing is moving. Coherent optical processing, either independently or in combination with digital processing, represents a specialized tool in the sense that the tool works best for tasks that have a special set of properties that involve rapid to near-real-time processing of signals with large space bandwidth products. It is in this context that applications of coherent optical processing technology to problems in mapping show potential as well as demonstrated promise.

Maps represent a display of spatial information on a geographic basis. Topographic maps, for example, represent the optical distribution of a selected class of terrain information such as hyplography, hydrology, vegetation, cultural features, place names, and reference grids. Cartography is the art and science of map making. The mapping process, in the broadest sense, involves the acquisition of data through remote sensing, processing the data to extract the required information, and storage and display of mapping information. In most cases, the acquisition of data takes the form of aerial photography. Hence, the basic sources of data for the preparation of topographic maps represent very high space bandwidth images of terrain surfaces recorded as aerial images. The requirement for near-real-time processing of this type of data source is a natural attraction for coherent optical processing techniques because the digital processing techniques are not yet efficient in this domain.

It is the intent of this chapter to provide an overview of the demonstrated as well as potential applications of coherent optical processing techniques in mapping. The major areas addressed are photogrammetric data reduction, synthetic aperture radar image reconstruction, and feature extraction. An evaluation of the application of coherent optical processing technology to mapping is presented to outline operational advantages and disadvantages. The material presented in this chapter is by no means exhaustive or complete and the reader is referred to the references cited for further details.

II. PHOTOGRAMMETRIC DATA REDUCTION

A. INTRODUCTION

Photogrammetry is defined as the art, science, and technology of obtaining reliable information about physical objects and the environment through the process of recording, measuring, and interpreting photographic images. The photogrammetric data reduction process uses instruments to measure terrain profiles and elevation contours from stereomodels reconstructed using the aerial stereotransparencies. Since the early 1960s, coherent optical processing techniques have been applied to photogrammetric data reduction. This section reviews application of optical data processing to stereophotogrammetry.

1. *Principles of Photogrammetric Stereo Compilation*

The spatial relationship between the two photographs forming the stereopair, and the three-dimensional terrain, is illustrated in Fig. 1. This figure refers to the simple and ideal case of vertical photographs in which the camera axes are parallel to each other and perpendicular to the reference plane. The transparency is a positive print of the negative and is located symmetrical to the negative about the perspective center (the camera lens). This representation avoids the problem of inversion of coordinates caused by the lens. The principal distance f is the distance between the perspective center and the image plane, and for large object distance it is equal to the focal length of the lens. The distance between the perspective center and the reference plane is defined as height H, and the separation of the camera axes is defined as the stereobase B. The axes of the photographs are parallel to the base-line axis. The various coordinate systems are defined in the figure. The height of the terrain is defined

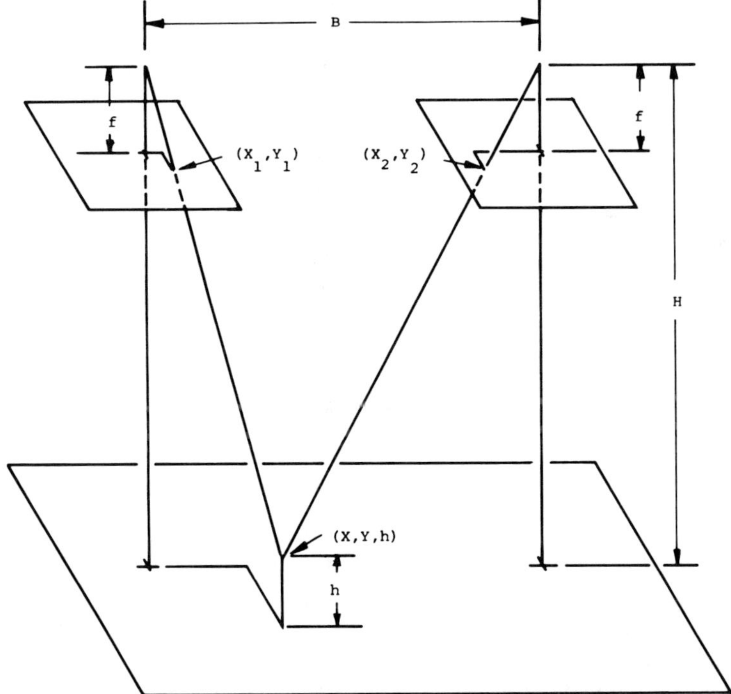

FIG. 1. Stereoscopic parallax for vertical photographs.

with respect to the reference plane. Using similar triangles, it is easy to show from Fig. 1 that

$$x_1 = fX/(H - f)$$
$$x_2 = f(X - B)/(H - h)$$
$$y_1 = y_2 = fY/(H - h)$$

The differential x parallax p_x is defined as

$$p_x = x_1 - x_2 = fB/(H - h) \tag{2.1}$$

As the terrain height h varies from object point to object point, the x parallax also changes, and the x parallax differences are the principal cause of stereo perception. The y parallax, defined as $p_y = y_1 - y_2$, is zero in the case of vertical photographs, but has a definite value whenever tilt is involved between camera axes (departure from the vertical case) which hinders stereoscopic vision.

In current practice the data reduction from stereophotographs is achieved in two ways: by direct measurement of the reconstructed spatial model of the terrain (double projection system) or by using a mathematical model defining the relationship between points in image space, object space, and perspective centers (analytical photogrammetry).

The principle of obtaining the stereomodel represents the reverse of the photographic process. The transparencies are placed in two projectors that are oriented so that the positive transparencies bear the exact relative angular relationship of the negatives in the cameras at the instants they were exposed. Light is projected through the transparencies, and where the rays from the corresponding images on the left and the right transparencies intersect below, they create a stereomodel. The concept is illustrated in Fig. 2. The stereomodel is measured by bringing into coincidence a reference mark with the stereomodel point. The measurement of the planimetric position of the reference mark then provides a means of measuring the stereomodel. In contrast to the direct measurement techniques, in the analytical approach, the coordinates of points in the photograph are measured independently and the planimetric position of the model point is computed analytically.

2. *Optical Processing—Its Relevance to Stereocompilation*

As discussed earlier, parallax differences provide for stereoperception and it is through the measurement of these parallax differences that the photogrammetric data reduction is achieved. All automated stereocompilation instruments are based on the ability to match conjugate images automatically and determine the parallax difference from object point to object point. The matching of the conjugate images involves examination of the similarity of the spatial density variation corresponding to the conjugate images. This is the process that is

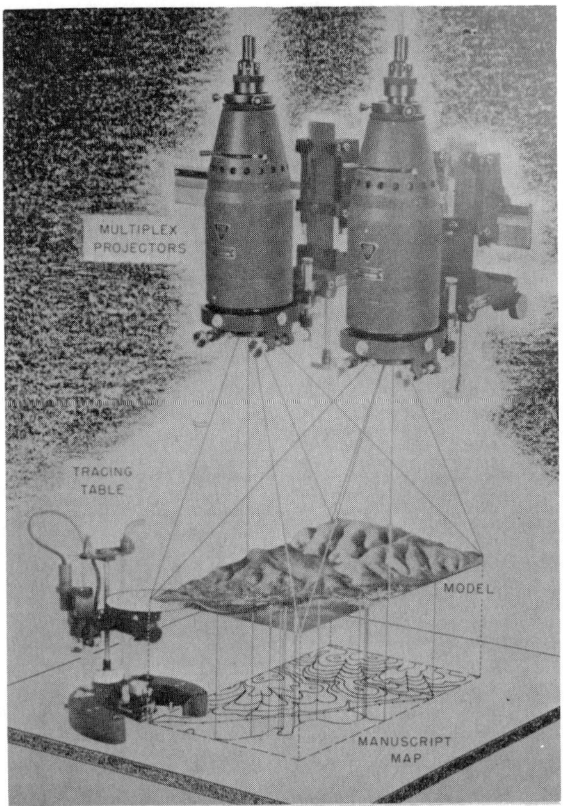

FIG. 2. Double projection plotter.

referred to as cross correlation. Mathematically, the cross correlations can be expressed as follows. The correlation function $\rho_{12}(x_0, y_0)$ is given by

$$\rho_{12}(x_0, y_0) = \frac{1}{A}\int_A T_1(x_1, y)T_2(x + x_0, y + y_0)\, dx\, dy \tag{2.2}$$

where A is the area over which the correlation is being defined, $T_1(x, y)$ and $T_2(x, y)$ represent the intensity transmittances of the two transparencies forming the stereopair, and x_0 and y_0 represent the displacement of the image bound by area A in transparency 2 with respect to its conjugate in transparency 1, in x and y directions. The two-dimensional correlation function $\rho_{12}(x_0, y_0)$ is generally a symmetric figure with its maximum at $x_p = y_0 = 0$, and ideally $\rho_{12}(x_0, y_0)$ decreases monotonically as x_0 and y_0 are increased. When x_0 and y_0 are zero, the conjugate images in the two transparencies are matched, hence, the measurement of the correlation coefficient permits the detection of the conjugate image coincidence. The measurement of the displacement of transparency 2 with respect to transparency 1 to obtain maximum correlation for different sampled areas on the transparencies permits measurement of the parallax for those sample areas.

This correlation process can be accomplished using electronic, optical, or digital correlation systems. In an electronic correlator, the optical transmittances of the two transparencies are converted to electrical signals by the use of a CRT flying-spot scanner. If the axis of the scan coincides with the x axis, the one-dimensional correlation is given by

$$\rho_{12}(t_0) = \tfrac{1}{2}T \int_{-T}^{T} f_1(t) f_2(t + t_0)\, dt \tag{2.3}$$

where f_1 and f_2 are proportional to the intensity transmittance of the two transparencies, t_0 represents the parallax x_0, and $2T$ represents the length of the scan. The basic principle of the electronic correlator is illustrated in Fig. 3. In the digital correlator, the individual transparencies are digitized independently and the correlation is performed in the computer. However, in all the optical correlation systems, the correlation is performed on a two-dimensional space without requiring transfer to an intermediate domain. It is in this context that the optical processing system plays an important role in automated stereo compilation.

B. Optical Correlators

The investigation of the application of optical systems to achieve correlation as an alternative to electronic correlation systems started in early 1960. Along with the progress in optical data processing methods and techniques came the

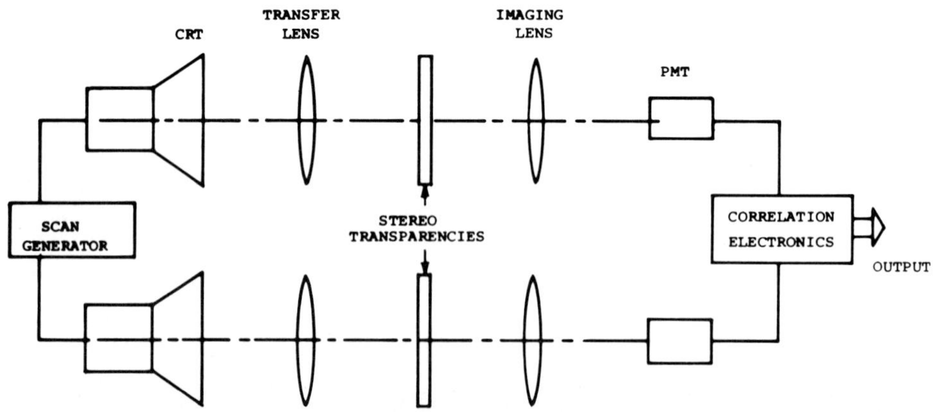

Fig. 3. Principles of electronic correlator.

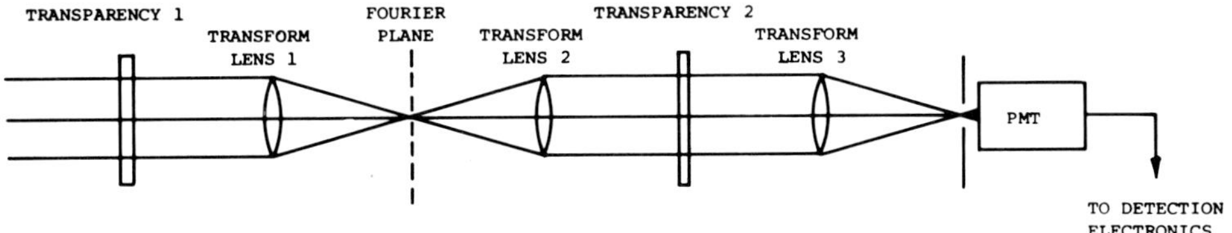

FIG. 4. Image–image correlation system.

multitude of proposals of optical systems to achieve automation of the stereocompilation process. While at the outset the optical schemes appeared to be simple, efficient, and fast, the actual implementation of the optical correlation methods have given rise to many problems and limitations. All of the methods utilize either coherent or incoherent optical correlation to obtain detection of image matching. They can be divided into three broad areas: image-to-image correlation, Fourier plane correlation, and correlation by interferometric methods. In the following pages a brief description of the basic concepts of these different correlation systems is presented and two of the experimental prototype optical correlation systems are described.[1]

1. *Basic Concepts*

The basic optical configuration of a coherent optical image-to-image correlator is shown in Fig. 4. Collimated light from a laser source illuminates a section of the transparency 1. The transform lens 1 produces an amplitude distribution at its back focal plane which is an exact Fourier transform of the transparency 1. A dc block placed on the optical axis at the back focal plane permits removal of the average transmittance of the transparency 1, permitting only the structure information to pass through. The transform lens 2 images the transparency 1 on the transparency 2 with unit magnification (−1), while transform lens 3 produces the Fourier transform of the amplitude distribution emerging from the transparency 2. The correlation signal appears on the optical axis in the Fourier frequency plane of the transform lens 3, and this signal is discriminated from other light distributions through the pinhole. For two identical images (the ideal case), the correlation signal is maximum when two images are exactly superimposed. In practice, however, the conjugate images on the two transparencies are never identical. In electronic correlation systems this problem is solved by properly shaping the scan pattern to generate maximum correlation signal. In an optical system this can be achieved by suitably modifying one transparency using spherical and anamorphic systems before it is imaged on the other.

---

[1] N. Balasubramanian, Coherent optics in photogrammetry, *in* "Topics in Applied Physics," (D. Casasent and H. J. Caulfield, eds.), Vol. 23, Applications of optical data processing, p. 119. Springer-Verlag, Berlin and New York, 1978.

In the image-to-image correlation systems just described, the parallax measurement was made by selection of a section of imagery in one transparency and detection of the location of the conjugate imagery on the second stereotransparency by physical translation. The correlation signal provided the means of detection. In the frequency plane (or Fourier plane) correlation systems, as the name suggests, the correlation detection is performed in the Fourier transform plane. The basic optical configuration describing the concept of frequency-plane optical correlation is illustrated in Fig. 5. First, a matched filter corresponding to transparency 1 is made as shown in Fig. 5a. The transparency 1 is then replaced by transparency 2 and is selectively illuminated section by section as shown in Fig. 5b. The x and y coordinates of the output signal then directly represent the x and y parallax of that part of the imagery on transparency 2 with respect to the conjugate imagery on transparency 1. The focal spot at the correlation plane due to the original reference beam used for making the matched filter provides a convenient origin against which the

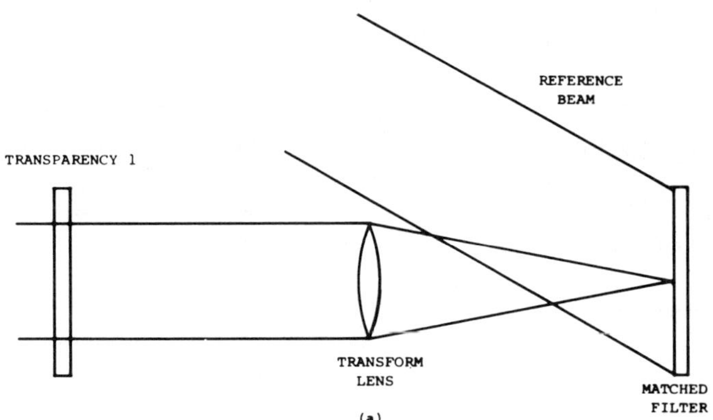

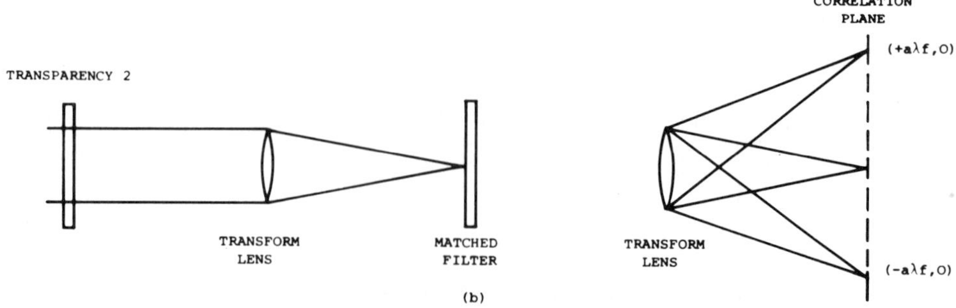

FIG. 5. Frequency plane correlation system.

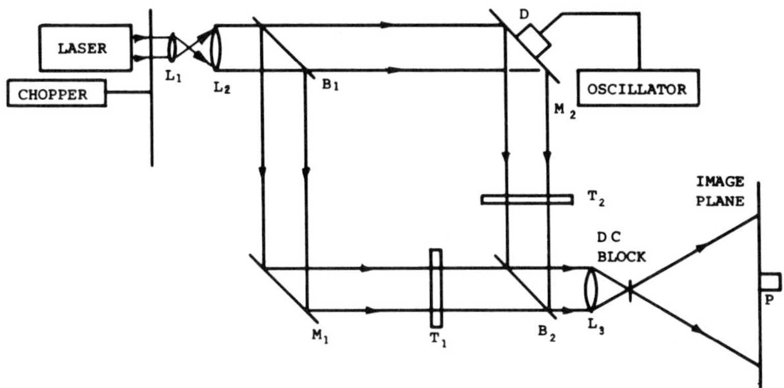

FIG. 6. Interferometric correlator; where M_1, M_2 are mirrors, B_1, B_2 are beam splitters, L_1, L_2 are beam expanding optics, T_1, T_2 are transparencies, L_3 the imaging lens, P the photodetector, and D the piezoelectric drive.

displacement of the correlation signal can be measured. An experimental prototype system based on this concept is discussed in detail later in this chapter.

In all the optical correlation systems described so far, the basic requirement to achieve correlation detection has been the ability to perform the multiplication of the amplitude transmittance of the two transparencies and integrate the product over the area of interest. Consider now the coherent addition of the two amplitude transmittances $t_1(x, y)$ and $t_2(x, y)$. The superimposed intensity distribution is given by

$$I(x, y) = |t_1(x, y)|^2 + |t_2(x, y)|^2 + 2t_1(x, y)t_2(x, y)\cos\Delta(x, y) \tag{2.4}$$

where $\Delta(x, y)$ is the phase difference between the two coherent beams at the image plane (Fig. 6). The independent detection of the third term containing the product of the amplitude transmittances of the two transparencies forms the basis of interferometric correlation. The measurement of the average fringe modulation carried by the cosine term over the correlation area permits independent detection of the third term as the correlation signal. An experimental prototype system based on the extension of the interferometric correlation concept is described in detail later in this chapter.

Over the years, many systems based on modification of the fundamental concept presented here have been theoretically proposed and experimentally demonstrated. However, none of them has gone beyond the laboratory demonstration stage. It is beyond the scope of this chapter to cover them all, and the reader is referred elsewhere for the details.[2]

---

[2] N. Balasubramanian and V. Bennett, Investigation of Techniques to Generate Contours from Stereopairs. U.S. Army Engineer Topographic Laboratories, Final Rep. ETL-0029 (1975).

2. *Experimental Prototype Systems*

While many different investigators have demonstrated, within the basic research environment, the potential applications of the optical correlation to the stereocompilation process, the real-world applications of the concepts are yet to be realized. Two different systems have been constructed for test and evaluation. They are:

(a) IMF (image matched filters) correlator, and
(b) HOC (heterodyne optical correlator).

The IMF correlator system is based on frequency plane correlation and the HOC system is based on interferometric correlation. In this section, brief reviews of the two systems are presented along with a discussion of the results obtained using them.

3. *Image Matched Filter Correlator*

The IMF system is a coherent optical processor designed for the purposes of parallax measurement in stereoscopic aerial photographs. The concept is based on frequency-plane correlation and the detailed discussions relating

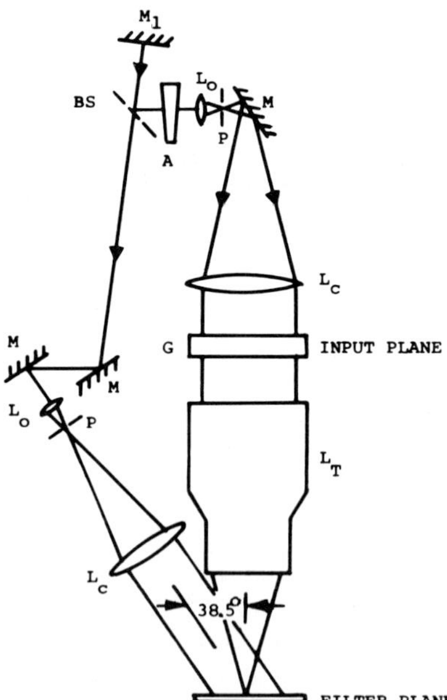

FIG. 7. IMF matched filter generation system; where M_1 is the beam routing mirror, *BS* the beam splitter, *A* the adjustable attenuator, L_O the microscope objective, *M* the mirror, *G* the liquid gate, L_C the collimating lens, *P* the pinhole, and L_T the Fourier transform lens.

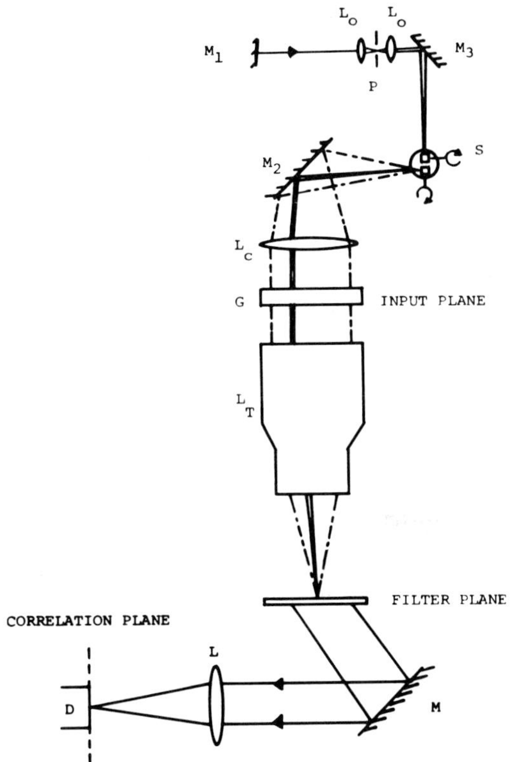

FIG. 8. IMF parallax measurement system; where the symbols are the same as Fig. 7 and S is the galvonometer scanning unit, M_2 the scanning mode mirror, L the correlation lens, and D the correlation detector.

to the system can be found in Rotz.[3] The optical configurations for the IMF system are illustrated in Figs. 7 and 8. The optical arrangement used to generate the matched filter is shown in Fig. 7 and the parallax measurement system using the matched filter generated is shown in Fig. 8. The input transparencies are held in a liquid gate and the instrument has an aperture of 100 mm × 125 mm. The Fourier transform lens has a focal length of 617 mm and a clear aperture of 217 mm. Variable attenuators placed in the unexpanded regions of the laser beam allow the adjustment of the ratio of reference- to signal-beam intensity for optimum filter modulation in spatial frequencies of importance.

For parallax measurement, all of the laser power is directed into a telecentric scanner consisting of a folded telescope and two mirror galvanometers whose axes are at right angles. Beam-forming optics focus a narrow cone of light into the first galvanometer mirror, which is in turn imaged by the telescope

---

[3] F. B. Rotz, Parallax measurement using coherent optical processing, *Proc. Int. Opt. Comput. Conf., 3rd* IEEE Cat. 75-CH0941-5C, p. 162 (1975).

onto the second galvanometer mirror. This device produces a narrow, scanning cone of light which appears to originate at a point on the second galvanometer mirror. A mirror, which is not present during filter generation, directs the scanning beam into the signal-beam collimator. The scanner and mirror are positioned so that the scanning beam appears to emanate from the same point as the signal beam did during the filter-making operation.

The scanning beam is driven to any point in the input gate by the signals from a pair of D/A converters coupled to a minicomputer. This computer performs the necessary corrections so that the operator can specify the scanning-beam location directly in photocoordinates. It is also used to automatically step through a sequence of positions.

The correlation plane is formed by a 500 mm $f/4$ lens; the system is folded by a mirror for compactness. The correlation peak is observed by means of a small vidicon camera mounted on a horizontal and vertical translation device. A pair of electronic crosshairs are superimposed on the television display. In order to measure the location of a correlation peak, the camera is translated until the correlation peak falls exactly under the crosshairs. The position of the camera is sensed by a pair of linear potentiometers. This method, while somewhat awkward, was expedient in this circumstance and avoided problems of calibrating the vidicon scan. During the test and evaluation of this system, care was taken to optimize the filter exposure to emphasize the spatial frequency band of importance. The lowest frequency of the band was determined by the aperture of the scanning beam, and the highest frequency was determined by the resolution of the imagery used. The input photograph was preprocessed with a high-gamma copy process to enhance the edge contrast present in the imagery. The evaluation of the parallax measurement process involved the use of Canadian photogrammetric test model, scale 1:16,000 for which the ground control and elevation data were available. The parallax measurements were made along several scan lines orthogonal to the stereobase. The scans, 100-mm long, were made with a 1-mm beam in 0.5-mm steps. The parallax measurements were made manually using the vidicon camera and the electronic crosshairs. Using control points in the model, the measured parallax data was converted from the photocoordinates to the ground coordinates so that an evaluation of the data can be made. The comparative results of the data obtained using the IMF correlator and the AS-11B automatic electronic stereocompilation instrument are illustrated in Fig. 9. A least-mean-square adjustment of the data permitted the removal of the rotational and translational errors in initial relative orientation of the transparencies. Considering the breadboard nature of the experimental IMF correlator system, the results clearly demonstrate the potential capabilities of this optical correlation system.

4. *Heterodyne Optical Correlation System*

The heterodyne optical correlation system has significant advantages in its application to stereocompilation. In the HOC system the two transparencies

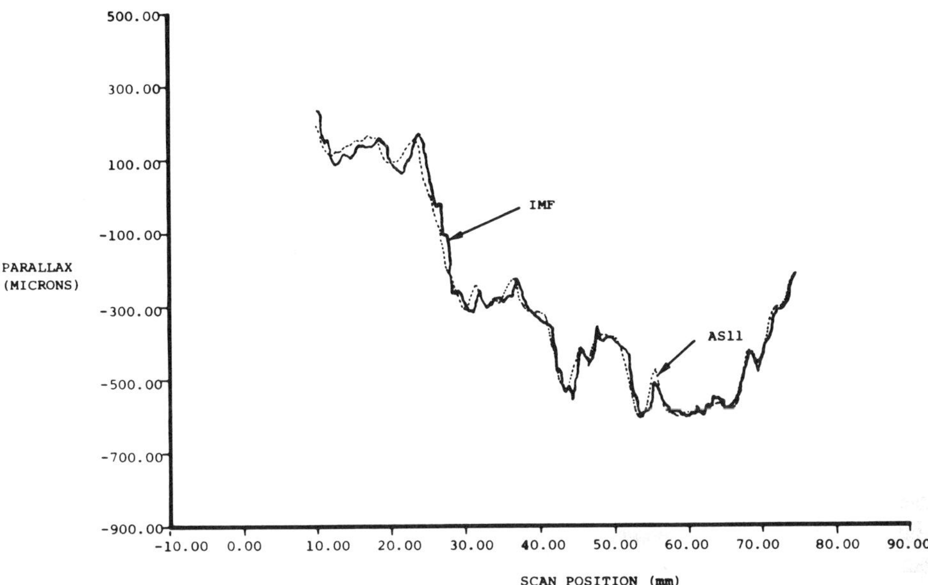

FIG. 9. IMF and AS11 parallax data.

forming the stereopair are relatively oriented and projected onto a common image plane where coincidence is detected. By means of heterodyne optical techniques, a normalized correlation coefficient is measured at each element of a photodiode array in the common image plane. The normalized correlation values are then used to define regions of conjugate image coincidence for a given orientation between transparencies. The principle of operation of the heterodyne optical correlator can be found in Balasubramanian.[4]

Consider a double-projection, direct-viewing photogrammetric plotting instrument (Fig. 10). It represents a simple and direct solution to the problem of forming an analogic, three-dimensional, accurate stereomodel of the earth's surface from two-dimensional aerial photographs exposed as stereopairs. The basic components of this system include a powerful illumination source, a precise projection system, a discriminating viewing system for accurate observation of dual projected images, and a system for precise measurement and delineation of the images. To form the stereomodel in the system, it is necessary to construct the same perspective relationship between the pair of transparencies in the projectors as that existing in the aerial cameras at the times of exposure. This process of relative orientation is followed by mutual adjustment of the projectors to bring the model to the correct scale and relationship to a datum

---

[4] N. Balasubramanian, Experimental Heterodyne Optical Correlator. U.S. Army Engineer Topographic Laboratories, Final Rep. ETL-0071 (1976).

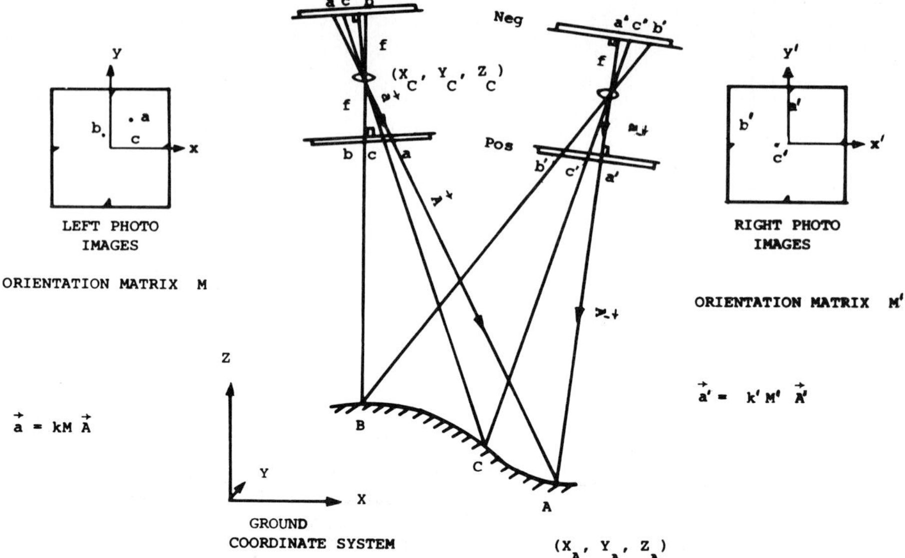

FIG. 10. Double projection plotter configuration.

plane. Measurement within the projected stereomodel usually involves determining coincidence or zero parallax for a given point in the model by placing a floating mark in contact with a point on the apparent surface of the model as viewed on the platen. The horizontal and vertical position of the point can then be measured at the scale of the projected model.

Now consider the conceptual implementation of the basic HOC as illustrated in Fig. 11 to be analogous to the double-projection direct-viewing photogrammetric plotting system discussed above. The basic components of this system include a powerful illumination source (the laser), a precise projection system consisting of two image projectors, a discriminating sensing system for accurate optoelectronic viewing of selected aspects of the dual projected images, and an electronic photodiode array detector serving as a precise measurement system. A stereomodel is not formed in the HOC, however. Each projector system serves as a rectifier to correct for scale, tip, and tilt, and the combined rectified images from both the projectors are superimposed in the image plane. Relative orientation of the transparencies in the projectors results in the removal of y parallax. Then x parallax associated with each point in the image plane is determined by measuring the relative x displacement necessary to obtain the maximum value of the normalized correlation coefficients. The x parallax values can then be converted to represent terrain elevations.

In order that the two images be mutually coherent with each other, the two projectors are illuminated in an optical system configured as a Mach–Zehnder

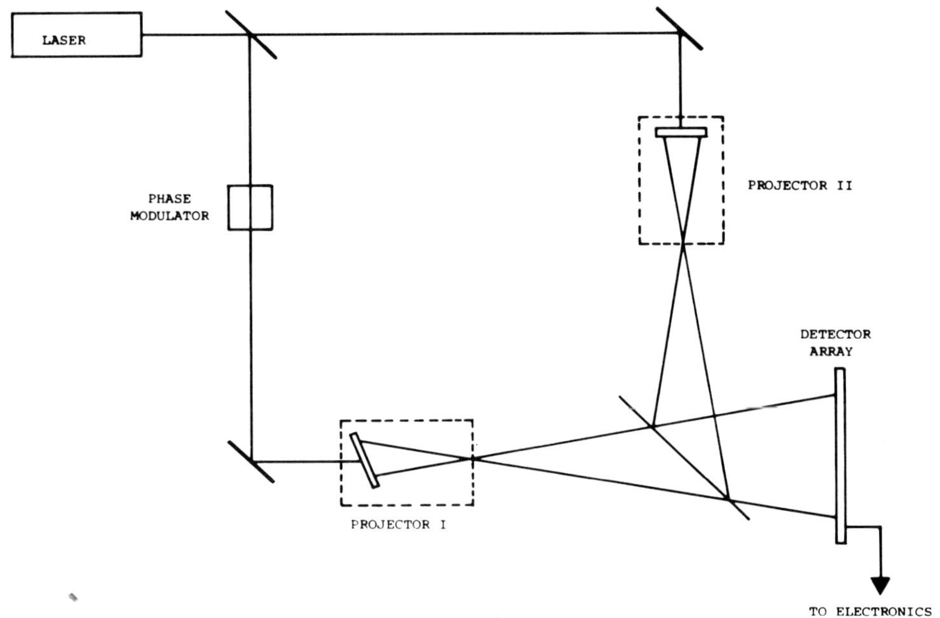

FIG. 11. Conceptual design of HOC.

interferometer. The effect of this arrangement, when the optical system is slightly detuned, is to create a stationary fringe modulation of the superimposed images. The fringe spacing should be larger than the detector aperture to make $\Delta_0(x, y)$ constant over the aperture so that at points where the correlation is maximum, the fringe modulation has a maximum value. Measurement of this fringe modulation is accomplished in the HOC by translating a mirror in the Mach–Zehnder interferometer, as shown in Fig. 11, and simultaneously measuring the time-varying intensity function in the image plane. Detection of the modulated signal at any point in the superimposed image plane forms the basis for the heterodyne optical correlator, and effectively increases the signal-to-noise ratio in detection by several orders of magnitude.

Changes in the normalized correlation value are caused mainly by the image structure associated with the two photographs, whereas the average transmittance has a constant bias level contribution to the correlation value. The ability to locate the maximum value of the normalized correlation coefficient can be enhanced by removing this bias level contribution. In the electronic correlation systems this is accomplished by dc filtering of the video signals. In the HOC, the bias level is removed by placing a small, nontransmitting light block on the optical axis of the two projection systems to remove the undiffracted light from the transparencies. Then diffracted light, representing the image structure, is allowed to pass through to the final image plane.

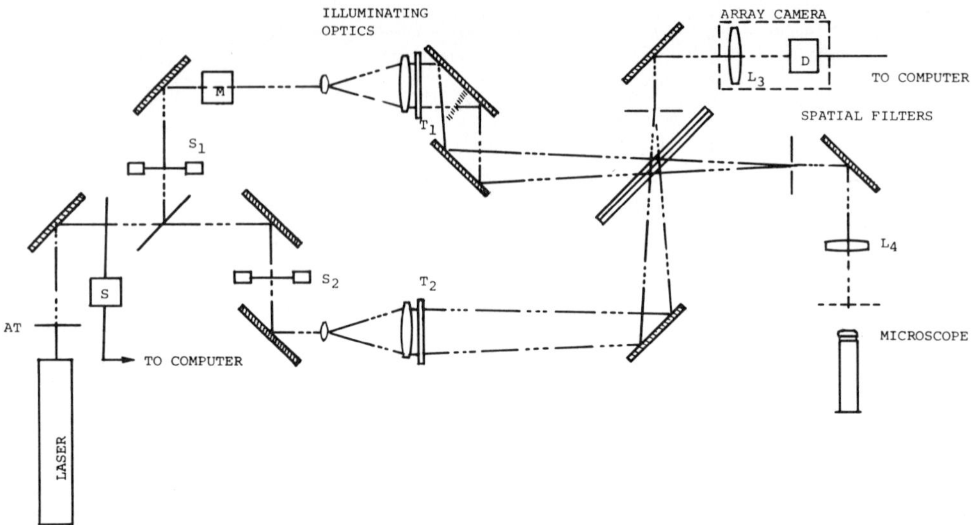

FIG. 12. Optical configuration of experimental HOC; where S_1, S_2 are shutters, T_1, T_2 are transparencies, M the phase modulator, D the detector array, L_3, L_4 the imaging lenses, S the solenoid controlled ND filter, and AT the linear wedge ND filter.

The optical configuration of this experimental HOC system is schematically shown in Fig. 12 and a picture of the system is shown in Fig. 13. The basic optical configuration still remains that of a two beam Mach–Zehnder interferometer; however, the two-projector optical system has been replaced by a single imaging system. This change has been necessitated by the relative sizes of the illuminated transparencies and the detector array. The modification, however, has resulted in simplicity and greater geometric accuracy during measurement.

Light from the laser (Spectra Physics Model 125, power 50 mW) is directed into the two channels of the interferometer with the 50–50 dielectric beam splitter. A linear wedge neutral-density filter permits attenuation of the laser power before it enters the interferometer. Electronically operated shutters S_1 and S_2 permit selective illumination of the two transparencies. The modulator M is used to modulate the optical path difference between the two channels of the interferometer. The modulator consists of a 5 × 5 × 1 mm glass plate attached to the spindle of a galvoscanner (General Scanning G-100). The oscillation of the glass plate in the path of the beam results in the time modulation of the optical path difference between the two beams.

The illuminating optics in each channel of the interferometer consists of a beam diverger and a condenser. The standard microscope objective (20×) and a spatial filter placed at its front focal point make up the diverging optics. The condenser consists of a combination of two collimating objectives mounted

FIG. 13. Experimental HOC system.

back to back so as to minimize wave front distortion. The objectives have focal lengths of 220 and 540 mm, and both lenses have a clear aperture diameter of 50 mm. The two transparencies T_1 and T_2 are mounted on translation stages and are placed in the converging beam, as close to the condenser optics as possible to obtain maximum coverage of the area of illumination at the plane of the transparency. The circular illumination area obtained in the experimental HOC has a diameter of approximately 40 mm.

The final imaging lens L_3 (EL Nikkor Projection Lens, 85 mm $f/2.8$) produces superimposed images of the transparencies at the common image plane occupied by the detector array. The beam splitter, acting as a tilted plane parallel glass plate, introduces astigmatic aberration in the imaging process. However, since the beam splitter is located at the long conjugate side of the imaging lens (the f-number of the cone of rays being large), the aberration introduced is negligible. The beam splitter is formed by cementing together two identical plane parallel glass plates with the beam-splitting coating at the cemented interface. This symmetric nature of the final beam splitter also assures that the aberrations have the same effect on both images. Thus, the aberrations produced have little effect on the correlation value.

The dc block is a circular opaque spot (0.25-mm diameter) formed with India ink on a microscope slide, and is placed at the point of convergence of the illuminating beam. The dc block is located between the beam splitters and the imaging lens L_3 and hence is common to both images. Because of this fact, once again the small aberration contribution of the microscope slide has no effect on the correlation value. It is to be emphasized here that the minimum number of optical components has been introduced between the transparencies and the final imaging lens. This simplicity in optical configuration not only enhances the resolution performance of the final imaging optics, but also significantly reduces geometric distortions during imaging. Hence, it is possible to maintain the geometric fidelity of the transparencies during the measurement process.

The final beam splitter has two output beam paths that are normal to each other. One of the beam paths, as described earlier, is directed to the array detector. Lens L_4 (EL Nikkor Projection Lens 85 mm $f/2.8$), placed symmetrical to lens L_3 in the path of the other beam, also produces superimposed images of the transparencies in the common image plane. The common image is examined under white light illumination with a $100\times$ microscope mounted on a three-axis stage. The visual examination of the common image plane permits manual relative orientation of the two transparencies.

Because of the extreme geometric fidelity of the detector array, each elemental area defines a precise location on the transparency. By using a grid plate for initialization, it is possible to relate precisely the coordinate system of the array elements to that of the photo stages. The procedure for compiling a stereomodel can be listed as follows: with the use of the microscope, the two transparencies representing the model are relatively oriented, and the x parallax direction is made to correspond to the x axis of the translational stages. During the relative orientation procedure, the two transparencies are adjusted essentially for tilt, tip, and scale. Also, the relative orientation of the array on the photocoordinate system is determined. The instrument is adjusted for the coincidence of the control imagery representing the datum. The x parallax is introduced in incremental steps, and the normalized correlation value at each element of the array for each step is determined. The resultant data can be represented in a three-dimensional form shown in Fig. 14. It is clear, then, that by examining the histogram of the normalized correlation coefficient as a function of x parallax steps, it is possible to define uniquely the x parallax associated with any given element of the array. The final output of the processor is a two-dimensional matrix of x parallax values representing two-dimensional elemental areas defined by the array (Fig. 15). It is to be noted here that these x parallax values relate to the photocoordinate system of the stationary transparency, but they can easily be transformed to elevation in the ground coordinate system. The effects of residual y parallax and perspective slope (within limits) are to reduce the maximum value of the normalized correlation value; however, these do not introduce errors in the x parallax measurement.

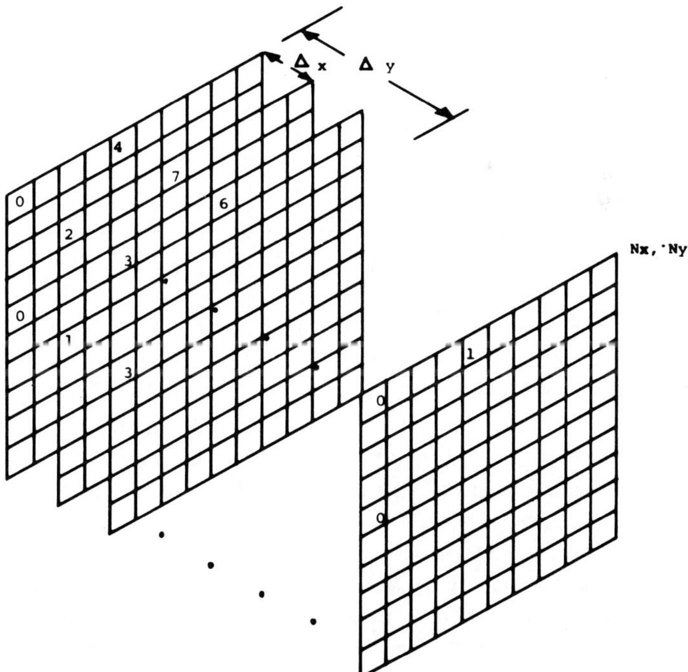

FIG. 14. Histogram of HOC data.

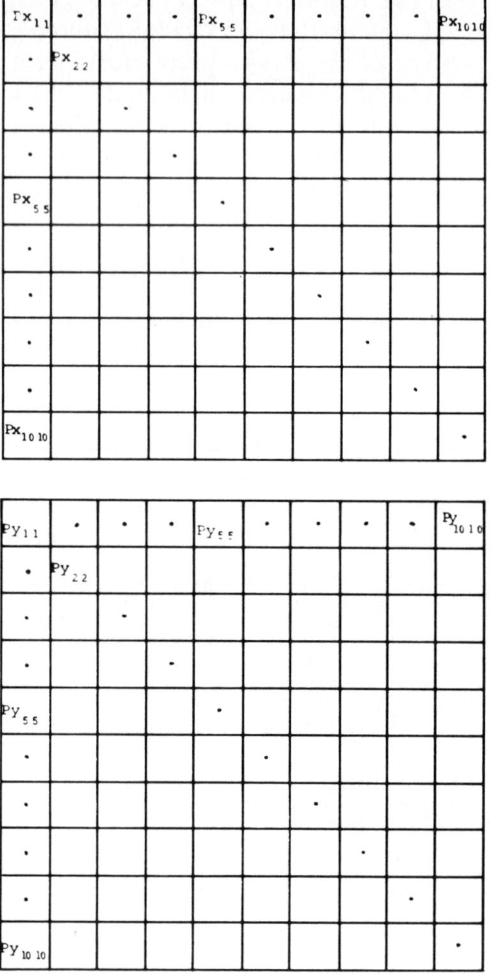

FIG. 15. HOC output matrices.

The effects of y parallax can be minimized by extending histogram analysis procedure to include several y parallax steps. This data reduction approach in HOC not only enables one to overcome operational limitations such as slope effects and residual y parallax, but permits potentially high speed operation, compared to electronic correlators.

For test and evaluation of the experimental HOC system, the Phoenix test range model 134/135 (scale 1:47,000) (Fig. 16) was used. On the reference photo 135, 25 image points within the area covered by the detector array were marked and their coordinates measured using a precision stereo comparator. Table I

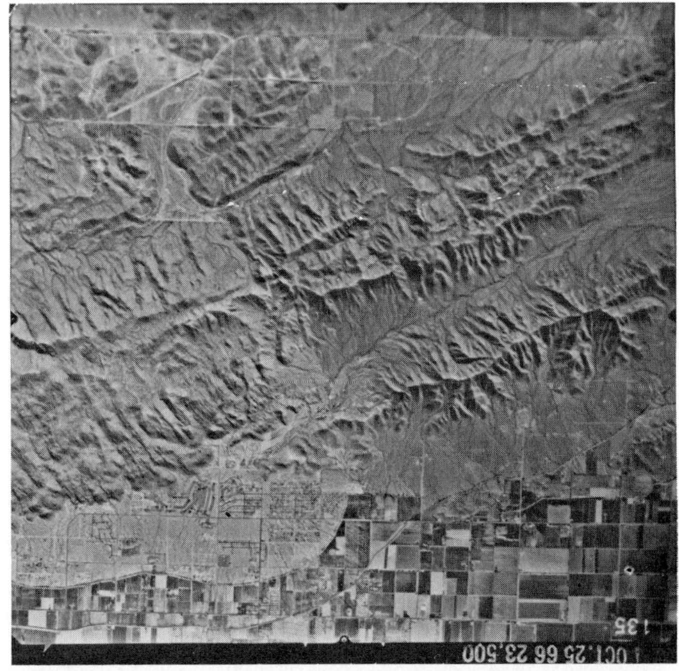

FIG. 16. Stereopair, Phoenix Test Range, 134/135 model.

TABLE I
HOC TEST RESULTS—MODEL 134/135

Point	Computed parallax	Measured parallax
1	0.459	0.457
2	0.472	0.482
3	0.495	0.482
4	0.487	0.495
5	0.579	0.596
6	0.597	0.584
7	0.575	0.584
8	0.529	—[a]
9	0.660	—[a]
10	0.470	0.470
11	0.693	0.686
12	0.469	0.457
13	0.484	0.457
14	0.455	0.370
15	0.432	0.381
16	0.453	0.444
17	0.831	0.609
18	1.441	1.420
19	1.415	1.420
20	0.634	0.571
21	0.463	0.431
22	0.464	—[a]
23	0.473	—[a]
24	0.475	0.431
25	0.569	0.558

[a] Points not in the field of view.

illustrates the comparison between the HOC parallax measurement values and the computed parallax measurement values. Figure 17 shows the contour representation of the HOC output for the same area. The test results indicate that the accuracy of parallax measurement is about 25 to 50 μm without any adjustment to the raw HOC data. Once again the test results have shown the potential of an optical correlator using real-world imagery.

C. SUMMARY

In this section a review of the applications of coherent optical correlation techniques to the measurement of parallax from stereophotographs has been presented. Two experimental systems are described in detail and their test and evaluation show the potential of coherent optical techniques in stereocompilation. Based on these results, it is reasonable to expect the routine use of coherent optical processors at least for specific tasks in the not too distant future.

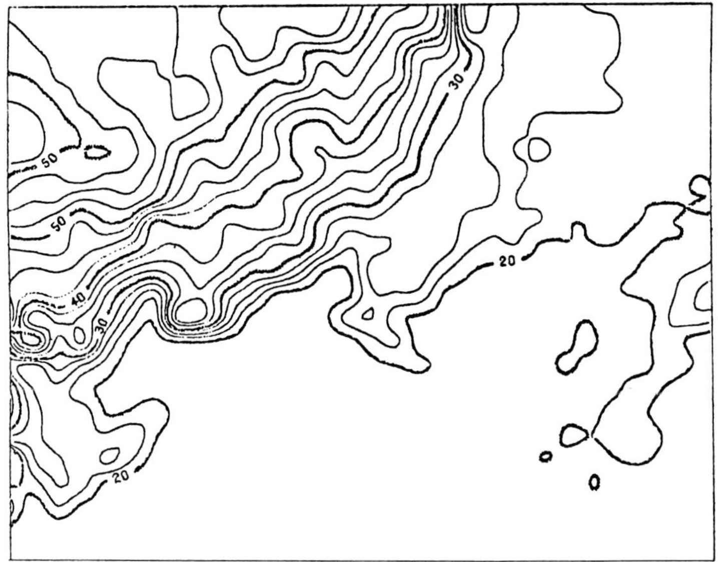

FIG. 17. Contour representation of HOC output data.

III. SYNTHETIC APERTURE RADAR SIGNAL PROCESSOR

A. INTRODUCTION

One of the most practical and successful implementations of optical data processing technology has been the development of the optical signal processor that is used to generate high-resolution terrain maps from the data film of a synthetic aperture radar. It is beyond the scope of this section either to provide a detailed description of the concept of the synthetic aperture radar or to discuss in detail its performance characteristics. It is the intent of this section to briefly review the application of optical data processing techniques to processing the synthetic aperture radar data.[5,6]

B. BASIC CONCEPTS

The synthetic aperture radar systems, often referred to as coherent side-looking radar, permit fine-resolution radar images to be generated at long ranges by the use of signal processing techniques. In the case of a side-looking airborne

[5] L. J. Cutrona *et al.*, On the application of coherent optical processing techniques to synthetic aperture radar. *Proc. IEEE* **54**, 1026 (1966).

[6] J. W. Goodman, "Introduction to Fourier Optics." McGraw-Hill, New York, 1968.

radar (SLAR) that is used to generate an image of the microwave reflectivity distribution of terrain strip that parallels the flight line of the aircraft, the azimuthal resolution obtainable at range R is limited by the linear extent of the antenna. This problem is overcome in the synthetic aperture radar (SAR) systems by creating an effective large extent antenna by proper signal recording and processing techniques. In the SAR systems radar pulses are transmitted from a sequence of positions along the flight path of the aircraft. The amplitude and phase of the radar returns are recorded at these positions and the recorded data is regarded as the signal obtained from a single element of a large array. By properly combining the amplitude and phase information, an effective aperture that is several hundred times longer than the antenna used can be synthesized.

In order to understand the signal-processing requirements and the application of coherent optical processing to such a task, it is necessary to review briefly the signal collection and recording process. There is a one-to-one analogy, between the SAR system and optical holography. In holography, the amplitude and phase of the scattered radiation from a coherently illuminated scene or object is recorded on a photographic film as an interference pattern resulting from interference with a reference beam. When the film is reilluminated after processing, the interference patterns reconstruct the original set of scattered waves to form the three-dimensional image of the original scene. In the SAR system the scattered radiation received is combined with a train of reference signals to form interference patterns that are recorded on data film.

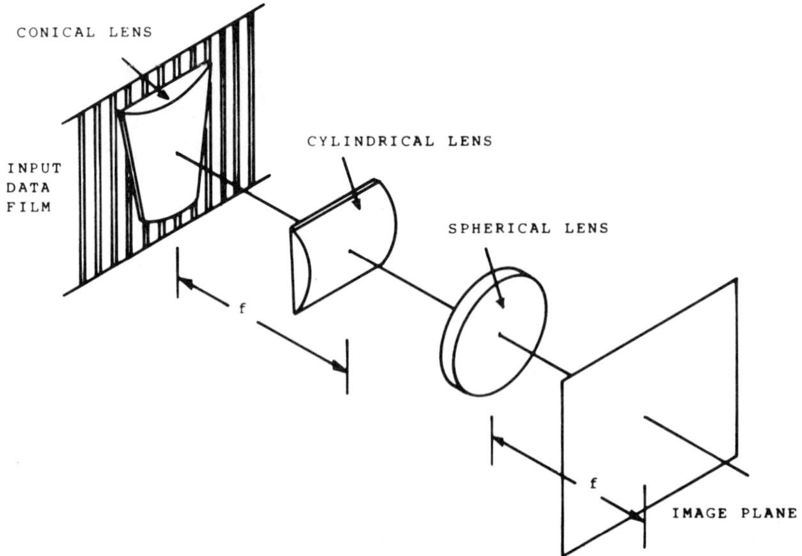

FIG. 18. Basic concept of radar signal processors.

The scattered radiation from each scatterer on the terrain is recorded as a narrow broken line parallel to the edge of the data film. The range of the scatterer is recorded as the distance of this line from the near edge of the data film.

When the data film is illuminated by a beam of coherent light, the pattern along each line gives rise to a diffraction pattern characteristic of scattered radiation from the original scatterer. The diffracted waves can be brought to focus to generate the microwave reflectivity distribution of the terrain strip that was originally illuminated. However, the focusing has to be accomplished only along the data-line direction since the scatterers are already resolved along the range direction. This problem is corrected using a cylindrical lens. However, optics is also required that correct for the range-dependent quadratic phase so that the final distribution can be obtained on a plane parallel to the data film. A conical lens is inserted immediately following the input film and its focal length depends linearly on the range coordinate. Thus, a cylindrical lens and a conical lens in combination achieve this complex data processing operation. The basic principle illustrating the optical reconstruction system is shown in Fig. 18. The spherical lens is only used to form the real image on the final output plane.

C. Radar Signal Processors

The basic elements of the coherent optical radar signal processor were first demonstrated in the late 1950s and field tested in the early 1960s.[7] A rapid evolution of the coherent optical processors then occurred resulting in significant modifications, the most important of which was the development of the processor configurations that are referred to as tilted plane and tilted cylinder processors.[8] In this configuration the use of spherical and cylindrical telescopes eliminated the need for the complex and unwieldy conical lens. This processor configuration represents the state of the art, in the sense that it is still being used to process SAR data.

Goodyear Aerospace Corporation developed an improved Universal Radar Signal Processor for the U. S. Army Engineer Topographic Laboratories, and delivered an advanced model to the Defense Mapping Agency Topographic Center for the production of planimetric maps. This coherent optical processor integrates the phase histories of the ground scene from the holographic image (data film) by the tilted plane processing approach to produce radar images at a scale of 1:50,000 and 1:250,000 for map compilation at a rate of an inch per minute. Detailed discussion on the theory and optical system characteristics of the tilted plane processors can be found elsewhere.[8]

---

[7] L. J. Cutrona *et al.*, A high-resolution radar combat surveillance system, *IRE Trans. Milit. Electron.* p. 127 (1961).

[8] A. Kozma *et al.*, Tilted plane optical processor. *Appl. Opt.* **11**, 1766 (1972).

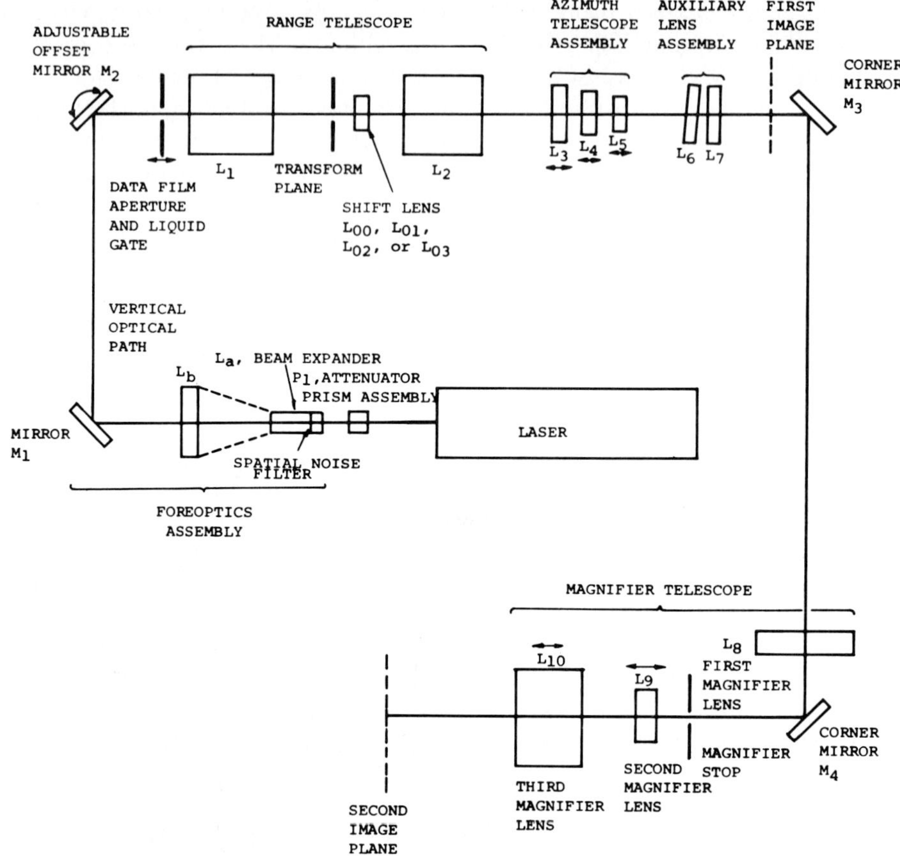

FIG. 19. Base plant coherent optical radar signal processor. Note: Arrows (↔) indicate adjustability and except for "vertical optical path" segment, diagram layout represents an on-top view.

The optical systems of the base plant coherent optical processor are illustrated in Fig. 19. The coherent light source for the system is a Spectra Physics model 125 He–Ne laser with over 50-mW power output. The rotation of a polarization prism in the path of the beam gives control of the output power from the laser. The beam-expanding telescope comprising lenses L_a, the spatial filter pinhole, L_a, and L_b help expand the laser beam diameter to the required size. The main optical system of the processor consists of two afocal or telescope lens systems. The expanded collimated beam is directed at the data film by the offset mirrors. The offset mirror M_2 adjusts the angle at which the collimated beam is incident upon the data film. This mirror provides an offset range of $\pm 4°$ in both range and azimuth directions.

The range telescope consists of two 11-in. diameter spherical lens assemblies which are identified as L_1 and L_2 in Fig. 19. Each lens assembly is composed of five individually corrected elements. The central shift lens of this telescope is powered only in the azimuth direction and is planoconvex with its convex vertex placed at the transform plane.

The azimuth telescope, which consists of three cylindrical lens assemblies (L_3, L_4, L_5), is positioned in the image space of the range telescope. The azimuth image is demagnified by varying separations of these assemblies in accordance with the three-lens telescope equation. Longitudinal positioning of the azimuth telescope, in association with the demagnification, will focus the azimuth image at the image film in coincidence with the range image.

The auxiliary lens assembly, L_6 and L_7, contains two cylindrical lenses. The lens L_6 has an adjustable tilt with respect to the plate on which L_7 is mounted. This lens assembly enables the signal processor to operate in a tilted-lens mode, as compared to a tilted-plane operation. The lenses L_8, L_9, and L_{10}

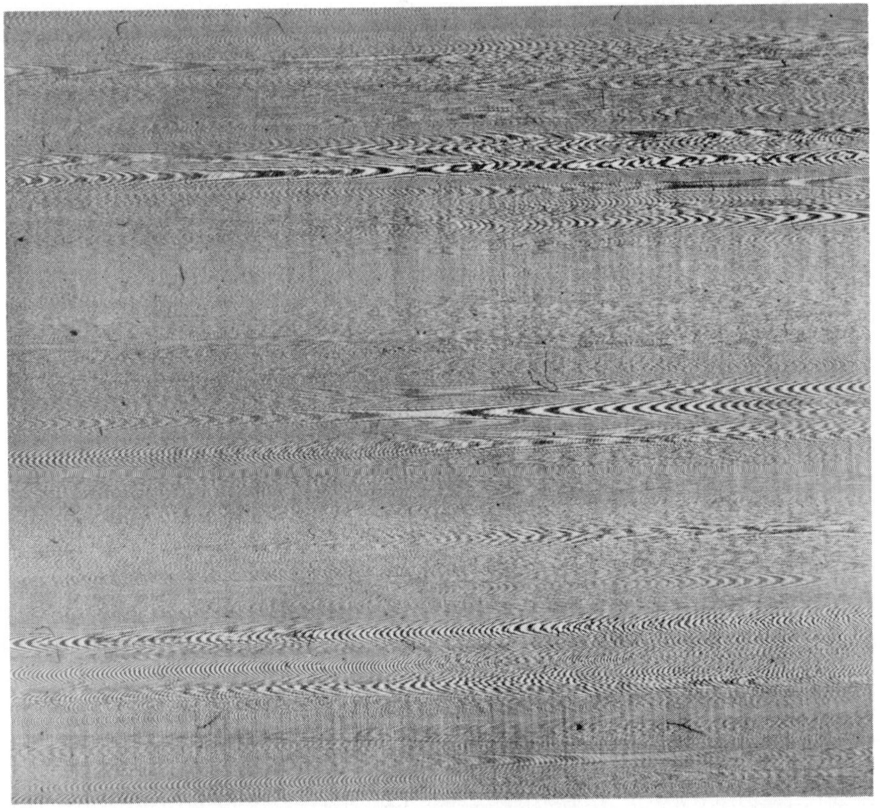

FIG. 20. Section of a data film.

FIG. 21. Reconstructed image; an example.

represent the magnifying telescope. They are a set of three spherical lens assemblies located as shown in Fig. 19. The spacing between lenses L_9 and L_{10} and their distances from L_8 determine the magnification of the telescope. A range of magnifications of $\frac{1}{2}$–3 is used in the base plant coherent optical processor. An example of the data film is shown in Fig. 20 and the radar image produced from it is shown in Fig. 21.

D. DISCUSSION OF RESULTS

In spite of the initial enthusiasm for the application of synthetic aperture radar system topographic mapping, the fervent activity of the late 1960s has slowed down considerably. Several factors have contributed to this decline. For the purposes of map compilation, the radar images have to be rectified and a high level of positional accuracy is needed on the images. The demands on the processor and subsequent image processing systems were severe to meet the required map accuracies. A simple overview of the activity in synthetic aperture radar research will show that while considerable attention has been focused on the construction of images from the radar data film, little has been done either to further exploit the images from a display point of view or to extract quantitative information from them. Also, recent trends in image restitution research are directed more towards digital approaches rather than coherent processor approaches.

IV. FEATURE EXTRACTION

A. Introduction

Feature extraction is defined as the detection, classification, and extraction of useful information from aerial imagery for mapping purposes. Here interest is in automated approaches to feature extraction, as opposed to manual or interactive approaches, wherein the human selects and applies an image transformation to produce a presentation from which he can make a feature extraction decision or a decision to apply further transformations leading to one or more feature extraction decisions. The automated approach deals with closed loop systems that are capable of command, control, and decision making. Optical systems *per se* possess none of these capabilities. Therefore, we must consider hybrid systems which employ optics to provide efficient image transformations while electronic or digital components provide the command, control, and decision-making functions. Since this discussion is mainly concerned with the optical transformations available for useful automated feature extraction, much less attention will be given to the command, control, and decision-making aspects which in many cases are poorly defined and await further research and development.

In this section three basic application areas are outlined where coherent optics can be employed in automated feature extraction systems of the future. These are in the areas of: (a) image assessment, (b) image handling, and (c) cartographic feature extraction. Each of these application areas will be defined along with the potentially useful coherent optical techniques that may be employed. Advanced concepts are shown for integrating optical and digital feature extraction capabilities which will hopefully be of advantage in the solution of many nontrivial feature extraction tasks of the future.

B. Application Areas

1. *Image Assessment*

Proper and efficient utilization of aerial imagery for various mapping-related tasks involves task-related image assessments. This is particularly important when the tasks are to be performed by automated systems that function efficiently only within narrow operational ranges. In this respect, image assessment relates to the sampling and evaluation of aerial imagery for use in selection and scheduling of the most efficient techniques to extract useful mapping information from the aerial images.

Humans are very flexible and are able to accommodate to changes in the quality of input data for mapping operations. Automated electronic and optical systems, on the other hand, cannot be expected to be as flexible. The UNAMACE is an example of an automated programmetric system designed to optically sample aerial stereotransparencies in a manner which allows electronic cross correlation of the sampled signals for determination of stereoparallax that is reducible to a measure of terrain relief in the stereoimage. It so happens that the performance of this system is severely limited by the characteristics of its input imagery in terms of image contrast and spatial frequency content. This will be used as an example of the nature of image assessment.

Proper utilization of the UNAMACE would involve scheduling its operation only to process those images or areas of image in which the attributes are within a range that will produce acceptable results efficiently. Image assessment here would involve measurement and evaluation of the image attributes (modulation and spatial frequency) at a spatial resolution commensurate with the UNAMACE sampling capabilities. Adverse areas of the image could be delineated: (a) to indicate areas for image processing to achieve acceptable performance, (b) to indicate areas prohibited to the UNAMACE for which the data must be extracted by other systems, possibly human, or (c) to indicate areas prohibiting data extraction. For the first class, detection of over- or underexposed image areas can possibly lead to chemical and/or electronic image processing for image improvement. In the second case, detection of natural and cultural terrain features having very low spatial frequency content, such as water bodies, snow-covered areas, some sand areas, etc., will enable their delineation and scheduling for manual data extraction techniques. In the third case, cross correlation is impossible due to terrain obscuration from thick clouds, fog, terrain relief-shadowed areas, or the terrain pattern is spatially similar, such as the periodic patterns of orchards, certain agricultural patterns, or forest canopies.

Another facet of image assessment could involve aspects of automated aerial image cataloging wherein the image quality is evaluated and rated for applicability to future processing tasks such as mensuration and cartographic feature extraction as indicated above. Additionally, automated image cataloging could include mission plotting where the boundaries of each photographic frame are referenced or plotted on a geographic reference base. This leads to a capability for assessing area coverage, for delineation of "holidays" that are unphotographed areas, or for indicating areas containing stereoscopic photography.

2. *Image Handling*

There are a number of image handling tasks associated with automated feature extraction that could lead to useful coherent optical capabilities, but in themselves do not produce information that is directly represented on a topographic map. An important point to bear in mind for feature extraction applied to mapping is that correct recognition and classification are only sufficient when accompanied by references to geographic coordinates within the required positional accuracies.

In automated image processing and feature extraction systems, recognition of photographic frame fiducial marks, annotated image control points, or ground control image points, could serve to position the image correctly and register the various coordinate systems for the processing which follows. The necessity for correct positioning and registration for mensuration tasks is evident from the discussion in previous sections.

Optical feature extraction techniques may be of value in automated selection of image control points for analytical block adjustment or the photogrammetric bridging of ground control points. In the first case, the recognition and identification of a number of ground image features for which geographic positions might be determined allows photogrammetric determination of the imaging camera's position and attitude at the time of exposure. This data for a block of photos can lead to an analytical determination of mapping control necessary in photogrammetric data reduction. In the second case, where ground control is sparsely distributed, intermediate photo control points strategically located can provide the required control to accomplish bridging.

3. *Cartographic Feature Extraction*

Optical cartographic feature extraction tasks are currently being accomplished by trained individuals. These tasks are labor intensive; techniques which reduce time, effort, and, hence, costs, are of the highest priority in the mapping community. There are needs to develop techniques leading to systems which will automatically recognize and classify natural and cultural image patterns for cartographic symbol encoding in map preparation. When one investigates map symbolization he will find only a few symbol classes for natural features associated with vegetation, hydrology, terrain conditions, and cultural features associated with structures, built-up areas, transportation facilities, etc. However, the range and variation of these patterns and their natural backgrounds lead to nontrivial pattern-recognition problems. This is attested to by the lack of progress to date. Automated cartographic feature extraction systems which operate on a large class of general problems will not be realized for many years. However, special-purpose systems that recognize and classify subsets of the required symbolization problems will be of value and lead to the overall capabilities of the future.

C. Basic Optical System Concepts

There are two fundamental approaches to optical feature extraction: optical power spectrum analysis and optical correlation.

1. *Optical Power Spectrum Analysis*

The principles of diffraction pattern phenomena have been known since 1873; however, the background for optical power spectrum analysis began with the processing of seismic data in 1965[9,10] when the laser light source was introduced. The potential for optically processing aerial photography was illustrated by Pincus and Dobrin[11] in 1966. In 1970, Lendaris and Stanley[12] illustrated the utility of optical power spectrum sampling for automatically classifying patterns in photographic transparencies. Since then, there have been a number of papers related to various aspects of the subject.[13–19]

To those familiar with coherent optics, the basic phenomena being investigated may be referred to as the diffraction pattern, the Wiener spectrum, and the intensity spectrum or the optical power spectrum. The latter will be used in this discussion. For the associated pattern-recognition system it is assumed that the optical samples from aerial photographic patterns will have optical power spectra with attributes representative of the patterns which will be electronically detected and classified by digital computer software into the desired information categories. This process will be termed optical power spectrum analysis (OPSA).

There are two commercial OPSA systems: the Recording Optical Spectrum Analyzer (ROSA) that is available from Recognition Systems, Inc., Van Nuys, Calif., and the Diffraction Pattern Analyzer, available from EIKONIX Corporation, Burlington, Mass. The basic components of both systems are a He–Ne light source, a spatial filter, an input plane for transparencies, a transform lens, and a detector with associated electronic signal processing capability. The similarities and differences of these systems will be outlined. Both systems may produce the same OPS data; however, their approaches are entirely different.

---

[9] M. B. Dobrin *et al.*, Velocity and frequency filtering of seismic data using laser light, *Geophysics* **30**, 1144 (1965).

[10] P. L. Jackson, Analysis of variable density seismograms by means of optical diffraction, *Geophysics* **30**, 5 (1965).

[11] H. L. Pincus and M. B. Dobrin, Optical processing of geological data. *J. Geophys. Res.* **71**, 4861 (1966).

[12] G. G. Lendaris and G. L. Stanly, Diffraction pattern sampling for automatic pattern recognition, *Proc. IEEE* **58**, No. 2 (1970).

[13] M. B. Dobrin, Optical processing in earth sciences, *IEEE Spectrum* **5**, (Sept. 1968).

[14] H. L. Pincus, The analysis of remote sensing displays by optical diffraction. *Proc. Int. Symp. Remote Sensing Environ., 6th, Univ. of Michigan* p. 261 (1969).

[15] S. Nyberg *et al.*, Optical processing for pattern properties, *Photogram. Eng.* **37**, 547 (1971).

[16] R. Groce and G. Burton, Techniques for high data rate 2-D optical pattern recognition, *RCA Rev.* **32**, 610 (1971).

[17] M. J. McCullagh and J. C. Davis, Optical analysis of 2-D patterns, *Ann. Assoc. Am. Geog.* **62**, 561 (1972).

[18] K. Tanaka and K. Ozawa, A new type of feature extraction of patterns using coherent optical systems, *Patt. Recog.* **4**, 251 (1972).

[19] N. Jensen, High-speed image analysis techniques, *Photogram. Eng.* **39**, 1321 (1973).

a. ROSA System. A picture of the ROSA system is shown in Fig. 22. Schematically, the ROSA system can be divided into three logical units: the coherent optical subsystem; the electronic detector, processing, and recording subsystem; and the computer processing and output with the FACEL pattern-recognition software developed by Recognition Systems Inc. These subsystems are shown in Fig. 23.

FIG. 22. ROSA system.

In the coherent optical subsystem, the 5-mW He–Ne laser provides a coherent beam which is scaled in amplitude by a variable attenuator prior to filtering with a pinhole spatial filter, and expansion to spherical wave fronts by the objective lens. The collimating lens converts the diverging spherical wave fronts into parallel wave fronts that illuminate one of ten circular sampling apertures ranging in diameter from $\frac{1}{8}$ to 3 in. Light transmitted by the aperture illuminates the aerial transparency and is spatially modulated and diffracted in transmission through the emulsion which varies spatially or patternwise in optical density. The transform lens processes the diffracted and undiffracted light to image the diffraction pattern on the detector in the back focal plane. The segmented detector, shown in Fig. 24, with 32 concentric ring elements and 32 wedge-shaped elements, integrates the intensity distribution of the diffraction-pattern incident on each element separately. (The large dark wedges are scratch blocks to inhibit false detection due to scratches in the photographic emulsion. However, Recognition Systems, Inc. also sells a detector without scratch blocks.) Electrical signals from each detector element are processed through individual amplifiers and then multiplexed to an autoranging amplifier and a binary-coded decimal converter where the data are converted into three significant digits and a decimal characteristic. The encoder appends a six-digit numerical descriptor to the data and formats the output for a set of eight punched cards or

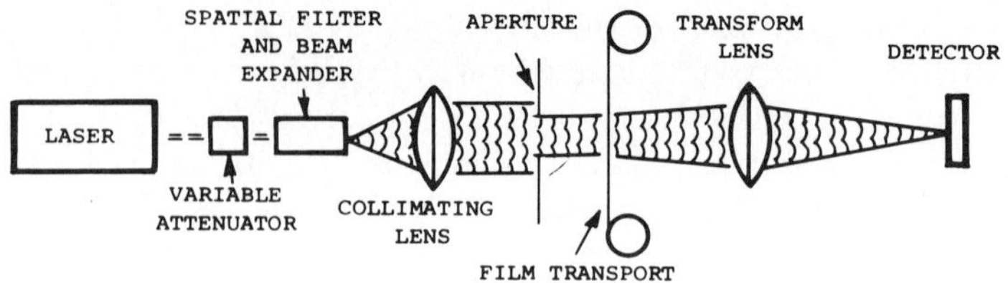

(a) COHERENT OPTICAL SYSTEM CONFIGURATION

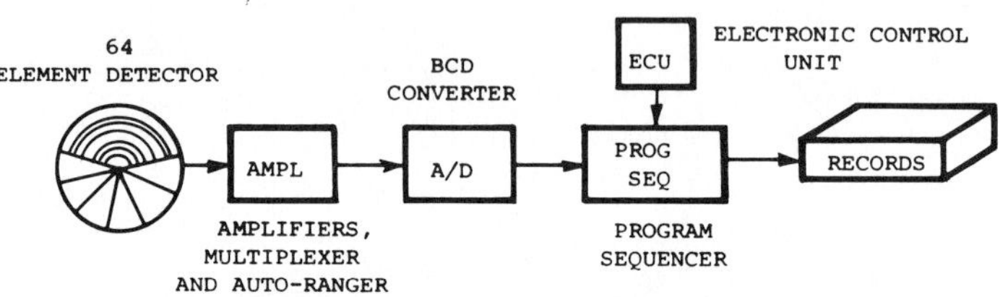

(b) ELECTRONIC PROCESSING AND RECORD SYSTEM

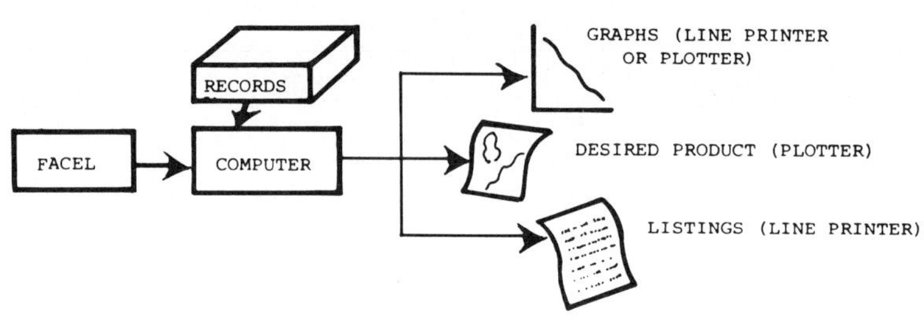

(c) SOFTWARE PROCESSING AND OUTPUT

FIG. 23. ROSA subsystems.

card images recorded on magnetic tape. Leighty and Lukes[20] have analytically characterized the ROSA optical and electronic processing, and the results are summarized in Fig. 25.

Each time a portion of an aerial photographic transparency is sampled, a vector of 64 scalar values is produced along with its descriptor. The FACEL software provides data manipulation and statistical classification subroutines for processing training samples or classifying unknown samples. Each recognition algorithm is partitioned into (a) a learning algorithm that trains on a set of labeled sample vectors and produces a file of decision parameters, and (b) a decision algorithm which applies these parameters to unknown sample

[20] R. D. Leighty and G. E. Lukes, Cloud screening from aerial photography, *Proc. ASP Ann. Meeting, 40th* p. 309 (1974).

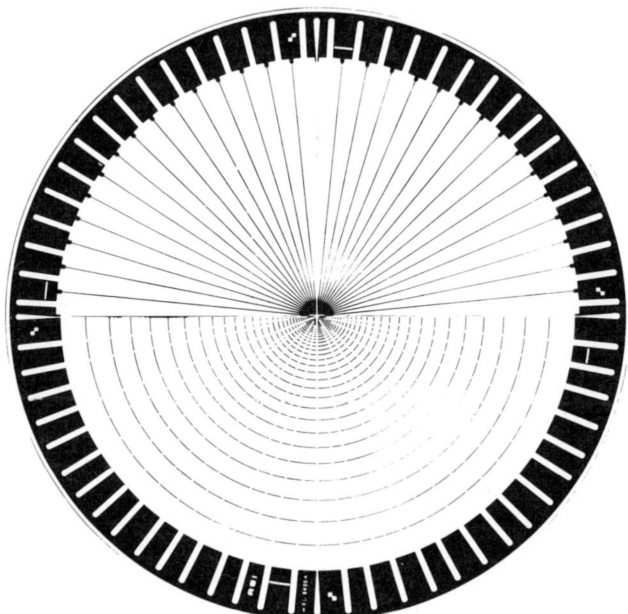

FIG. 24. ROSA segmented detector.

vectors and assigns class labels. Applying statistical pattern-recognition techniques, a set of labeled sample vectors for which the classes are known is used to train a recognition algorithm that can then classify unknown sample vectors. The segmented detector reduces the infinitely dimensional optical power spectrum into a 64-dimensional sample vector, and further reduction of dimensionality is accomplished during the statistical training to reduce core storage and processing requirements on the unknown vectors.

Although any statistical classification algorithm can be incorporated in FACEL, two of the algorithms included in the software are of very general value in feature extraction. One algorithm is a nonsupervised multiclass algorithm which develops multiple hyperplane-separated regions in the learning

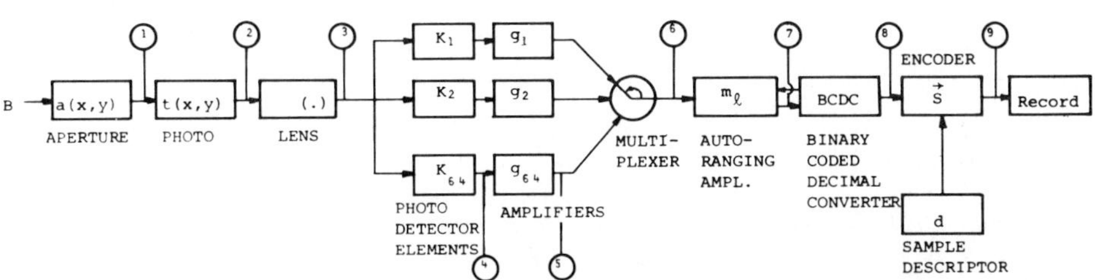

FIG. 25. Analytical characterization of ROSA.

phase and assigns a decision class to each unknown sample based on projections onto these multiple hyperplanes in the decision phase.[21] Another algorithm, described by Kasdan,[22,23] is a nonparametric supervised learning algorithm for multiclass decisions with abstention action that develops decision regions based on statistically equivalent blocks. The decision algorithm assigns a class or abstention action to each unknown sample based on the region in which the sample falls.

b. AOPSA System. The EIKONIX Corporation's Advanced Optical Power Spectrum Analyzer, Model EC-742, hereafter referred to as AOPSA, is shown in Fig. 26. As with the ROSA system, the AOPSA system can be divided into three logical units: the coherent optical subsystem; the detector, processing, and recording subsystem; and the computer processing and output subsystem. Here we want to indicate how the AOPSA differs from the ROSA in its optical and electronic processing of OPS data.

A block diagram of the AOPSA is shown in Fig. 27 and provides the clues to the differences between it and the ROSA system shown in Fig. 23. Firstly, the AOPSA is designed to be used by an operator evaluating film, roll, or chip transparencies on a light table. A zoom microscope with camera is available for detailed inspection of the images, and when an OPS measurement is desired, that portion of the image is moved to the optical axis of the AOPSA sampling unit. Eleven sampling apertures ranging in size from 1.0 to 30.0 mm in diameter are located between the film and the transform lens. To compensate for large variations in beam energy when changing from one image sampling aperture to another, a compensating neutral-density filter is automatically placed in the sampling beam. The subsequent OPS sampling in the AOPSA is more aptly described as a spatial frequency filtering operation, because a set of annular masks are sequentially positioned in the spatial frequency plane, and the energy passing each filter is collected and detected by a photomultiplier (PMT), then amplified and recorded. The output of the PMT is amplified by five cascaded amplifiers that produce an effective dynamic range of at least 10^6 with an rms accuracy of 2×10^{-6}. The AOPSA system monitors laser energy from the 15 mW He–Ne laser to provide feedback to maintain this measurement accuracy.

The AOPSA automatically sequences and positions up to 20 annular masks specified by the operator in a period of 30 sec or less for the OPS measurement

---

[21] R. L. Mattson and J. E. Dammann, A technique for determining and coding subclasses in pattern recognition problems, *IBM J.* **9**, 294 (1965).

[22] H. L. Kasdan, A distribution-free classification procedure with performance monitoring capabilities, *in* "Techniques of Optimization" (A. V. Balakrishnan, ed.), p. 21. Academic Press, New York, 1972.

[23] H. L. Kasdan, Nonparametric feature analysis methods for sampled diffraction patterns, *Proc. SPIE Annu. Tech. Meeting, 16th*, 187 (1972).

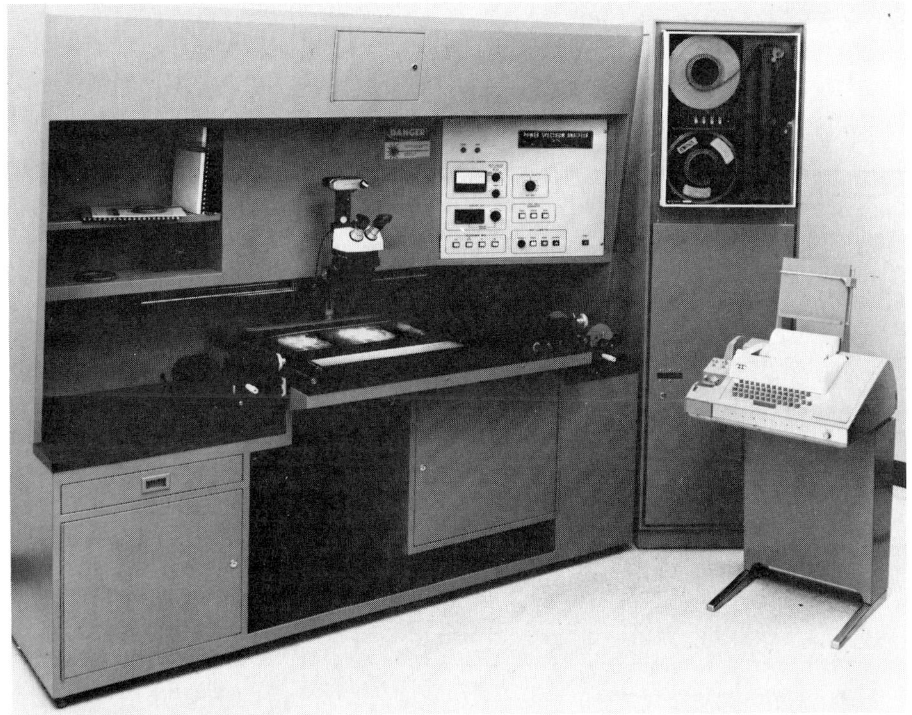

FIG. 26. AOPSA system.

at each sample point in the input image. All the data acquired during the measurement cycle are then stored in the computer memory for analysis by resident algorithms.

It should be noted that one-dimensional OPS distributions can be sampled by the operator positioning a rotatable slit in the spatial frequency plane. The OPS distributions can then be sampled at any desired angle.

c. OPS System Considerations. There are a number of decisions and trade-offs to be made when selecting or assembling an OPS system for feature extraction purposes. The first consideration involves selection of a transform lens that approaches diffraction-limited performance in the band of spatial frequencies of interest. The focal length of this lens, together with the wavelength of the laser light, determine the scaling factor for the diffraction pattern in the spatial frequency plane. When sampling aerial imagery of low resolution, a long focal-length lens might be used, while for high-resolution aerial imagery a short focal-length transform lens may be needed to properly scale the diffraction pattern in the spatial frequency plane.

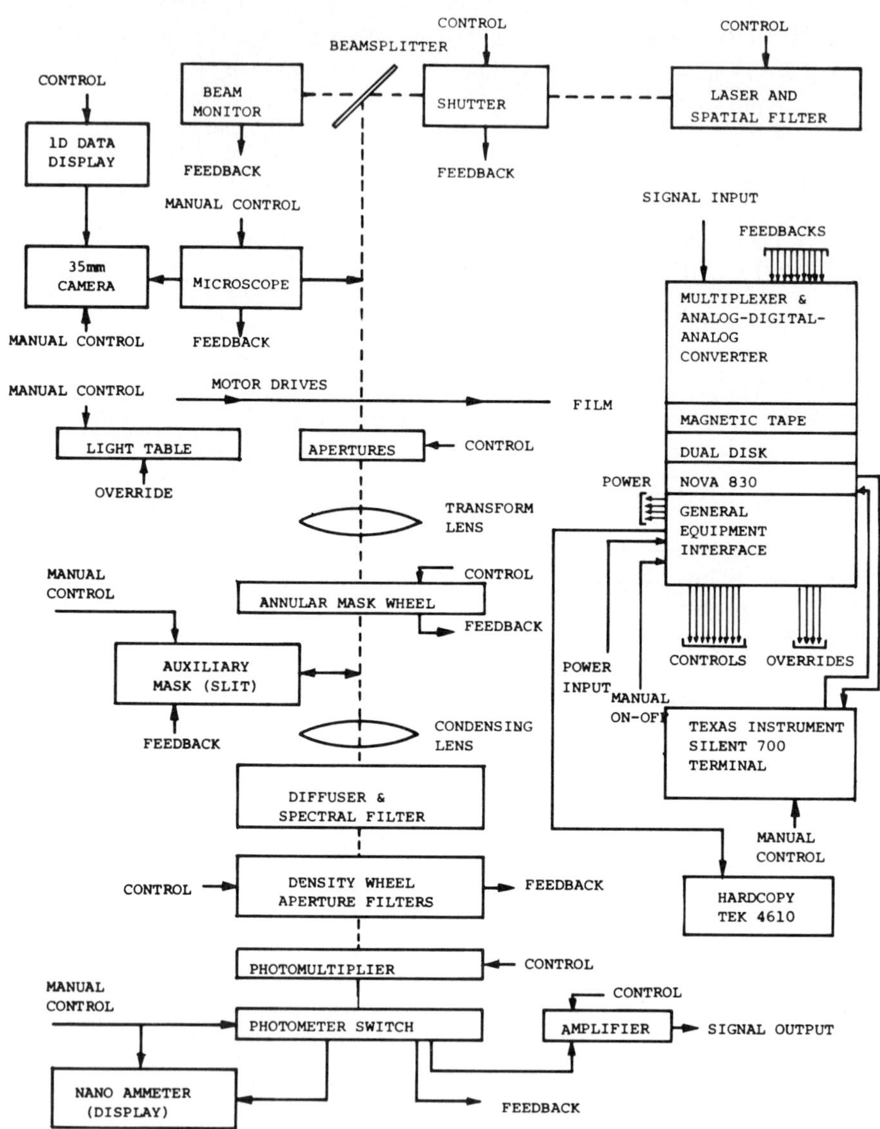

FIG. 27. Block diagram of AOPSA system.

Next, there are sampling aperture-size considerations. From a theoretical point of view, the amplitude transmission function of the aerial transparency is sampled by (multiplied by) the circular sampling aperture in the object plane, and this is represented in the spatial frequency plane as the convolution of the respective Fourier transforms of the aperture and limited-image functions. In a linear system, the transform of object patterns in the sampling aperture are added in the spatial frequency domain. Phase information is lost along with positional information. Thus, as the aperture diameter is made smaller, the convolutional smoothing increases, but the positional information is obtained with more accuracy. Also, with smaller apertures there will be less statistical smoothing due to additive transforms and sample-to-sample variation will be rather great, making recognition difficult. The sampling aperture, as a rule, should not be smaller than the smallest object to be detected.

The ROSA system lends itself to automated sampling by computer-controlled translation of the input transparency, and this leads to considerations for selection of a geometric space sampling strategy in the object plane. The general approach is to select an imaginary two-dimensional rectangular grid overlaying the transparency which calls for OPS samples at the grid intersections. For a given sampling aperture diameter, the grid spacings would then dictate the portions of the transparency sampled. If sampling speed is of major importance, then a coarse sampling grid with large sampling apertures may be indicated, while if object detection, classification, and delineation are of major importance, a relatively small sampling aperture is indicated with a sampling grid which causes some overlap of the adjacent sample spaces.[24]

For the AOPSA system one has some flexibility for selection of the annular apertures for the spatial frequency plane filtering. Depending upon the type of objects to be detected or analyzed in the input transparencies, there may be an optimum set of filter annuli that would make subsequent OPS data analysis simpler.

Statistical algorithms for data analysis can be very complex and require long execution times. No classification algorithm can provide an optimum generalized solution and, therefore, no rigorous guidelines exist for establishing an OPS classification algorithm. The goal in algorithm selection is to obtain the simplest algorithm that separates the information classes of interest. The selection process is accomplished by sampling representative patterns and applying different algorithms until the results are maximized in some sense. Kasdan[23] provides a comprehensive overview of the data handling and feature analysis methods associated with the FACEL software of the ROSA-system, and this discussion can be translated to apply to the AOPSA system as well.

---

[24] G. E. Lukes, Cloud screening from aerial photography applying coherent optical pattern recognition techniques, *Proc. SPIE* **45**, 265 (1974).

d. Advantages and Limitations of OPSA. The prime advantages of the OPS analysis for feature extraction from aerial transparencies lie mainly in the speed of sampling and decision process. In a sense, the 64-element vector for each ROSA sample or the 20-element vector for the AOPSA sample represents considerable bandwidth reduction over the number of digital bits associated with a similar aperture size. The OPS decisions are generally made from a single vector, whereas digital classification requires computing many differences to establish a spatial pattern of value. Other advantages of OPS systems lie with the simplicity of the optical system, and the simplicity of digital control logic and classification algorithms once selected.

Limitations of OPSA for feature extraction pertain to the limited amount of information available in the power spectral domain, the spatial position ambiguity associated with the use of large apertures, and the optical noise from scratches, dust, finger prints, etc. Liquid gating can be employed to reduce the noise somewhat; however, the problems associated with use of a volatile liquid in systems where speed is important usually outweigh the potential advantages.

2. *Image-to-Image Correlation*

The basic concepts of image-to-image correlation related to photogrammetric data reduction have been described in Section II. Here we will have additional comments related to the general problem of feature extraction. Image-to-image correlation techniques are best applied to image cross correlation where the images have a high degree of similarity. Thus, the technique is of value for locating standard symbols in photoimages such as reseau marks or fiducial marks, but difficult to apply to scenes of natural and cultural terrain objects for reasons which will be reviewed presently.

The general procedure in applying image-to-image correlation techniques is first to register the images in the optical subsystem; next, the images are mechanically or optically translated and rotated until a maximum signal is detected. In general, this signal is not normalized and this can lead to detection of false correlation peaks. Application of thresholds in the signal space are usually invalid also, because of the lack of signal normalization. The usual correlation strategies involve a general search for a maximum signal with a coarse or low sensitivity scheme using some *a priori* knowledge about the general vicinity where the maximum is to be expected. This coarse search could involve the use of large sampling apertures and relatively large shifts in position and rotation. When a coarse peak of interest is located, sampling aperture and electronic gain changes can be effected and the search becomes one of small nutations in position, rotation, and scale around the previous found peak. The correlation signal is sampled and a digital interpolation algorithm selects the location of the peak.

As stated, image-to-image correlation is particularly effective when the image to be detected is highly constrained, as in the case of special symbols such as reseau marks. Masks can be made of the symbols and employed in a relatively simple system. To allow all degrees of freedom in terms of position, rotation, and scale for two arbitrary images involves a system of prohibitive complexity. Additionally, natural and cultural terrain scenes of the same areas exhibit considerable differences due to aspect angle, season, sky condition, sun angle, etc.

Interferometric correlation, as applied in the heterodyned optical correlator (HOC), approach to image-to-image correlation and, as discussed in Section II, has not been applied to feature extraction. This approach might provide a very sensitive and fast capability for the fine correlation phase because the normalized correlation could allow efficient search strategies. However, another approach would be required for the coarse correlation phase if the heterodyned correlator could not be desensitized.

3. *Image-to-Matched-Filter Correlation*

Image-to-matched-filter correlator concepts applicable to photogrammetric data reduction were introduced in Section II. In this section we extend these general concepts to feature extraction.

Matched filtering may be performed by using a reference function generated from the pattern of interest and detecting the correlation peaks in the output data to determine the location of matching patterns in the input image. The general procedures employed involve filter preparation, sampling, and detection. As in image-to-image correlators, image-to-matched-filter correlators operate best on input image patterns that are highly constrained. For highly constrained patterns, such as reseau marks, the reference masks can be optimized to obtain a high signal-to-noise ratio (SNR) with a relatively simple optical system. However, for detection of natural and cultural image patterns of interest where there is a high degree of variability, and where image scale, rotation, and aspect angle are also variables, the SNR can be drastically reduced when these parameters have only small differences between the reference and input images.

A definitive study of image-to-matched-filter correlation that will be quoted often in the remainder of this chapter, was prepared by Rotz *et al.*[25] while with Radiation, Inc. (now Harris Corporation). Figure 28, from that reference, shows the cross-correlation SNR versus rotational difference in degrees between the matched filter and an aerial image of a rural landscape in the input. The SNR is reduced by 13 dB with a rotation of only 1° in this illustration, while

[25] F. B. Rotz *et al.*, Mission Review Screener, RADC-TR-72-129 Final Rep. (June 1972).

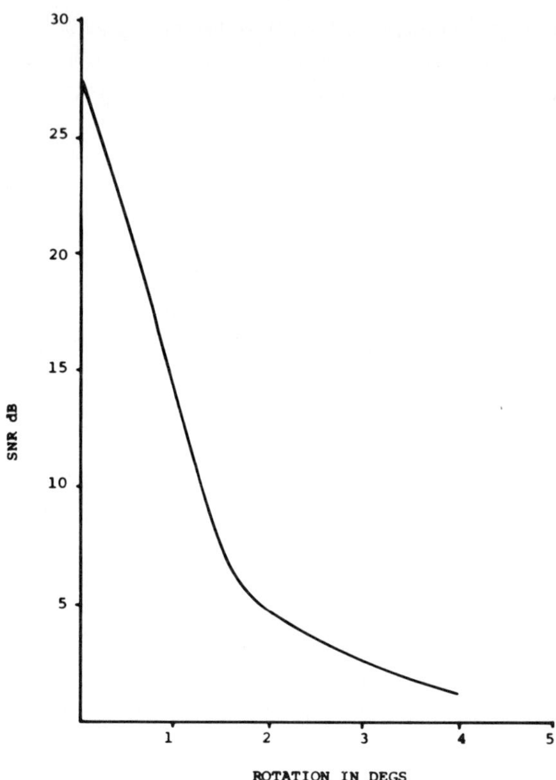

FIG. 28. Cross correlation SNR versus orientation of matched filter, Rotz *et al.*[25]

their illustrations of rotation effects on the autocorrelation of other terrain types showed even larger SNR reductions. They point out, however, that scale and orientation sensitivity are reduced if the size (not scale) of the input is reduced. Their results show that for a scale change of 1%, there is an SNR loss of 16 dB for a 100-mm diameter input image, while for a 20-mm diameter apertured input image, the loss is only 3 dB for a 1% scale change. They observed that image quality had a significant effect on correlation. Sharp, high-contrast aerial images produced strong, well-defined correlations, while low-resolution images produced poor SNR results.

As in image correlation, coarse and fine correlation strategies can be employed to locate and optimize discrimination of the correlation peaks. A large sampling aperture will reduce the discrimination between true signal correlations and noise, but it allows simplified initial detection of correlations resulting from scale and orientation mismatch. Once a potential correlation peak is located, the sampling aperture can be reduced, and the scale and orientation of the input data can be varied by zoom lenses and rotating *K* mirrors to optimize correlation.

The main objection to the use of image-to-matched-filter correlators in an operational environment has to do with the lack of available read/write/erase optical modulators having operating characteristics for near-real-time preparation and the use of the complex matched filters. Nisenson and Sprague[26] discuss the Itek Corporation PROM device in real-time optical correlation, which will lead to improved capabilities when the devices are developed to handle filters with large space bandwidth products and improved response at higher spatial frequencies.

Casasent and Psaltis[27] have described new optical transformations, e.g., Mellin transforms, Fourier–Mellin transforms, and combinations of generalized Fourier–Mellin operations, which show in theory and laboratory experiments that images having scale and rotational differences can be correlated with no loss in SNR. They further show that the scale and rotational differences between the objects can be determined by the location of the correlation peaks.

D. Applications to Image Assessment

Four applications of coherent optical feature extraction techniques pertaining to the area of image assessment will be discussed: cloud screening, cloud tracking, mission plotting, and image quality evaluation.

1. *Cloud Screening*

Clouds in aerial images, as previously stated, obscure the terrain patterns of interest. When evaluating newly acquired aerial imagery, detected cloud masses could necessitate additional aerial missions to obtain cloud-free photography. In this application, the evaluation must be rapid and as soon as possible after the aircraft returns to base. Additionally, if the photography has been accepted with some cloud cover, the locations of obscured areas are important for scheduling the automated mapping systems and searching the image data base for alternate photography of the obscured areas. Cloud screening can be viewed as a simple two-class pattern recognition problem where samples are to be classed as either "cloud" or "noncloud." The positive transparencies of cloud image patterns are typically highly transmissive and without structure. Therefore, the power spectra of cloud images typically have a significant portion of the incident energy centered on the optical axis and the energy distribution decreases rapidly to insignificant values within the noise level.

---

[26] P. Nisenson and R. A. Sprague, Real-time optical correlation, *Appl. Opt.* **14**, 2602 (1975).

[27] D. Casasent and D. Psaltis, New optical transforms for pattern recognition, *Proc. IEEE* **65**, 77 (1977).

Here we present a simple, automated approach to rapid and accurate cloud screening tasks after the OPS research, providing the technological base for the screener.

a. OPS Research System. The ROSA OPS system previously discussed has been used to conduct cloud screening research, and these results have been described by Leighty and Lukes[20] and by Lukes.[24] For these experiments, a square 16 × 16 element grid with intersections spaced at $\frac{1}{4}$ inch was used to sample 1:100,000 nominal scale photography obtained with a KA-30 aerial camera. A 40-in. transform lens focal length was used. The results summarized here are mainly those of Lukes.[24]

Data normalization was investigated. This was necessary because the ROSA system operates with an unstabilized or unmonitored laser and, also, significant variations in transmission of aerial images can be experienced due to changes in exposure conditions at the time of imaging, or to a lesser extent, film emulsion and film processing variations. Two types of data normalization were tested: (a) dc normalization where the value of the innermost ring was scaled to a fixed value and sample values from all other detector elements were adjusted proportionally; and (b) total energy normalization where the sum of the rings was scaled to a constant and the sample values of all detector elements were adjusted accordingly. In comparative tests on cloud detection data sets, both procedures performed satisfactorily, but the total energy normalization demonstrated slightly higher performance.

Feature selection algorithms were tested to reduce the dimensionality from the 64-element vector to a much smaller set, yet retain the information necessary to properly partition the decision space. Dimensionality reduction further served to reduce the size of the training set required, as well as the required computer memory and processing time. One approach to feature selection involved a somewhat arbitrary choice of four ring components from the sample vector, on the premise that clouds could be distinguished on the basis of spatial frequency bands. However, the four ring components could have just as well been those having the least variance over all cloud samples in the training set. Another approach synthesized a new three-dimensional vector from the original vector components. The first synthesized component represented the maximum difference between adjacent wedges in the sample vector to determine if a pattern containing edges existed in the sampling aperture. The second component of the synthesized vector was computed from the sum of the absolute ring-value differences between those of a mean cloud and the sample. The third component resulted from computing the maximum absolute difference between the ring values of a sample vector and the vector for the mean cloud. Thus, the synthesized vector has elements computed from algorithms that operate on all or part of the sample vector. The two feature selection approaches were tested for a given classification algorithm with the sample data collected with $\frac{1}{2}$ and $\frac{1}{8}$ in. diameter sampling apertures. The results indicated little difference between feature selection approaches for the smaller sampling aperture, while for the larger aperture the synthesized vector produced better results that seemed to indicate that specialized software could improve performance with larger aperture data. In an additional set of experiments, Lukes[24] selected eight algorithms to produce different synthesized vectors; four components were based on ring data and four on wedge data. Subsets of two- and three-component vectors were selected from the eight possible, and were tested with a training set for statistical significance as described by Kasdan[23] to evaluate class separation provided by the four classification algorithms in the FACEL library. When these subsets were run against the test data, numerous subsets demonstrated classification accuracies in excess of 95% with any of the four classification algorithms. This similarity in performance would seem to indicate that a classification algorithm could be selected on such secondary criteria as storage requirements and execution time. The majority of false classifications in these experiments occurred where the sampling aperture was located at a cloud boundary and included both "cloud" and "noncloud."

The ROSA system is well suited for research in OPS applications but in general is not suited for an operational environment. Thus for operational cloud screening, a system was designed which better suited the operational needs.

b. Operational OPS System. The objectives of the operational cloud screening system were to have the capability of screening a 9.5-in. roll of film in lengths up to 200 ft at the rate of about 50 ft per minute, with 40 samples of a $\frac{1}{4}$-in. aperture per scan line. The system was to use existing technology. The subsequent preliminary design integrated four subsystems:

(a) a coherent optical subsystem consisting of a rotating multifaceted mirror to scan a telecentric laser beam using parabolic mirrors for Fourier transform optics,
(b) an electronic subsystem consisting of a ROSA detector with high-speed signal processing capabilities,
(c) a minicomputer digital subsystem utilizing selected parts of the FACEL software, and
(d) a continuous film transport system. Figure 29 shows a sketch of the optical subsystem.

From a collimated source the first parabolic mirror focuses the beam onto a rotating multifaceted mirror. As the facets rotate, the reflected diverging beam is scanned across the collimating section of the first parabolic mirror and is reflected as a collimated beam parallel to the optical axis. Scanning the transparency with the collimated telecentric beam is necessary so that the diffraction pattern will not translate on the detector. The second parabolic mirror serves

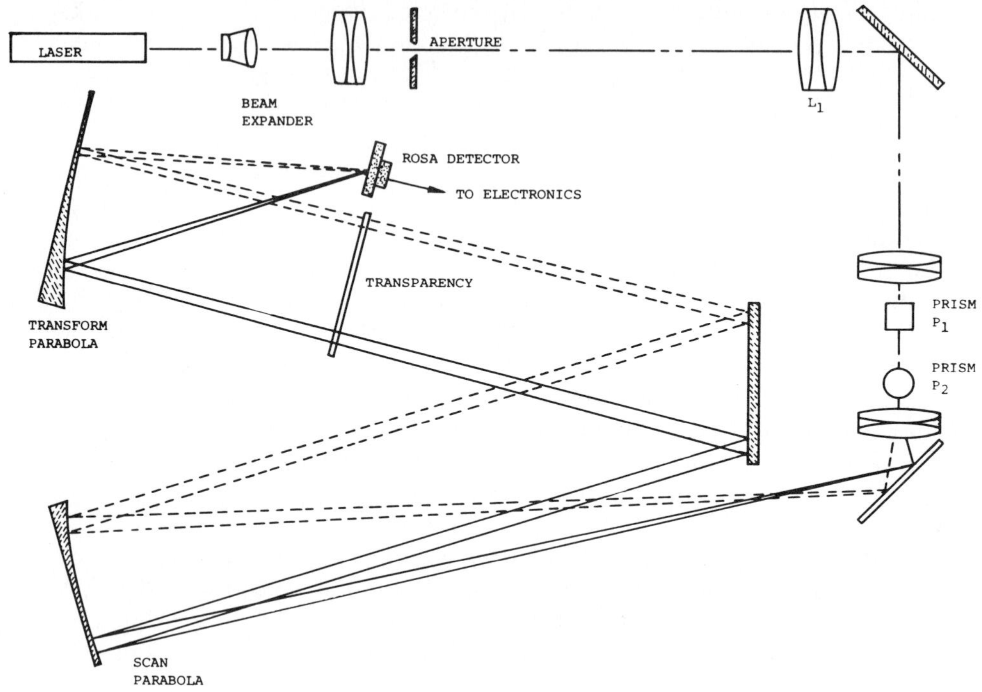

FIG. 29. Large format optical scanner based OPS system.

to collect the diffraction energy and focus it on the detector. To meet the time requirement, a diffraction sample had to be obtained and processed in 1 msec. This controlled the selection of the scan rate and the film transport requirements. Only eight elements of the ROSA detector were to be used, each having an individual amplifier.

2. *Cloud Tracking*

The tracking of cloud images from satellite photography by image-to-matched-filter correlation has been demonstrated by Rotz and Greer.[28] They utilized pictures obtained at half-hour intervals from the geosynchronous ATS III satellite to obtain cross correlation of local regions illuminated by a telecentric scanning beam. The results were used to estimate velocity vectors for prediction of local wind pattern of the imaged areas, and they found that in most cases it was possible to obtain good correlation with images taken 1 hr apart.

[28] F. B. Rotz and M. O. Greer, Photogrammetric and reconnaissance applications of coherent optics, *Proc. SPIE* **45**, 139 (1974).

3. *Mission Plotting*

The increasing capabilities for acquiring aerial imagery at faster and faster rates has led to a study of image-to-matched-filter correlation feasibility to automated mission plotting. This study, reported by Rotz *et al.*[25] and Rotz and Greer,[28] indicated two basic modes of cross correlation for mission plotting: (a) combined use of frame-to-frame correlation and pass-to-pass correlation, and (b) correlation of images from different missions. The results of this study will be described.

a. Frame-to-Frame and Pass-to-Pass Correlation. Suppose orthophoto maps do not exist of the entire geographic area of interest, but that the area is crossed by long, separated strips of imagery obtained at some recent time. This existing photography has been plotted and represents the reference images. It is required to plot the location of the new mission photography obtained along flight lines normal to the older photography. Locating ground reference points on at least one photo of the new photography and then correlating it with the reference photography, represents frame-to-frame correlation. Between the strips of old photography, referred to as passes, a frame-to-frame correlation of the new photography is used to bridge by boot strapping until the next pass is located. This results in cumulative errors that can be distributed to the positional locations of photographic frames lying between the reference frames of the old imagery.

b. Mission-to-Mission Correlation. In mission-to-mission correlation, new imagery is correlated with reference imagery from previous missions that could have been obtained many years previously.

c. Study Results. The experiments were conducted with aerial photography of different terrain types, i.e., industrial, urban, rural, and mountainous regions. The frame-to-frame correlation peaks were typically 15 to 30 dB above the rms noise, while the mission-to-mission SNRs were typically greater than 8 dB. The occurrence of cultural features (man-made structures) in the images produced stronger correlation than the natural scenes. Mountainous regions were the most difficult to correlate due to the relief displacement, and the terrain slopes in images obtained from different perspectives.

The study concluded that the accuracy of the cross-correlation operation exceeded 1/200 of the mission altitude and that an automated mission plotting capability could have an average throughput of 50 frames/min, and plot to an accuracy of 1/600 of the frame width in both directions.

4. *Image Quality Evaluation*

While cloud screening, cloud tracking, and mission plotting are important image assessment operations, the subject of image quality evaluation is assuming a role of prime importance in the mapping community. Brock,[29] in his appraisal of techniques for aerial photographic image evaluation, summarizes the history as exhibiting an intriguing counterpoint on the theme of objective and subjective image quality. He states that variations appear, disappear, and reappear according to the shifting emphasis in the appreciation that there can be no progress in accuracy without taking the human being out of image evaluation, and no confidence in its significance without putting him in again. Here, we will discuss image evaluation from two viewpoints: (a) image evaluation for human application, and (b) image evaluation for machine applications. This is not a commonly appreciated distinction but we believe it demonstrates a very important partition for coherent optical applications in the future.

a. Image Evaluation for Human Applications. Subjective image quality implies the value of the aerial photography for conveying information to a human observer about the ground scene. For many years, the only numerical scale to photographic image quality dealt with resolving power, and it is still more widely recognized than any other criterion in the rating of aerial photography. Objective techniques have tended to present some form of quality index based solely on physical variables such as MTF or granularity. For simplicity and historical reasons, the users of aerial imagery desire a single image quality index. Thus, the objective and subjective scales should show correlation. Nill,[30] in a recent paper, demonstrated a high degree of correlation between image quality assessments by human observers and a normalized low-order moment of the optical power spectrum. He used the EIKONIX Optical Power Spectrum Analyzer to sample the optical power spectrum with 20 discrete annular apertures, each with a bandwidth of 10 cycles/mm from 0 to 200 cycles/mm. A quantity approximating a low-order moment of the optical power spectrum is obtained by first normalizing the discrete value detected through each annular aperture by the total power, then multiplying this result by the square of the spatial frequency associated with the midradius of that annular aperture, and, finally, summing the products over the frequency range desired. The result, which he terms the "merit factor," was experimentally tested from 54 frames of imagery originally exposed on EK3414 aerial black-and-white film and then duplicated on EK30-192 duplicating film. The images had a scale of about 1/30,000 and the scenes represented airports, clusters of trucks, naval facilities, etc. A wide range of quality conditions were obtained due to varying image smear, focus, and atmospheric haze and turbulence

[29] G. C. Brock, "Image Evaluation for Aerial Photography." Focal Press, London, 1970.

[30] N. B. Nill, Scene power spectra: the moment as an image quality merit factor, *Appl. Opt.* **15**, 2846 (1976).

conditions. Each test image was rated by an average of six experienced observers on a quality scale of 0.0 (low quality) to 0.8 (high quality), according to whether specific tasks of detection and recognition could be performed at any observer-preferred image magnification. The duplicate positive film transparencies were measured with a sampling aperture of either 0.1, 0.2, 0.4, 1.0, or 2.00 mm depending upon the relative sizes and clustering of the objects of interest in a given scene. The common logarithm of the merit factor was then plotted against the observed image quality with correlations of 0.75 and 0.88, depending on whether or not aperture compensation was used.

b. Image Evaluation for Machine Applications. Subjective image quality does not necessarily agree with the picture quality sought for machine processing. For example, it is often desirable to print aerial photographs with automatic dodging systems at a much lower contrast than would be optimum for machine processing, in order to retain information in highlights and shadows by compressing the dynamic range. The need for image evaluation for automated electronic optical image processing systems was emphasized in Section A,1 of this chapter and will not be repeated here. Rather, consider an image processing system in the general context of systems engineering wherein each component or subsystem has a transfer function related in some way to the information to be processed. Think of the image as the input signals or data to be processed. If areas of the input image have measured signals that fall in the poor performance zone of the systems transfer function, it could alert a supervisor of pending problems and he should take appropriate action to minimize costs and maximize results in some sense. This rationale will lead to objective image evaluation related to classes, types, and individual pieces of equipment for a particular output product.

A major difference between image evaluation for human and machine use has to do with the nature of the evaluation. In image evaluation for human use, a single image quality measurement is usually acceptable for a photograph or an entire roll of photographs. Whereas for machine processing, the measurements must be made over the entire area of each frame at the resolution commensurate with the subsequent processing instrument so as to define adverse areas.[31]

This points to another major difference between the two approaches to evaluation: the adverse areas may not be related to conventional image quality. For example, the adverse areas may represent images of natural and cultural objects having inherent low spatial frequency patterns (water, snow, sand, clouds, etc.), low contrast patterns or patterns of excessive contrast, overexposure (sun glint), underexposure (shadows), etc. In the case of the

[31] N. Balasubramanian, High-resolution Optical Power Spectral Analysis. Final Rep. to USAETL (1978).

UNAMACE automated mapping system, conjugate stereo aerial images are cross correlated to extract measurements of terrain elevations and adverse areas for this system including terrain patterns which have auto and cross-correlation functions with potential selecting a false correlation peak. Such images, for example, are orchards and plowed fields with periodic patterns, or uniform forest canopies with quasi-periodic.

Lukes[32] is studying image evaluation applied to the UNAMACE with the ROSA and the large format optical scanning OPS system similar to that shown in Fig. 29. The objective of this research is a high-speed image evaluation system that can be applied to programming the UNAMACE system to avoid adverse areas where low efficiency or failure would normally result.

We believe the future image evaluation systems will be components of hybrid optical/digital image processing systems which have the capability to detect adverse areas, apply optical or digital image processing to selected adverse areas to improve performance potential to an acceptable level, and to program the actions of the subsequent automated systems.

E. Applications to Image Handling

If aerial images are to be used in completely automatic processing systems of the future, techniques must be developed to automatically register the image coordinate system with the instrument coordinate system, and, in turn, automatically register the image coordinate system with the geographic reference system of the real world. In general, the optics community has concerned itself with aberrations in the output images of its image processing systems, but now we see an extra dimension: the processed data must be relatable to the geometry of the real world to be of value. It can be argued that without effective registration capabilities, automated image processing by optical or digital means will not exist outside of the laboratory in mapping applications. Many of the subjects previously discussed in this chapter inferred a registration capability, e.g., cloud screening, mission plotting, and cloud tracking.

Here we wish to mention briefly only two types of optical techniques which will serve as tools for handling images in automated systems of the future. It is not within our scope to detail the photogrammetric concepts of interior and exterior orientation in which these tools will be of value, but let it suffice to say that an optical system incorporating a digital computer for command, control, and decision making can automate a number of the image handling and image processing tasks now accomplished only manually or semiautomatically.

---

[32] G. E. Lukes, Personal communication.

1. *Reseau Mark Detection*

This discussion addresses capabilities associated with registration of image and measurement coordinate systems with reseau marks. Reseau marks in aerial images are precise geometric symbols, usually in the shape of a small cross, which are exposed as part of the image format at the time of aerial imaging. Spatial arrays of reseau marks were originally used to study the geometric deformations of images due to shrinkage and stretching of the emulsions and film base. Development of improved materials have negated their value for this purpose, but reseau-type symbols are still used to mark the center and other positions within the aerial photographic format. These marks, along with fiducial marks, annotate the mid or end points of the photographic image, and are of great value for registration.

Krukikoski *et al.*[33] investigated the feasibility and measurement of reseau marks with coherent matched filter correlators. The matched filter for these standard symbols is easily prepared, and one would think there would be no problem in detecting the location of the correlation peaks. However, consider that the reseau marks are embedded in the terrain image to the point that it is often difficult to locate the marks when looking at the image. Krukikoski *et al.* found that the matched filter-detection process effectively isolates the reseau from the surrounding imagery in a manner which makes it possible to perform automated measurements of reseau position. They incorporated an automatic position-detection capability in their laboratory correlator, and determined that position could be determined with accuracies ranging from 0.1 to 0.4 μm, depending primarily on the sharpness of the reseau transmission profile and on the contrast between the reseau and the surrounding imagery. They pointed out the possibility of encountering terrain backgrounds, such as plowed fields, which may have similar characteristics to the reseau, and remarked that further sophistication could be added to the detector system of the matched filter correlator so that it could recognize and flag such situations at the expense of the system's simplicity.

2. *Image Control Point Selection*

This discussion addresses the potential for automatic registration of image and ground coordinate systems. We have discussed the boot-strap matched filter approach to bridging between controlled photography for mission plotting. Locally unique image patterns were selected for their value in conjugate image point correlation. It is believed that the manual process of selecting unique

---

[33] S. J. Krulikoski *et al.*, Coherent Mapping Techniques. Bendix Research Lab. RADC-TR-70-62, Final Rep. (1970).

image points could be automated with optical power spectral analysis search strategies. Once located, matched filters could be prepared with the use of real-time optical modulators.

F. Applications to Cartographic Feature Extraction

A complete capability for automated extraction of cartographic information from aerial imagery is many years from realization in a production environment. Here we will outline the nature of the problems and some research concepts that are undergoing investigation or planning at this time. We treat only two large problem areas: cartographic data class extraction and change detection analysis, but the discussion must convey the state of the technology.

Before beginning these discussions, we must clarify a few basic terms often misused in pattern recognition or feature extraction literature, at least from our perspective. So as not to encroach on definitions of others we will define our terms. The first term, "DETECTION," will mean the realization of a difference in signals that in some way relates to differences in objects of interest and their backgrounds. We can talk of designing detectors to optimize our signal values, as with film–filter combinations, threshold detectors, etc. The second term, "RECOGNITION," will mean signals can be separated from their backgrounds. The term, "IDENTIFICATION," indicates that we can label the recognized objects with names and descriptors. And the last term, "ANALYSIS," will mean that we can analyze the pattern of objects in the scene and judge their significance in the context of a problem to be solved which is currently under consideration. These definitions provide a hierarchy for future automated feature extraction systems and we will use them in the following discussions.

1. *Extraction of Cartographic Data Classes*

Maps contain selected information of the earth's surface. This information is coded or symbolized and is called cartographic data by the mapping community when applied to topographic mapping. Some examples of cartographic data classes are roads, bridges, structures, utilities, urban areas, rivers, intermittent streams, terrain elevations, geographic grids, geographic names, and political boundaries. Our attention here is directed to the information classes that can be automatically extracted from aerial imagery for map symbolization. Semiautomated systems exist for extracting terrain elevation data and interactive and automated cartographic data of interest in our discussion, once extracted and symbolized. To date, cartographic feature extraction is a time-consuming and labor-intensive process that represents a major bottleneck in present day mapping operations. Interactive approaches are under development to speed data extraction but the ultimate goal is to replace the human by automated extraction systems, at least up to the stage where the first edit of the extracted data is scheduled. The solution to this problem will not be realized for many years.

The following discussion will first indicate the complexity of the problem, then indicate research concepts under consideration, and, finally, two research systems presently under investigation will be outlined.

a. Complexity of the Problem. We assume that the present day aerial cameras and scanning reconnaissance systems provide the capability to detect terrain features of interest to the mapping community. Mapping cameras generally utilize high-resolution black-and-white emulsions; however, detectors and emulsions exist for detection from the near-uv, visible, ir, and microwave portions of the electromagnetic spectrum. This wealth of detection capability has not been exploited in automated pattern recognition systems, mainly because of the complexity of the image registration problems presented by differing formats, imaging dates, etc. Variability of the real-world terrain images, however, provides the main reason for lack of progress in automated terrain feature extraction for all but the simplest of pattern types. Where signals are relatively invariant with respect to their backgrounds, a system can be designed for their recognition, delineation, and identification. Variability is associated with the sensing systems, their recording and processing approaches, the environment, seasonal conditions, the geographic area, the complexity of natural features such as vegetation type or terrain materials, and the complexity of the structures of man and his modifications to the environment.

A study of the literature on the subject of automated pattern recognition will indicate that almost any problem has a solution if the input data are optimized. Optical pattern recognition literature is no exception when given the optimum input image. This has led to over-confidence and over-selling because the literature presents isolated instances of feasibility without relating to the failures. If we are to solve the general problems, we must know where the failures occur with respect to special case solutions. And the best approach to obtaining a knowledge of pattern variability is to understand the variability of the objects under study. There are no simple solutions to complicated real-world problems.

b. Research Concepts. Our perspective of automated pattern recognition leads us to the conclusion that more robust pattern recognition capabilities are required that will lead to more sophisticated equipment, with complex software to control the equipment. We will now outline the rationale for this conclusion. The discussion will mainly concern the following items.

Recognition, Identification, and Analysis Sophistication

(1) Develop structural descriptions of pattern classes.
(2) Develop analysis rules and techniques for their application.
(3) Develop approaches to use of existing information in data bases.

Present pattern recognition approaches assume a flat-surfaced earth, i.e., terrain elevations and terrain configurations are neglected. Yet if a photo interpreter is asked to study terrain patterns he will usually request stereophotographs so that his analysis is in the three-dimensional context of the real world. It is believed that terrain elevations must be a type of data for future terrain-pattern recognition systems. Digital pattern recognition approaches based only on radiometric or tone data are doomed to failure for general solutions until we understand the natural and physical causes of the variations and are able to correct for the variations by some normalization approach. The multispectral pattern recognition approaches have enjoyed popularity in digital processing because of processing simplicity relative to approaches involving spatial patterns.

In the real world objects have size, shape, and arrangement characteristics according to some set of natural laws or actions of man. For example, road systems in a given geographic area will have patterns that are dependent on the terrain and on the engineering policies of the area, giving uniformities in width, grade, separation, junctions, etc. Additionally, most natural and cultural terrain patterns can be divided into subpatterns, where the subpatterns have a higher degree of internal uniformity. For example, in urban areas, residential, commercial, and industrial subareas will each have their distinguishing characteristics. This suggests a set of structural descriptions of patterns and how they relate to associated patterns, and of subpatterns and how they relate to their associated subpatterns. These structural descriptions must be integrated into a set of analysis rules which dictate how the descriptions and rules should be applied. Leighty[34] has developed an approach for automated terrain analysis involving predicate logic that has not been tested for generality, but indicates that artificial intelligence applications to pattern recognition, as developed at Stanford Research Institute,[35,36] will provide the intelligent pattern analysis procedures. Future automated pattern recognition systems must include provision for incorporating known information stored in a data base such as the cartographic data bases presently under development by the U.S. Geological Survey and the Defense Mapping Agency.

c. Research Systems. Optical processing systems have speed and bandwidth advantages for future feature extraction systems, but digital systems will be required for nonlinear processing and for overall command, control, and decision making. Two hybrid research systems currently under preliminary

[34] R. D. Leighty, A Logical Approach to Terrain Pattern Recognition for Engineering Purposes. Ph.D. Dissertation, Ohio State Univ. (June 1973).

[35] H. G. Barrow, Interactive Aids for Cartography and Photo Interpretation. Stanford Research Institute, Semiannual Technical Rep. DAAG29-76-C-57 (November 1976).

[36] E. H. Shortliffe, Mysin: A Rule-Based Computer Program for Advising Physicians Regarding Anti-Microbiological Therapy Selection. Stanford Univ., STAN-CS-74-465 (October 1974).

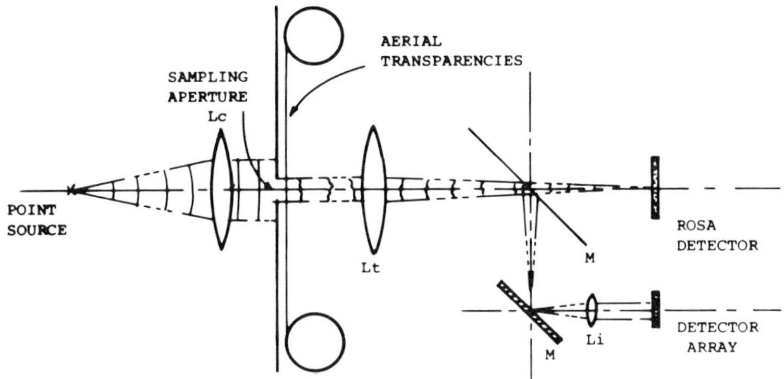

FIG. 30. Hybrid optical/digital system for both frequency and image sampling.

investigation at the U.S. Army Engineer Topographic Laboratories will now be discussed.

The first research system, shown schematically in Fig. 30, rotates a mirror into the ROSA system so that the sampling aperture is imaged on a Reticon detector array. This provides data from the spatial frequency domain and the space domain for recognition and identification purposes. Hierarchical pattern recognition strategies will be employed to choose when and what to sample from each domain. Sampled data in the spatial frequency domain will provide suggested pattern classes, the space domain will be sampled according to information extracted from the spatial frequency and directional samples, and provide information to select classes from those suggested by frequency-domain detection and recognition. The system of Fig. 31 also provides increased spatial resolution for screening operations discussed in Section C,1.

The optical and mechanical system associated with Fig. 30 has been implemented, but serious experimentation awaits development of software for control and decision making.

The second research system, shown schematically in Fig. 31, involves the Heterodyned Optical Correlator described in Section II for extraction of terrain elevations from stereoimages, a ROSA detector, and a second space domain detector array to view the apertured image without the dc filter. The integral detector array in the correlator provides a pseudoderivative image of value, but the original image is desirable for many reasons. It is anticipated that the concepts of Leighty[35] can be implemented with this system, and the experience gained with the first system will provide a background for the more sophisticated system of Fig. 30.

2. *Change Detection Analysis*

Suppose we have two aerial photographs of the same area that were exposed at different dates. Can we detect the differences between the two images? Of

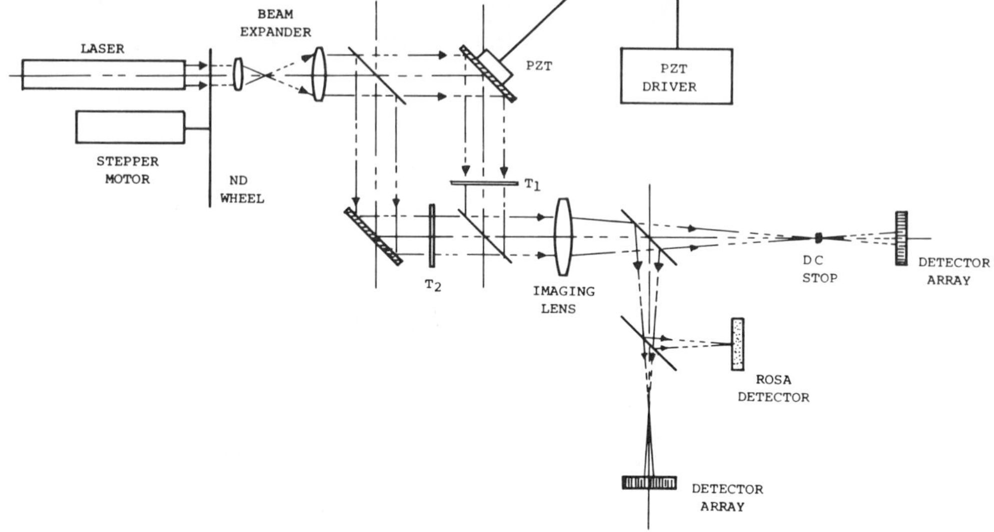

FIG. 31. Hybrid optical/digital system for sampling terrain elevations, image, and spatial frequencies.

course, a rather simple system could be designed to indicate where the two images differed, but this system, as such, would not be useful in automated operations in change detection analysis, as we shall soon see. What we need for change detection analysis is a system which will detect, identify, and analyze the differences and select only those of value.

The problem concerns the maps of our rapidly changing world. Most of the civilized world has maps which are out dated because man has been feverishly modifying the landscape with roads, dams, buildings, etc. Remapping is very costly and necessary only where major modifications to a map sheet are needed. In most cases, only small modifications are necessary.

Image-to-image correlation has many of the characteristics needed to dectect the differences in two images of the same area at different dates. Also, Lukes[37] investigated the feasibility of applying optical power spectrum techniques to the detection of natural and cultural changes. While both approaches can detect differences, neither by itself can separate the signals from the noise. The difference in noise could be due to sources of variability discussed above, in addition to registration problems. If the new or changed cartographic feature can be detected in this noise, then the features of interest must be identified, and the significant changes over the entire map format would be used to decide

[37] G. E. Lukes, Integration of optical power spectrum analysis and projective sampling for land use change assessment, *ASP Annu. Meeting* (March 1977).

whether or not to revise and reprint the map. It is believed that the signal-to-noise enhancement techniques required are nonlinear and require digital or hybrid optical/digital processing systems.

Thus we see that change detection analysis presents a problem more severe than feature extraction because it requires analysis in addition to identification. It is believed that useful approaches to change detection analysis will emerge after capabilities for integration of data bases into pattern recognition systems are developed.

V. EVALUATION OF COHERENT OPTICAL PROCESSORS IN MAPPING

In evaluating the feasibility and applicability of any technical concept to a particular application area, the system parameters relevant to the performance characteristics must be defined, and a set of criteria must be established. It is important, particularly in the application of coherent optical processors to mapping, since the system requirements are unusual and are not representative of the typical application. The size, shape, and resolution of the input and output requirements are unconventional, and the operating environment is hostile from the optical system point of view. In this section, the operational limitations associated with the application of coherent optical processors is first discussed, and the major operational advantages are then outlined. The general system parameters considered in evaluating the performance characteristics of the coherent optical processors are (a) signal-to-noise ratio, (b) space bandwidth product of the system, (c) precision and accuracy of measurement, (d) complexity of the overall system, and (e) the ability of the system to withstand rugged environmental conditions.

A. Operational Limitations

1. *Input Formats*

In most mapping applications, the input photographic transparency to be processed is large compared to the normal demands on optical systems, and they range from 70 mm to 9 in. formats. The large size of the transparency requires large and bulky optical components which represent a serious drawback. Depending upon the aerial camera, interior, and exterior geometries, the imagery is distorted and often needs space-variant, nonlinear, and anamorphic distortion corrections to be used in optical processing systems. While it is true that off-line operations such as size reduction, differential rectification, and image enhancement can be accomplished to make it feasible, as well as improve the viability of the optical processing systems, the inherent lack of geometric control during the reproduction processes forms the basis for objections from the mapping community. Also, the optical systems should be capable of handling inputs of different forms such as films, plates, negatives, positives, etc. In some instances, the use of liquid gates to overcome phase distortion associated with the input substrates causes severe material handling problems resulting in considerable reduction in operating speed. While progress is being made with regard to the limitations of optical components and processor systems in handling large input transparencies, the size and diversity of input transparencies represent a major drawback in the application of coherent optical processor systems to mapping.

2. *Optical System Quality*

The number of resolvable elements needed along a line can often be used as a good merit factor to describe the demands on the optical system. In most mapping applications, the optical system must be capable of providing about 50 lines/mm over a 300 mm field. Thus the optical system must possess 15,000 resolvable elements along any radial direction in its field, and systems capable of delivering such performance are complex and expensive. Apart from this, the tolerances associated with distortion requirements are very stringent (of the order of 0.1 to 0.01 %). Since the geometric fidelity of the process data is of extreme importance in mapping applications, while it is possible to calibrate and remove system distortion, such processes require time-consuming and laborious operation. Optical systems capable of meeting such performance requirements in coherent optical processors applied to mapping require special design and fabrication tasks resulting in high cost.

B. Limitations Due to the Nature of Coherent Processor Output

The output from a coherent optical processor takes mainly two distinctive forms. In the first case, it represents a quasi-final product such as restituted radar imagery or the differentially rectified output of an optical correlator. In the second case, the output of the coherent optical processor represents an intermediate signal that is not directly associated with the final product. The examples for this case are correlator outputs, signals associated with image feature extraction systems, etc. The limitations caused by the output requirements in the first case are direct and are well documented. The degradation in cosmetic quality caused by the nature of the coherent illumination includes ghost images and stray reflections as the most common. The output of the class covered by the first case often requires postprocessing to remove system distortions. This necessitates an effective interface to convert the output signal to an electronic form. The use of scanning photodetectors in most cases not only reduces the bandwidth of the output, but also removes the operational speed advantages gained by resorting to optical processing techniques. In those cases in which the output of the processor represents only a small portion that is sequentially assembled to form the final product, the registration errors and grey-scale variations give rise to unacceptable cosmetic defects. These problems are being slowly overcome by efficiently combining postdigital processing techniques with optical processing, and this approach holds considerable promise.

The above-mentioned output problems are not present in the class of systems represented in the second case in which the output represents only an intermediate signal. The signal-to-noise ratio in detection is the major problem area in this case. By resorting to unconventional geometries for signal detection and measurement, the problems of signal discrimination, dynamic range of the signal detected, and volume of the data that need to be measured are being overcome. Effective editing of the intermediate signal using digital processing techniques has also minimized many of the problems associated with noise and spurious signals. Coherent processors of this class are slowly taking their fast hold and are more likely to succeed.

C. Coherent Noise Limitations

The problems of coherent illumination in producing a cosmetically acceptable final output product have been alluded to earlier. In particular, the diffraction patterns caused by dusts, scratches, and striations, the ringing of sharp edges, and the fringe patterns caused by film substrates, are some of the predominant cosmetic defects. A good review of the various coherent noise problems is presented elsewhere.[38]

D. Environmental Limitations

From an optics point of view, the mapping environment is harsh and hostile. The demands for extreme component stability in such environments in the midst of mechanical vibrations caused by other equipment and instruments often represents a major problem. While any instrument based on near and emerging high technology requires special handling and delicate maintenance, the optical systems often represent an extreme of this case. Environmental control such as airconditioning and dust control are not unique only to optical systems, but they still remain a hindrance in the psychological acceptance of coherent optical systems in the "real world."

---

[38] J. DeVelis and G. O. Reynolds, Noise in coherent optical systems, *Proc. SPIE* **52**, 94 (1974).

E. Operational Advantages

In spite of all the operational limitations outlined above, the research and development of coherent optical processors in mapping applications continues. This is mainly because of the many significant operational advantages offered by coherent optical processing technology. Optical processing systems are capable of handling inputs that have a large space bandwidth product (10^8 and over) without compromising the speed of processing. The requirements on comparative digital and analog electronic systems are too severe to be either economical or practical today. It is the parallel processing capability of the optical process that provides the inherent speed advantages in coherent optical processing systems. With the widespread and commercial exploitation of lasers and laser systems over the horizon, many of the fabrication and system design problems are bound to be solved in the near future. Hence, considering the complexity and the volume of the data processing need of the mapping applications, the sophistication and the special needs of coherent optical processing systems cannot be considered unusual. In comparison to digital systems capable of handling some tasks, the optical processors are simple, compact, and can become very economical once some of the environmental limitations are overcome. Potentials for combining the strengths of coherent optical processors with those of digital techniques in hybrid approach are being demonstrated. Hence, while the limitations of coherent optical processors outlined earlier are many, the superior performance characteristics offered by coherent optical processing technology (at least in theory) seem to continuously overcome initial setbacks and problems in practical implementation.

CHAPTER 8

Infrared Detectors

DONALD E. BODE

Santa Barbara Research Center, Goleta, California

I. INTRODUCTION

The purpose of this chapter is to present a basic understanding of infrared detectors to the optical engineer. In this treatment, the various detector types are classified and the physics of photosensitivity of important detectors is discussed in a simple manner with a minimum of mathematics. The different parameters associated with the performance of infrared detector types are defined. The concept of background-limited performance that establishes the theoretical limit to the signal-to-noise ratio from an infrared detector is described. A summary of present infrared detector technology which includes packaging and cryogenic concepts is presented along with a discussion of future trends.

Infrared detectors fall into two major classes based on their energy conversion mechanism; namely, thermal detectors and photon detectors. Thermal detectors respond to incident infrared radiation when this irradiation causes an increase in temperature, which is generally sensed by a physical parameter. If the detector has a truly "black" receiver, the ideal detector will have a response dependent on the energy and independent of wavelength. In other words, the response per watt will be flat with wavelength. Very few practical detectors have

ISBN 0-12-408606-3

a response that is independent of the wavelength of the incident irradiation. One device that comes close to meeting this criterion is the pneumatic cell (Golay cell) where the absorbing membrane has an electrical impedance that matches the impedance of free space. In this case, the membrane has a maximum absorption of incident irradiation that is also independent of the wavelength. For this reason, this device has been accepted as a standard detector in far-infrared spectrometers operating well beyond 100 μm.

The other class of infrared detector is the photon detector. In this case, the detector responds to the individual photons (quanta) of incident electromagnetic irradiation. Two basic principles can be used for photon detection. One detector uses the external photoeffect. This is the effect where an incident photon "kicks" an electron out of a selectively coated cathode into the vacuum interface where it can be collected by an anode, as in the simple photocell, or accelerated to the first of a series of dynodes, as in the photomultiplier tube, to achieve gain enhancement. The external photodevice is generally used in the ultraviolet and visible regions of the spectrum. The *S*-1 photocathode surface (Ag–O–Cs) responds into the near infrared to about 1.1 μm. However, the *S*-20 photocathode (Sb–K–Na–Cs) is far superior for most of the visible ranges. To date, no external photoeffects respond beyond about 1.2 μm (i.e., no materials have been discovered with a work function below about 1.0 eV). To achieve photon response beyond about 1.2 μm it is necessary to resort to an internal photoeffect principle. In this case, the absorbed photon generally excites an electron to a higher energy state within the solid material. This excitation gives rise to a change in electrical conductance for the case of a photoconductor, or to a generation of electric current for the case of a photovoltaic detector. The ideal photon detector, either external or internal, has a flat quantum response out to its characteristic cutoff wavelength where it theoretically drops to zero. The cutoff wavelength depends on the photocathode work function for the external photoeffect device. The cutoff wavelength of an internal photoeffect device depends on the forbidden energy gap in an intrinsic semiconductor material, or on the energy difference (optical activation energy) between an impurity state and an energy band of an extrinsic semiconductor material.

The distinction between the basic detector differences is shown in the diagrams in Fig. 1. A similarity is seen here between the simple photocell (external) and photodiode (internal) because the effective gain is unity in both cases. In the photocell the photoejected electron leaving the cathode reaches the anode, and in the photodiode the photoexcited electron and/or hole is separated at the barrier (crosses the depletion layer) and produces a gain of unity. For the case of the photomultiplier tube (external) the photoelectric electron is accelerated to the first dynode where it produces a large number of secondary electrons. These secondary electrons are further accelerated to the second dynode producing more secondary electrons. This process is repeated until a very high gain is achieved. A gain exceeding one million is possible.

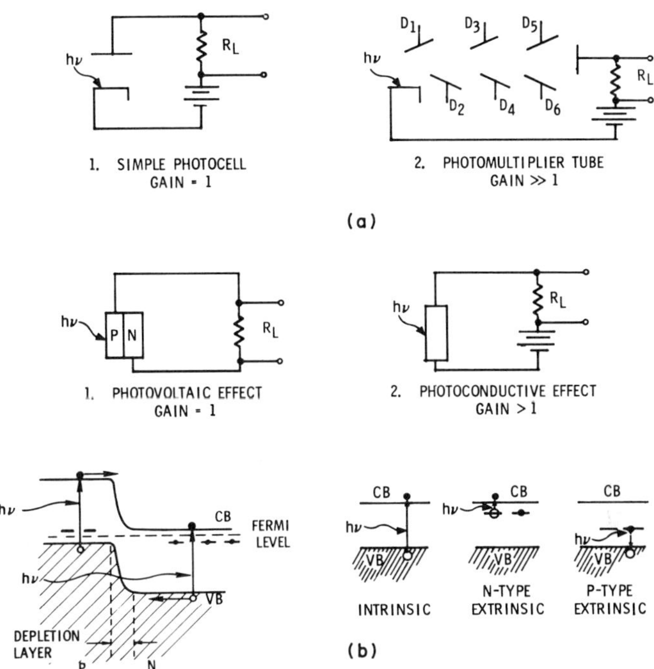

FIG. 1. Diagrams comparing (a) external and (b) internal photoeffects.

The closest analogy to this type of device in the internal photoeffect class is the avalanche photodiode. In this case, the device is designed and operated at a potential where the photoexcited carrier is accelerated by the applied electric field through the depletion layer so fast that it acquires sufficient energy to excite other electrons by collision with the valence electrons in the crystal lattice, resulting in an avalanche process that produces gain. The use of these devices to date has been limited to high-speed and high-gain laser radiation detection, and primarily limited to intrinsic germanium and silicon materials with their characteristic spectral limitation. The main disadvantage in avalanche photodiodes involves extremely accurate control of biases and operating temperature.

In Fig. 1, photoconductive detectors are used as an example of an infrared detector that demonstrates the principle of a widely used infrared detector exhibiting gain, and is compared to the photomultiplier which is widely used in the ultraviolet and visible regions. In the case of the photoconductor, a gain greater than unity can be achieved when the effective carrier lifetime exceeds the carrier transit time between electrodes. The photoconductive gain in this case is the ratio of the carrier lifetime divided by the transit time, and represents the number of times the photoexcited carrier circulates in the external circuit. In many infrared photoconductors this gain might be limited to a range between

10 and 100 for a variety of reasons. However, if the carrier lifetime is sufficiently long, and carrier mobility sufficiently large to produce a very short transit time, a photoconductive gain of a million or more is possible. This is comparable to that achieved in photomultiplier tubes.

Two types of photoconductivity are shown in Fig. 1: intrinsic and extrinsic. Intrinsic photoconductivity occurs when an electron is excited from the valence band of a semiconductor into the conduction band of the semiconductor by a photon of sufficient energy ($h\nu$). In this manner, an electron–hole pair is created and free to move under the influence of the applied electric field producing a photocurrent in the external circuit. Extrinsic photoconductivity occurs when a photon of sufficient energy excites an electron from a donor impurity atom into the conduction band for the case of an n-type semiconductor, or excites a hole from an acceptor impurity atom into the valence band for the case of the p-type semiconductor.

II. DETECTOR PERFORMANCE PARAMETERS

A. Responsivity

One of the most fundamental performance parameters of an infrared detector is responsivity. Responsivity is a parameter that describes the magnitude of signal response from a detector, resulting from a unit of radiant power on the detector element. Responsivity can be expressed in terms of amperes per watt or volts per watt, depending on preference. The most fundamental way to report responsivity is to treat the detector as a short-circuit current generator and report the short-circuit current responsivity in amps per watt of incident radiation power. For this case, the short-circuit current responsivity ($\mathscr{R}_{sc}$) can be calculated by the following relation

$$\mathscr{R}_{sc} = \frac{\eta e \lambda G}{hc}(1 + \omega^2\tau^2)^{-1/2} \tag{2.1}$$

where η is the quantum efficiency, e the charge on the electron, λ the wavelength of irradiation, h is Planck's constant, c the velocity of light, G the photoconductive gain, $\omega = 2\pi f$, where f is the frequency, and τ the carrier lifetime. A convenient way to calculate theoretical responsivity values mentally is to note that e/hc is approximately equal to 0.8×10^6 C/J m, consequently at low frequency (where $\omega^2\tau^2 \ll 1$)

$$\mathscr{R}_{sc} \cong 0.8\eta\lambda G \qquad \text{A/W} \tag{2.2}$$

if λ is expressed in micrometers. Equations (2.1) and (2.2) use G to represent the photoconductive gain. The two major photon devices being considered here are the photovoltaic detector (photodiode) and the photoconductive detector. The photodiode has a gain of unity. The gain of a photoconductor depends on a number of factors, but the fundamental concept is that the photoconductive gain is equal to the effective carrier lifetime (τ) divided by the carrier transit time (T_{tr}). Expressed mathematically

$$G = \tau/T_{tr} = \mu\tau E/S \tag{2.3}$$

where μ is the carrier mobility, E is the electric field strength, and S is the spacing between electrodes.

At times an engineer is interested in knowing how to calculate the magnitude of signal voltage he might achieve under a given load condition. In this case, if the photoconductor with resistance R_D is connected in series with the load resistor R_L as shown in Fig. 1b2, one can apply the Norden transformation, treat the detector as a short-circuit signal current generator, and consider that the load resistor and detector resistance are in parallel. In this case

$$\frac{V_s}{P} = \mathcal{R}_{sc}\left(\frac{R_D R_L}{R_D + R_L}\right) \tag{2.4}$$

where V_s is the measured signal voltage and P is the radiant power on the detector element in watts.

B. Noise

Four types of noise can be present in an infrared detector. They are Johnson–Nyquist noise, thermal noise, photon noise, and $1/f$ noise. The characteristics of these different types of noise will be discussed here.

1. *Johnson–Nyquist Noise*

Johnson–Nyquist noise is the type of noise that is present in any ordinary conductor. This noise is caused by the random motion of charge carriers within the solid. The noise voltage is given by the relation

$$V_{n_J} = (4kTR_D\Delta f)^{1/2} \tag{2.5}$$

where k is Boltzmann's constant, T the absolute temperature, R_D the resistance of the detector, and Δf the frequency bandpass.

The short-circuit noise current per root Hz (i_{n_J}) can be obtained by dividing by the resistance and square root of the bandpass of the detector to obtain

$$i_{n_J} = (4kT/R_D)^{1/2} \tag{2.6}$$

The frequency spectrum of this noise is flat.

2. *Thermal Generation–Recombination Noise*

In a semiconductor material, such as a photoconductor, charge carriers can be generated within the material by thermal excitation. For example, in the case of an intrinsic semiconductor, electrons can be excited from the valence band into the conduction band by thermal means, and conduction can occur by electrons and holes. A statistical fluctuation of carriers is associated with this excitation. In addition, a statistical fluctuation of carriers is associated with recombination. The combination gives rise to a noise known as thermal generation–recombination noise. The noise is flat with frequency in the low-frequency region, but will roll off in the high-frequency region at 6 dB/octave since the noise is proportional to $(1 + \omega^2\tau^2)^{-1/2}$, where τ is the recombination time constant.

3. *Background Generation–Recombination Noise*

If the temperature of the photoconductor described above is reduced to a sufficiently low level, the magnitude of background-generated carrier fluctuations will then dominate over thermal generation–recombination noise. In this case, if the background generation–recombination noise (g–r noise) also exceeds the Johnson noise, the $1/f$ noise, or any other noise, one has the condition of a background-limited infrared photoconductor (BLIP detector). In this case, the signal and noise will both have the same frequency response. The signal and noise will both be flat up to a frequency determined by the effective recombination time constant (τ) because both the signal and noise are proportional to $(1 + \omega^2\tau^2)^{-1/2}$. Both the signal and noise will drop off at 6 dB/octave in the high-frequency domain. Because both the signal and noise roll off in the same manner, the signal-to-noise (S/N) ratio can remain flat to a higher frequency than the responsive time constant would dictate. The S/N ratio will start to fall after the noise approaches the Johnson–Nyquist noise or the $1/f$ noise, whichever occurs first. If the first limiting noise is $1/f$ noise, D^* will fall at 3 dB/octave. When the g–r noise (or $1/f$ noise as the case may be) approaches the Johnson noise, the D^* will approach a frequency roll off of 6 dB/octave.

4. *$1/f$ Noise*

The least understood of the four noise types is $1/f$ noise. This type of noise is frequently referred to as current noise because it is generally proportional to the current raised to a power of about unity. The noise power is inversely proportional to the frequency. The noise is generally associated with trapping and untrapping fluctuations occurring at the surface and/or it can be caused by the contacts or even defects in the crystal lattice. An analytical expression for the noise is

$$i_{n_{1/f}} \propto i^a/f^b \tag{2.7}$$

where $a \geq 1$, $b \sim \frac{1}{2}$, f is the frequency in Hertz, and i the bias current.

While $1/f$ noise generally degrades the performance of a detector in the low-frequency domain, it is interesting to note that for a sufficiently slow detector $1/f$ noise can also degrade performance in the high-frequency domain. In other words, since g–r noise rolls off at 6 dB/octave and $1/f$ noise rolls off at 3 dB/octave, the components can cross over and the detector can be $1/f$-noise limited at low and high frequencies and background g–r-noise limited (BLIP) at intermediate frequencies.

5. *Other Noises*

The noise characteristics in photovoltaic detectors are somewhat different from those in the photoconductive detector. This main difference is that a photodiode exhibits a background-generated shot noise; but, since the recombination occurs at or near the contacts, the detector does not exhibit a recombination noise. For this reason, it is possible to achieve a theoretically ideal photovoltaic detector with a detectivity (S/N ratio) that is the square-root-of-two greater than a photoconductive counterpart.

There are other types of noise that are frequently associated with an infrared detector that are not necessarily basic to the mechanism within the detector. One type of noise is thermal fluctuation noise where thermal fluctuations in cooling cause temperature instabilities that produce a noiselike effect. Another phenomenon that can produce spikes in the noise output of detectors is the interaction of a high energy particle, such as a gamma ray, with the detector. Another type of degradation can be microphonism. Microphonism is generally not caused by the detector element, but is associated with the packaging of the element and the manner in which the leads are brought out. Similarly, radio frequency interference is another degrading influence that must be guarded against in the design of the appropriate package for the detector.

C. Signal-to-Noise Ratio Parameters

1. *Noise Equivalent Power (NEP)*

The noise equivalent power is the most fundamental parameter describing the performance of an infrared detector. It is defined as that incident radiant power on the detector element that produces a signal level equal to the noise level (i.e., $S/N = 1$). It is fundamental to the system user since it represents (neglecting optical losses) the power collected by the primary optics that produces a signal equal to the noise of a detector-noise limited system.

When discussing an NEP value of a detector, it is important to specify the detector area and electrical frequency bandwidth as well as wavelength (or blackbody temperature) and operating frequency. In BLIP-type detectors, it is also important to specify the effective field of view (FOV) and background temperature or effective background photon flux in the spectral region used.

2. *Detectivity* (D, D^*, *and* D^{**})

Three figure-of-merit values have evolved over the years to eliminate some of the confusion and ambiguities in comparing detector performance data. The first concept was D which was introduced by Jones to provide a convenient manner of describing a detector in that D was directly proportional to the S/N ratio. D is defined by the relation

$$D = 1/\mathrm{NEP} \tag{2.8}$$

While D was convenient compared to NEP in that better performance meant higher D values, it did not remove the confusion about detector area (A) and electrical bandpass (Δf). Consequently, Jones later defined D^* (pronounced Dee star) as

$$D^* = (A\Delta f)^{1/2}/\mathrm{NEP} \tag{2.9}$$

This figure of merit is the most commonly used parameter for comparing both infrared detectors and infrared systems.

When dealing with background-limited detectors, the concept of D^* still leaves uncertainties when comparing detectors between laboratories. The ambiguities are caused by not specifying the background irradiance present during the measurements. To solve this problem, the concept of D^{**} (pronounced Dee double star) was introduced to eliminate some of the confusion by standardizing on a detector viewing a background temperature of 295°K through a 2π-steradian FOV. If the detector was not measured under these conditions, then appropriate corrections must be made to the data considering that the

$$D^* \propto 1/Q_{\mathrm{B}}^{1/2} \tag{2.10}$$

where Q_{B} is the effective background photon flux in photons per square centimeter per second. For a cold conical FOV aperture stop

$$D^* \propto 1/\sin \tfrac{1}{2}\theta \tag{2.11}$$

where $\frac{1}{2}\theta$ is the half angle of view determined from the center of the detector element. Consequently,

$$D^{**} = \sin \tfrac{1}{2}\theta (D^*) \tag{2.12}$$

Compensating for a different background temperature or cold optical bandpass filter in addition to the FOV correction requires the use of Planck's radiation law integrated over the wavelength passband from λ_1 to λ_2 for the background temperature (T_{B}) using the relation

$$Q_{\mathrm{B}} = 2\pi c \sin^2 \frac{\theta}{2} \int_{\lambda_1}^{\lambda_2} \frac{1}{\lambda^4} \left[\frac{d\lambda}{\exp[(hc/\lambda k T_{\mathrm{B}}) - 1]} \right] \tag{2.13}$$

If the quantum efficiency of the detector/filter combination is not flat between λ_1 and λ_2, a more complex compensation is required to accurately evaluate D^{**}.

D. Time Constants

For the photon-detector user there are four major time constants of importance, and the system will be limited by the longest of these four.

1. *Responsive Time Constant*

This time constant is generally defined as the time it takes for a square-pulse signal response to decay to 37% ($1/e$) of its equilibrium value. For a detector having an exponential rise and decay characteristic, this will relate to the frequency where the responsivity has fallen 3 dB below the low-frequency value (i.e., 0.707 level). The responsive time constant (τ_r) can be calculated from $f_{0.707}$ using the relation

$$\tau_r = 1/2\pi f_{0.707} \tag{2.14}$$

Some detectors have double or multiple time constants because of competing mechanisms. In this case, interpretation of either pulse-response or frequency-response characteristics is more complex.

2. *Detective Time Constant*

For a BLIP detector whose generation–recombination noise level far exceeds the amplifier noise level or any other noise level, it is possible to operate the detector over a much wider bandwidth than that determined by the responsive time constant. In this case, the frequency roll-off of the noise will have the same characteristics as the frequency roll-off of the signal. In other words, the S/N ratio can be flat to a much higher frequency, as previously discussed. By designing high-frequency boost into the amplifier, the system user can frequently achieve a much wider bandwidth out of a comparatively slow detector such as PbS. At the frequency intercept of the responsive roll-off of 6 dB/octave and a flat noise such as Johnson noise, the S/N value will be down to 0.707 of the low-frequency value. The detective time constant (τ_{D^*}) is defined from this 0.707 frequency value which in itself has been called f^* (pronounced Eff star)

$$\tau_{D^*} = 1/2\pi f^* \tag{2.15}$$

An important concept for rating detector performance is the concept of the D^*f^* product. In this case the D^* is the low-frequency or maximum value on a D^* versus frequency plot. The D^*f^* product concept is similar to the gain $\times$ bandwidth product that is very familiar to most electrical engineers. Frequently, it is possible to trade off bandwidth for D^* and vice versa. For example, by reducing the background photon flux incident on a BLIP detector (such as a

Ge:Hg detector) it is possible to greatly enhance the low-frequency D^* value. However, it is necessary to use a higher-value load resistor because of the higher resistance value of the detector. The enhancement of high-frequency detectivity becomes limited by the RC time constant of the circuit and/or the dielectric relaxation time constant of the detector material, as the electrical resistivity of the material increases with the reduction of background-generated carriers.

Some of the highest D^*f^* products observed to date have been for photoconductive HgCdTe detectors that have a high photoconductive gain and a low electrical resistance. D^*f^* products as high as 10^{18} cm $\mathrm{Hz}^{3/2}$ W^{-1} have been reported for 0.1 eV HgCdTe when operated at a sufficiently low temperature.

3. *Circuit Time Constant*

In many detector applications the inherent time constant of the detector is not necessarily the dominant time constant. Frequently, the time constant of the amplifier may dominate. As discussed in Section II,D,2, an extrinsic semiconductor such as Ge:Hg or Si:As when operated under very low background and operating temperature may have an extremely high electrical resistivity. In this case, operating with a load resistor of 10^{10} Ω might be required. Under this condition a total input capacitance (detector + lead + FET input) could be 5 pF. The RC time constant in this case is 50 msec. This is at least 6 orders of magnitude greater than the carrier time constant which is generally less than 50 nsec.

In the case of a photovoltaic detector, such as PV InSb or PbSnTe, the time constant is generally limited by the product of the junction capacitance and external load. The RC time constant needed to achieve a high D^* generally far exceeds the fundamental limitation, which is the photogenerated carrier diffusion time. Obviously, one can reduce the external load resistance and increase the bandwidth, but this will result in a reduced D^* as the Johnson noise current exceeds the background photon generated shot noise current.

4. *Dielectric Relaxation Time Constant*

As was mentioned in Section II,D,2, the dielectric relaxation time (τ_{rel}) constant of a high-resistivity material can be the dominant time constant. The dielectric relaxation time constant is given by the relation

$$\tau_{\mathrm{rel}} = \varepsilon\varepsilon_0 \rho$$

where ε is the dielectric constant of the material and ε_0 the permittivity of free space. If one considers any photoconducting insulator, the "dark" resistivity

(ρ) may well exceed 10^{14} Ω cm. In the case of extrinsic germanium where the dielectric constant is 16 and the resistivity is 10^{14} Ω cm

$$\begin{aligned}\tau_{\text{rel}} &= (16)(8.85 \times 10^{-14}\ \text{F/cm})(10^{14}\ \Omega\ \text{cm}) \\ &= 141.6\ \text{sec}\end{aligned}$$

In this situation, circulation of carriers through the external circuit with any practical time constant is not possible and an ac gain greater than unity is not possible. The ac gain for a dielectric relaxation-limited case saturates at a gain of 0.5 for most extrinsic semiconductors to date (such as Ge:Hg, Ge:Cu, Ge:Cd, and Si:Ga). However, Si:As appears to have a dielectric relaxation-limited gain that saturates closer to unity, but the reason for the apparent difference is not well understood. In any event, the higher responsivity of Si:As compared to other extrinsic semiconductor materials when operating in a preamplifier-limited mode makes it the favorite candidate if adequate cooling is possible.

5. *Thermal Time Constants*

The four types of time constant just discussed were those generally associated with photon detectors. Thermal detectors, such as thermistor bolometers and thermopiles, are limited by the thermal time constant of the device. To approach radiation-limited performance in this type of device, it is necessary to isolate the detector element from thermal losses. However, this action would result in an extremely long time constant. Consequently, an engineering compromise is usually made to achieve the desired time constant and its associated D^*. Here, again, one is making a $D^* \times$ bandwidth tradeoff. The design of thermal detectors is quite adequately discussed by Smith *et al.*[1] Proper adherence to the design criteria described in this text can yield detectors with a performance close to the Havens[2] limit.

E. Background-Photon-Limited Performance

As discussed in the section on noise, a background-limited infrared photoconductor's dominant noise is caused by statistical fluctuations of carriers created by the background photon flux (Q_B). For a photoconductor there is also a statistical fluctuation in the carrier recombination rate. This generation–recombination noise sets an upper limit on D^*. The D^* at maximum wavelength

---

[1] R. A. Smith, F. E. Jones, and R. P. Chasmar, "Detection and Measurement of Infrared Radiation." Oxford Univ. Press, London and New York, 1957.

[2] R. R. Havens, *J. Opt. Soc. Am.* **36**, 355 (1946).

(λ_{max}) for a BLIP detector is given by the relation

$$D^*(\lambda_{max}) = \frac{\lambda_{max}}{2\,hc}\left(\frac{\eta}{Q_B}\right)^{1/2} \tag{2.16}$$

where η is the quantum efficiency, and Q_B the effective background photon flux. For a photovoltaic detector, one observes only generation noise. Recombination takes place outside the depletion layer; consequently, there is no recombination noise. The D^* of a background-limited photovoltaic detector at maximum wavelength is given by the relation

$$D^*(\lambda_{max}) = \frac{\sqrt{2}}{2}\,\frac{\lambda_{max}}{hc}\left(\frac{\eta}{Q_B}\right)^{1/2} \tag{2.17}$$

Figure 2 shows dependence of $D^*(\lambda_{max})$ on wavelength for both ideal photoconductors and ideal photovoltaic detectors while viewing a 295°K background radiation through a 2π-steradian FOV (by definition this is D^{**}). Since the quantum efficiency of these ideal detectors is assumed to be 100%, no detector (having a flat quantum response to its cutoff wavelength, λ_{max}) can exist above its respective curve.

However, as the effective background radiation is reduced, the curves in Fig. 2 will move up inversely as $Q_B^{1/2}$, as shown in Eq. (2.10) with an assist from Eq. (2.13), until the effective photon noise no longer dominates. In some cases, this can be orders of magnitude improvement in D^*.

F. Preamplifier-Limited Detectivity

When a detector is no longer background limited as described above, but is limited by the preamplifier, one has the condition of a preamplifier-limited infrared photoconductor (PLIP rather than BLIP operation). For the case of a PLIP detector, Eqs. (2.16) and (2.17) no longer hold for photoconductors and photodiodes, respectively. In the PLIP case

$$D^*(\lambda) = \frac{\sqrt{A}}{\text{NEP}} = \frac{\eta e \lambda G \sqrt{A}}{hc\, i_{n_p}} \tag{2.18}$$

where G is the photoconductive gain, and i_{n_p} the short-circuit noise current of the preamplifier for a 1-Hz bandwidth. The photoconductive gain in a photovoltaic detector is unity. The photoconductive gain of a photoconductor is given by Eq. (2.3). If the preamplifier noise is dominated by the Johnson–Nyquist noise in the load resistor (R_L), then

$$i_{n_p} = (4kT/R_L)^{1/2} \tag{2.19}$$

and

$$D^*(\lambda) = \frac{\eta e \lambda G}{2hc}\left(\frac{AR_L}{kT}\right)^{1/2} \tag{2.20}$$

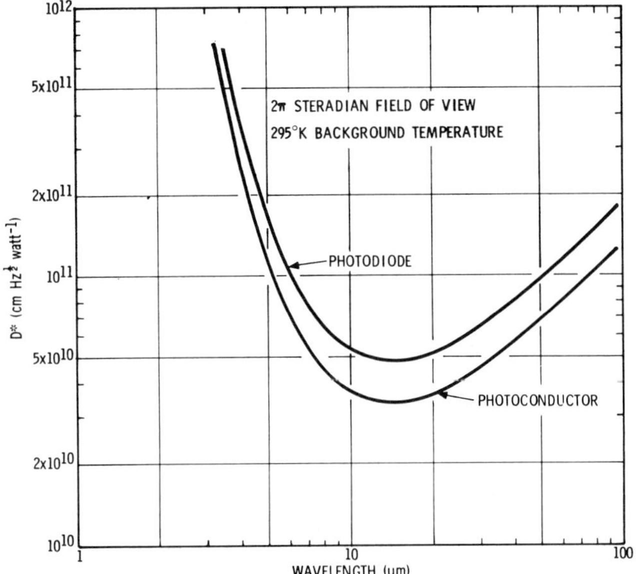

FIG. 2. The dependence of the detectivity of an ideal background-limited detector on cutoff wavelength.

Equation (2.20) shows that in order to achieve a high S/N in a PLIP mode, one must maximize the load resistor. The limits of doing this are set by the desired f^* as discussed earlier. Reduction of load-resistor temperature and reduction in capacitance and microphonics require the preamplifier to be cryogenically cooled next to the detector element.

It is also interesting to note that D^* is proportional to the square root of the area. Of course, this is not surprising since Jones defined D^* to normalize out the area. In the PLIP case, it would be better to use Jones' original D which is the reciprocal of the NEP, or just the NEP itself. The driving reason for this is to use a figure of merit that is area independent.

G. Performance Limit for Thermal Detectors

Jones[3–5] has derived a thermodynamic limit for thermocouples and bolometers where the minimum detectable power (H_m) is given by the relation

$$H_m = 2.76 \times 10^{-12}/\tau^{1/2} \quad \text{W}$$

[3] R. Clark Jones, *J. Opt. Soc. Am.* **37**, 879 (1947).
[4] R. Clark Jones, *J. Opt. Soc. Am.* **39**, 327 (1949).
[5] R. Clark Jones, *J. Opt. Soc. Am.* **39**, 344 (1949).

for a detector having a 1-mm^2 receiving area. This can be arranged to yield a

$$D^* = 1.81 \times 10^{10}\tau^{1/2} \quad \text{cm Hz}^{1/2}\ \text{W}^{-1}$$

Earlier, Havens defined a limit based on empirical examination and extrapolation from experimental devices. This minimum detectable power (H_m) was expressed as

$$H_m = 3 \times 10^{-12}/\tau \quad \text{W}$$

for a 1-mm^2 detector. This yields a

$$D^* = 1.67 \times 10^{10}\tau \quad \text{cm Hz}^{1/2}\ \text{W}^{-1}$$

If a detector has a 1-sec time constant, the Havens limit and Jones limit are effectively the same.

The difference in the dependence on the time constant comes from the fact that Jones was considering a temperature–noise-limited detector and Havens was evaluating bolometers limited by Johnson noise.

III. TYPES OF INFRARED DETECTORS

A. Classification of Detectors

Basically, two classes of optical radiation detectors are used in systems today. The external photoeffect devices, such as the simple photocell, photomultipliers, and channeltrons, are generally used in the ultraviolet, visible, and very near infrared (less than 1.2 μm) regions of the spectrum. Since this chapter concerns infrared detectors, a discussion of external photoeffect devices will not be given here. The diagram shown in Fig. 3 helps to sort out the different types of infrared detectors and modes of operation. Examples given are those commonly used and are not meant to be all inclusive.

Figure 3 shows a classification of infrared detectors into two major classes—thermal detectors and photon detectors. These two classes of detectors will be discussed in Sections III,B and III,C, starting with the thermal detectors.

B. Thermal Detectors

Thermal detectors are divided into two groups as shown in Fig. 3. The mechanical group includes those detectors relying mainly on a mechanical property to detect and/or measure the magnitude of infrared radiation received. The second group of detectors relies on an electrical property of the materials and/or structure composing the detection device.

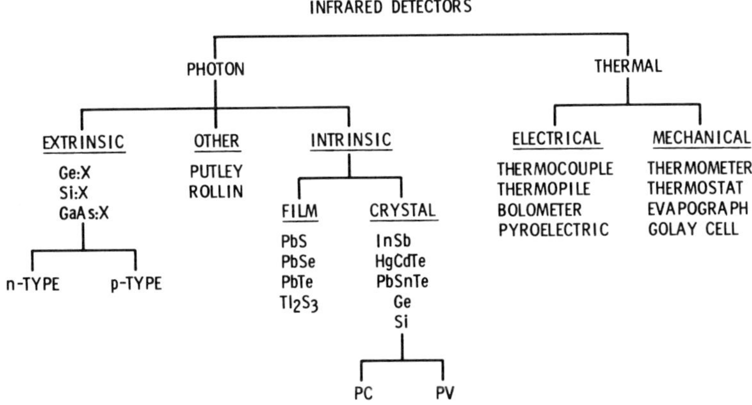

FIG. 3. Infrared detector classification tree.

1. *Mechanical Devices*

One of the earliest types of infrared detector was that demonstrated by Sir William Herschel in 1800 when he exposed two thermometers, one with the bulb blackened and the other with the bulb silvered, to infrared radiation and observed the different temperatures. Since a thermometer depends on the expansion of a column of mercury, alcohol, or some other liquid, it is classed as a mechanical device.

Another device that is commonplace in the home and industry for measuring temperature in place of the thermometer is a thermostat. This relies on the difference in expansion coefficients of two metals in a bimetallic strip. While a thermostat is generally used for measuring and controlling temperatures in homes, furnaces, ovens, and factory processing equipment, it is not generally used as a detector of radiant energy.

a. Evapograph. One of the first devices used for infrared imaging employed a thermal receiver and was called an evapograph. In this case, a thin film of oil is coated on the absorption membrane. This is then exposed to the infrared scene. Where the temperature is hottest, the oil evaporates more. Visible light is then reflected off the absorber membrane and a scene appears on a screen that reproduces the original infrared scene with visible light.

b. Pneumatic Cell (Golay Cell). If one were to blacken a balloon and expose it to radiant energy, it would expand significantly as the absorbed energy heats the air inside the balloon and increases its volume according to the relation

$$PV = NRT$$

where N is Avogadro's number, R the ideal gas constant, T the absolute temperature, P the pressure (in this case, atmospheric), and V the volume.

This principle is the basis of the pneumatic cell invented by Golay. The Golay cell is essentially a container filled with gas and having a blackened receiver membrane on one side. When the membrane receives radiant energy, the gas in the cavity is heated and expands, causing a displacement of a flexible optically reflecting membrane on the other side of the cavity. An optical system is used to sense this displacement in much the same way as one senses the displacement of a mirror in an old-fashioned galvanometer. If the receiver membrane has the impedance of free space (376.7 Ω), the device will have a very high absorption coefficient that is very flat with wavelength over a wide spectral range.

Golay cells are the standard in the laboratory for far-infrared spectral measurements, where a limited number of detectors are available and none with as flat a spectral behaviour with energy. While the Golay cell is useful in the laboratory where one has a stable, benign environment, it is not very useful in the active world where vibration is generally present (as in airplanes, missiles, tanks, and other moving vehicles).

2. *Electrical Devices*

a. Bolometers. Another device that is quite suitable for application to moving vehicles and a vibrating environment is the thermistor bolometer.[6,7] This device utilizes a change in electrical resistance when the element is heated by the incident radiation. The sensitive element is generally made by sintering a mixture of oxides of manganese and nickel. Lead attachment and mounting of the "thermistor flakes" are proprietary techniques of the manufacturer. It is possible to immerse these elements on the backside of a lens with a high index of refraction, and in this way gain an improvement in performance over a nonimmersed detector using conventional optics. In general, room-temperature bolometers have a D^* in the low 10^8 cm $\mathrm{Hz}^{1/2}$ W^{-1} range, and have a time constant greater than 10 msec. However, they do respond to long wavelengths at room temperature. Although one would like to think that these thermal detectors have a flat response to energy with wavelength, they generally do not have a flat response. The response of a bolometer generally falls off with increasing wavelength and frequently shows structure depending on the coatings and blackening technique used.

b. Thermocouples and Thermopiles. The next device of interest is the thermocouple. For many years thermocouples have been used as the principal detectors in infrared spectrometers. They are generally made with fine wires of bismuth and antimony to sense the temperature of a receiver that has been blackened with a gold black. The output signal voltage is generally a few microvolts per

[6] W. H. Brattain and J. A. Becker, *J. Opt. Soc. Am.* **36**, 354 (1946).

[7] E. M. Wormser, *J. Opt. Soc. Am.* **43**, 1, 15 (1953).

microwatt. But here again, one should not assume that the thermocouple responsivity is flat with wavelength. The thermocouple output can drop between 10 and 40% going from 2 to 30 μm depending on the individual detector and manufacturer. The response of the thermocouple falls quite rapidly with frequency. The thermocouple is usually operated at a very low chopping rate such as 13 Hz; in this case, a synchronous amplifier is generally used to reduce the bandwidth and noise output.

When a large-area receiver and high-responsivity output are desired, a number of thermocouple junctions may be coupled in series to raise the output voltage. These devices are called thermopiles. A more recent technique of fabricating very sensitive thermopiles with a smaller time constant than the wire device just described is to use evaporated thin-film couples formed by vacuum evaporation of bismuth and antimony. In this case, an extremely thin substrate, receiver pad, and black coating are used to minimize the thermal inertia and optimize the frequency response.

c. Pyroelectric Detectors. Another detector that has become more and more interesting in recent years is the pyroelectric detector. This detector is generally made from a material having a high pyroelectric coefficient. Examples of such materials are triglycine sulfate, lithium tantalate, and polyvinyl fluoride. In these materials, the impinging electromagnetic radiation displaces ions in the structure of the material to create polarity changes within the material that are proportional to the rate of change of energy with time. The ionic displacements can be sensed in the external circuit by the use of an electrometer type of circuit.

A major difference exists between this detector and conventional thermal detectors such as a bolometer and a thermopile. A bolometer's resistance can change from one dc level to another. The thermopile's output voltage can change from one dc level to another. The pyroelectric detector is not able to operate in a dc mode (the output is zero), but it responds to transients. This is an advantage when used as a moving target indicator or intrusion sensor, but is a disadvantage in an absolute dc radiometer or a radiometer with an uncertain relative velocity to the object being measured. The alternating current requirement is also a disadvantage for the case of a pyroelectric vidicon camera tube where panning, chopping, or rotation techniques must be used to create the necessary dP/dt per pixel. The primary advantage of a vidicon over a scanner is that it has no moving parts.

C. Photon Detectors (Extrinsic)

As shown in the classification tree in Fig. 3, there are two major categories for internal photon detectors; namely, extrinsic and intrinsic. The extrinsic detectors can be operated in only one major mode—the photoconductive mode.

The standard host crystals for extrinsic detectors are usually germanium or silicon and occasionally gallium arsenide. Various impurities are used to provide the desired cutoff wavelength and operating temperature. These impurities in Ge and Si are acceptors if they are from Group III of the periodic table, such as B, Al, Ga, or In. The impurities are donors and form an *n*-type semiconductor if they are from group V of the periodic table, such as P, As, Sb, and Bi. The fundamental mechanism of photoconductivity in both *n*-type and *p*-type photoconductors can be described with the aid of Figs. 4a and 4b, respectively, using Si:As as an *n*-type example, and Ge:Hg as a *p*-type example.

For the case of Si:As detectors as shown in Fig. 4a, the single crystal is generally grown by a float zone method using As in the vapor phase to achieve an As concentration around 3×10^{16} arsenic atoms/cm^3. Acceptor compensation is generally present in the form of residual boron to compensate any residual donor impurities that have an activation energy less than As (e.g., P or Sb). One desires the highest possible operating temperature; a donor shallower than As would reduce the required operating temperature and extend the cutoff wavelength. In general, it is extremely difficult to accurately control compensation concentrations below about 10^{13} cm^{-3}. The concentration of compensated As impurities (i.e., N_{As^+}) will have a direct effect on the photoelectron lifetime. When a photon is absorbed by the impurity within the crystal,

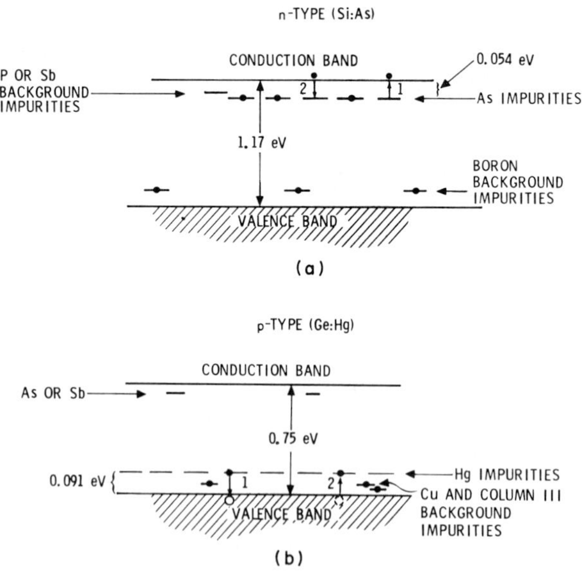

FIG. 4. Energy band diagrams comparing (a) an *n*-type infrared photoconductor with (b) a *p*-type infrared photoconductor.

it excites an electron from the orbital bound state of the arsenic center into the conduction band of the silicon host lattice (transition 1 in Fig. 4a). The electron will then be free to move within the conduction band of the crystal under the influence of the electric field (E) to attain a mean drift velocity ($\bar{v}_{d_e}$) until it encounters an empty As center (As^+) whose capture cross section is given by σ_{As^+}. The probability for capture is proportional to the concentration of As^+ centers (N_{As^+}), the cross section for capture (σ_{As^+}), and the speed with which they encounter centers ($\bar{v}_{d_e}$). The carrier lifetime (τ_e) is inversely proportional to the probability for capture, consequently 3.1

$$\tau_e = [N_{As^+}\sigma_{As^+}\bar{v}_{d_e}]^{-1} \tag{3.1}$$

To achieve a high responsivity from the detector, one wants a high photoconductive gain. This means one wants to maximize the $\mu_e \tau_e E$ product and use the smallest interelectrode space possible for the system requirement. The highest quality crystals will achieve the highest mobility possible. The maximum electric field strength is set by impact ionization of the impurity centers. The only other factor left to optimize is the lifetime, and about the only way to do this is to decrease the compensation. However, controlling the compensating impurity concentration at less than one part per billion with accuracy and uniformity of a few percent of this level to achieve element-to-element uniformity in an array is a severe problem. Consequently, an engineering compromise is normally made.

For the case of the *p*-type extrinsic detectors, Fig. 4b considers the case of Ge:Hg as an example. In this case, everything is inverted. A photon excites a hole into the full valence band (actually a valence electron is raised into the acceptor state leaving a hole behind in the valence band). The hole is now free to drift toward the negative electrode under the influence of the electric field. The hole ultimately encounters a compensated Hg center and recombines. As just described for Si:As, the lifetime is given by

$$\tau_h = [N_{Hg^-}\sigma_{Hg}\ \bar{v}_{d_h}]^{-1} \tag{3.2}$$

Here again one must optimize performance by adjusting the compensation to adjust the $\mu_h \tau_h E$ product.

Occasionally, a system user may desire a very short time constant for a laser detector to exhibit a flat response beyond a gigahertz. In this case, one can achieve such performance by "killing" the lifetime through compensation with a donor such as Sb for the case of Ge:Hg. It must be recognized that although the responsivity will be flat with frequency, the magnitude will be down proportionately because the gain and corresponding responsivity has been reduced by lowering the $\mu_h \tau_h E$ product. Here, again, a gain × bandwidth tradeoff has been made.

D. Photon Detectors (Intrinsic)

There are five basic modes for the operation of intrinsic semiconductors as infrared detectors; namely:

Photoconductive mode
Photovoltaic mode
Avalanche diode mode
Phototransistor mode
Photoelectromagnetic effect mode

Only the two most popular modes will be discussed here—the photoconductive and photovoltaic modes.

1. *Intrinsic Photoconductors*

The basic model for intrinsic photoconductivity is shown in Fig. 1b2. In this case, a photon with sufficient energy creates an electron–hole pair. If one were able to achieve the case where the lifetime is limited by radiation recombination, the electron would recombine wth the hole, and the electron and hole lifetime would be equal. This is the situation where the laws of a bimolecular recombination in a chemical sense are obeyed, and the rise and decay characteristics to a square pulse of radiation are different (hyperbolic tangent rise and hyperbolic decay) and intensity dependent. This is not the case for any known infrared photoconductors. Because of a general lack of purity the photoconductor is usually *n*-type or *p*-type. Recombination is a monomolecular process for these infrared photoconductors, and the rise and decay characteristics to a square pulse of radiation are exponential. The majority carrier lifetimes and mobility are generally different from the minority carrier lifetime and mobility. In addition, the effect of trapping can frequently lead to multiple time constants. A detailed discussion of these complexities is beyond the scope of this chapter.

Intrinsic photoconductors such as HgCdTe and InSb can exhibit a gain greater than unity produced by the circulation of majority carriers through the external circuit. However, increasing the electric field does not increase the gain up to impact ionization as it does for the extrinsic photoconductor. For the case of the intrinsic photoconductor, gain saturation occurs when photogenerated minority carriers are swept out of the photoconductor. When minority carrier sweepout occurs, the electron and hole recombine at the contacts and the time constant is reduced and the gain saturates. The value of the gain at saturation obeys the relation

$$G = \tfrac{1}{2}(1 + b)$$

where $b = \mu_e/\mu_h$, which is the mobility ratio of electrons to holes. For 0.1 eV HgCdTe, the mobility ratio b is about 300 so that a gain of about 150 can be achieved before minority carrier sweepout occurs. There is a means of achieving

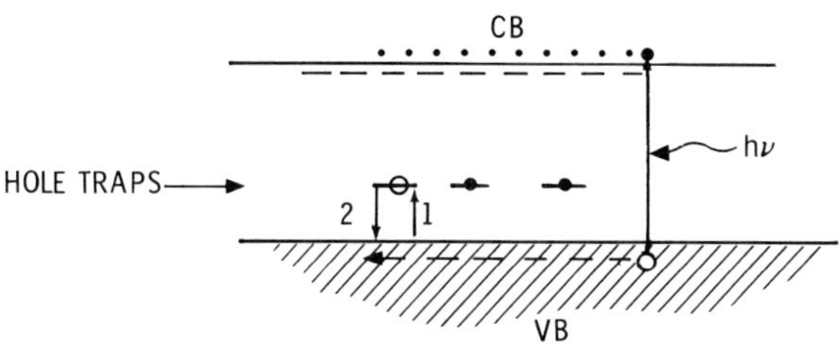

FIG. 5. Energy band diagram illustrating minority carrier trapping in n-type HgCdTe operating in the low-temperature trapping mode.

a higher photoconductive gain out of a material like n-type HgCdTe if cooling to sufficiently low temperatures and operating at sufficiently low backgrounds are possible. In this case, one can operate in the "trapping mode." This situation can be explained with the aid of the energy band diagram shown in Fig. 5.

This diagram shows that a photon creates an electron-hole pair. As the hole drifts toward the negative electrode, it is trapped by a hole trap (path 1). It will sit in this trap until freed thermally or by a background photon. At very low temperature and background photon flux it may sit there a very long time. In the meantime, the electron that was generated by the signal photon can circulate in the external circuit and generate a very high gain producing a very high responsivity. Since the noise is fixed by the comparatively large number of electrons constantly in the conduction band, a very high S/N and correspondingly high D^* can be achieved. But, here again, a $D^* \times$ bandwidth tradeoff has been made. It should be pointed out that the technology for introducing and distributing these "hole traps" uniformly throughout the crystal is not very well established at the present time.

2. *Photovoltaic Detectors*

The energy band diagram of a photovoltaic detector is shown in Fig. 1b1. Although some of the first diodes ever made were PbS, Ge, and Si point-contact diodes, a photovoltaic detector of today uses a p–n junction to operate. The junction can be formed by a variety of different techniques; namely:

(a) junction grown during crystal growth,
(b) impurity diffusion,
(c) stoichiometric imbalance by diffusion of a compound semiconductor (self-diffusion),
(d) ion implantation,
(e) proton bombardment,
(f) surface metal diode (Schottky diode),
(g) liquid phase epitaxy,
(h) molecular beam epitaxy, and
(i) alloy junction.

Depending on the method of formation, a junction can vary from an abrupt junction with a narrow depletion layer to a PIN structure with a very wide depletion layer. The PIN (*P*–Intrinsic–*N*) structures are used when a very low capacitance is needed to achieve the smallest *RC* time constant possible for laser detector applications.

The capacitance of a photodiode is determined by the dielectric constant of the material and width of the depletion layer. As the detector is biased into the reverse direction, the depletion layer will widen and the capacitance will roll off. The manner in which the capacitance depends on voltage depends on the grading of the junction. For an abrupt junction

$$C \propto V^{-1/2}$$

but for a diffused junction with a linearly graded impurity distribution, the capacitance

$$C \propto V^{-1/3}$$

Photovoltaic detectors are generally made as *p*-on-*n* or *n*-on-*p* structures, depending on the material type and many sophisticated engineering tradeoffs. They are generally made as mesa-type structures; however, recently planar structures are being more widely exploited. The processes used for fabrication are compatible with wafer processing by photolithographic techniques. Many high-density arrays can be made on one wafer, then probe tested and sliced into detector-array chips to be mounted into a Dewar package.

After antireflection coating, photovoltaic InSb, HgCdTe, PbSnTe, InGaSb, InAsSb, PbTe, InAs, and PbGeTe can all have a quantum efficiency between 50 and 80% at their respective spectral peaks. The choice of the detector material (and composition in the case of the ternary materials) depends on the desired operating spectral range and/or operating temperature. The most maturely developed and widely used infrared photodiode is InSb. It was first explored in the mid-1950s and has been under continual development and improvement for the past 20 years.

E. Infrared Photon Detector Performance

Table I represents an example of the performance values for a variety of infrared detectors that are available today. This table is arranged in approximate order of increasing cutoff wavelength. Conventional operating temperatures for use under normal background conditions are indicated. Most of the detectors in this table can have a BLIP-type performance in that the D^* will increase with the reduction in background radiation. Some of the detectors shown are non-BLIP because the indicated operating temperature is too high for BLIP behavior.

TABLE I
STATE OF THE ART OF A VARIETY OF INFRARED DETECTORS (AUGUST 1976)[a]

Detector type	Mode	Recommended spectral range (μm)	Temperature (°K)	λ_m (μm)	$D^*(\lambda_m, f_m)$ Range (cm $Hz^{1/2}$ W^{-1})	Remarks
PbS	PC	1–3.0	295	2.4	$0.8–1.5 \times 10^{11}$	Non-BLIP
PbS	PC	1–3.5	195	2.8	$4–8 \times 10^{11}$	
PbS	PC	1–4.0	77	3.2	$1.5–2.5 \times 10^{11}$	
InAs	PV	1–3.8	295	3.4	$3–5 \times 10^{8}$	Non-BLIP
InAs	PV	1–3.5	195	3.2	$5–8 \times 10^{11}$	
InAs	PV	1–3.2	77	3.0	$0.5–1 \times 10^{12}$	
$Hg_{0.7}Cd_{0.3}Te$	PC	1–5	192	4.6	$1–2 \times 10^{11}$	
Insb	PV	1–5.4	77	5.3	$1.5–2 \times 10^{11}$	
PbSe	PC	1–4.4	295	3.8	$0.3–1.5 \times 10^{10}$	Non-BLIP
PbSe	PC	1–5.1	195	4.8	$1–5 \times 10^{10}$	
PbSe	PC	1–7.0	77	5.0	$1–2.5 \times 10^{10}$	
Ge:Au	PC	2–10.6	77	5.5	$3–7 \times 10^{9}$	Non-BLIP
$Pb_{0.2}Sn_{0.8}Te$	PC	1–12	77	11.0	$2–6 \times 10^{10}$	
$Hg_{0.8}Cd_{0.2}Te$	PC	1–13	77	12.0	$2–6 \times 10^{10}$	
Ge:Hg	PC	2–14	30	11.0	$1–2 \times 10^{10}$	Performs at 77°K (very low D^*)
Si:Ga	PC	2–17	23	16.0	$1–2 \times 10^{10}$	
Ge:Cd	PC	2–22	20	21.0	$2–3 \times 10^{10}$	
Si:As	PC	2–23	18	22.0	$2.5–4 \times 10^{10}$	
Si:P	PC	2–28	15	23.0	$1.5–3 \times 10^{10}$	
Ge:Cu	PC	2–30	14	24.0	$1.5–3 \times 10^{10}$	
Ge:Zn	PC	2–40	8	38.0	$1–2 \times 10^{10}$	

[a] 60° Field of view; 295°K background temperature.

One may note that the cutoff wavelengths of PbS and PbSe move to longer wavelengths as the operating temperature is reduced. The shift of the spectral edge with temperature is a characteristic of detectors made from intrinsic semiconductors. This is caused by the fact that the forbidden energy gap in an intrinsic semiconductor is temperature dependent. The energy gaps of PbS, PbSe, PbSnTe, and HgCdTe increase as the temperature is increased. The energy gaps of Si, Ge, InAs, and InSb decrease as the temperature is increased.

F. PHOTON DETECTORS (OTHER)

As shown in Fig. 3, two types of photon detectors cannot be classed in a simple manner as intrinsic or extrinsic, but are worthy of mention. Both of the detector types described here are primarily used as far-infrared detectors in the spectral range from 100 μm to several millimeters. These infrared detectors extend into the microwave region. Named after their inventors, the devices are the Rollin detector and the Putley detector. Both detectors utilize InSb as the semiconductor material and are best operated at liquid helium temperature (4.2°K) or lower.

1. *Rollin Detector*

The Rollin detector[8] is effectively an electronic bolometer. It is normally made from very high purity (below 10^{13} electrons/cm^3) *n*-type InSb. In this device a long-wavelength photon excites electrons in the very bottom of the conduction band to a somewhat higher energy state (i.e., creating a "hot" carrier) within the conduction band, and produces a change in mobility. This change in electron mobility by the absorption of far infrared photons produces a change in electrical conductivity. This device is different from all the other photon devices described here because in those cases the carrier concentration is changed. This device is not very suitable for operation at shorter wavelengths (less than about 50 μm) because the probability for free carrier absorption is proportional to λ^2.

2. *Putley Detector*

This detector was first described by Putley[9] in 1960 and is an improvement of the Rollin detector. The principle of operation is the application of a magnetic field to InSb to separate the impurity levels from the conduction band. The freeze-out of carriers into these impurity levels produces an increasing resistance with increasing magnetic field. This produces a higher responsivity than would be present in the simple Rollin detector. In addition, a very low energy photon can then excite electrons from the impurity levels into the Landau subband, formed from the conduction band by the magnetic field, and produce an enhanced photoconductivity with a field-dependent spectral behavior. Because the energy required to excite an electron into the Landau subband can be extremely small (less than 1 meV), response beyond 1000 μm is possible. An additional enhancement in photoconductivity can also be produced by cyclotron resonance with the aid of a high magnetic field. This phenomenon produces structure in the spectral response.

The Putley detector has limited applicability because of the requirement for a high magnetic field, and a limited need at the present time. Since liquid helium is necessary for detector operation, the high magnetic field problem is generally handled by the use of a small superconducting electromagnet. To date, application of the device has been limited to far-infrared spectroscopy.

---

[8] B. V. Rollin, *Proc. Phys. Soc.* **77**, 1102 (1961).

[9] E. H. Putley, *Proc. Phys. Soc.* (*London*) **76**, 802 (1960).

IV. DETECTOR ARRAYS

Single-element infrared detectors and straight linear, staggered linear, and dual line arrays are being used today in a large variety of optical system designs. These detector arrays are generally prepared with relative ease for modest numbers of elements up to 200 or so. However, there has been an increasing trend toward larger and larger multielement focal-plane arrays. The reason for this is that if one wants to cover a large FOV with proper spatial resolution in a reasonable observation time, one must share the FOV with an adequate number of detectors to do the job. In addition, a proper system analysis which considers the noise of the required amplifier bandpass, as well as the frequency response limitation of realistic detectors, will show that the S/N of the system (D^*) will be proportional to the square root of the number of elements in the array. This is the driving force toward the ever-increasing number of elements in the focal plane. Figure 6 shows a plot prepared by the author in 1972 and presented at a conference[10] in 1976. This curve demonstrates this growing demand for more detectors in the focal plane array by showing the history between 1958 and 1972, and an extrapolation of this to 1978. The data points shown represent approximate times for actual infrared systems that went operational. The 1958 data point for a single-element detector is for the Sidewinder and Falcon missiles. The top end of the "gray area" extrapolation represents an extrapolation of existing data. The bottom end of the "gray area"

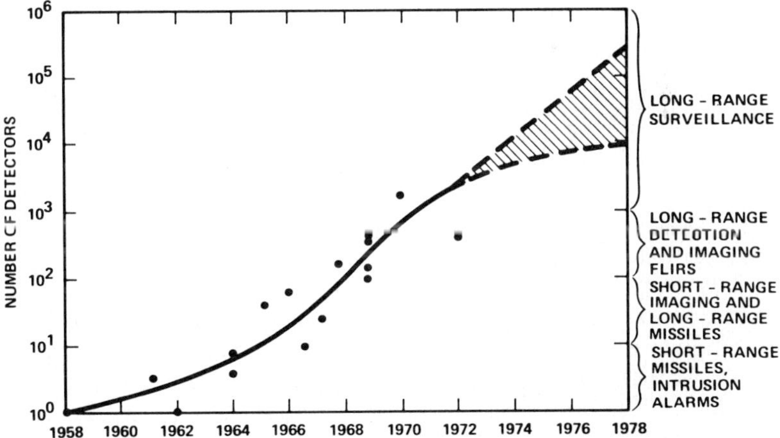

FIG. 6. Growth in the number of detector elements in the focal plane of operational sensors since 1958.

[10] D. E. Bode, *Proc. Electro-Opt. Syst. Design Conf., New York* (1976).

represents the author's conservative viewpoint and prediction of things to come. More optimistic and "bullish" system analysts are inclined to swing the extrapolation up rather than down. In any event, the desire is for an extremely large number of elements on the focal plane.

For a nonscanning detector array, a mosaic or large matrix of detector elements with an extremely high focal-plane density is needed. The manner for getting the leads out of the focal plane and maintaining a large packing density has been a severe problem. There are techniques under development that show promise for solving this problem. The other challenge is that of addressing the mosaic array and reading out the information in a low-cost solid state manner.

To date, the nearest approach to the focal plane being discussed is the silicon diode vidicon. In this case the focal plane (or retina) is made by photolithographic delineation of a diffused silicon wafer. A conventional retina uses 2000 × 2000 elements/in.^{2}. The array is addressed and read out with an electron beam. The spectral limitation of intrinsic silicon does not permit operation very far into the infrared region. There are other infrared image converters responding to longer wavelengths that use a variety of different photoconductive or thermally sensitive materials on their retinas for electron beam readout. These devices have not been widely used to date.

What is needed at the present time is a low-cost, compact, two-dimensional infrared detector focal-plane array (FPA) with solid state readout. Many ideas are on the drawing boards and under development at the present time, but nothing has become a reality to date.

The closest approach to the ideal sensor is the human eye. The optical system is compact. The light input is automatically controlled by the iris diaphragm to achieve AGC. The retina is curved properly to match the focal plane. The sensing elements (rods and cones) are distributed to provide both high resolution at moderate sensitivity, or high sensitivity with lower resolution. The lead attachment problem (nerve endings) has been solved along with the interconnect to the computer (the brain). The dark-adapted eye has the ability to almost sense individual photons. This corresponds to an NEP of about 10^{-18} W. Mother nature can make the most ingenious engineers feel humble with their inferiority and shortcomings.

In addition to nonscanning sensors, two-dimensional arrays are also needed for scanning sensors. The scanning sensor offers an easier solution to nonuniformity problems or background discrimination problems than the nonscanning sensor. Of course, the scanning system does not have the advantage of continual integration on each pixel to achieve higher response. Another disadvantage of the scanning system is that of missing an event that occurs between dwell times. However, the primary disadvantage of a scanning system compared to a nonscanning system is the reduced reliability by bearing wear.

The primary use of a matrix FPA in a scanning system is to use a parallel scanned array of detector channels to cover the FOV in the cross-scan direction

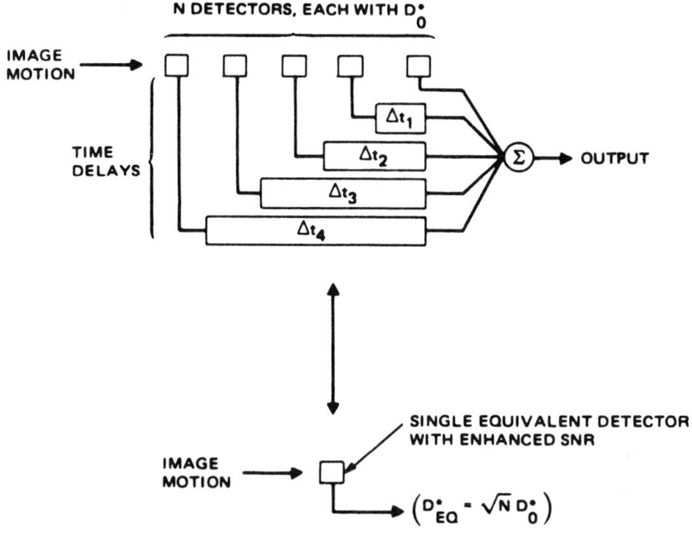

FIG. 7. Concept of time delay and integration (TDI).

and to use a serially scanned array in each channel in a manner that employs time delay and integration (TDI) to achieve an improved S/N ratio in the scan direction. The diagram in Fig. 7 explains the principle of TDI. In this case, the signal from each detector element in the scan direction is delayed and added in phase to achieve a pulse enhancement by the number of elements in the serially scanned line. Of course, the noise adds as the square root of the sum of the squares. Consequently, the S/N ratio is increased approximately by the square root of the number of elements. The appropriate delay-line network to match the scan speed can be built from lumped circuit components of the proper parameters, as has been done in the past, for the case of serially scanned FLIRs. However, the upcoming and latest technology under development at the present time is to use charge-coupled devices (CCDs) to perform both TDI in the scan direction and multiplexing of channel outputs in the cross-scan direction. Two approaches are being used for these developments, one hybrid and the other monolithic.

The serial scanning of each line offers two advantages beyond the S/N improvement by the $\sqrt{n}$. Firstly, serial scanning provides an averaging effect that smooths out element-to-element nonuniformities and provides better statistically averaged uniformity (on a channel-to-channel basis) than a straight parallel scan array will afford. The other advantage is that a dead element or weak element in the string is held up (a dead element in a simple parallel arrangement will cause a black line of missing information) by the others in the string. Obviously, the more elements serially coupled, the better. However, to achieve these three advantages more complexity and increased cost of the FPA are imposed. With the advent of LSI techniques in the detector industry, these additional costs may not be excessive, and the improved performance may well result in reduced cost at the system level. For example, improved D^* can permit a reduction in the size and cost of the optics, providing a smaller and, perhaps, lower cost system. Another example is that an improved D^* can permit operation of systems at higher operating temperatures where the detector performance is not maximum (i.e., non-BLIP) and lost D^* can be offset by the use of TDI. In this case, the combination of reduced cooler costs, size, power requirements, noise, and maintenance requirements may well offset the increased complexity and detector/electronics cost of the FPA.

There are basically two approaches for making two-dimensional detector focal-plane arrays using CCD processing on the cooled focal plane as briefly described earlier. The one method is a hybrid method and the other monolithic (single crystal material). The hybrid method can be accomplished in one of three basic ways at present, namely: (a) a sandwich structure with bump interconnects, (b) a sandwich structure using the "island approach," and (c) stacked boards.

Figure 8 shows the approach using indium bump interconnects. In this case the detector array is "flip-chipped" onto the silicon CCD array. Figure 9 shows a diagram of the "island approach," where evaporated leads are brought down over the edge of the detector array chip, over a glue line, and onto the

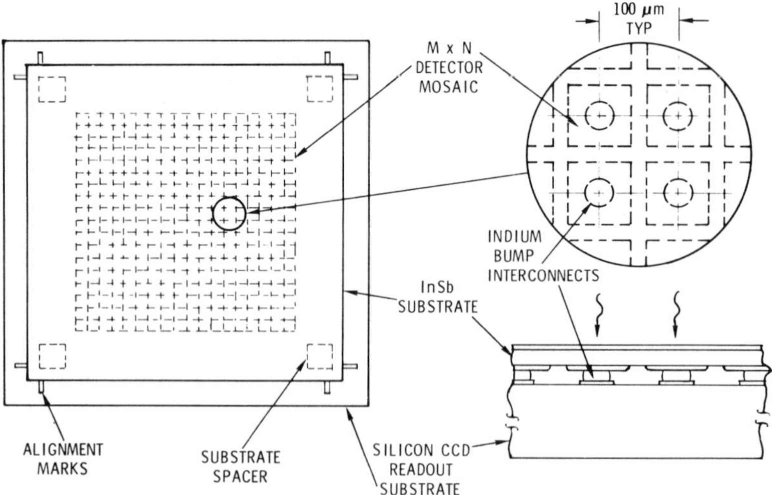

FIG. 8. Diagram illustrating the concept of a backside illuminated sandwich structure of a two-dimensional infrared diode array coupled to a silicon CCD readout chip using indium bump technology.

silicon wafer. Because of the spacing needed between islands, this approach has a lower packing density than the above mentioned sandwich structure that uses indium bump interconnects. The third method that can be used is shown in Fig. 10. In this method, linear arrays or dual lines of linear arrays, are deposited or cemented to the edge of ceramic "boards." In this case, the necessary circuit elements, CCDs, resistors, capacitors, FETs, and so forth can be arranged on the circuit board using a large amount of real estate down in the third dimension. The boards can then be stacked with thin spacers between to achieve a matrix array with a large packing density. The advantage of this configuration is that more real estate is available for circuitry than the two methods described above. The disadvantage of this method involves use of more volume plus the difficulty of achieving alignment and clamping accuracy along with the additional labor involved. The bump and island interconnect methods achieve accurate alignment with a minimum amount of labor by using photolithographic techniques similar to standard LSI processing.

The monolithic method involves the incorporation of CCD registers and associated active and passive circuit elements along with the detector array on a single chip of infrared semiconductor. These detector/CCD array chips are made by wafer processing techniques that use multilayer techniques in combination with accurately registered photolithographic masks.

The technology for preparation of monolithic focal-plane arrays (MFPA) is fairly well established for visible detectors utilizing intrinsic silicon. The technology for the preparation of high-density monolithic focal-plane arrays using extrinsic silicon or intrinsic infrared semiconductor materials (such as InSb or HgCdTe) for infrared detection is in its infancy. However, the future holds promise for MFPA chips containing many thousands of detector elements (pixels) per chip at a low cost, providing that adequate support for development is maintained.

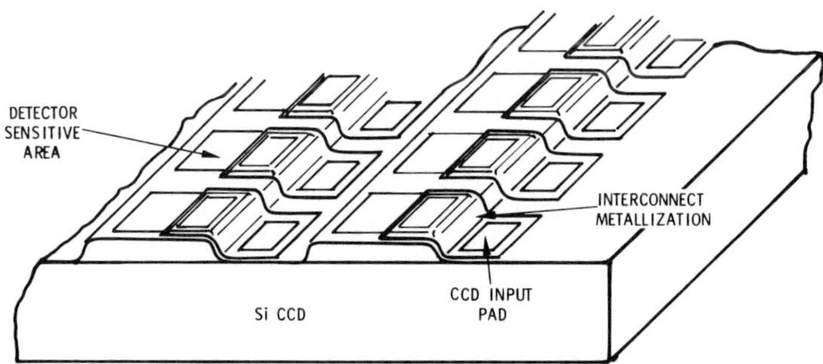

FIG. 9. Diagram of a two-dimensional sandwich structure that employs photoetched evaporated metal interconnects.

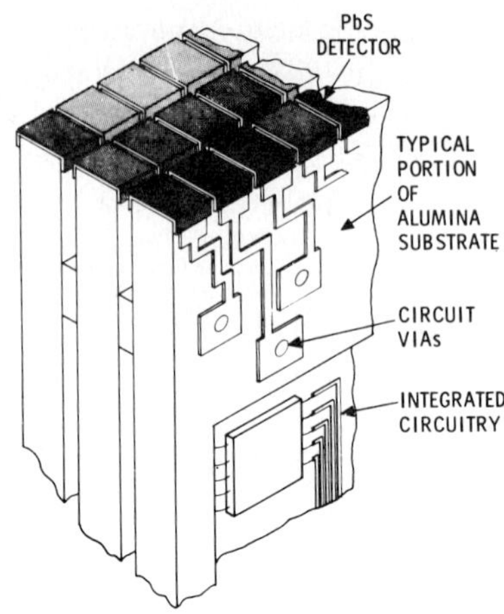

FIG. 10. Diagram illustrating the concept of stacking ceramic circuit boards to achieve a two-dimensional infrared detector array with readout circuitry.

V. INFRARED DETECTOR PACKAGING

Ambient infrared detectors can be packaged for system application into a modified TO-5 or TO-18 style transistor package that is fitted with an infrared transmitting window. Ambient infrared detector arrays are mounted into transistor flatpacks containing multilead outputs and fitted with an appropriate infrared transmitting window.

However, most infrared detectors must operate at low temperatures to achieve BLIP performance. The further out in wavelength that a detector goes, the lower the operating temperature must be to achieve BLIP performance. A photovoltaic InSb detector responding to about 5.5 μm has to be cooled below about 120°K to operate in a BLIP manner. A Ge:Hg detector responding to about 14 μm has to be cooled below about 35°K for BLIP operation under normal background conditions. On the other hand, a Ge:Ga detector responding to 120 μm must be cooled below about 3°K to operate in BLIP fashion. It should be mentioned that as one reduces the background below the normal 2π steradian, 295°K background, one must also reduce the detector operating temperature to ensure that the thermally generated carriers remain well below the background generated carriers to achieve the ultimate BLIP performance.

The packaging of infrared detectors and infrared detector arrays must give every consideration to the method of achieving cryogenic operation and heat dissipation without introducing microphonics or electromagnetic interference.

A simple method of cooling an infrared detector is to use a liquid cryogen such as liquid nitrogen in a simple glass Dewar package fitted with an infrared-transmitting window such as that shown in Fig. 11. The infrared window

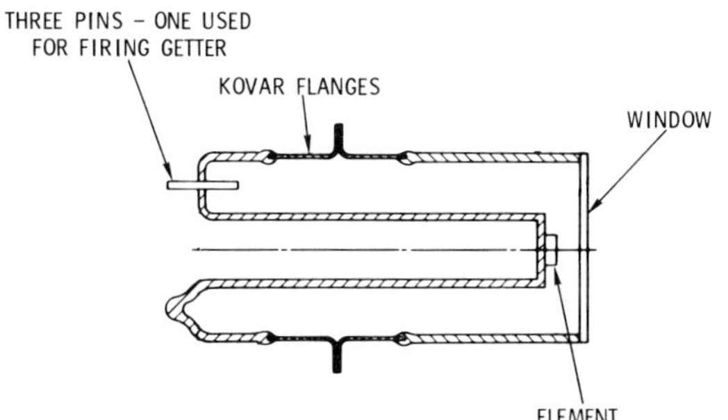

FIG. 11. Cross-sectional drawing of a simple glass Dewar.

FIG. 12. FIG. 13.

FIG. 12. Example of Joule–Thomson cooling showing spray of liquid nitrogen droplets emanating from Joule–Thomson orifice.

FIG. 13. Example of a glass Dewar package with radial feedthroughs for high-density detector arrays. (Courtesy SBRC.)

chosen is usually sapphire or antireflection-coated silicon, depending on the application. A convenient method of cooling an infrared detector is to use Joule–Thomson expansion of a gas such as nitrogen or argon. A simple example demonstrating this type of cooling is shown in Fig. 12. In this case, compressed nitrogen gas is expanded out of the Joule–Thomson orifice to regeneratively cool the incoming gas. Ultimately, liquid nitrogen sprays out of the orifice as shown in the photograph.

A more complex glass Dewar package than that shown in Fig. 11 is shown in Fig. 13. This package was developed for use with linear arrays of HgCdTe with as many as 200 elements. The number of radial leads is tailored to fit the number of elements in the package.

A convenient metal Dewar package is shown in Fig. 14. This package has the advantage of being able to change detector elements, filters, and windows to suit experimental requirements. The Dewar is more rugged and versatile

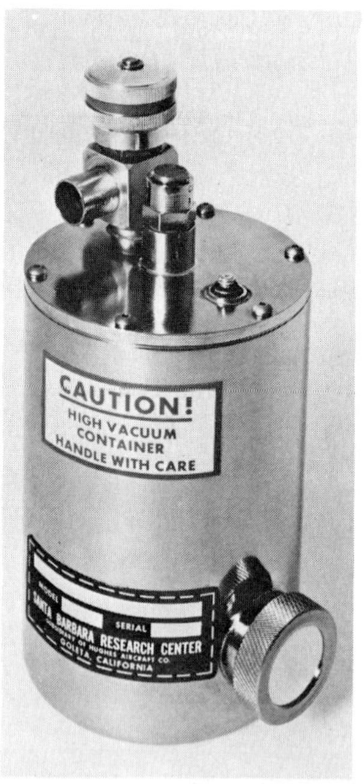

FIG. 14.

FIG. 15.

FIG. 14. Metal Dewar package for use in laboratory.

FIG. 15. 180-Element Ge:Hg detector array package for use with a closed-cycle refrigerator. (Courtesy SBRC.)

than that shown in Fig. 11, but has the disadvantage of larger size, and the requirement to evacuate periodically.

For specific applications, metal Dewars can be engineered to operate for many years with a permanent vacuum. Figure 15 shows an example of a 180-element Ge:Hg array operating in a metal Dewar for use with a closed-cycle refrigerator. In this case, the operating temperature is about 25°K and the infrared-transmitting window is coated germanium.

Another type of detector array package that has recently found wide acceptance is the common detector module for he U.S. Army Common Modular FLIR equipment. Figure 16 shows an example of such a package containing a 180-element HgCdTe detector array.

An alternate method for cooling an infrared detector array operating below 5 μm, such as PbS, PbSe, or $Hg_{0.7}Cd_{0.3}Te$, is to utilize a thermoelectric cooler (Peltier effect). Figure 17 shows a 64-element $Hg_{0.7}Cd_{0.3}Te$ detector array mounted onto a four-stage Bi_2Te_3 thermoelectric cooler which in turn is soldered to the copper heat sink of an evacuated 70-pin radial feedthrough package outfitted with an antireflection-coated sapphire window. Present thermoelectric coolers can be made to operate at about 195°K in a normal ambient using about 5 W of electrical power.

Another important method of cooling infrared detectors is to use radiation-to-space cooling. This method has been employed in geosynchronous orbiting satellites to achieve temperatures well below 100°K in order to operate photoconductive $Hg_{0.8}Cd_{0.2}Te$ detectors in the 8- to 14-μm atmospheric window.

FIG. 16. Common detector module with a 180-element HgCdTe detector array.

FIG. 17. 64-Element $Hg_{0.7}Cd_{0.3}Te$ detector array mounted on four-stage thermoelectric cooler. (Courtesy SBRC.)

In conclusion, the following methods for cooling infrared detectors and detector arrays are available for consideration:

(a) liquid cryogen,
(b) solid cryogen,
(c) Joule–Thomson expansion,
(d) Peltier effect,
(e) Gifford–McMahon cycle,
(f) Stirling cycle,
(g) Solvay engine,
(h) Vuilleumier refrigerator, and
(i) radiation-to-space cooling.

The choice of the proper cooling method and packaging technique must consider a combination of mission profile, performance specifications, engineering tradeoffs, logistic requirements, reliability, and cost.

CHAPTER 9

Principles and Applications of Optical Holography

B. J. THOMPSON

College of Engineering and Applied Science, University of Rochester, Rochester, New York

I. INTRODUCTION

It is now 30 years since the invention of holography by Gabor[1–3] and a very considerable literature has been developed that covers the fundamental concepts, the basic techniques, the properties of appropriate recording materials, and the various applications of holography. The field is mature enough that a variety of monographs and texts are now available which both

[1] D. Gabor, *Nature* (*London*) **161**, 777 (1948).
[2] D. Gabor, *Proc. R. Soc. London Ser. A* **197**, 454 (1949).
[3] D. Gabor, *Proc. Phys. Soc.* **B64**, 449 (1951).

ISBN 0-12-408606-3

introduce the concepts and survey the field (Stroke,[4] DeVelis and Reynolds,[5] Smith,[6] Kock,[7] Caulfield and Lu,[8] Lehmann,[9] Vienot *et al.*,[10] Collier *et al.*,[11] Butters,[12] Cathey,[13] Ostrovsky[14]). Tanaka and Nakajima,[15] under the auspices of the Japan Optical Engineering Research Association, have prepared, on a continuing basis, an excellent bibliography on the published literature on holography that is an invaluable resource to workers in the field. Major conferences held on an international basis provide an excellent source of information as well as an historic perspective to the subject of holography and how it is applied (Thompson,[16] Robertson and Harvey,[17] Vienot *et al.*,[18] Barrekette *et al.*,[19] Camatini,[20] Thompson and DeVelis,[21] Aprahamian,[22] Erf,[23] Stroke *et al.*,[24] Greguss,[25] Marom *et al.*[26]). A series of specialized conference proceed-

---

[4] G. W. Stroke, "An Introduction to Coherent Optics and Holography," 2nd ed. Academic Press, New York, 1969.

[5] J. B. DeVelis and G. O. Reynolds, "Theory and Applications of Holography." Addison Wesley, Reading, Massachusetts, 1967.

[6] H. M. Smith, "Principles of Holography." Wiley, New York, 1969 (see also 2nd ed., 1975).

[7] W. E. Kock, "Lasers and Holography." Doubleday, Garden City, New York, 1969.

[8] H. J. Caulfield and S. Lu, "The Applications of Holography." Wiley, New York, 1970.

[9] M. Lehmann, "Holography: Technique and Practice." Focal Press, London, 1970.

[10] J.-Ch. Vienot, P. Smigielski, and H. Royer, "Holographie Optique, Developpements-Applications." Dunod, Paris, 1971.

[11] R. J. Collier, C. B. Burkhardt, and L. H. Lin, "Optical Holography." Academic Press, New York, 1971.

[12] J. N. Butters, "Holography and Its Technology." Peter Peregrinus, London, 1972.

[13] W. T. Cathey, "Optical Information Processing and Holography." Wiley, New York, 1974.

[14] Yu. I. Ostrovsky, "Holography and Its Applications" (Translated from the Russian by G. Leib). Mir Publ., Moscow, 1977.

[15] S. Tanaka and T. Nakajima, "Bibliography on Holography." Japan Optical Engineering Research Association, Tokyo.

[16] B. J. Thompson (ed.), Holography, *Proc. SPIE* **15** (1968).

[17] E. R. Robertson and J. M. Harvey, "The Engineering Uses of Holography." Cambridge Univ. Press, London and New York, 1970.

[18] J.-Ch. Vienot, J. Bulabois, and J. Pasteur, "Applications of Holography." Univ. de Besançon, 1970.

[19] E. S. Barrekette, W. E. Kock, T. Ose, J. Tsujiuchi, and G. W. Stroke (eds.), "Applications of Holography." Plenum Press, New York, 1971.

[20] E. Camatini (ed.), "Optical and Acoustical Holography." Plenum Press, New York, 1972.

[21] B. J. Thompson and J. B. DeVelis (eds.), Developments in holography, *Proc. SPIE* **25** (1972).

[22] R. Aprahamian (ed.), Engineering applications of holography, *SPIE Symp. Proc.* (1972).

[23] R. K. Erf (ed.), "Holographic Nondestructive Testing." Academic Press, New York, 1974.

[24] G. W. Stroke, W. E. Kock, Y. Kikuchi, and J. Tsujiuchi (eds.), "Ultrasonic Imaging and Holography—Medical, Sonar, and Optical Applications." Plenum Press, New York, 1974.

[25] P. Greguss (ed.), "Holography in Medicine." IPC Science and Tech. Press, Guildford, England, 1976.

[26] E. Marom, A. A. Friesem, and E. Wiener-Avnear (eds.), Applications of holography and optical data processing, *Proc. Int. Conf., Jerusalem, August 23–26, 1976*. Pergamon, Oxford, 1977.

ings on acoustical holography is also available.[27–32] Finally, a series has recently appeared entitled "Advances in Holography"[33] that has now reached four volumes with many more promised. Each of these volumes contains one or more specialized review articles.

The purpose of this chapter is to discuss the basic concepts of holography and describe the fundamentals of the various methods of recording holograms in both transmission and reflection. The important parameters associated with these methods will be given, and the characteristics of the resultant images described. A discussion of some of the major application areas will be included.

A hologram is formally described as the recorded intensity distribution of the interference pattern produced by the interaction of an unknown field with a known or reproducible background or reference field. The fields referred to here could be optical, acoustical, or microwave (however, only optical fields will be considered in this chapter). The recorded intensity pattern, the hologram of the unknown field, may then be illuminated with the background or reference field to reproduce the unknown field. It will be recalled that under normal circumstances, when the intensity of a field is recorded, only the intensity of that field can be reproduced; that is, the amplitude is known but the phase is lost. The interference term provides a method of storing the phase information as positional information of the interference fringes; the amplitude information is the envelope on the fringes.

Gabor's invention of holography consisted of a realization that the classical Fresnel diffraction pattern of an opaque object in an otherwise transparent field, can be thought of as an interference pattern between the light diffracted by the opaque object and the undeviated light passing through the transparent part of the field. A record of this interference pattern is a hologram. Gabor recognized for the first time that a record of this particular interference pattern (a specific type of Fresnel diffraction pattern) did indeed contain all the necessary information to uniquely determine the cross section of the diffracting object. Furthermore, he noted that merely illuminating the hologram with the original beam that illuminated the opaque object would recreate the complex amplitude of the original field in the plane in which it was recorded and that the field would propagate as if the object was still there; hence an image of the original object could be formed.

---

[27] A. F. Metherell, H. M. A. El-Sum, and L. Larmore (eds.), "Acoustical Holography," Vol. 1. Plenum Press, New York, 1969.

[28] A. F. Metherell and L. Larmore (eds.), "Acoustical Holography," Vol. 2. Plenum Press, New York, 1970.

[29] A. F. Metherell (ed.), "Acoustical Holography," Vol. 3. Plenum Press, New York, 1971.

[30] G. Wade (ed.), "Acoustical Holography," Vol. 4. Plenum Press, New York, 1972.

[31] P. S. Green (ed.), "Acoustical Holography," Vol. 5. Plenum Press, New York, 1974.

[32] N. Booth (ed.), "Acoustical Holography," Vol. 6. Plenum Press, New York, 1975.

[33] N. H. Farhat (ed.), "Advances in Holography," Vols. 1–4. Dekker, New York, 1975.

Once the basic concepts had been understood in detail, then a variety of methods of forming the hologram in both transmitted and reflected light were designed. The major groups of these holographic methods will be discussed in Section III. It is worth noting that the concepts and original implementation of the holographic process were developed long before the laser was invented and developed; however, the laser did enable considerable versatility to be achieved in the design of holographic systems.

The motivation for Gabor's invention of the holographic process was a specific application in electron microscopy. The development of holographic methods and systems has been largely driven by the desire to achieve new techniques that could be applied to the solution of real problems. Hence applications of holography are of particular importance and a short review of the major applications will be given in Section IV.

II. PRINCIPLES OF BASIC TYPES OF HOLOGRAMS

There appears to be an almost infinite variety of methods and optical configurations for producing holograms. This apparent array of possibilities is often confusing to workers who seek to apply holographic techniques to the solution of technological problems. A systematic approach to the methods of recording is worthwhile, since it helps establish the basic principles and, hence, the fundamental basis for the various methods. The approach taken here will be to first consider holograms formed in transmitted light, since historically these were the first areas developed, and a discussion of these types of holograms will clearly illustrate the basic principles. Holography in reflected light can then be considered as a separate and, of course, very important class.

A. In-Line Holograms Formed in Transmitted Light

To begin this discussion, we may consider Fig. 1a. Here the opaque letters AO located in an otherwise transparent field are illuminated with a collimated beam of coherent light. We are interested in the field at a distance z from the object. In purely classical optical terms, this problem would be considered to be a problem in Fresnel diffraction—that is, that the field in the x–y plane is the Fresnel diffraction pattern associated with the input distribution in the ξ–η plane. Analyzing the problem in this way is certainly correct; however, it misses the important insight that is the essence of holography. Gabor recognized that the propagation problem from the ξ–η plane to the x–y plane could be stated somewhat differently. He considered the resulting field in the x–y plane to be made up of two components; the light diffracted by the object itself and the undiffracted light. These two components are, of course, coherent with respect

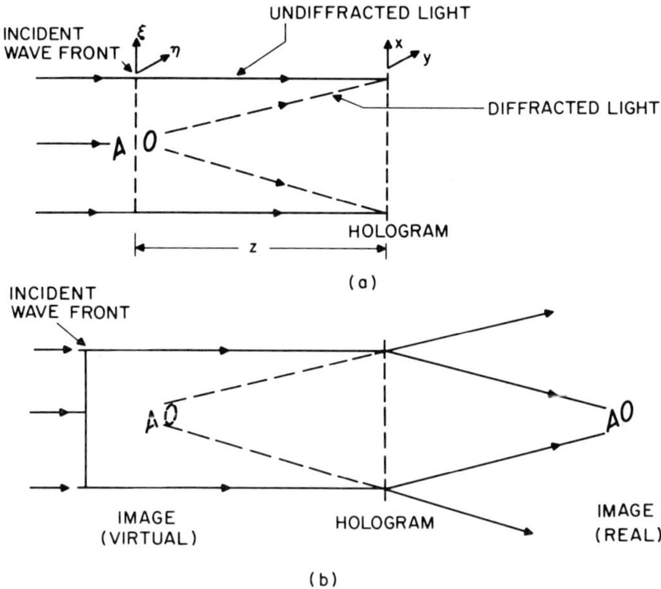

FIG. 1. The in-line holographic process. (a) Formation of the hologram. (b) Formation of the image from the hologram.

to each other and will interfere. This recorded interference pattern is the hologram.

The process can be represented formally by considering the field diffracted by the object to be given by $\psi_0(x)$ and

$$\psi_o(x) = a_o(x) \exp[i\phi_o(x)] \tag{2.1}$$

where $a_o(x)$ is the amplitude and $\phi_o(x)$ is the phase. For simplicity, Eq. (2.1) has been written with only the x variable, although, of course, in general the functions are two-dimensional. The undeviated field also has a complex amplitude in the x plane given by

$$\psi_r(x) = a_r(x) \exp[i\phi_r(x)] \tag{2.2}$$

The resultant field $\psi_h(x)$ is then given by

$$\begin{aligned} \psi_h(x) &= \psi_o(x) + \psi_r(x) \\ &= a_o(x)e^{i\phi_o(x)} + a_r(x)e^{i\phi_r(x)} \end{aligned} \tag{2.3}$$

When this field is recorded, it is the intensity that is detected thus

$$\begin{aligned} I_h(x) &= \psi_h(x)\psi_h{}^*(x) \\ &= a_o{}^2(x) + a_r{}^2(x) + a_o(x)a_r(x)\{\exp[i(\phi_o(x) - \phi_r(x))] \\ &\quad + \exp[-i(\phi_o(x) - \phi_r(x))]\} \end{aligned} \tag{2.4}$$

1. *In-Line Fresnel Holograms*

The field $\psi_o(x)$ is determined by the standard propagation calculation. Thus when $z < (2a)^2/\lambda$ (where $2a$ is the maximum dimension of the opaque object and λ is the wavelength of the illuminating light) and z is large enough that small angles can be assumed, then the field is a Fresnel diffraction pattern and

$$\psi_o(x) = C\frac{e^{ikz}}{z} e^{ikx^2/2z} \int_{-\infty}^{\infty} O(\xi) \exp\left(\frac{ik\xi^2}{2z}\right) \exp\left(\frac{-ikx\xi}{z}\right) d\xi \tag{2.5}$$

where $O(\xi)$ is the amplitude transmittance of the object and C is a constant. Under this condition the hologram is often referred to as a Fresnel hologram—the name being determined by the nature of the diffracted field $\psi_o(x)$.

2. *In-Line Fraunhofer or Far-Field Holograms*

If $z \gg (2a)^2/\lambda$, then the recording plane is said to be in the far field of the object, and the diffracted field associated with the opaque object is that traditionally referred to as Fraunhofer diffraction. Equation (2.5) then becomes

$$\psi_o(x) = C\frac{e^{ikz}}{z} e^{ikx^2/2z} \int_{-\infty}^{\infty} O(\xi) \exp\left(\frac{-ikx\xi}{z}\right) d\xi \tag{2.6}$$

The integral in Eq. (2.6) reveals that the field $\psi_o(x)$ is the Fourier transform of $O(\xi)$, multiplied by a quadratic phase term in x. A hologram formed under these conditions is usually called a Fraunhofer or far-field hologram—again, the name is determined by the nature of the diffracted field.

B. The Image Formed from the Hologram

The basic concept of holography can be understood by writing Eq. (2.4) in a simple form by assuming that the undeviated field is just a constant in both amplitude and phase. Thus $\psi_r(x)$ can be written as unity and hence Eqs. (2.3) and (2.4) become

$$\psi_h(x) = 1 + a_o(x)e^{i\phi_o(x)} \tag{2.7}$$

$$I_h(x) = 1 + a_o^2(x) + a_o(x)\{e^{i\phi_o(x)} + e^{-i\phi_o(x)}\} \tag{2.8}$$

We will now assume that the intensity given in Eq. (2.8) is recorded linearly so that when the recording is coherently transilluminated, the resultant amplitude transmittance is proportional to the recorded intensity. Further, we will assume that the illumination is the constant field $\psi_r(x) = 1$; then the field immediately after the hologram is $\psi_h'(x)$ given by

$$\psi_h'(x) = [1 + a_o^2(x)] + a_o(x)e^{i\phi_o(x)} + a_o(x)e^{-i\phi_o(x)} \tag{2.9}$$

Equation (2.9) contains three terms, the second and third of which are of particular interest. The second term represents a wave propagating from the original object as if it was still there, and hence is associated with a virtual image; the third term is the so-called conjugate wave that is 180° out of phase with the second term and is a wave that converges to a real image of the object at a conjugate location. This process is illustrated in Fig. 1b. The light associated with the first term in Eq. (2.9), as well as the light associated with the two image terms, is all propagating in the same direction and, hence, if a screen is placed in the plane containing the real image, then the other image is out of focus in that plane; specifically, this out-of-focus component is another hologram of the original object formed at a distance equivalent to $2z$. The presence of this hologram causes some degradation of the image—as does the presence of the term $[1 + a_o^2(x)]$.

Holographic imaging is then a two-step imaging process that can be carried out without the use of a lens. However, a focusing element must have been automatically built into the hologram, otherwise the real image could not be formed. This focusing element is often referred to as a zone lens and is formed by the interaction (interference) between the two fields when the hologram is recorded. The information about the object is the envelope function on this interference term. Further insight into the process can now be gained by pursuing this idea further. We may inquire about the nature of the image of a point object. The hologram of a point object will be produced by the interaction of a spherical wave generated by the point and the undiffracted plane wave (see Fig. 2a). Thus, from Eq. (2.7) and Eq. (2.8)

$$\psi_h(x) = 1 + b\exp(ikx^2/2z) \tag{2.10}$$

$$I_h(x) = (1 + b^2) + 2b\cos(kx^2/2z) \tag{2.11}$$

where b is a constant that will be less than unity. Figure 2b gives a representation of a hologram of a point object. The intensity distribution is reminiscent of a classical zone plate but is a continuously varying function rather than a binary distribution. Because of its functional form, this distribution is sometimes called a sine-wave zone plate. Figure 2c shows a similar distribution in a one-dimensional case which is the intensity distribution in the interaction of a cylindrical wave with a plane wave. If the recorded distributions of Figs. 2b and 2c are illuminated with a collimated beam of light, then they would produce two beams; one diverging from a virtual focus and the other converging to a real focus (see Fig. 2d).

For an extended object, the hologram is formed by a superposition of the complex amplitude described by Eq. (2.10) for every point in the object. Thus every point in the object will have its own zone lens associated with it and when the hologram is illuminated, two images will be formed.

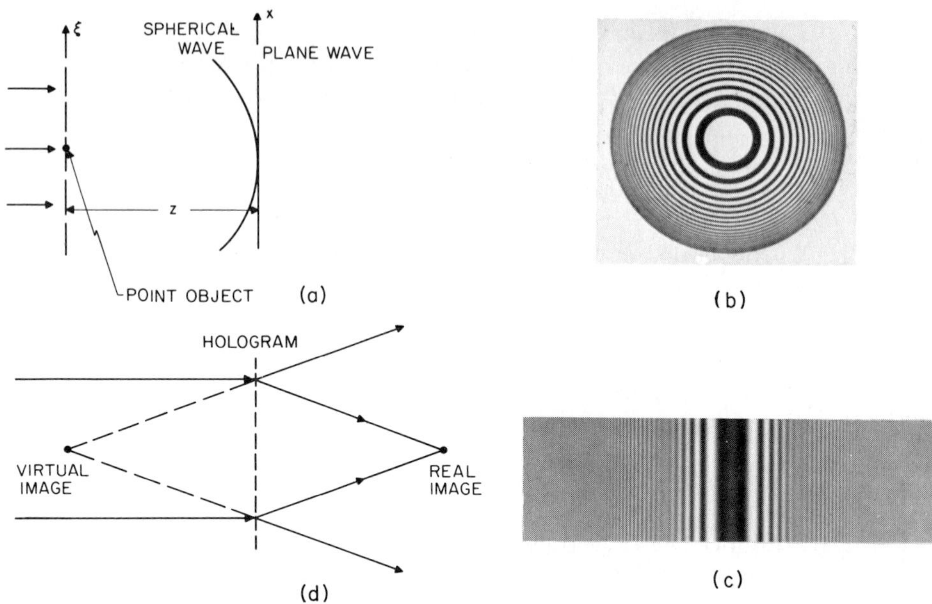

FIG. 2. Holographic process with a point object. (a) Formation of the hologram by the interference of a plane wave and a spherical wave. (b) A representation of the hologram of a point object, i.e., the zone lens. (c) The hologram of a line object—the one-dimensional zone lens. (d) Images formed from the hologram.

Figure 3 illustrates the result for an in-line Fresnel hologram.[34,35] The example chosen here is the hologram of a wire of diameter 160 μm illuminated with a collimated beam of coherent light of wavelength 5461 Å. Figures 3a,b,c show the intensity profile across the hologram; since the problem is essentially one-dimensional, the distribution is constant in the perpendicular direction. The intensity profile of the real images formed from these holograms is shown, for comparison, in Figs. 3d,e,f. The image is present but is, of course, degraded by the presence of the hologram associated with the other image. It is of some consequence to note that if the hologram is recorded on film and then the film is normally processed as a negative, the image of the wire is bright on a dark background. Hence the profile of these images, shown in Figs. 3d–f, is shown with the wire image bright. Naturally, if the hologram had been recorded as a positive, then the image of the wire would have been dark on a bright background.

[34] G. A. Tyler, Far-Field Holography: A Reassessment and New Applications. Ph.D. Thesis, Univ. of Rochester, 1978.

[35] G. A. Tyler and B. J. Thompson, to be submitted.

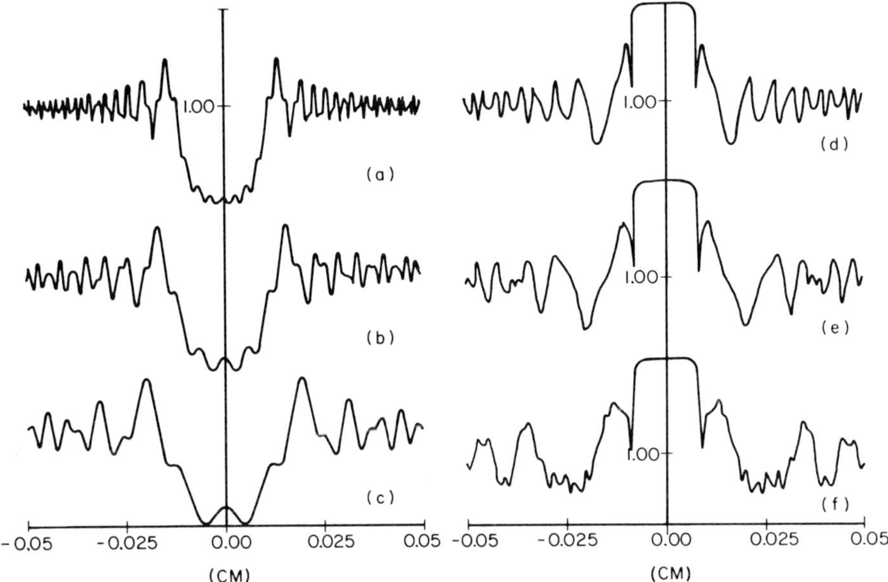

FIG. 3. An illustration of the in-line Fresnel hologram. (a), (b), and (c) illustrate a trace across the hologram of a 160-μm diameter wire illuminated with a collimated beam of light of wavelength 5461 Å. (a) $z \simeq 1.0$ cm. (b) $z \simeq 2.0$ cm. (c) $z \simeq 4.0$ cm. A trace across the intensity profile of the image formed from these holograms is shown in (d), (e) and (f) [after G. A. Tyler, Ph.D. Thesis, Univ. of Rochester (1978)].

Figure 4 illustrates the results for an in-line far-field hologram of the same wire. The hologram is now a more slowly varying function but is extended over a much larger distance. The component parts of this hologram are very readily seen in Fig. 4a; the rapid variation is, of course, the zone-lens structure and the envelope on that zone lens is the diffraction pattern of the cross section of the wire which is a function $\mathrm{sinc}(kdx/z)$, where $2d$ is the width of the wire. A photograph of this hologram is shown in Fig. 4b, which, of course, exhibits the features described above. Figure 4d shows a photograph of the image produced from the hologram of Figs. 4a and 4b. It will be noted that the image of the wire is quite good and that a far-field hologram is seen corresponding to a distance $2z$, but this hologram is well removed from the image. It is this important property that makes far-field holograms of great value in particle size analysis (see Section IV,A,2). Finally, in Fig. 4c, for comparison with the image shown in Figs. 3d–f, the intensity profile across the image is formed from a far-field hologram of the same wire used in Fig. 3, and is shown in the same scale. This comparison shows rather nicely the advantage of a far-field hologram when circumstances permit that type of hologram to be recorded.

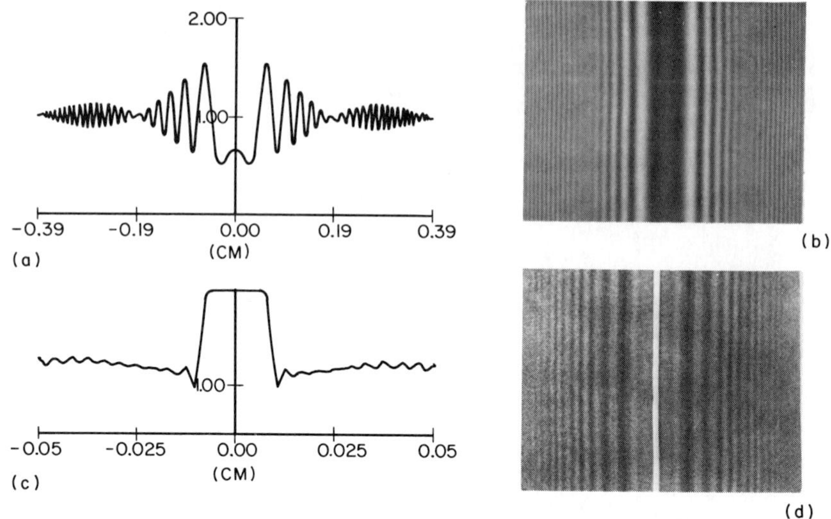

FIG. 4. An illustration of in-line far-field holograms. (a) Profile of the intensity distribution in the far-field hologram of a wire. (b) Photograph of the hologram. (c) Image formed from holography" (H. J. Caulfield, ed.), Chapter I. Academic Press, New York, 1979.

In-line holograms are excellent for illustrating the basic principles of holography, but apart from the very significant specific application of the far-field type to the analysis of small objects in dynamic situations,[36,37] they have been superseded by other methods that allow the two images to be separated.

C. THE OFF-AXIS REFERENCE BEAM HOLOGRAPHIC PROCESS

While Gabor had indicated that it might be possible to separate the two images by a suitable optical arrangement, it was left to Leith and Upatnieks[38,39] to actually devise and implement the method. This work provides the cornerstone of most of modern holographic technology. In principle, the method is very simple and hence elegant. The idea is to use a separate controllable reference beam for producing the hologram. Figure 5a illustrates this technique in one of its many forms. The object is again the two letters AO that are opaque in an otherwise transparent field. Collimated illumination is again used but part of the incident beam is passed through the prism to illuminate the plane in which the hologram is to be recorded. This is the reference beam and it is inclined at an

[36] B. J. Thompson, J. H. Ward, and W. R. Zinky, *Appl. Opt.* **6**, 519 (1967).

[37] B. J. Thompson and J. H. Ward, *Sci. Res.* No. 10, 37 (1966).

[38] E. N. Leith and J. Upatnieks, *J. Opt. Soc. Am.* **52**, 1123 (1962).

[39] E. N. Leith and J. Upatnieks, *J. Opt. Soc. Am.* **53**, 1377(1963).

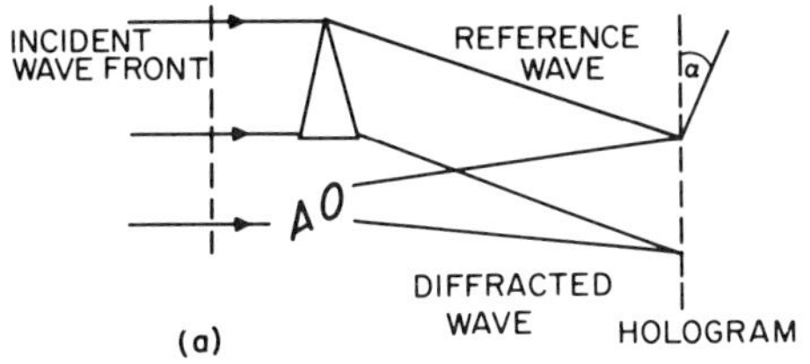

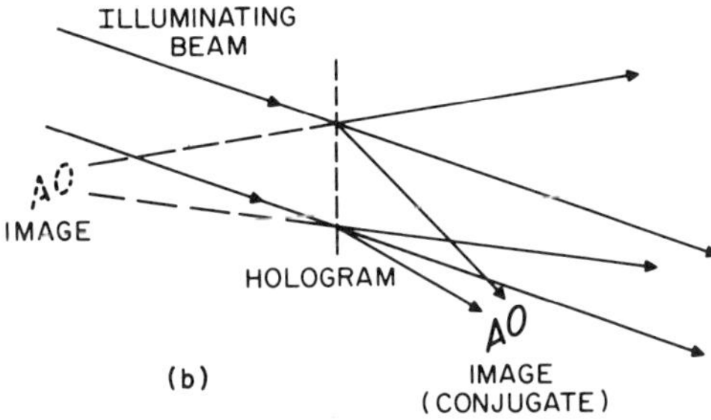

FIG. 5. The off-axis reference beam holographic method. (a) Formation of the hologram. (b) Formation of the images.

angle to the forward diffracted beam. The hologram is then the recorded interference pattern between the forward diffracted beam and this separate reference beam. In terms of Eq. (2.7), the field in the hologram plane, is now

$$\psi_{\mathrm{h}}(x) = e^{ik\sin\alpha} + a_{\mathrm{o}}e^{i\phi_{\mathrm{o}}(x)} \tag{2.12}$$

where α is the angle between the reference beam and the forward beam. When this hologram is recorded and reilluminated, then again two images are formed as illustrated in Fig. 5b. An image (virtual in this particular geometry) is formed in the location of the original object, a second image (real in this particular geometry) is formed on the other side of the hologram, and the light propagating to this image is at an angle determined by a term $\exp(ik2\sin\alpha)$. The zone-lens concept discussed earlier is still appropriate, but now the zone lens is actually an off-axis section of that lens.

Thus if α is correctly chosen, then the light associated with the two images is propagating in different nonoverlapping directions; furthermore, the forward diffracted light is also nonoverlapping with the image beam. This separation of the light associated with the two images is, of course, of extreme importance in holography and produced the surge on work in the field starting in 1962. It will also be noted that the object no longer has to be mainly transparent, since the required reference beam is separately controlled. Thus the letters AO could be transparent letters in an otherwise opaque field.

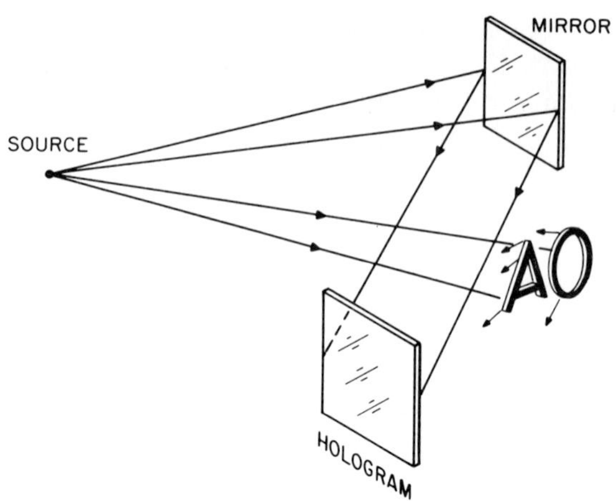

FIG. 6. Formation of the hologram in reflected or scattered light.

This off-axis reference beam technique, together with the use of the laser with its relatively long coherence length, allows holograms to be formed in reflection. This process is illustrated in Fig. 6. The light is scattered by the object onto the holographic recording plane and the required reference beam is provided by the mirror. If the object is removed and the hologram illuminated with the original reference beam, then a virtual image is located at the position of the original object as if it was still there. The image, of course, has all the properties associated with the object as viewed through the window which is determined by the hologram.

Coherent systems are notorious for the noise problems associated with dust and minute flaws in various components. A much better cosmetic quality of the image can be achieved by using a diffuser in the beam that illuminates the object. Effectively, this produces a spatial redundancy and the hologram can be considered to be a superposition of a multiplicity of holograms just as if multiple coherent sources had been used.

D. Parameters of the Holographic Process

Once the hologram is formed and recorded, most of the properties are already determined. For example, the resolution of the final image is determined by the extent of the hologram. That is, the half-width of the impulse response of the two-step imaging process is given by $1.22z\lambda/2h_e$, where z is the distance from the object to the hologram and h_e is the effective width of the hologram. This effective width might be the actual width of the recording material if the hologram is actually recorded to the edge of the recording material. Often, however, other factors will limit the extent of the hologram; the most important of these is the actual resolution limit of the recording material, since the interference fringes which produce the hologram get closer and closer together as the extent of the fringe field increases. DeVelis and Reynolds[5,40] have used the inverse of the half-width of the impulse response as a resolution criterion, and show that the resolution limit in object space, RL_{ob}, is given by

$$RL_{ob} = 1.64[l - (\sin \alpha/\lambda)] \tag{2.13}$$

where l is the resolution of the film in line pairs per millimeter and α is the angle of the reference beam (see Fig. 5a).

The effective width of the hologram also determines the field of view since the hologram is acting as a window through which the object is being viewed.

In this discussion so far, plane-wave illumination has been assumed, in which case the magnification is unity. Clearly a spherical wave may be used in both the formation of the hologram and in the subsequent image formation. Let the radius of these spherical waves be R_1 and R_2, respectively. Then Eq. (2.13) becomes

$$RL_{ob} = \frac{1.64(1 - \sin \alpha/\lambda)}{1 - z_1/R_1} \tag{2.14}$$

where z_1 is the distance from the object to the recording plane. The magnification also changes as is given by

$$n = [1 - (z_1/R_1) - (z_1/z_2)]^{-1} \tag{2.15}$$

where z_2 is the distance from the hologram to the image.

A useful parameter that is often quoted for holographic systems is the space–bandwidth product, which defines the number of resolution elements of the holographic recording. Since the hologram is not an image, the size of the hologram can sometimes be extended so that the resolution requirements of the recording medium can be reduced.

The brightness of the image is nearly always of concern to holographers, particularly those working on displays. An important parameter is the diffraction efficiency of the hologram, which is defined as the power diffracted into one of the image beams (which is equivalent to a first-order wave of a diffraction grating) to the power illuminating the hologram. Usually, holograms are illuminated with a uniform beam and hence the diffraction efficiency becomes the ratio of the intensity diffracted into the first order to the input intensity. Holograms recorded on film to produce the normal density as a function of exposure will have a diffraction efficiency of about 6%. Bleached photographic film has been used to increase the efficiency since the density hologram is converted to a phase hologram. It is the refractive index difference between the silver halide crystallites

[40] J. B. DeVelis and G. O. Reynolds, Fresnel holography, *in* "Handbook of Optical Holography" (H. J. Caulfield, ed.), Chapter I. Academic Press, New York, 1979.

(e.g., silver chloride has a refractive index of just over 2) and the gelatin (refractive index 1.54) that produces the phase record. Several bleaching techniques have been discussed in the literature and required formulations listed (see, e.g., Lamberts and Kurtz[41] and Zeck[42]). The basic process consists of exposing the film (e.g., say Kodak 649F). It is then developed, fixed, and hardened. The film is then refixed, bleached, stabilized, and dried. Most of the methods involve liquid bleaching; however, Graube[43] has successfully used a bromine vapor method. With these bleached films, the diffraction efficiency can be improved to better than 30%.

The variety of recording materials that have been considered for holography is almost endless and it is not possible to discuss them here; however, a number of reviews do exist that can be referred to with confidence.[42,44–46]

A further increase in diffraction efficiency that can approach 100% can only be achieved if so-called "thick holograms" are used. The hologram is now formed within a volume of recording material.

III. SPECIALIZED HOLOGRAPHIC METHODS

The fundamental types of holograms were discussed in Section II, together with their properties. Naturally, as the subject has developed, a variety of rather specialized techniques have been used for specific purposes. A short discussion here will help with this relatively large array of terminologies.

A. IMAGE PLANE HOLOGRAMS

Holograms can be formed in an image plane[47,48] but the part of the object not in focus in that plane will actually be a Fresnel diffraction pattern plus the actual reference or background beam. For the image plane, however, the hologram can be formed with very limited spatial coherence and, hence, reduced speckle. Application of image plane holograms have been limited to some problems in microscopy[49] and in holographic subtraction.[50]

---

[41] R. L. Lamberts and C. N. Kurtz, *Appl. Opt.* **10**, 1342 (1971).

[42] R. G. Zeck, Data Storage in Volume Holograms. Ph.D. Thesis, Univ. of Michigan, 1974.

[43] A. Graube, *Appl. Opt.* **13**, 2942 (1974).

[44] R. L. Kurtz and B. Owen, *Opt. Eng.* **14**, 393 (1975).

[45] J. Gladden, Review of Photosensitive Materials for Holographic Recordings, ETL0128. Eng. Top. Lab., Fort Belvoir, Virginia (in press).

[46] H. M. Smith (ed.), "Holographic Recording Materials." Topics in Applied Physics, Vol. 20, Springer-Verlag, Heidelberg, 1977.

[47] L. Rosen, *Appl. Phys. Lett.* **9**, 337 (1966).

[48] G. W. Stroke, *Phys. Lett.* **23**, 325 (1966).

[49] M. E. Cox and K. J. Vahala, *Appl. Opt.* **17**, 1455 (1978).

[50] K. Bromley, M. A. Monahan, J. F. Bryant, and B. J. Thompson, *Appl. Opt.* **10**, 174 (1971).

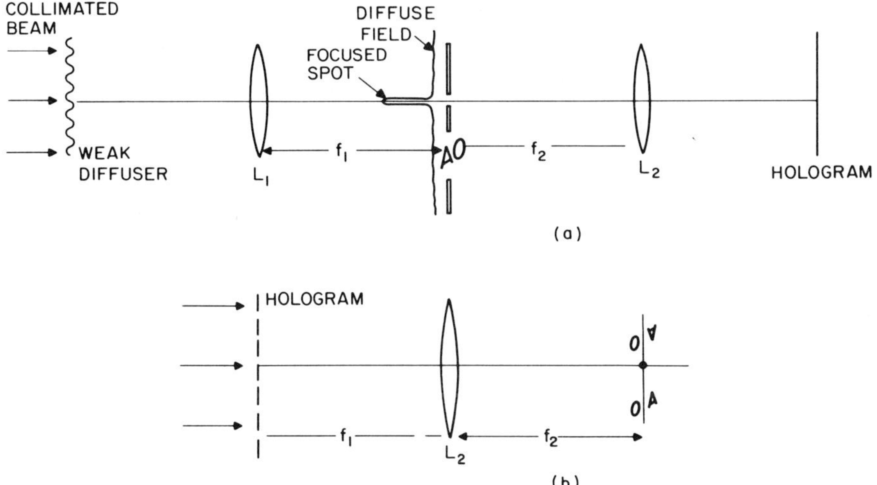

FIG. 7. Method of producing a Fourier transform hologram with diffuse object illumination. (a) Formation of the hologram. (b) Formation of the twin images [after E. N. Leith and J. Upatnieks, *J. Opt. Soc. Am.* **53**, 1377 (1973)].

B. FOURIER TRANSFORM HOLOGRAMS

In Section II,A,2, holograms were described that used a Fraunhofer diffraction pattern of the object formed in the far field. Classically there is another important method of producing a Fraunhofer diffraction pattern using a lens. When the far-field approximation is used, it is merely an approximation to infinity, since the Fraunhofer pattern of an object illuminated with a collimated beam is actually produced at infinity. This field at infinity can be imaged by a lens to produce a real image in the focal plane which is the Fraunhofer diffraction pattern of the object. If a reference beam is now added to the diffraction pattern, then a hologram is formed. Figure 7a shows one configuration for this purpose.[51] A collimated beam of light is incident on a weak diffuser; the lens L_1 produces a focused spot of light on axis and a diffuse illumination surrounding it, part of which is used to illuminate the object. The hologram is recorded in the Fraunhofer plane of that object produced by lens L_2. Since the object is placed in the front focal plane of lens L_2, the focused spot in that plane may be isolated with a small aperture, and then produces a collimated reference beam. The difference in lateral position of the small aperture and the object position produces an effective off-axis reference beam. To produce an image from the hologram, it is reilluminated with a collimated beam (Fig. 7b) and a lens produces the two

[51] E. N. Leith and J. Upatnieks, *J. Opt. Soc. Am.* **54**, 1295 (1964).

images as illustrated. Initially, this type of hologram was called a generalized hologram[38] and later a Fraunhofer diffraction hologram[51] or Fraunhofer hologram.[52] The name Fourier transform hologram,[4] however, seems to have become the standard accepted terminology.

The importance of these Fourier transform holograms is in optical processing applications in which the hologram is used as a spatial filter in the Fourier transform plane of a given input.[53]

C. Local Reference Beam Holography

Since it is not always possible to have complete control over the beam illuminating the object, it is sometimes advantageous to be able to produce the reference beam from the light scattered by the object. If a portion of the light from the object is focused onto a small aperture, then the light diverging from this aperture can act as the reference beam.[54] This method has been used in the study of laser beam parameters[55] and this application can be used as an example of this particular type of hologram.

Figure 8a shows a schematic diagram of a local reference beam system used initially for the study of a He–Ne laser wave front and is designed for use with a low-carrier frequency. In this particular example, the laser output to be studied was produced by a hemispherical cavity (30 cm) and could be run in a number of low-order transverse modes. The system is a modified Mach–Zehnder interferometer so that the path difference between the two beams could be made very small. The local reference beam is provided by the beam passing through the beam splitter BS, mirror M_1, pinhole P, and collimating lens L. The beam to be studied reaches the hologram plane via mirror M_1. The two beams are combined with the hologram plane by a beam combiner BC.

This same system has been used for the study of pulsed outputs of both ruby and Nd-glass lasers. Figure 8b shows the intensity pattern of a ruby laser pulse reconstructed from a local reference beam holographic record. The reconstructed field can be tested interferometrically, and Fig. 8c shows fringes crossing the field indicating the relative phases of the various parts of the output. For this illustration, the ruby cavity was purposefully misaligned to produce some higher-order transverse modes.

Because image plane holograms can be produced with very much relaxed coherence conditions, local reference beam holography can be most successfully used in an image plane configuration.

---

[52] A. W. Lohmann and D. P. Paris, *Appl. Opt.* **6**, 1739 (1967).

[53] A. Vander Lugt, *IEEE Trans. Inform. Theory* **IT-10**, 139 (1964).

[54] H. J. Caulfield, J. L. Harris, H. W. Hemstreet, and J. G. Cobb, *Proc. IEEE* **55**, 1758 (1967).

[55] C. Roychoudhuri and B. J. Thompson, *Opt. Eng.* **13**, 347 (1974).

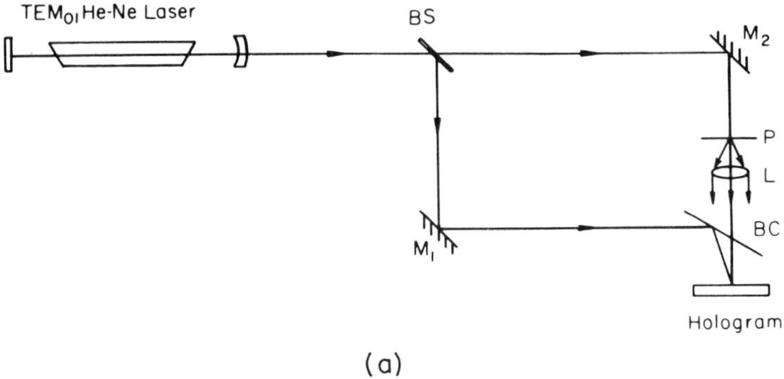

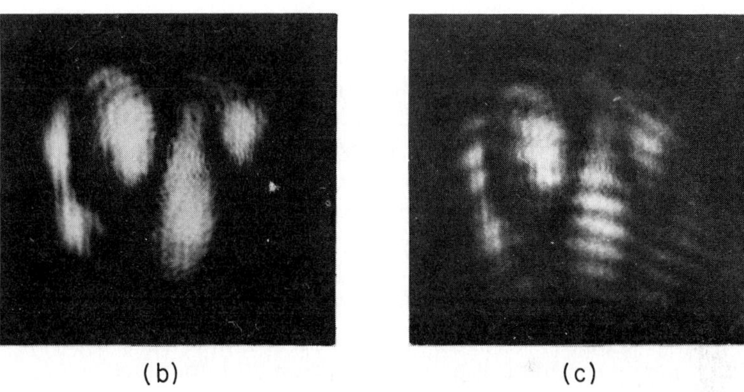

FIG. 8. An example of local reference beam holography. (a) Schematic diagram of system. (b) Intensity formed from reconstructed output of pulsed ruby laser. (c) Interferogram formed from reconstructed output of pulsed ruby laser [after C. Roychoudhuri and B. J. Thompson, *Opt. Eng.* **13**, 447 (1974)].

D. WHITE-LIGHT HOLOGRAMS

For many technological purposes, such as particle size measurement, nondestructive testing, and vibration analysis, the use of the laser in the final image-forming process is not particularly objectionable. However, for display purposes, there is a great desire to have the image formed with ordinary thermal light. Clearly, if this can be achieved with good image brightness, then it can be more convenient, more pleasing to the viewer, and remove the possible safety hazards associated with the laser radiation. It is not surprising, then, that considerable effort has gone into trying to achieve systems that can be used, at least in the reconstruction stage, with ordinary thermally produced light.

1. *Lippmann–Bragg Holograms*

Denisyuk[56] used the basic principles of the Lippmann color photographic method to produce both monochrome and full color images with white light. In the Lippmann color imaging process, the photographic emulsion is backed by a reflective layer so that when the image is formed in the emulsion, some of the image light passes through the emulsion and is reflected back out through the emulsion. The dimensions of the emulsion are such that within that thin layer, the incident light passing into the emulsion can interfere with the light reflected back through the emulsion. A standing wave interference pattern is produced which results in a series of developed silver layers in the emulsion parallel to the surface. When the developed emulsion is reilluminated, the image is seen in the original color, since reflected light from the planes in the developed emulsion adds constructively at the appropriate angle in accordance with Bragg's law. The multiple layers within the emulsion act as a filter for the reflected light to select the appropriate color.

Denisyuk used this same method but recorded a scattered field from an object rather than an image field. The basic system is shown in Fig. 9a. The light illuminating the object passes through the emulsion and illuminates the object on the other side. Light is then scattered by the object back into the emulsion to interfere with the incident beam to produce a hologram. Naturally this requires considerably more spatial and temporal coherence than the Lippmann color image process. The resulting interference fringes are recorded in the depth of the emulsion and parallel to the emulsion surface in a way related to the Lippmann process, except that it is a hologram that is formed rather than merely an image. When the hologram is illuminated with an incident white-light beam, an image is formed as illustrated in Fig. 9b, the appropriate color being selected by the planes of silver within the emulsion. This process is related to the diffraction of x rays by a crystal first discussed by Bragg, and to the use of the term "Bragg diffraction" when a light beam is diffracted by an acoustical standing wave in a thick medium. Hence, the Denisyuk method gives holograms that are sometimes referred to as Lippmann–Bragg holograms. They have also been called white-light holograms, since the image is formed with white light and the hologram acts as its own selective wavelength filter.

Naturally, the basic method introduced by Denisyuk can be implemented with different configurations but most particularly with a separate reference wave. The white-light reconstruction is, of course, retained (see, e.g., Stroke and Labeyrie[57]).

---

[56] Yu. N. Denisyuk, *Sov. Phys.-Dokl.* **7**, 543 (1962). Yu. N. Denisyuk, *Opt. Spectrosc.* **15**, 279 (1964). Yu. N. Denisyuk, *Opt. Spectrosc.* **18**, 152 (1965).

[57] G. W. Stroke and A. E. Labeyrie, *Phys. Lett.* **20**, 368 (1966).

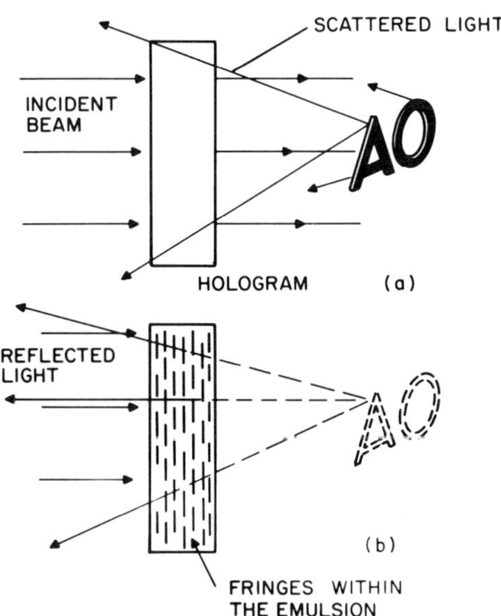

FIG. 9. The white-light hologram formed using the Denisyuk method (Lippmann–Bragg hologram). (a) Formation of the hologram. (b) Formation of the image.

2. *Rainbow Holograms*

An interesting new method was pioneered starting in 1969 by Benton[58] which is often referred to as rainbow holography. A hologram of the object of interest is made in the usual way. Since the major thrust of this work was for displays, the objects of interest were, of course, three-dimensional scattering objects. This hologram is then illuminated and a real image produced (see Fig. 10). A second hologram is made using the light from the real image and a new reference beam. Before the second hologram is recorded, a horizontal slit is placed across the first hologram. Hence, the second hologram contains complete image information about the object which was a diffuse object but the parallax in the vertical plane will be lost. The second hologram is illuminated from the opposite direction (180°) from the original reference beam direction. When the second hologram is illuminated, a virtual image is produced of both the object and the slit with all the light actually passing through the spatial region containing the vertical image of the slit. The image of the object will be seen only if the eyes are placed in the location of the virtual image of the slit. If white light

[58] S. A. Benton, *J. Opt. Soc. Am.* **59**, 1545A (1969). S. A. Benton, *Opt. Eng.* **14**, 402 (1975). S. A. Benton, "Holography in Medicine" (P. Greguss, ed.). IPC Science and Technology Press, London, 1975.

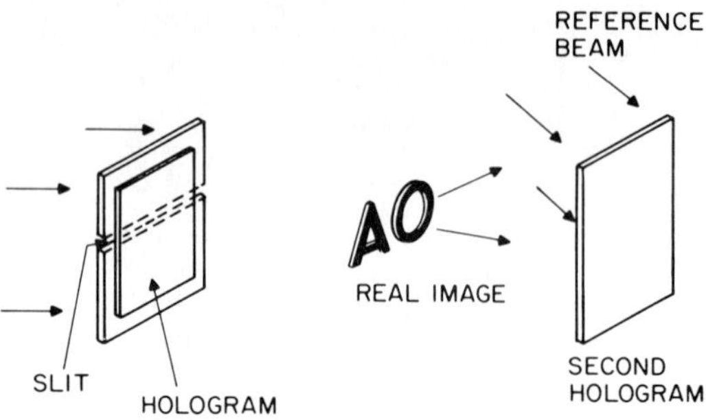

FIG. 10. Formation of the rainbow hologram.

is used to illuminate the second hologram, then by the normal process of diffraction, the virtual image of the slit will be displaced vertically, and will, in fact, be a white-light spectrum. Hence the color of the image will be determined by the actual location of the eyes. Thus as the eyes are moved, the color will change through the spectrum; hence the name rainbow hologram. The vertical parallax is, of course, not restored with the white light illumination. Figure 11 shows a white-light transmission hologram. The hologram is illuminated with a microscope illuminator shown in the upper part of the figure.

3. *Multiplex Holograms*

A slightly different method has been perfected by Cross and his co-workers[59] at The Multiplex Company in San Francisco. This process starts by making a short movie film of a subject that is located on a rotating table. The illumination is ordinary studio lighting. The approximately 1000 frames from this movie are then made into individual holograms. The first frame is illuminated coherently and a large cylindrical lens is used to allow the hologram to be formed in a vertical line which fills a slit placed over the recording material. A hologram of this first frame is then recorded on the recording film and is about 20-cm long and 0.1-mm wide. The recording film is then moved and the second frame recorded as a hologram. This process is repeated until the 1000 frames are each recorded as holograms. The result is a series of 2000 individual holograms (since each frame is actually printed twice), each representing a different view of the object. The hologram is arranged into a cylindrical shape (or an arc of a cylinder) and illuminated with white light. On looking through the hologram, a virtual image is seen inside the cylinder that looks very three-dimensional. The eyes are actually seeing a number of images produced by the section of the hologram that is being looked through. Again, the vertical parallax has been discarded in the process.

[59] The Multiplex Company, 454 Shotwell Street, San Francisco, California.

FIG. 11. An example of the image formed from a rainbow hologram [after S. A. Benton, *Opt. News* **3**, No. 2, 16 (1977)].

A variety of such holograms have been made and are available from The Multiplex Company. Figure 12 shows one view of perhaps the most famous of these holographic images. It is entitled "Kiss II" and was made by Cross and Brazier. In this hologram, as the observer moves around the cylinder of film, a young lady first winks and then blows a kiss to the observer. The effect is very realistic and, of course, fascinating.

The method just described uses a series of ordinary two-dimensional images to build up a synthetic hologram. Related techniques have been developed that are sometimes called holographic stereograms. The individual frames of a movie are illuminated coherently and a lens is used to spread the light over a diffuser. A hologram is recorded on a film with a slit placed in front of the recording material. Each strip is a hologram. The total series of holograms then produces an image of different views of the object similar to stereo but with a multiplicity of views rather than just two.[60]

Holograms made up of a composite of individual recordings were first suggested by Pole in 1967.[61] Other workers have used a fly's-eye lens to produce a multiplicity of views of the object. This so-called "integral photography" can be illuminated with a collimated beam to produce a three-dimensional image.

[60] D. J. DeBitetto, *Appl. Opt.* **8**, 1740 (1969).

[61] R. V. Pole, *Appl. Phys. Lett.* **10**, 20 (1967).

FIG. 12. An example of multiplex holography (The Multiplex Company, San Francisco, California).

E. INCOHERENT HOLOGRAMS

Since a hologram has been described as an interference pattern, it would appear to be an essentially coherent process. However, under the correct conditions, the coherence conditions can be relaxed as, for example, in the case of image plane holograms. The terminology "incoherent hologram" has been used in the literature to describe a variety of processes. Conventional holograms can be produced by light from an incoherent source since the light gets more spatially coherent as it propagates.

An object incoherently illuminated can be used to produce a hologram if the field is caused to interfere with itself, since there is point by point coherence between the two fields.[62,63] The image from this type of hologram is formed using coherent illumination in the usual way.

---

[62] A. W. Lohmann, *J. Opt. Soc. Am.* **55**, 1555 (1965).

[63] G. W. Stroke and R. C. Restrick, *Appl. Phys. Lett.* **7**, 229 (1965).

The temporal coherence requirements can be relaxed if achromatic fringes can be formed. Several techniques have been suggested[62–66] and implemented, but not a great deal of use has been made of the particular method.

1. *Coded-Aperture Imaging*

Synthetic holograms can be produced by an incoherent shadow-casting method. This technique is usually called coded aperture imaging. Each point in the object produces a shadow of an aperture. (The aperture may have a number of forms but a zone-plate aperture will illustrate the method.) Thus each point in an object will produce a shadow of the zone plate, whose position and scale will be dependent upon the particular geometry. If these shadows are recorded, then there will be an intensity distribution on the film produced by the incoherent superposition of the shadow zone plates. This record can now act in a manner similar (but not identical) to a hologram and an image can be produced. This method was first suggested by Mertz and Young[67] for use in astronomy. Considerable effort has now been extended to develop the technique for x ray and nuclear tomography as a medical diagnostic.[68–70] More recently, the method has been extended to x-ray imaging of laser fusion plasmas.[71]

The developments in coded-aperture imaging require a separate discussion on their own; and since they are, strictly speaking, not holograms, no further description is appropriate here.

A final form of synthetic hologram can be mentioned—the computer-generated hologram. The two-dimensional hologram is actually calculated and then written out on film, usually as a binary distribution. This hologram is then treated as if it had been recorded in the usual way.

IV. APPLICATIONS OF HOLOGRAPHY

The fascination of holography has produced a wide variety of possible applications. Unfortunately, not all of these applications have seen the light of

---

[64] E. N. Leith and J. Upatnieks, *J. Opt. Soc. Am.* **57**, 975 (1967).

[65] R. H. Katyl, *Appl. Opt.* **11**, 1241 (1972).

[66] E. N. Leith and B. J. Chang, *Appl. Opt.* **12**, 1957 (1973).

[67] L. Mertz and N. O. Young, *Proc. ICO Conf. Opt. Instrum. London* (K. J. Habell ed.), p. 305, Chapman and Hall, London, 1965. See also, L. Mertz, "Transformations in Optics." Wiley, New York, 1962.

[68] H. H. Barrett, W. W. Stoner, D. T. Wilson, and G. D. DeMeester, *Opt. Eng.* **13**, 539 (1974).

[69] W. L. Rogers, L. W. Jones, and W. H. Beierwalters, *Opt. Eng.* **12**, 13 (1973).

[70] H. Weiss, E. Klotz, R. Linde, G. Rabe, and U. Tiemens, *Opt. Acta* **24**, 305 (1977).

[71] N. M. Ceglio, X-ray imaging, *Proc. SPIE* **106**, 55 (1977).

day outside the research laboratory in which they were conceived. Some of the applications were even ill-conceived in the first place! Nevertheless, it is important for the proper development of a field that each and every reasonable idea be pursued to the point where it is shown to be feasible. The actual use of these results depends not only on their feasibility but also on whether there is a need. It still remains true that if the problem under study can be solved with standard techniques, then that will probably be easier and cheaper than attempting to solve the same problem holographically. However, if the problem cannot be solved by conventional methods, then maybe holography is worth a try.

Applications of holography fall into three major areas. The first of these is in imaging for display purposes, high-speed photography, specialized microscopy, etc. The second category is concerned with holographic interferometry and contour generation. The third area involves the use of a hologram as an optical element in its own right, either as a lens or grating, and as a filter in coherent optical processing. Some applications have already been mentioned briefly in an earlier section of this chapter, where special types of holograms were discussed.

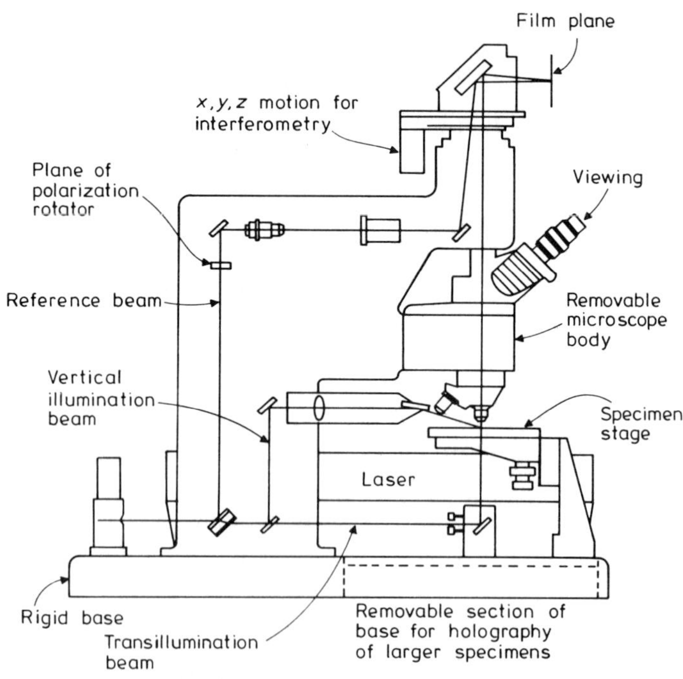

FIG. 13. A schematic diagram of the holographic microscope [after R. F. Van Ligten, Holography, *Proc. SPIE* **15**, 75 (1968)].

A. Holographic Image Formation

The image formed by most optical systems is, in fact, three-dimensional; but once the image is recorded, it becomes a two-dimensional representation with a depth of field determined by the actual system used and, of course, no parallax can be recovered. Holography is a two-step imaging process which allows the three-dimensional information to be stored and a true three-dimensional image to be recreated.

1. *Photographic Applications*

Holographic methods of recording three-dimensional information are useful for data gathering when it is not possible to record the total information with a conventional movie camera system. Photography of a transient event when the exact location of that transient event is not known in advance is a difficult technological problem. This problem can be overcome by holographic imaging. However, many of the examples of this require pulsed lasers to achieve the short exposure times necessary. Many papers in the literature discuss the specific application of holography to the recording of high-speed transient events[72] with pulsed ruby lasers. (A variety of papers can be found in the various Proceedings of the International Congresses on High Speed Photography.[73–76]) Holography can be combined with some of the standard techniques of high-speed photography to yield a series of holograms that provide detailed information about changes in the object. For example, high-speed holocinematography has been used with spatial multiplexing to produce a series of holograms and then, of course, a series of images.[77]

2. *Microscopy*

The range of applicability of these methods extends from quite large objects several centimeters in size, down to the smallest resolvable objects of a few microns. At this smaller scale, the application is, of course, in microscopy. The conventional microscope has been modified to become a holographic microscope. Van Ligten[78] has pursued the work and has developed a commercial holographic microscope that is shown schematically in Fig. 13. The sample on

---

[72] R. E. Brooks, L. O. Heflinger, R. F. Wuerker, and R. A. Briones, *Appl. Phys. Lett.* **7**, 92 (1965).

[73] O. Helwick (ed.), *Int. Kongr. Kurzzeitphotogr., 7th, Zurich, 1965*. Verlag, Dr. Othmar Helwick, Darmstadt, 1967.

[74] N. R. Nilsson and L. Hogberg (eds.), "High Speed Photography." Wiley, New York, 1968.

[75] P. J. Rolls (ed.), High speed photography, *Proc. Int. Congr., 11th.* Chapman and Hall, London, 1975.

[76] W. G. Hyzer and W. G. Chace (eds.), *Proc. Int. Congr. High-Speed Photography, 9th.* SMPTE Publ. 25 (1970).

[77] K. J. Ebeling and W. Lauterborn, *Opt. Commun.* **21**, 67 (1977).

[78] R. F. Van Ligten, Holography, *Proc. SPIE* **15**, 75 (1968).

the microscope stage is illuminated with light from a helium–neon laser which then passes up through the body and lenses of the microscope in the usual way. The reference beam for producing an off-axis hologram is brought to the hologram plane by a path outside the body of the microscope itself. Once the hologram has been recorded, a permanent record of that sample has been obtained for future study. The advantage here occurs when a transient event has to be studied in a relatively thick sample. Resolution can be obtained at about 1 μm and is limited by speckle unless some spatial or temporal redundancy techniques are used. Application of this type of microscopy has been limited, but useful applications appear to have been found in the study of polymers.[79]

Holographic microscopy has been pursued in a number of other ways that do not use a conventional microscope. One of the most successful of these other methods has been the application of inline far-field holography for a variety of particle and droplet size analysis tasks. While the method was first discussed in 1963,[80] the reader is referred to a recent detailed reassessment of the method[81] and a review of the various applications.[82] The method was first applied to the study of naturally occurring fog in order to measure the droplet size distribution and to determine the evolution with time of this distribution. After the successful application of this technique to fog particle measurements, a variety of other problems have been usefully studied. These applications include cloud chamber droplets, explosively generated aerosols, marine plankton, rocket engine exhaust, two-phase flow, glass fibers (see Thompson[82] for a more detailed discussion). Current applications include inhalation studies of dust particles, blood cell measurements, wind tunnel and dust erosion studies, cloud studies, and snowflake measurements.[83]

Figure 14 shows an example of the use of in-line far-field holograms. Figure 14a shows a portion of the hologram of a field of particles. The image of just a small portion of the total field is shown in Fig. 14b. The images are seen in dark field and the individual pollen grains are approximately 20 μm in diameter.

Off-axis holographic methods have also been used in particle size analysis work, and in some systems both techniques are used.[83]

3. *Displays*

The magic of the holographic image has caught the imagination of scientists, engineers, and artists alike and, hence, the idea of a holographic display has received considerable attention. Commercial-type displays can be successfully

[79] R. F. Cournoyer, M. B. Rhodes, and S. Siggia, *J. Polym. Sci.* **13**, 1023 (1975).

[80] B. J. Thompson, *J. SPIE* **2**, 43 (1963).

[81] G. A. Tyler and B. J. Thompson, *Opt. Acta* **23**, 685 (1976).

[82] B. J. Thompson, *J. Phys. E. Sci. Instrum.* **7**, 781 (1974).

[83] J. D. Trolinger, *Opt. Eng.* **14**, 383 (1975).

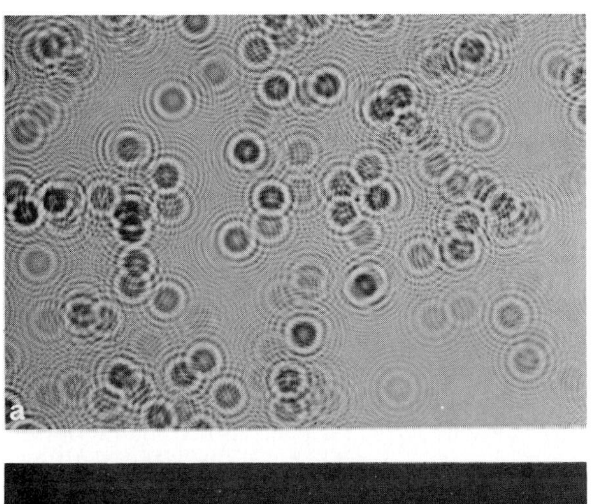

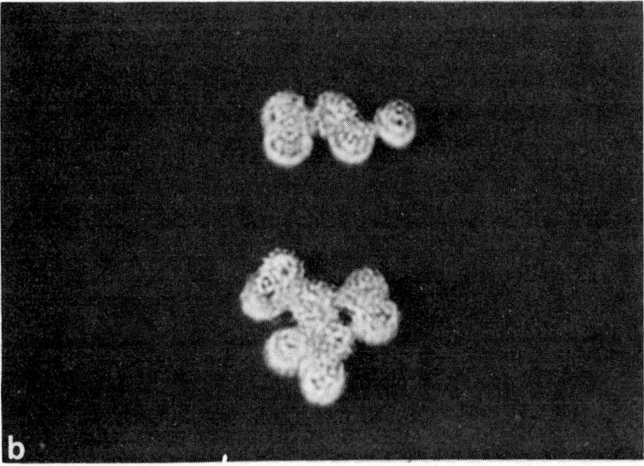

FIG. 14. Particle size analysis using far-field holography. (a) Hologram of a field of pollen particles. (b) Image formed from a portion of the hologram.

produced by the techniques devised by Benton and by The Multiplex Company. Before these methods were developed, Jeong[84] had made 360° holograms that were viewable with white light.

Holography can be used as a new medium for artists, and considerable progress has been made with this medium. The Museum of Holography in New York City is now well established and has a variety of permanent and special exhibits, both at the headquarters and in their traveling shows.

The desire to make good displays which accurately portray the original object has some other important applications. A permanent record can be made

[84] T. H. Jeong, *J. Opt. Soc. Am.* **57**, 1396 (1967).

of a variety of works of art so that as these objects are eroded or damaged, accurate repairs can be made.[85] Holograms of rare objects can be extremely useful for a student in studying those objects without the necessity of travel to a remote museum. For this latter purpose, it is essential that very accurate color reproduction is obtained. Further work is necessary on this aspect of holographic displays.

4. *Information Storage*

Since a hologram is actually a method of storage of information in a unique way, it can be thought of as a method of coding information. This idea has not escaped notice, and the idea of using holography as a means of storing information (the information can be an image, a diffraction pattern, a binary code sequence, etc.) has many intriguing aspects. Since the holograms can easily be recorded so that any one piece of the hologram contains the total image, this record is much less susceptible to dust and scratches. A scratch will not eliminate a part of the object but merely limit slightly the overall resolution. This built-in redundancy is, of course, only obtained at a price. The price is the required resolution of the recording material on which the hologram is stored.

Leith and Upatnieks[51] first illustrated that two separate images could be stored on the same holographic plate and be recovered independently. In this first result, the two objects were two transparencies placed at different distances from the holographic recording material. In general, a number of inputs can be stored by varying the angle and orientation of the reference beam. This holographic method is related to the earlier method of optical multiplexing, but now the holograms rather than the image inputs themselves are being multiplexed. It is not surprising that the holographic storage of a multiplicity of images suffers from the same type of problem as the earlier multiplexing methods. The limiting problem is noise.

Most of the current work on holographic storage, or holographic memories as they have been called, uses individual small holograms rather than overlapping holograms. Recent reviews give an excellent overview of this application.[86,87] The emphasis is on storing information in digital formats so that the memories can be readily coupled into large computers. Two approaches are being worked on. The first of these is called a block or page-organized memory and is intended to provide fast random access. The second type of memory is sequentially organized and is designed for a slower access. Figure 15 shows a system designed by Nelson *et al.*[88] that stores information in both machine-

---

[85] J. F. Asmus, G. Guttari, L. Lazzarini, G. Musumeci, and R. F. Wuerker, *Stud. Conservat.* **18**, 49 (1973).

[86] G. R. Knight, *Opt. Eng.* **14**, 453 (1975).

[87] B. Hill, "Advances in Holography" (N. Farhat, ed.), Vol. 3. Dekker, New York, 1976.

[88] R. H. Nelson, A. Vander Lugt, and R. G. Zeck, *Opt. Eng.* **13**, 429 (1974).

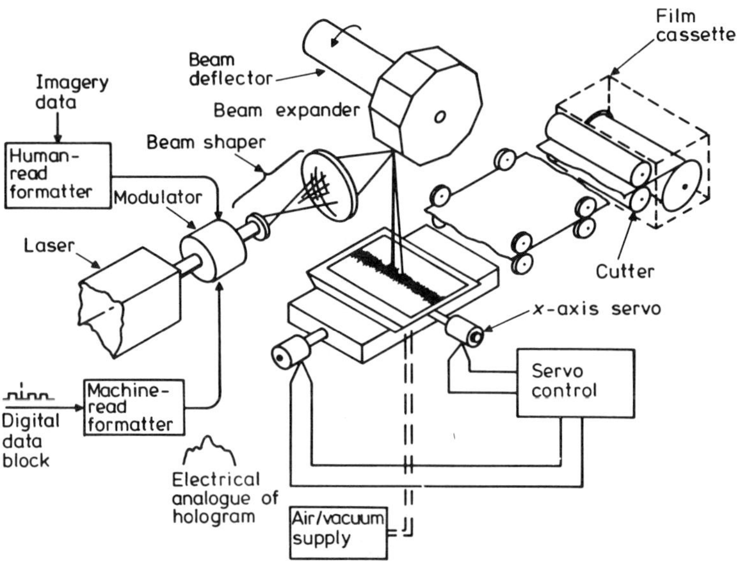

FIG. 15. Schematic diagram of a sequentially stored holographic memory system [after Nelson *et al.*, *Opt. Eng.* **13**, 432 (1974)].

readable (digital) and human-readable (image) formats. An initial digital input is converted into a synthetic Fourier transform hologram and is recorded on film with a modulator and laser scanning subsystem. This Fourier transform hologram is the machine-readable information. The direct image information is also written on film with the same laser scanning system to provide the human-readable storage. In use, this system stores the information on a 4×6 in. microfiche format that contains five rows of 12 images for a total of 2.5×10^6 bits. The system can also operate in a different format to fill the entire microfiche with holograms. This particular system has been built and used for the storage of cartographic information.

It is still too early to tell whether holographic memories have a significant future or not.

B. Holographic Interferometry

The most widespread use of holography appears to be in nondestructive testing using interferometric methods. Holographic interferometry allows interference methods to be applied to situations that would not normally be considered suitable for interferometric testing.

The concept of holographic interferometry occurred to several groups of workers all at about the same time, and since they were working independently, they deserve equal credit.[72,89–94] The basic concept is to produce two or more three-dimensional images of the same object under different conditions. If these two three-dimensional images are coherent with respect to each other, then any changes that have taken place between the first image and the second image will produce a set of interference fringes across the image which relate to these changes.

There are various methods of implementing this idea. One method requires that a hologram of the object under test be made and then the virtual image is formed directly on the object itself; the object is, of course, coherently illuminated by the virtual image. If the object is now moved slightly, fringes will result. In a second method, two holograms can be recorded on the same photographic plate and the two virtual images produced. Again, if the object has been moved slightly after the first hologram has been recorded, the overlapping image will contain fringes.

There is no reason why more than two holograms cannot be recorded in this way and a multiplicity of coherent images formed. However, it is not obvious that the resulting fringes could have any easy interpretation in general, since the fringes will be related to the difference between each pair of holograms. There is a special case, however, when this particular method has validity. The particular case is in the study of vibrating objects. In this application, a multiplicity of holograms are recorded by making a time-averaged exposure in the holographic recording material while the object is in periodic motion. When the composite image produced from this time-averaged hologram is formed, fringes are observed which contain loci of equal amplitude of vibration of the object.

1. *Holographic Interferometry with a Single Hologram*

The object under test is illuminated coherently and an off-axis hologram recorded with the light scattered by the object. After processing, this hologram is repositioned in its original location and illuminated. On looking through the holographic plate, a virtual image will be seen at the location of the original object and, if the object is still there, will be identical with the object. If the hologram is not relocated properly, the image and the object will be shifted relative to each other, and fringes will be observed. These fringes are an aid to precise relocation of the hologram. With the hologram in the correct location, the object can now be changed and fringes will result in the viewed image; the

[89] J. M. Burch, The 1965 Viscount Nuffield memorial paper, *Prod. Eng.* **44**, 431 (1965); J. M. Burch, *Z. Angew. Math. Phys.* **16**, 111 (1965).

[90] R. J. Collier, E. T. Doherty, and K. S. Pennington, *Appl. Phys. Lett.* **7**, 223 (1965).

[91] K. A. Haines and B. P. Hildebrand, *Phys. Lett.* **19**, 10 (1965).

[92] M. H. Horman, *Appl. Opt.* **4**, 333 (1965).

[93] R. L. Powell and K. A. Stetson, *J. Opt. Soc. Am.* **55**, 1593 (1965).

[94] C. Reid and N. R. Wall (see Burch[89] who gives credit to these authors).

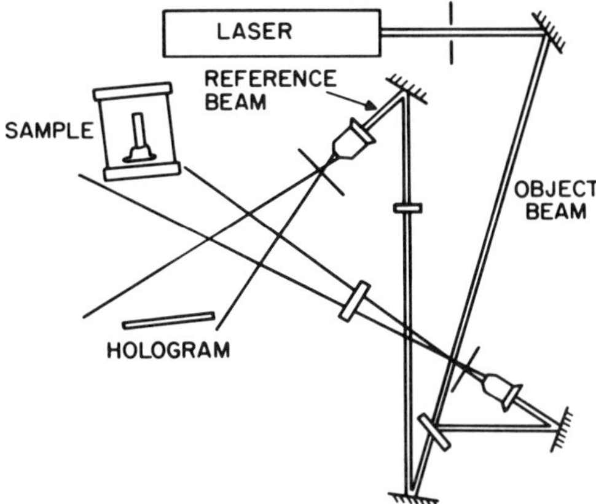

FIG. 16. Schematic diagram of a holographic interferometer for the study of magnetostriction [after S. Ramon, M.S. Thesis, Univ. of Rochester (1971)].

fringes being produced by interference between the illuminated object and the image from the hologram. The detailed analysis and interpretation of these fringes has been carried out for a variety of circumstances.[95–98] For small displacements of the object, the amplitude of the two interfering terms can be considered to be equal and the fringes are cosine fringes, where the argument of the cosine is the phase difference between the two interfering terms.

Since the fringes are viewed directly and they can be followed as the object is changed, this particular method is sometimes referred to as real-time holographic interferometry. Figure 16 shows a schematic diagram of a system for using holographic interferometry for the study of magnetostriction.[99] Light from a helium–neon laser illuminates the sample via the optical path shown and the reference beam falls on the same side of the holographic plate as the light scattered by the sample. The sample is a bar of ferromagnetic material and a hologram is made of the face of this sample. A ring is positioned around the sample to act as a reference surface. The image from the hologram is then formed back onto the sample with a deliberate tilt error so that vertical fringes are produced (Fig. 17a). A field is then applied to the ferromagnetic material and the real-time fringes observed as the field is increased. Figures 17b–e show examples of these fringes. The fringes have rotated, which is a combination of bending (Guillemin effect) and longitudinal contraction (Joule effect). The

---

[95] E. B. Aleksandrov and A. M. Bonch-Bruevich, *Sov. Phys.-Tech. Phys.* **12**(2), 258 (1967).

[96] N. Abramson, *Appl. Opt.* **8**, 1235 (1969).

[97] G. M. Brown, R. M. Grant, and G. W. Stroke, *J. Acoust. Soc. Am.* **45**, 1166 (1969).

[98] K. Stetson, *Optik* **29**, 386 (1969). K. Stetson, *J. Opt. Soc. Am.* **64**, 1 (1974).

[99] S. Ramon, The Measurement of Magnetostriction by Real-Time Holographic Interferometry. M.S. Thesis, Univ. of Rochester (1971).

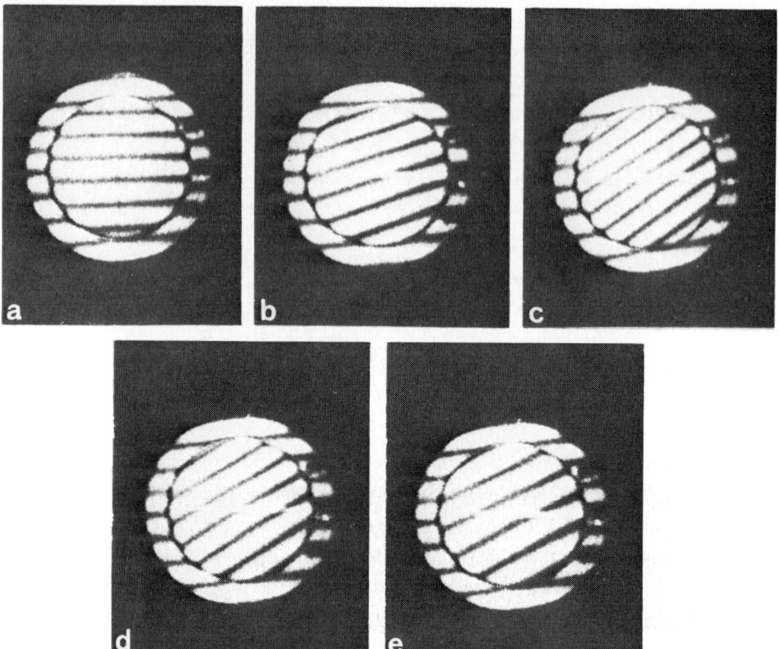

FIG. 17. Interference fringes produced by the holographic interferometer of Fig. 16. This sample is a ferromagnetic material placed in a radial magnetic field. (a) Fringes formed with no field applied but with a deliberate tilt between the holographic image and the object. (b–e) Fringes formed as the field is increased; fringe rotation is observed [after S. Ramon, M.S. Thesis, Univ. of Rochester (1971)].

rotation of the fringes increases with increasing field. The various magnetostrictive effects have been examined with this method.[99]

The use of a deliberate repositioning error to create a set of fringes can be extremely useful, as seen above. This idea has also been used to study vibrations. The image formed by the hologram is again displaced slightly with respect to the object to produce a set of straight-line fringes. Now, if the object is caused to vibrate, the observer will see the fringes disappear in the region of nodes and remain at antinodes.

2. *Holographic Interferometry with Two Holograms*

Since the illustrations in Fig. 17 are of particular situations, these same results could be obtained by making two holograms. The first hologram is made as described in Section IV,B,1. The magnetic field is then applied and a second hologram made, preferably on the same photographic plate, after a tilt has also been introduced. When the two holograms are illuminated, the two images will be superpositioned and the result will be the same as that illustrated in, for example, Figs. 17b–e. This technique has some slightly different properties than the single hologram method. It is usually employed without the deliberate repositioning errors and is often called lapsed-time or frozen-fringe holographic interferometry and has been used in such classic work as the tire testing experiments and in the study of debonding in laminated structures.

The possible applications of both real-time and the frozen-fringe methods are many, varied, and include strain measurement, photoelasticity studies, and gas flow analysis, to name but a few. The techniques draw their strength from being nondestructive testing methods.

3. *Time-Averaged Holographic Interferometry*

The multiple exposure or time-averaged method of holographic interferometry has found considerable use in the nondestructive testing of vibrating objects. This particular method was first reported in the literature in 1965.[93] The holographic recording system is the basic off-axis reference beam method that is illustrated, for example, in Fig. 16. The difference is that while the hologram is being recorded, the object is caused to vibrate in one of its modes, and will undergo many complete oscillations during that exposure. When the hologram has been developed and is reilluminated, the image is seen to contain fringes of equal amplitude of vibration of the object. Figure 18 shows the image formed of an acoustical transducer vibrating in its first harmonic.[100] This particular example is quite simple to illustrate the idea, but the method can and has been used for studying quite complex objects, including turbine blades, musical instruments, quartz crystals, and disk brakes. Understandably, there are a number of variations of this method (see, e.g., Archbold and Ennos[101]), but space does not permit a discussion of them here.

C. Holographic Optical Elements

It is possible to consider using a hologram as an optical element in its own right, particularly in a coherent optical system, since the hologram can modify the amplitude and phase of the field passing through it in a predetermined way. Since a lens-like structure is automatically built into a hologram, the idea of making use of the structure as a lens is readily apparent. It is also possible to consider the interference term that is the hologram as a grating-like structure; it is only a small step from that idea to the suggestion that holographic gratings could be designed and fabricated. Gratings and lenses are conventional optical elements; however, holography has a very important role to play in providing an important method of fabricating spatial frequency filters for coherent optical processors.

---

[100] M. A. Monahan and K. Bromley, *J. Acoust. Soc. Am.* **44**, 1225 (1968).

[101] E. Archbold and A. E. Ennos, *Nature (London)* **217**, 942 (1968).

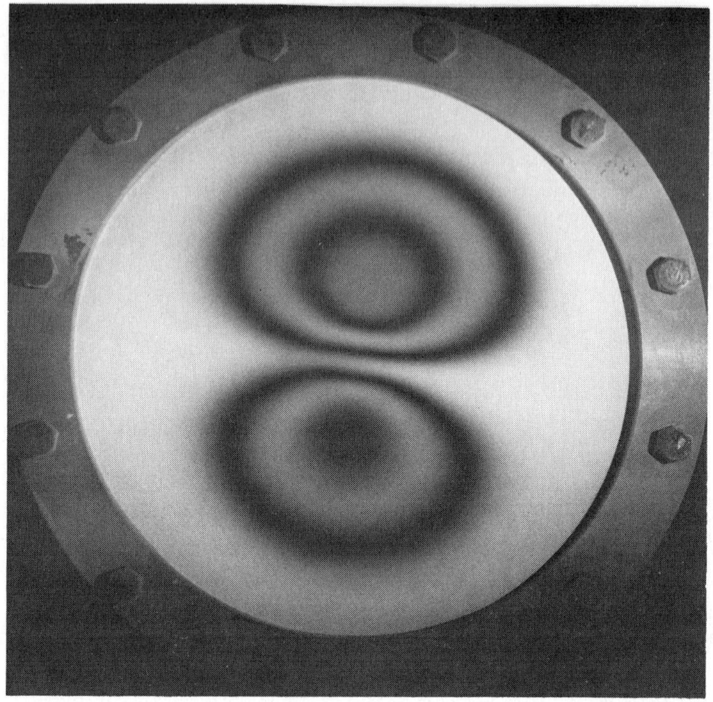

FIG. 18. Image of an acoustic transducer using the method of time-averaged holographic interferometry [after M. A. Monahan and K. Bromley, *J. Acoust. Soc. Am.* **44**, 1225 (1968)].

1. *Holographic Lenses, Mirrors, and Gratings*

Rogers' pioneering papers[102] on the zone-lens concept in holography provide the foundation not only for the image-forming process in holography but also for the idea of a holographic lens. The concept is straightforward, but the implementation is more difficult. Ray tracing programs have been developed that can be used for holographic lens design.[103] The production of high-efficiency devices requires very careful fabrication, usually in materials like hardened dichromated gelatin.

The most important possible use of holographic lenses occurs when it is necessary to locate a lens on an unusual surface. An example of this application is in the field of head-up displays.[104]

Since the lenses being discussed are holographic, of course, they can be used in a reflective mode and become a replacement for a mirror.

---

[102] G. L. Rogers, *Proc. R. Soc. Edin.* **A63**, 193 (1950). G. L. Rogers, *Proc. R. Soc. Edin.* **A63**, 313 (1951).

[103] J. N. Latta, *Appl. Opt.* **10**, 599, 609, 666, 2698 (1971).

[104] D. H. Close, *Opt. Eng.* **14**, 408 (1975).

Holographic gratings have already achieved some considerable commercial success since serious work started in 1969.[105] The advantages of holographic gratings are in their efficiency and accuracy. In addition, the scattered background light can be reduced significantly over conventional ruled gratings. Gratings with 10,000 grooves/mm can be obtained with widths up to 60 cm, and plane and concave configurations can be made.

In coherent optical processing systems, it is often necessary to fabricate a filter that will modify both the amplitude and phase of an optical field in a continuous manner over the two-dimensional plane. The fabrication of such filters had been a major problem for many years until Vander Lugt suggested the use of the holographic filter.[53] In the initial work, a matched filter was prepared which required a complex filter that was the inverse of the Fourier transform of the signal to be recognized. Thus the final output of the coherent optical processing system was a delta function. This type of filter is performing an autocorrelation. The elegance of Vander Lugt's method is that the filter can be generated directly on a hologram (usually a Fourier transform hologram) without having to calculate the actual filter function required. The matched filtering method using holographic filters has found a number of applications and paved the way for the development of hybrid optical processing systems.

The use of holographic filters has been extended to a much broader class of optical processing methods than the example given above, with particular attention being given to aberration balancing and image enhancement. Any detailed discussion of this particular application of holography would require a much more detailed discussion of optical processing methods; the interested reader is referred to the extensive literature on this subject (see, e.g., Stroke *et al.*[106] and Thompson[107]).

V. CONCLUSIONS

The basic concepts of holography are quite straightforward. However, the methods of implementation are quite diverse and often the specific methods are devised with applications in mind. In this chapter, most of the major methods currently in use in holography have been described and discussed.

The search for useful applications of holography continues, as does the study of new techniques for implementing the basic ideas. The brief survey of applications should only be considered as examples of the type of applications that have attracted attention; the discussion should in no way be considered complete.

---

[105] A. Labeyrie and J. Flamand, *Opt. Commun.* **1**, 5 (1969).

[106] G. W. Stroke, M. Halioua, F. Thon, and D. H. Willasch, *Proc. IEEE* **65**, 39 (1977).

[107] B. J. Thompson, *Proc. IEEE* **65**, 62 (1977).

CHAPTER 10

Image Intensifiers

HERBERT K. POLLEHN

U. S. Army Night Vision and Electro-Optics Laboratory, Fort Belvoir, Virginia

I. INTRODUCTION

Man's visual senses enable him to orient himself in this world. They are also the prerequisite for gathering the information that leads him to new discoveries, inventions, and innovations. It is, therefore, not surprising that much effort has been expended to extend the capabilities of the human eyes. Until only a few decades ago all the efforts were mainly directed toward overcoming the resolution limitations. Telescopes and microscopes allowed the observation of objects that were too far away or too small to be resolved by the human eye. For all these observations it was always realized that the intensity of the light entering the eye had to be sufficiently high. At lower light levels the resolution capability decreases until it becomes too dark to even resolve the largest objects.

ISBN 0-12-408606-3

Over the past three decades imaging devices have been developed with sensors not only significantly more sensitive than the human eye, but also extending imaging capabilities into wavelength regions outside the visible. For this article only those imaging devices will be considered where the detection of the light is based on the photoelectric effect. These devices are commonly called image intensifiers, or sometimes image converters if the main purpose of the device is to convert an image formed by radiation of one wavelength region into an image at another wavelength region—normally the visible. No distinction will be made in this chapter.

Today, image intensifiers are being used from the x ray through the visible into the near-infrared wavelength regions. Although most of the development has been provided by the military, which still is the largest user, image intensifiers have found applications in many other areas. In astronomy they extend the capability of telescopes when images of weak radiation or short duration are to be observed. Law enforcement and the study of nocturnal animals are aided by image intensifiers. They allow a faster recording in spectrometers. In medicine, x-ray intensifiers are widely used to reduce dosage rate and to observe dynamic processes.

In the following sections the main emphasis will be placed on image intensification as used for night vision applications. For many other applications, the material given in the following sections can be used either directly or with some small modifications. In Section II the principal parameters determining the performance of image intensifier systems will be introduced. Sections III and IV are devoted to two main components, the photocathode and the microchannel plate, respectively. In Section V the operation and performance of image intensifier tubes will be discussed.

II. PERFORMANCE OF IMAGE INTENSIFIER SYSTEMS

A. NIGHT VISION SYSTEMS

The discussion of the performance of image intensifier systems will be limited to systems used for night vision applications, sensitive in the visible or near-infrared wavelength region. For many other applications the same performance criteria can be used with only slight modifications.

Night vision intensifier systems are used for target detection, recognition, and identification, and also for various types of operations such as driving, helicopter piloting, maintenance work, map reading, or walking and general orientation. For all these tasks a different scenery and a different set of environmental conditions might exist in any specific use. In addition, several of the above tasks might have to be performed for a specific mission using the same system.

The design of a system is further complicated since no good general quantitative performance model exists for any of the above tasks, and weight, size, and cost also have to be taken into consideration. On the other hand, the importance of a certain system parameter on the performance for a certain task can generally be described in qualitative form, and even some quantitative analysis might be possible. For example, for driving applications a relatively wide field of view is necessary, but generally with a wider field of view the size of the image will become smaller or the size of the system has to be increased. Knowing the size of the object that must be seen at a certain distance and the field of view for safe driving, other system parameters, especially size and weight, can be determined. In the following the most important system and environmental parameters will be introduced. For a static detection task quantitative detection criteria have been developed. Even though some of the assumptions made in this model are questionable, some good insight can be gained into the ultimate performance limitations of image intensifier systems. A description of the model will be given at the end of this section.

B. System Parameters

Figure 1 shows a typical night vision image intensifier system. Radiation from a distant object is collected and imaged onto the photocathode by the objective lens. The power P reaching the image plane (photocathode) is a function of the irradiance at the entrance pupil or aperture of the objective H (watts per square meter), the open area of the aperture A, and the transmission of the lens T_{tr}. Thus

$$P = HAT_{tr} \qquad \mathrm{W} \tag{2.1}$$

or expressed through the T-number of the objective lens,

$$P = H\tfrac{1}{4}\pi(f_o/T\text{-no})^2 \qquad \mathrm{W} \tag{2.2}$$

with f_o the focal length of the objective lens.

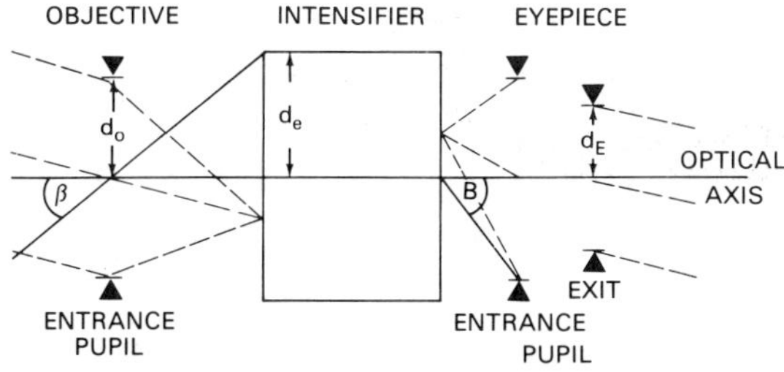

FIG. 1. Image intensifier system.

The irradiance H at the entrance pupil is dependent on the object radiance N (watts per square meter per steradian), the size of the object A^x as seen by the system, the distance between the system and the object R (meters), and the atmospheric absorption,

$$H = (NA^x/R^2)e^{-\sigma R \times 10^{-3}} \tag{2.3}$$

where σ is the atmospheric absorption coefficient. If the object is uniformly illuminated by the scene incident irradiance H_s—neglecting shadowing effects—the object radiance can be written as

$$N = (H_s \rho)/\pi \tag{2.4}$$

where ρ is the reflection coefficient.

If expressed in photometric units, P is replaced by F (lumens) and H by E (lumens per square meter or lumens per square foot).

In Fig. 2 the spectral distribution of the scene incident irradiance for some nighttime conditions are shown. Figure 3 shows reflectance coefficients of some typical objects and scenes as a function of wavelength. Attenuation coefficients (per kilometer) in the visible (0.6 μm) range from 6×10^{-2} on a very clear day to over 1.0 in. haze or fog.

The field of view 2β, and the subjective magnification M, as defined by the ratio of the image size on the retina with and without the system, are important parameters. They are given by[1] (see Fig. 1):

$$\beta = \tan^{-1}(d_e/f_o) \tag{2.5}$$

$$M = m_1(d_o/d_E[1 + 4(f\text{-No})^2]^{1/2} \sin \theta \tag{2.6a}$$

or

$$M = \frac{m_1}{d_E}(d_o^2 + 4f_o^2)^{1/2} \sin \theta \tag{2.6b}$$

where m_1 is the magnification of the intensifier and (f-No) is the f-number of the objective lens. The other parameters can be seen from Fig. 1. The field of view and the subjective resolution can be chosen independently and made, at least in principle, as large or as small as desired. But practical considerations make it difficult to obtain a wide field of view and, at the same time, a large magnification, as would be desirable for systems used for surveying and target acquisition. The field of view can be increased [Eq. (2.5)] by either increasing d_e or decreasing f_o. An increase in d_e becomes very expensive. The standard diameter for image intensifier cathodes currently in production is 25 or 18 mm for $2d_e$. Decreasing the focal length has practical limitations—the last element of the objective

[1] A. D. Schmitzler, *in* "Photoelectronic Imaging Devices" (L. M. Biberman and S. Nudelman, eds.), Vol. 1, p. 89. Plenum Press, New York, 1971.

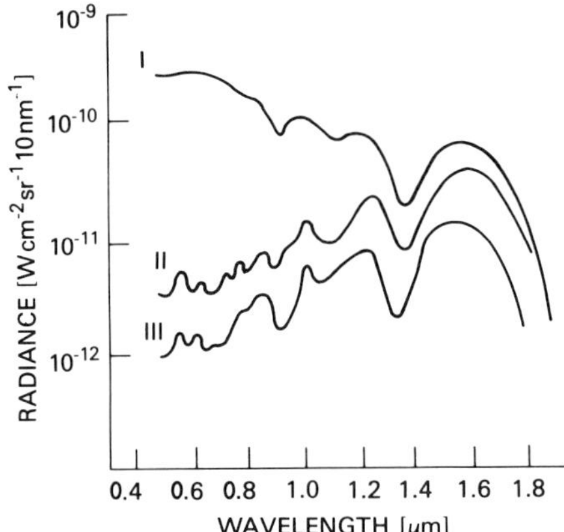

FIG. 2. Night sky radiance. (I) Full moon, no clouds. (II) No moon, no clouds, starlight. (III) No moon, overcast, starlight.

cannot extend into the vacuum envelope of the tube. It also decreases the magnification as can be seen from Eq. (2.6b). Without affecting the field of view, the magnification can be increased by increasing the magnification of the intensifier m_1 or the entrance pupil of the objective d_o, or by decreasing the exit pupil of the eyepiece. The exit pupil of the eyepiece should be at least equal to the entrance pupil of the human eye. A larger pupil ($d_e > 10$ mm) is very desirable to allow some eye movements and placement errors of the observer. Most image intensifiers currently in production have a magnification of $m_1 = 1$. Special tubes with different magnification will be discussed later. To increase the entrance aperture d_o is most desirable, not only to obtain higher magnification but more importantly to increase the collection efficiency of the system. From the above discussion it becomes obvious that some compromises have to be

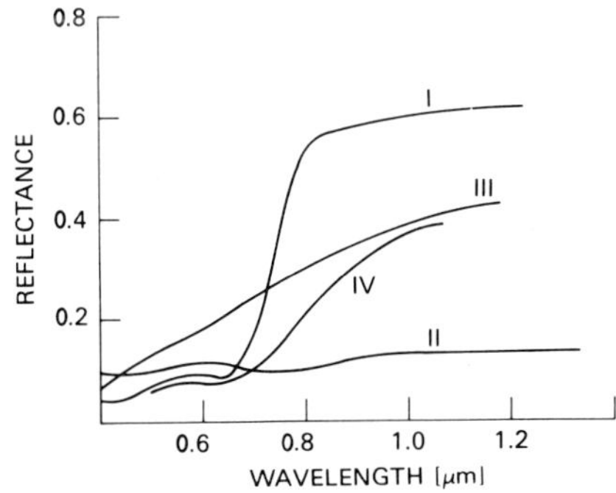

FIG. 3. Spectral reflectance. (I) Green grass, foliage. (II) Hard target, truck. (III) Dead grass. (IV) Man in green uniform.

TABLE I

SYSTEM PARAMETERS FOR NIGHT VISION GOGGLES AND SMALL STARLIGHT SCOPE

	Night vision goggles	Small starlight scope
Tubes	2–18 mm second generation wafer	25 mm second generation inverter
T-Number	1.58	1.7
Field of view 2β (deg)	40	15
Magnification M	1	3.5
Exit pupil (mm)	10	10
Weight (lb)	1.9	4

made in designing a night vision system, especially since cost, size, and weight also have to be considered very heavily. In Table I the parameters of two typical night vision systems are given. The night vision goggle system is a head-mounted device used as an aid for driving and patroling at night. For such a system an overall system magnification of 1.0 with a wide field of view is most appropriate.

C. STATIC PERFORMANCE MODEL

It has been indicated that the aperture size of the objective, the magnification of the system, and the field of view are important system parameters that determine its performance. The size of the aperture determines the intensity or the number of photons arriving from the scene that can reach the photocathode. At the cathode a certain percentage of these photons will release photoelectrons, thus forming an electron image. This electron image is transferred to the output, where the electrons, after multiplication by the gain mechanism, are normally used to generate radiation in the visible wavelength region. In an ideal intensifier, the output radiation in any instant of time and at each point is proportional to the intensity in the electron image at the corresponding point of the input. For the following considerations an ideal image intensifier is assumed. Deviations will be discussed later.

The ultimate limitation on the visual perception of images, or the detection of brightness differences, is set by the quantum nature of light. The photons arriving at the photocathode are distributed randomly in time. In discussing these final limitations it is convenient to divide the image plane into adjacent areas of equal sizes. The areas are first assumed to be large compared with the resolution limits of the system, so that image-degrading effects can be neglected. The average intensity in each of these areas can be determined by Eq. (2.4), but the intensity fluctuates with time. If N is the number of photons arriving at one area in a certain time Δt, then $N^{1/2}$ is the standard deviation from this average. The number of released electrons is

$$N_e = \int_0^\infty N(\lambda)\Theta(\lambda)\, d\lambda \tag{2.7}$$

where $\Theta(\lambda)$ is the quantum efficiency of the photocathode and λ is the wavelength of the light. Since an ideal image intensifier is assumed, the number of scintillations N_s in each area at the output of the intensifier is identical to the number of electrons in the respective input area. It can be easily shown that N_e or N_s is also distributed randomly in time. The perception of brightness differences between areas $N^{(1)}$ and $N^{(2)}$ is now dependent on the difference in the average number of scintillations within the integration time of the eye and the fluctuation in the difference σ_{12}. It is now assumed that a 50% probability of detection is given by

$$K_{50\%} = \alpha \frac{N_s^{(1)} - N_s^{(2)}}{\sigma_{12}} = \frac{\Delta N}{\sigma_{12}} \tag{2.8}$$

The factor α depends on perception parameters of the eye. For an ideal image intensifier,

$$\sigma_{12}^2 = N_s^{(1)} + N_s^{(2)}$$

then

$$K_{50\%} = \alpha(N_s^{(1)} - N_s^{(2)})(N_s^{(1)} + N_s^{(2)})^{-1/2} \tag{2.9}$$

Setting

$$C = |N_s^{(1)} - N_s^{(2)}|/N_s^{(1)}$$

$$K_{50\%} = \alpha(N_s^{(1)})^{1/2} C/(2 - C)^{1/2} \tag{2.10}$$

For Eqs. (2.9) and (2.10) it has been assumed that the contrast at the intensifier is the same as the scene contrast C. In reality, the contrast is reduced by scattering in the atmosphere and unavoidable veiling glare in the objective lens. The veiling glare of the objective is measured by illuminating the whole field of view with the exception of one degree in the center. The ratio of the illuminance in the central area to the illuminance in the evenly illuminated field is taken as a measure of the veiling glare, V_g. For all objective lenses used for night vision devices special care is taken in reducing the amount of veiling glare. The specification calls for $V_g < 0.02$. The contrast reduction due to scattering in the atmosphere reduces detection ranges in haze or fog, and under these conditions is of more significance than the decrease in the number of photons due to atmospheric absorption.

So far, adjacent large areas of equal size have been considered. In a more realistic scene for night vision devices an object is surrounded by a more or less evenly illuminated background. The principles and formulas developed above are then applied by assuming an adjacent area of equal size to the object, and an irradiance equal to the background. The areas of uniform illumination have been assumed to be very large compared to the resolution element of the system. For smaller objects or, better, image sizes, the image transfer characteristics of the system normally expressed through the modulation transfer function (MTF), will further reduce the contrast. The veiling glare can be considered as the low-frequency part of the MTF. For the simple case of a sinusoidal bar pattern of contrast $C(V)$, the contrast after passing through the system will be $H(V) \times C(V)$, where $H(V)$ is the MTF value at the spatial frequency V. To apply this concept for real objects the Johnson criterion has been proven useful. Under this criterion the contrast C of a real target is also reduced to $H(V) \cdot C$, where $H(V)$ is now the MTF value for a sine-wave frequency $V = N/S$ (lp/mm), where S is the smaller dimension of the target in millimeters, measured at the output of the intensifier. N depends on background features and on the amount of information needed for the target. The information content is generally categorized in detection, recognition, and identification. Detection ($N \sim 1$) just means that a "foreign" object is in the scene. By recognizing the target ($N \equiv 2$–3) a distinction can be made, for example, between a truck or a tank, whereas under identification ($N \equiv 2$–7) all essential features of the target can be seen. The wording "essential features" is very loose and may mean something different under different circumstances, which naturally implies then that N will be different.

The length s of the image A of a target of actual length s^* can be calculated

$$s = \frac{s^* \, d_e}{R \tan \beta} = \frac{s^* f_o}{R} \tag{2.11}$$

where β is half the field of view and R is the distance between the system and the target. Equation (2.11) is strictly true only for on-axis targets, but due to the normally large ratio between R and the focal length of the objective lens, the differences of s for off-axis targets are negligible.

For all applications where search functions are required, a large field of view is desirable, but as can be seen from Eq. (2.11), a large field of view results in a small image, which in turn results in a larger V, and therefore smaller $H(V)$, reducing the contrast. A theory giving a functional dependence between image size, field of view, and the time required to find a target, is still waiting for its discovery.

In all the above discussions no image-degrading effects due to the human eye or the image intensifier have been taken into account. The performance of the human eye under different conditions, especially threshold conditions, is covered in a different chapter in this series. The main degradations may be expressed as the eye MTF and eye "noise." Eye noise becomes negligible for sufficient gain in the intensifier. Technically, for all intensifiers, sufficient gain can be obtained, but cost consideration sometimes requires the use of a low-gain device. The eye MTF practically always has to be taken into account and may be lumped together with the system MTF. A high subjective magnification [Eq. (2.6)] decreases its influence.

Now taking MTF degradations into account, the contrast [Eq. (2.9)] can be rewritten as

$$K_{50\%} = \alpha \frac{N_s^{(1)} - N_s^{(2)}}{(N_s^{(1)} + N_s^{(2)})^{1/2}} H(v) \tag{2.12}$$

Now N_s is equal to the scintillation density D_s times the image area A, which is assumed to be equal to the background area. The scintillation density in our ideal intensifier is equal to the irradiance or illuminance at the photocathode times the cathode sensitivity Θ. Also taking into account that the image area $A = A^* f_o^2/R^2$, where A^* is the object area, combining Eq. (2.12) with Eqs. (2.3) and (2.2) results in

$$K_{50\%} = K \frac{d_o \int H_s(\lambda)\Theta(\lambda)e^{-\sigma(\lambda)R}[\rho_1(\lambda) - \rho_2(\lambda)]\, d\lambda\, H(v)}{R[\int H_s(\lambda)\Theta(\lambda)e^{-\sigma(\lambda)R}(\rho_1(\lambda) + \rho_2(\lambda))\, d\lambda]^{1/2}} \tag{2.13}$$

The constant K contains perception parameters but also the integration time of the eye or system, depending which one is longer.

In a real image intensifier not all photoelectrons will generate scintillation. In addition, the energy of the scintillation varies according to some distribution, which will be discussed later. These effects are taken into account by the noise figure N_F defined as the ratio of the signal-to-noise ratio at the input of the intensifier tube to the signal-to-noise ratio at the output. Also, the dark current I_D has to be taken into account. With these additions (2.13) becomes

$$K_{50\%} = K \frac{d_o}{RN_F} \frac{\int H_s(\lambda)\Theta(\lambda)e^{-\sigma R}[\rho_1 - \rho_2]\, d\lambda\, H(v)}{[\int H_s(\lambda)\Theta(\lambda)e^{-\sigma R}[\rho_1 + \rho_2]\, d\lambda + 2I_D]^{1/2}} \tag{2.14}$$

From this equation it is obvious that for optimum performance a photocathode should be chosen where the night sky irradiance H_S is high, the atmospheric attenuation coefficient $\sigma(\lambda)$ is low, and the difference in the reflection coefficients is large. All these factors improve with longer wavelength, especially beyond 0.7 μm. It is therefore not surprising that much effort is devoted to the development of photocathodes with high cathode sensitivities at longer wavelength. In Section III currently used photocathodes, together with current developmental and research efforts, will be described.

III. PHOTOCATHODES

A. DEFINITIONS

The photocathode is the most important component of an image intensifier. The intensity distribution in the light image is converted through the photocathode into a distribution of electrons where the number of electrons released into the vacuum is proportional to the intensity of the light. The energy and/or the number of electrons can be increased to a level where each photoelectron creates more photons than the average number of photons required to release one photoelectron from the cathode. This forms the basis for image intensification, and such an amplification process would not be possible without photoemission.

In the first part of this section the fundamentals of the photoemission process will be given, followed by a description of the most important photocathodes currently in use or in development. For a description of the photoemissive properties of various other material systems the reader is referred to Sommer,[2] and for a more quantitative, theoretical analysis of negative affinity photoemitters he is referred to Bell[3] and references in these books.

A photocathode consists of a thin film of material attached to a glass or fiber-optics substrate (Fig. 4a). If the cathode is illuminated through the substrate (Fig. 4b), the illumination is called front illumination and the cathode operates in the semitransparent mode. If illuminated from the vacuum side, the cathode is said to have back illumination and is a reflection cathode. Reflection cathodes are easier to fabricate and quite often exhibit higher efficiencies. But for image intensifiers, a semitransparent photocathode is much preferred for obvious reasons.

The two most important parameters that characterize a photocathode are the sensitivity and the dark current. For cathodes operating in the visible or near-infrared region the sensitivity is quite often expressed by a single number: the amount of current generated from the cathode when illuminated with 1.0 lumen of light at a color temperature of 2854 °C (microamperes per lumen). Besides its obsolete and incorrect nomenclature, as discussed by Biberman,[4] it can only be used as a meaningful performance criterion if cathodes having equal shapes of the spectral response curve are compared. The spectral response is generally given as a function of wavelength or as a function of photon energy. The response is defined as the current released from the photocathode into the vacuum for one watt incident radiation [milliamperes per watt], or as the ratio of the number of photons releasing an electron into the vacuum to the number of photons incident onto the photocathode, expressed in percent quantum efficiency. For these parameters the input light, be it in lumens, watts, or number of photons, is considered as the radiant or luminous flux at the input of the substrate for front illumination or at the vacuum–cathode interface for back illumination. Losses due to absorption in the substrate or reflections at the

[2] A. H. Sommer, "Photoemissive Materials." Wiley, New York, 1968.

[3] R. L. Bell, "Negative Electron Affinity Devices." (Clarendon), Oxford Univ. Press, London and New York, 1973.

[4] L. M. Biberman, *in* "Photoelectronic Imaging Devices" (L. M. Biberman and S. Nudelman, eds.), Vol. 1, p. 9. Plenum Press, New York, 1971.

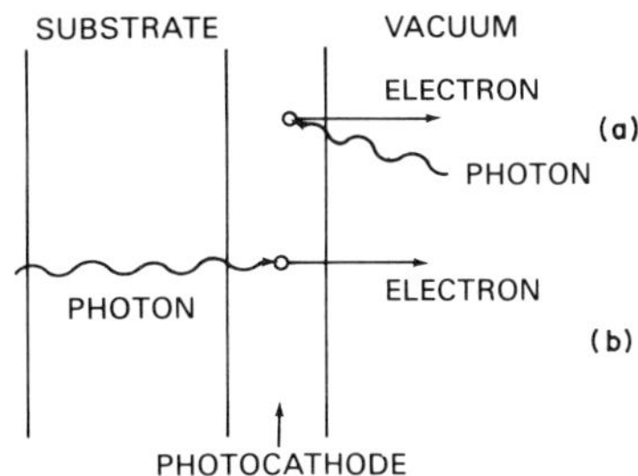

FIG. 4. Photocathode structure. (a) Reflection mode. (b) Transmission mode.

various surfaces are included in the above parameters. The dark current is normally expressed in amperes per square centimeter. For image intensifiers an "equivalent background input" (EBI) in lumens or watts per square centimeter is frequently used as a measure for the dark current.

B. SOLID STATE CONCEPTS

The physical mechanism of the photoemission process can be easier explained by introducing some concepts from solid state physics. In all solids electrons are confined and can only travel in certain energy bands. The band with the highest energy is called the conduction band, the next lower band is the valence band (Fig. 5). The two bands are separated by a band of energy E_g—the bandgap energy—with no energy states. In a semiconductor at zero degrees Kelvin all states in the conduction band are empty and all states in the valence band are filled. The conductivity is zero. At higher temperatures some electrons can acquire sufficient energy to transition from the valence into the conduction band. The number of electrons, and with it the conductivity, increases with increasing temperature, but decreases with increasing band gap. Insulators have band gaps of several electron volts. In a metal at zero degrees Kelvin, electron states in the conduction band are filled up to the so-called Fermi level, and are empty at higher energies. The Fermi level is defined as the 50% point in the energy distribution of the electrons in the solid. For this definition it is of no concern whether allowed energy states are present or not at the Fermi

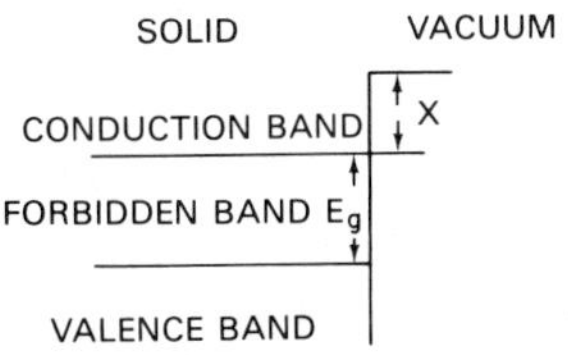

FIG. 5. Band structure of solid and vacuum interface. E_g is the band gap, X is affinity.

level. In a semiconductor the Fermi level is in the middle of the band gap. The energy between the Fermi level and the vacuum level is the work function ϕ, and the energy from the lower conduction band edge to the vacuum level is the affinity χ.

The model of a semiconductor pictured so far resembles a crystal with no surfaces and consisting of molecules of only one kind, as completely pure silicon, Cs_3Sb, or GaAs. First, completely pure crystals do not exist and, second, in all good semiconductor photocathodes molecules are added that have fewer valence electrons than the molecules of the host crystal. In alkalide–antimony photocathodes additional antimony is added and in GaAs, zinc is added. The additions, called dopants, create additional energy states in the forbidden band just above the upper edge of the valence band. These states are very close to the valence band, and electrons from the valence band can easily transition into these states, leaving empty states (holes) in the valence band. The semiconductor is called *p*-type and the Fermi level is close to the valence band. At the surface of the semiconductor the band structure of the bulk material is disrupted. Allowed electron states can be present at energies within the band gap of the bulk material. If such states are originally occupied, electrons from these states will transfer in *p*-type material into the available empty states of lower energy. This leaves an excess positive charge at the surface and excess negative charge in the bulk immediately next to the surface. This changes the Fermi level only slightly, but results in a significant bending of the valence and conduction bands as shown in Fig. 6b. The important effect for photoemitters, as will be discussed later, is the lowering of the effective affinity χ. In many materials, including metals, the affinity or work function (for metals) can be further lowered by depositing an electron-positive material such as Cs on the surface. An electro-positive material will easily transfer an electron to a lower level of the bulk material. With addition of oxygen—probably forming dipoles—the affinity can be lowered to negative values in some materials (Fig. 6b). See also Fig. 7.

C. The Photoemissive Process

Photoemission is a three-step process.

(a) Absorption of a photon, elevating an electron into a state of higher energy.
(b) The motion of the electron through the material.
(c) The escape of the electron into the vacuum.

In an ideal photocathode, all photons arriving at the photocathode will release an electron into the vacuum, giving a 100% quantum efficiency. The best quantum efficiency achieved so far is just above 50% at one specific wavelength, falling off at shorter and longer wavelengths. Losses occur in all three steps mentioned above.

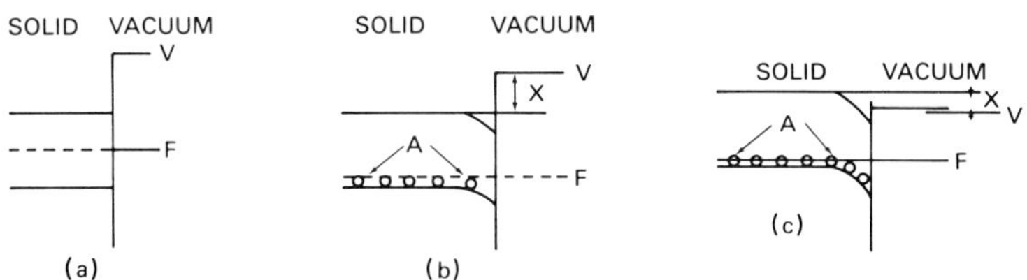

FIG. 6. Crystal–vacuum interface. (a) Undoped crystal. (b) Crystal with p-type doping, positive affinity. (c) Crystal with p-type doping, negative affinity. V is the vacuum level, F the Fermi level, and A the acceptor states in band gap.

First, not all the photons arriving at the photocathode will be absorbed. Some of the light will be reflected at the substrate input and substrate–cathode interfaces in semitransparent photocathodes, or at the cathode–vacuum interface in reflective photocathodes. These reflections are especially high for metals. For semiconductor photocathodes operating in the transmission mode, reflection losses can be made very low with antireflection coatings. But, as the mode of operation already indicates, significant losses occur due to transmission of light through the cathodes. This not only reduces the quantum efficiency but—quite often more importantly—is a main contributor to the veiling glare in image intensifiers. It can be reflected from other components of the intensifier back to the photocathode, releasing electrons at unwanted positions. The transmission losses can be reduced by increasing either the thickness or the absorption coefficient. An increase in thickness will increase losses in the other two steps of the photoemission process. For each cathode material a different optimum thickness exists. The absorption coefficient of any material is fixed; however, good cathodes must have a high absorption coefficient. Figure 8 shows the absorption coefficient for three frequently used photocathodes as a function of the photon energy.

In the absorption process a photon interacts with a lattice electron, elevating it to a level of higher energy. In a metal (Fig. 7a) the photon normally reacts with an electron from the partially filled conduction band, increasing its kinetic energy. Since allowed energy states are at any higher energy, a photon of any

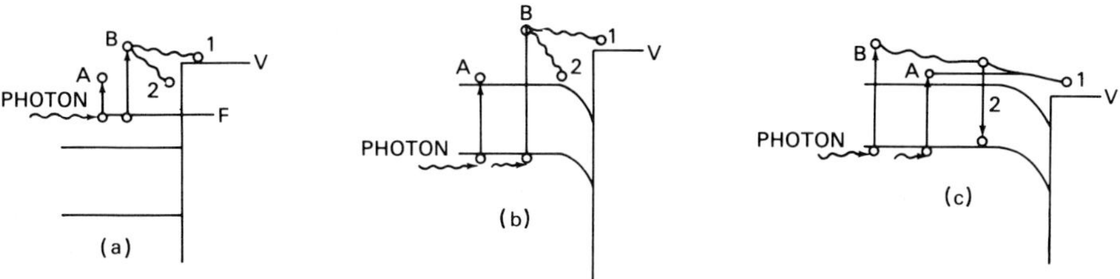

FIG. 7. Photoemissive process. (a) Metal. (b) Positive affinity semiconductor. (c) Negative affinity semiconductor.

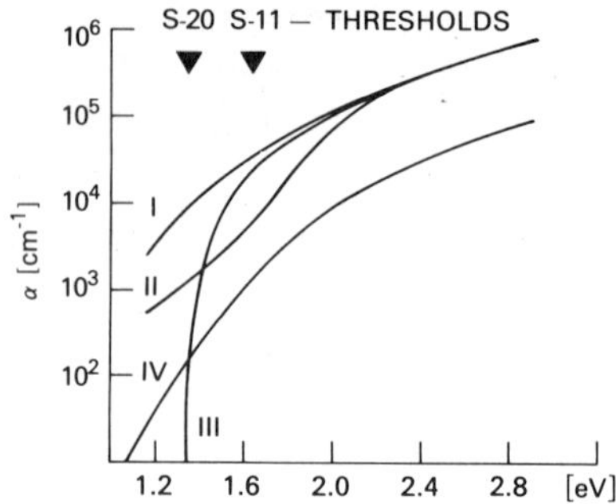

FIG. 8. Absorption coefficients for some photocathode materials. (I) (Cs) Na–K–Sb (S-20). (II) Cs_3Sb (S-11). (III) GaAs. (IV) Si.

energy may be absorbed. Neglecting impurity states and electrons elevated into the conduction band due to thermal excitation, a photon can react with an electron in a semiconductor only if it has sufficient energy to elevate the electron from the valence band into the conduction band (Figs. 7b and 7c). But only if the affinity of the cathode is negative, or at least zero, can a photon that only has the band gap energy E_g initiate photoemission (transition A in Fig. 7c). For positive affinity the photon has to have the additional energy χ (Fig. 7b), and for a metal the threshold energy is given by the work function. The probability of a photon actually reacting with an electron is dependent on the distribution of the energy states within the conduction and valence bands. The states are characterized by their energy and momentum. In each transition, both energy and momentum have to be preserved. A detailed analysis of the absorption process will go beyond the scope of this article and the reader is referred to the literature.[5] An electron excited above the vacuum level still has to travel through the material and might lose enough energy to have an energy below the vacuum level when finally reaching the surface (B-2 in Figs. 7a and 7b, and A-2 in Fig. 7c). The causes for energy losses in a semiconductor are inelastic scattering (production of phonons) with the crystal lattice or at crystal imperfections, grain boundaries, stress lines, or foreign particles. These reactions can occur only until the electron reaches the lower edge of the conduction band. At this point the electron has lost all its kinetic energy. Any motion is now governed by the diffusion equations[3] since no electric field exists in the cathode, with the exception of the band-bending region. For a positive-affinity photocathode this energy is already too low, but in a negative-affinity photoemitter, this electron still has a chance to leave the crystal. For any further energy losses an electron at the lower conduction band edge must transition to empty states in the valence

[5] R. K. Willardson and A. C. Beer, (eds.), "Semiconductors and Semimetals," Vol. 3. Academic Press, New York, 1968.

band or to states within the forbidden gap. Energy states in the band gap are again due to impurities, stress lines, distortion in the crystal, or, as stated before, can be found at crystal boundaries. The energy losses of the excited electron can be minimized by proper cathode preparation. First, ultraclean materials and working areas must be used. As mentioned before, all good photoemitters are *p*-doped. The dopants are impurities that should be avoided, on one hand, but are, on the other hand, required to lower the affinity through band bending. A good balance has to be found experimentally. The amount of stress lines, distortions, and boundaries within the photocathode is a function of the fabrication process. Cathodes made out of III–V components such as GaAs can be fabricated as a single crystal through epitaxial crystal growth techniques. The only boundaries in such a photocathode are, ideally, the vacuum and substrate interfaces. In order to avoid or minimize electron states within the forbidden gap at the cathode–substrate interface, another epitaxially grown layer with a wider band gap is used as a buffer layer between the substrate and the cathode. Such a layer not only minimizes electron transitions into lower energy states, but at the same time acts like a mirror for electrons traveling toward the interface (Fig. 9). Alkali compounds are currently the most used photocathodes. Single crystals cannot be grown from these materials. The material is of a multicrystalline structure with grain sizes dependent on the fabrication processes. Generally larger grain sizes, which means less boundaries, are obtained by evaporating small amounts of material over long periods of time. For metals, the main cause for energy losses is interaction with other electrons. In reactions with other electrons the energy loss per event is significantly higher than in scattering with the lattice. From the qualitative description of the photoemission above, we might deduce the requirement for a good photoemitter:

(a) The absorption coefficient should be high for all wavelengths where the cathode is sensitive.

(b) The affinity should be as low as possible, preferably negative.

(c) Energy losses should be minimized. Therefore, semiconductors should be preferred over metals.

(d) For low dark currents, a large work function is required.

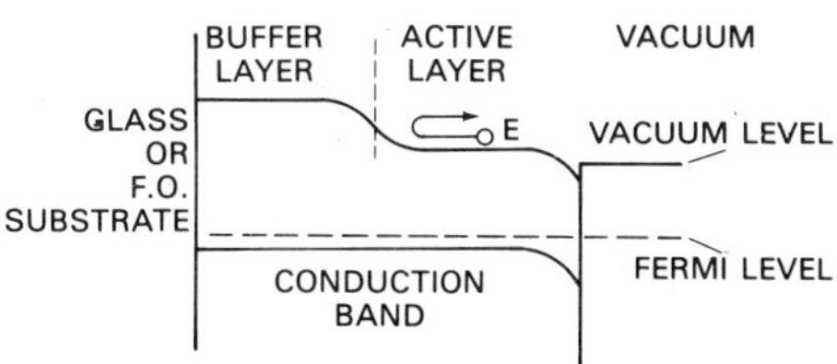

FIG. 9. Photocathode with buffer layer. Electron E will be reflected when traveling towards the buffer layer. Buffer layer is transmissive for photons with energies smaller than the band gap.

D. Special Photocathodes

1. *The Silver Cesium S-1 Photocathode*

The first cathode that will be discussed is a metal, violating the rules set up for good photocathodes. But the S-1 is the first photocathode applied to wide practical applications and is still the only commercially available semitransparent photocathode sensitive up to a wavelength of 1.1 μm. The S-1 was discovered in 1935. Figure 10 shows a typical spectral response curve. The sensitivity to 2854° light is typically 30 μA/lm and the dark current is high, between 10^{-13} and 10^{-11} A/cm^2. The high values of the dark current generally correspond to higher infrared sensitivity. The S-1 decays slowly (years) under storage conditions and much more rapidly in operation and increasing light levels. Improved life and dark current are obtained if the cathode is cooled. Image intensifiers with S-1 photocathodes are known as zero generation intensifiers. Much research has been conducted in order to understand the physical processes and towards improving the sensitivity of the S-1 photocathode. Both these efforts have had only marginal success. The S-1 photocathode probably consists of small islands of silver heavily coated with a thick layer of cesium oxide to reduce the work function below 1.0 eV. The limitations of this photocathode can easily be deduced from the general discussion of the photoemissive process. The absorption is low due to the high reflection of silver and to the thin layer, which is required since excited electrons rapidly lose their energy in electron–electron interactions.

2. *The Multialkalide S-20 Photocathode*

The S-20 multialkalide photocathode is the most widely used cathode for low-light-level application and is employed in all first and second generation image intensifiers. It was discovered accidentally by Sommer in 1955. The first S-20 cathodes had a luminous sensitivity around 150 μA/lm. Today typical sensitivities are around 350 μA/lm and sensitivities above 500 μA/lm are obtained quite frequently with dark currents of 10^{-14} A/cm^2.

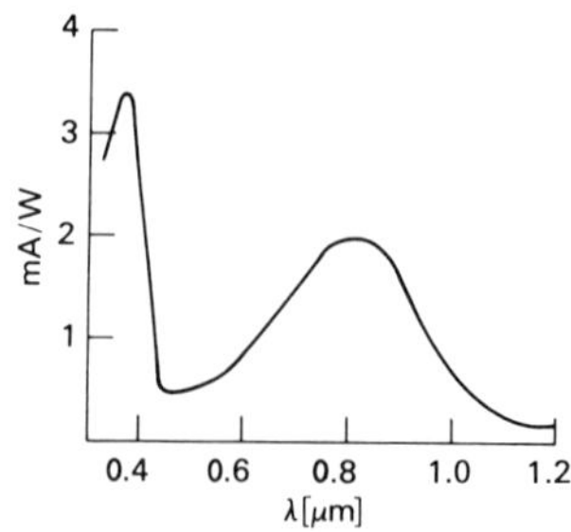

Fig. 10. Typical spectral response curve of Ag–O–Cs S-1 photocathode.

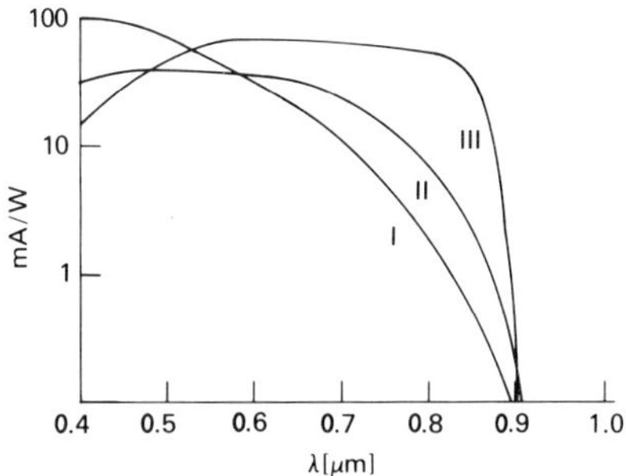

FIG. 11. Spectral response curves of (Cs) Na–K–Sb S-20 photocathodes with increasing substrate quality and thickness from I to III.

The S-20 employed in image intensifiers is a semitransparent polycrystalline semiconductor film. The approximate composition is Na_2KSb–Cs with some excess Sb to provide a *p*-type doping. Some Cs might be in the bulk material, but most of the cesium covers the surface to lower the affinity. The forbidden gap is 1 eV and with careful processing an affinity between 0.3 and 0.4 eV can be achieved. This gives a long-wavelength cutoff around 0.9 μm. Figure 11 shows three typical spectral response curves. The differences in the response curves from I through III demonstrate the achievements obtained over the last couple of years in photocathode fabrication techniques.[6]

Referring back to Figs. 2 and 3, which show the night sky spectral irradiance and reflectance from various objects, it is obvious that curve III must be preferred for night vision applications. The improvements were obtained through the development of sophisticated fabrication methods and extreme care in respect to cleanliness. This resulted in films with large stress-free crystal sizes and low impurity levels, giving large escape depth for excited electrons, which allows increasing the thickness of the photocathode. The thickness of photocathodes with a response curve III of Fig. 11 is above 1000 Å. A typical processing scheme for a good S-20 photocathode is as follows: First a thin layer of K_3Sb is formed by alternate or coevaporation of K and Sb at elevated temperatures. The photosensitivity of the layer is monitored continuously and the evaporation is stopped when a peak in the sensitivity is obtained. In the next step, Na_2KSb is formed by exposing the K_3Sb film to Na vapor and again alternate or coevaporation of K and Sb. This process is continued until the desired thickness is obtained. As a measure of thickness, the decrease in photosensitivity for short wavelengths is commonly used. In the last step, alternate evaporations of Cs and Sb are performed until peak sensitivity is obtained. All materials are deposited at a very slow rate to allow the formation of large-size stress-free crystals. The fabrication of a good S-20 photocathode takes about 6 hr.

[6] I. P. Csorba, *Proc. Elec. Opt. Syst. Design Conf.* p. 636 (September 1976).

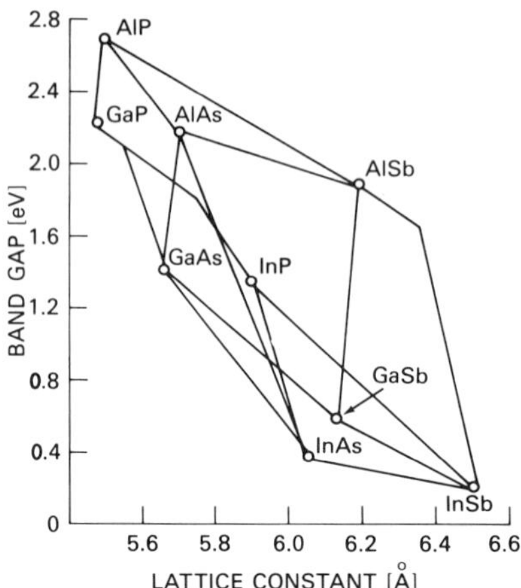

FIG. 12. Band gap and lattice constants for III–V compounds.

For even higher sensitivities, larger crystal sizes—ideally single crystals—and a further reduction in the affinity, ideally to negative values, would be required. Efforts in this direction for the S-20 cathode have not been successful. Both these requirements have been met with photocathodes composed of alloys formed from elements of the third with elements of the fifth column of the periodic system.

3. *III–V Photocathodes*

The first activation to negative affinity was achieved by Scheer and van Laar in 1965.[7] They activated a vacuum-cleaved single crystal of GaAs to a luminous reflection sensitivity above 500 μA/lm. Since that time a considerable amount of research and development has been devoted to understanding the photoemission process of III–V compounds and to devising fabrication techniques. Both efforts have been successful.

III–V compounds have the zinc blende crystal structure. Crystal constants and band gaps can be seen in Fig. 12. Binary compounds are represented by points, tertiary by lines connecting the points, and quaternaries by the areas enclosed by the corresponding lines. For example, the important compound $Ga_xIn_{1-x}As$ is represented by the line connecting the binaries GaAs ($x = 1$) and InAs ($x = 0$). The band gap varies continuously from 1.4 eV ($\lambda = 0.9$ μm) to 0.35 eV (4 μm) with the lattice constant varying from 5.65 Å to 6.05 Å.

Today cathodes are fabricated by epitaxial growth technique. In this technique the photoemissive layer is grown on a seed or substrate crystal at high temperature. The substrate crystal must have the same crystal structure, and

[7] J. J. Scheer and T. van Laar, *Solid State Commun.* **3**, 189 (1965).

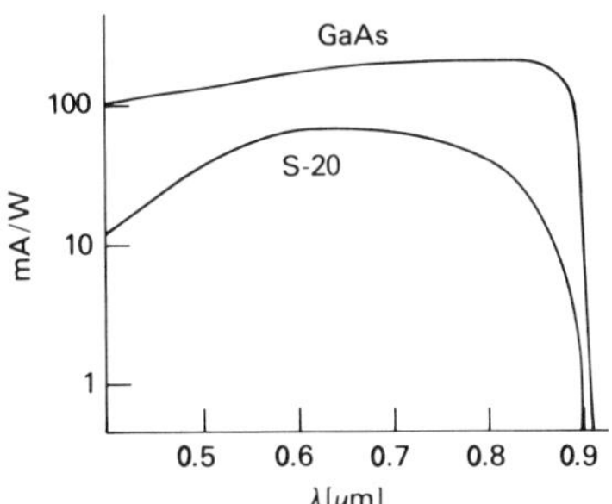

FIG. 13. Spectral response of GaAs in reflection, and very good S-20 photocathode.

for a stress-free growth should ideally have equal crystal constants. Substrate crystals grown by the boat or Czochralkyi method and commercially available are the binaries GaP, GaAs, InP, InAs, and InSb. For photocathode applications, InAs and InSb must be discarded due to their low band gaps of 0.35 and 0.2 eV, respectively. This leaves (Fig. 12) GaP, GaAs, and InP as possible choices for substrate material. To fabricate now a reflective cathode on, e.g., GaAs, an epitaxial layer of GaAs with the right doping can be grown directly on the substrate. In order to extend the sensitivity to longer wavelength, In can be added. The addition of In will change the crystal constant. A frequently used method is to start the growth with only a small amount of In and increase the percentage of In until the desired composition is obtained. With this method, sensitivities over 2000 μA/lm have been obtained.[8]

Figure 13 shows the spectral response of a reflection InGaAs photocathode and for comparison, the response of a very good S-20 photocathode. Another important photocathode fabricated in a similar way is $GaAs_{1-x}P_x$ grown on GaP.[9] Quantum efficiencies up to 50% at $\lambda = 0.53\ \mu$m have been reported. Referring to Fig. 12, ternary or quaternary compounds could be grown on InP substrates with perfect lattice match and bandgaps as low as 0.8 eV (1.7 μm). The most successful growth has been obtained with the quaternary $In_{0.88}Ga_{0.12}As_{0.23}P_{0.77}$.[10] Quantum efficiencies above 10% have been obtained at 1.1 μm.

All the above-mentioned photocathodes are not very useful in the semitransparent mode. For a GaAs photocathode grown on GaAs, all the light will be absorbed before reaching the sensitive layer. Only the GaInAsP cathode can be used to some degree in the semitransparent mode, but the bandwidth is very narrow. The long-wavelength cutoff is given by the quaternary at 1.1 eV ($\lambda \sim$ 1.1 μm) and the short-wavelength cutoff by the InP substrate.

[8] D. Jackson and E. M. Yee, *Proc. IEEE* **59**, 90 (1971).

[9] J. S. Escher and G. A. Antypas, *Appl. Phys. Lett.*, **30**, 314 (1977).

[10] J. W. James *et al.*, *Appl. Phys. Lett.* **22**, 276 (1973).

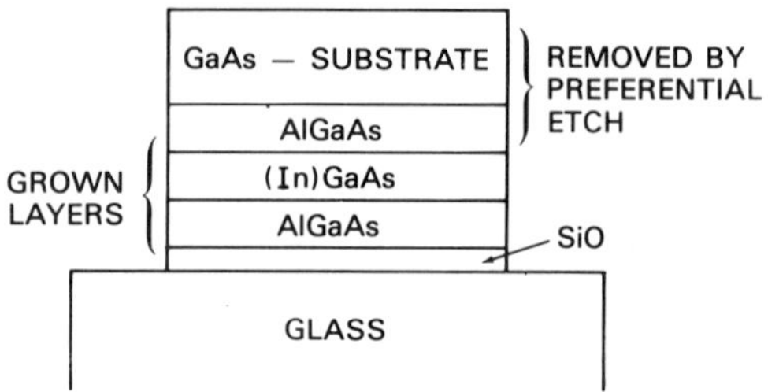

FIG. 14. Semitransparent (In) GaAs photocathode.

With the techniques described so far, the only substrate usable for semitransparent photocathode sensitive at shorter wavelength is GaP. No direct lattice match is achievable for any III–V compound. A possible solution is to grow a graded layer of GaInP onto the GaP substrate with an increasing amount of In until the lattice constant matches the desired photocathode material—GaAsP, GaAs, or GaInAs (Fig. 12). This has been tried extensively with moderate success.

The best semitransparent photocathodes fabricated today are of an "inverted" structure grown by the liquid phase epitaxial growth techniques (Fig. 14). In this technique AlGaAs is grown on GaAs. The lattice constants differ only slightly. For an even more perfect growth, a grading layer of InGaAs can be grown first. The next step is the growth of the active (In) GaAs layer. The indium content is chosen so that the lattice constant exactly matches the previous AlGaAs. Now another AlGaAs layer is grown and after coating with SiO_2, the whole structure is sealed to glass. The GaAs substrate and the first AlGaAs layer are removed with preferential etch methods exposing the active Zn-doped cathode. In this photocathode the short-wavelength cutoff is determined by the AlGaAs layer and, as can be seen from Fig. 12, is strongly dependent on the Al content. A similar method can be used to fabricate a broadband semitransparent GaInAsP cathode. Here the GaInAsP is grown with perfect lattice match onto an InP substrate. Then a very thin layer of InP is grown on the quaternary compound, the whole structure sealed to glass, and the InP substrate removed. This quaternary photocathode can be activated only to a very small negative affinity. At room temperature the cathode is unstable. In addition, a high dark current limits its use. Stability and dark current can be improved by cooling. For a III–V compound with band gaps lower than about 1.1 eV ($\lambda < 1.1$ μm), activation to negative affinity cannot be achieved. To extend the wavelength further into the long-wavelength region, "field assisted" photocathodes are currently being developed. In these photocathodes an external field is applied to lower the barrier at the substrate–vacuum interface. Several structures are under consideration. The development is still in its early research phase. Some success has recently been reported with the so-called "transferred electron" photocathode.[11] Photoemission up to a wavelength of 1.4 μm has been obtained.

4. *UV Photocathodes*

Photocathodes used in the UV do not differ in their mechanism from cathodes sensitive in the visible wavelength region. All photoemitters with good sensitivity in the visible region have normally also good sensitivity in the UV region if deposited on a suitable window material. In the 2000–3000 Å region quartz is usually used as a substrate material, and from 1000 to 2000 Å LiF is used. Below 1000 Å all materials become absorbent and only reflective-mode cathodes in windowless "tubes" can be used. Of considerable importance are so-called solar blind photocathodes. These are cathodes that do not detect the radiation of the sun and can therefore be used in daylight. A good solar blind photoemitter in the 1000–3000 Å wavelength region is CsTe. MgO and most alkali halides have good sensitivities in the LiF region and in the vacuum UV. LiF is a good emitter below 1000 Å.

IV. MICROCHANNEL PLATES

A. Description

A microchannel plate (MCP) consists of a large number of hollow glass tubes fused into a disk-shaped array. The input and output surfaces are coated with metal electrodes. The glass tubes have some conductivity and each might be considered as a continuous dynode electron multiplier. An electron entering one of the tubings or channels will create secondary electrons while striking

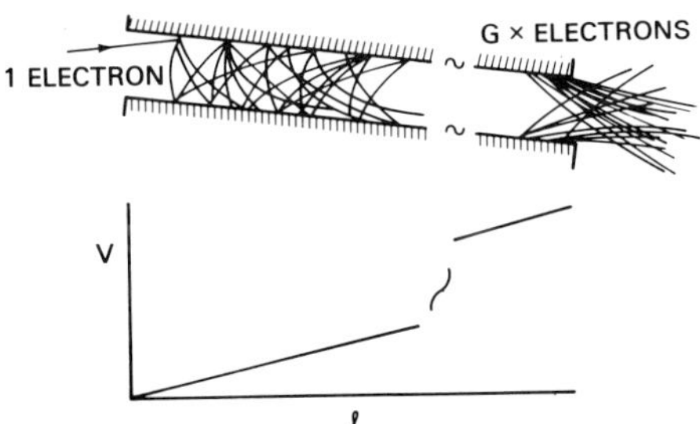

FIG. 15. Microchannel; and voltage as a function of channel depth *l*. *G* is the electron gain.

the channel wall (Fig. 15). If a voltage is applied to the MCP, a field exists within the channel accelerating these secondaries. They will eventually hit the wall again, creating new secondaries. This process will continue to the output of the

[11] J. S. Escher and R. Sankoran, *Appl. Phys. Lett.* **29**, 87 (1976).

channel. If the secondary emission coefficient is larger than 1.0, the number of electrons will increase through the channel. Since the input and output of the MCP is in direct registration, this device is ideally suited as a gain mechanism for electrooptical imaging devices. In the following, the main characteristics of the MCP, important for its operation in an imaging device, will be discussed.

B. Gain

The gain of the MCP is dependent on the average number of collisions N of the electrons with the channel wall and on the average secondary emission coefficient δ at each of these collisions. Neglecting the fact that some of the electrode material is penetrating into the channel, the gain g is given by

$$g \sim \gamma\delta^{N} \tag{4.1}$$

where γ is the secondary emission coefficient for the first impact of an electron in the channel. Gamma is normally greater than δ. For a constant length and diameter of a channel, δ will increase with increasing voltage. This results in a more than linear increase in the gain of the MCP. In current MCPs with 15-μm center-to-center spacing between the channels, the gain roughly doubles for every increase of 50 V. If the voltage is now kept constant but at constant diameter d, the length of the channel l is increased, the gain will initially rise. At small l/d ratios the effect of the increasing number of collisions is more effective for the gain than the reduction in the secondary emission coefficient. The secondary emission coefficient decreases since the impact energy for each collision decreases—the total voltage is assumed to be constant. At some l/d ratio, the secondary emission coefficient becomes so low that a gain reduction is obtained. An approximate formula for the gain has been derived by Eschard and Manley[12]:

$$G = [KV_0{}^2/4V\alpha^2]^{4V\alpha^2/V_0} \tag{4.2}$$

K is a constant and the assumption is made that the secondary emission coefficient δ is proportional to the accelerating voltage V_e; $\delta = KV_e$; V is the average initial energy of a secondary electron emitted normal to the surface; V_0 is the total voltage applied to the channel; $\alpha = l/d$ is the length-to-diameter ratio of the channel. Guest[13] has calculated the dependence of the gain on the l/d ratio by computer simulation (Fig. 16). These universal gain curves are of great importance. For each voltage, the gain has a maximum at a certain l/d ratio. In these maxima, small changes in the l/d ratio will result in only small changes in the gain. In an MCP the length of each channel is the same, but the

[12] G. Eschard and B. W. Manley, *Acta Electron.* **14**, 19 (1971).

[13] A. J. Guest, *Acta Electron.* **14**, 79 (1971).

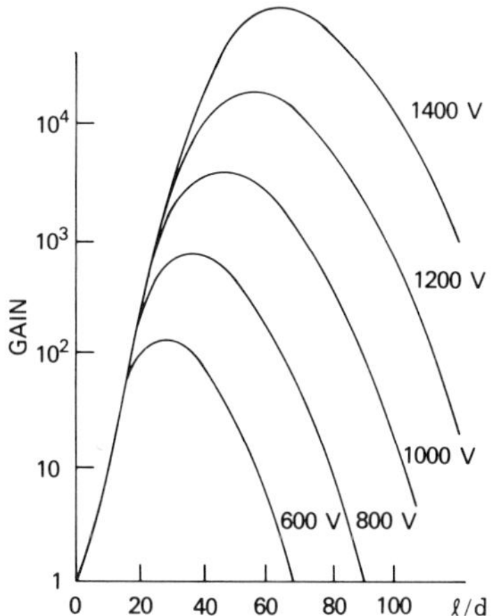

FIG. 16. Electron gain of MCP as a function of the length–diameter ratio (l/d), with the applied voltage as parameter. Universal gain curves.

diameter changes slightly from channel to channel, depending on manufacturing tolerances. If the operating conditions are not set (the MCP operates in or near the maxima of the gain curves) variations from channel to channel will cause fixed pattern noise. In most image intensifiers the gain required is between 100 and 1000. Therefore, MCPs for this application have an l/d ratio between 30 and 40 and the operating voltage is between 600 and 900 V. The secondary emission coefficient for MCP material has been measured by several authors. Since the glass used and the processing of MCPs differs from manufacturer to manufacturer, it is not surprising that differences in the secondary emission coefficient have been obtained. For a "fresh" MCP the secondary emission coefficient seems to vary between 2.0 and 3.7 at $V_{max} = 300$ V. Figure 17 shows the dependence of the secondary emission coefficient on the accelerating voltage

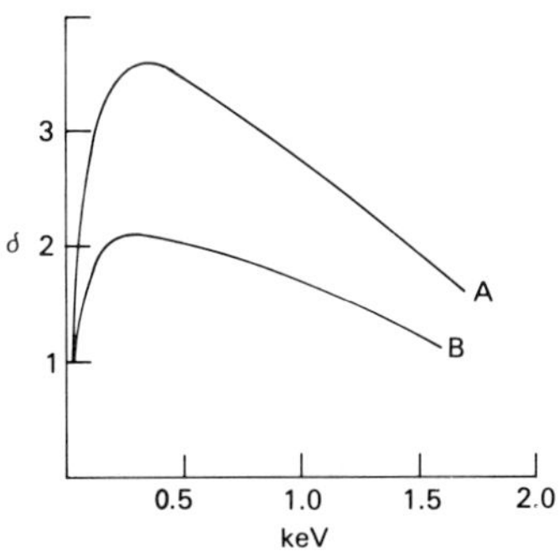

FIG. 17. Secondary emission coefficient δ of MCP glass as a function of electron energy for normal incidence. Reported values fall between curves A and B.

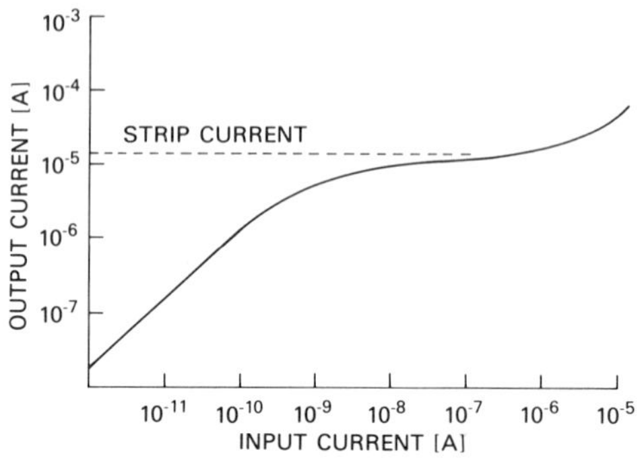

FIG. 18. Typical transfer curve of MCP. MCP voltage: 1000 V.

as measured by Hill[14] and Authinarayanan and Dudding[15]. In both references it is stated that the secondary emission coefficient is strongly dependent on the potassium content of the surface layer. With continuous electron bombardment, the potassium content and with it the secondary emission coefficient and the gain decreases.

Under continuous operation of the MCP, the output current will increase linearly with the input current, until the output current reaches about 10% of the standing current between the input and output of the MCP. With higher input currents the MCP shows saturation effects (Fig. 18) causing a drop in the gain. If the plate resistance would remain constant, the current would saturate at the standing or strip current. But in operation the resistance drops causing the gain to drop more gently. If the gain approaches 1.0, the output current again starts to become linear with the input current, and the gain stays constant. This saturation effect is only a function of the output current density. Each individual channel acts as its own independent multiplier. The strip current is carried by the channel wall and is therefore proportional to the radius of the channel for a given material. The number of channels within a given area is inversely proportional to the square of the channel radius. Therefore, higher total current outputs from an MCP can be obtained with smaller channels. Since this saturation is only dependent on the output current of each individual channel, this saturation effect may also be observed if the gain, instead of the input current, is increased.

For short pulses or individual electrons, the output current might be limited due to space-charge effects before the above saturation conditions are reached. For very high gains the negative charge, especially at the output of the channel, decreases the electric field, thus decreasing the accelerating voltage of the

[14] G. E. Hill, *Adv. Elect. Electron. Phys.* **40A**, 153 (1976).

[15] A. Authinarayanan and R. W. Dudding, *Adv. Elec. Electron. Phys.* **40A**, 167 (1976).

secondary electrons, and with it the secondary emission coefficient. The implications of these saturation effects on the operation and performance of image intensifiers will be discussed later.

C. Image Transfer Characteristics

The image transfer characteristics are normally expressed by the modulation transfer function (MTF) even though the criteria of spatial invariance and linearity are not met. The MTF, due to the structure of the microchannel plate, is dependent on the center-to-center spacing of the individual channels and can be calculated in the same way as for fiber-optic plates. It can easily be seen that the current saturation will decrease the contrast. For applications in image intensifiers two other effects are important. In an image intensifier, electrons are accelerated toward the input of the MCP. Some of these electrons will hit the closed area where some of them will be reflected or produce secondaries. Depending on the emission angle and energy and the strength of the electric field at the MCP input, these electrons will return to the MCP at some distance from the original impact point. These backscattered electrons are a major cause for the MTF reduction at low spatial frequencies. For the imaging of the output electrons of an MCP onto a phosphor screen, the energy and angular distribution is of prime importance. With larger spread, less favorable MTF characteristics are obtained. In order to constrain at least the angular variation, a material with low secondary emission coefficient is coated several diameters deep into the channel at the output side.

D. Noise Characteristics

In this section only signal-induced noise will be discussed. Fixed pattern noise, dark current, and emission points are other sources of noise. They are mainly functions of manufacturing techniques and control.

The signal-induced noise of the microchannel plate determines the noise figure of an intensifier tube. The noise figure is defined as the ratio of the signal-to-noise ratio at the input of the device to the signal-to-noise ratio at the output

$$N_F = \frac{S/N_{in}}{S/N_{out}} \tag{4.3}$$

The noise figure is a function of the area and the bandwidth over which the signal-to-noise ratios are determined. For the above definition the signal-to-noise ratios have to be measured over the same area and bandwidth. As mentioned before, electrons hitting the closed area can produce secondaries which might or might not be measured, depending on the measurement area. Since this is more of an image intensifier tube effect, it will be neglected for the following discussion.

Two factors contribute to the noise figure:

(1) Electrons accelerated toward the MCP might not produce an output pulse by hitting the closed area, or for a low secondary emission coefficient, might not produce secondaries even when entering a channel. The detection efficiency D is the ratio of the number of output pulses to the number of electrons accelerated toward the MCP.

(2) Single electrons will not always generate output pulses of equal size. The resulting distribution is characterized by the mean value m and the standard deviation σ. It can be shown that

$$N_F = D^{-1/2}[1 + (\sigma/m)^2]^{1/2} = D^{-1/2}N_{FG} \tag{4.4}$$

For reasons that will become more clear later, D is further divided into D_1 and D_2 with $D = D_1 D_2$. Here D_1 represents the electrons that enter a channel but still do not generate an output pulse. N_{FG} is called the gain noise figure.

Several attempts have been made to derive a formula for the noise figure from theoretical concepts. All these formulas can only be considered as approximate solutions, or contain some unknown parameters. Generally in all the derivations, the multiplication process is divided into two parts; the first impact of the electron in the channel, and the rest of the gain mechanism. For the first formula the only assumption made is that the secondary emission process can be described by Polya statistics. Polya statistics cover a continuous class of discrete distributions with the Poisson and the discrete exponential (Fury) distribution as limits. With γ as the secondary emission for the first strike of the electron in the channel, and $R(S)$ the unknown moment generation function for the gain process after the first strike, the moment generation function for the whole process is given by

$$Q(S) = [1 + b\gamma\{1 - R(S)\}]^{-1/b} \tag{4.5}$$

where b is a parameter in Polya statistics that determines the shape of the distribution. For $b = 0$, a Poisson, and for $b = 1$, a Fury distribution is obtained. With these assumptions, the noise figure can be calculated to

$$N_F = D_1^{-1/2}\left\{\frac{1}{\gamma}\left[1 + \left(\frac{\sigma^\chi}{m^\chi}\right)^2\right] + 1 + b\right\}^{1/2} \tag{4.6}$$

here σ^χ is the standard deviation and m^χ the mean for the multiplication process excluding the first strike. Bell[16] assumed for the secondary emission process a Poisson distribution and a linear birth or Fury process for the rest of the multiplication process. Independently, he obtained

$$N_F = D_1^{-1/2}\{(2/\gamma) + 1\}^{1/2} \tag{4.7}$$

[16] R. L. Bell, Private communications.

Equation (4.7) can be easily derived from Eq. (4.6). For a Fury process resulting in an exponential distribution $\sigma^\chi/m^\chi = 1$ and for a Poisson process $b = 0$. Escher[11] derived

$$N_F = D^{-1/2}\{(1/\gamma) + 2\}^{1/2} \tag{4.8}$$

The underlying assumption for deriving Eqs. (4.7) and (4.8) was that the multiplication process always starts at the same point in the channel. In an image intensifier, the electrons arrive at the MCP normal to the surface. The channels are biased between 5° and 10° in respect to the normal. Neglecting electro-optical "micro" lensing effects, the arriving electron may penetrate up to $18d$ for 5° bias and up to $7d$ for 10° bias before making the first impact. As discussed before, the length of the channel is chosen to be between $30d$ and $40d$. This means in the worst case the average gain difference for single electrons can be 30%. This certainly degrades the distribution and with it the noise figure. With a simplified calculation it can be shown that this can degrade the noise figure up to a factor of 1.3. To compare the formulas with measured data, only experimental data can be used where care has been taken that secondaries produced by electrons hitting the closed area cannot enter a channel. Not too many of these measurements exist. In addition, it is difficult to determine the secondary emission coefficient for the first strike. Table II shows measured noise figures for a well scrubbed regular MCP, and two MCPs where MgO or CsI has been deposited at the input of the channels. The given values for the secondary emission coefficient are approximate numbers. As can be seen from the table, the experimental data fit very well with Eq. (4.6) for $b = 0$ and $\sigma^\chi/m^\chi = 1.7$. Equation (4.7) gives too large values and Eq. (4.8) too small values. As mentioned before, too small—but not too large—values are expected from these formulas due to the variation in the location of the first strike.

From the discussions above, it is obvious that improvements in the noise figure can be obtained by increasing the open area ratio and the secondary emission coefficient for the first strike and by decreasing the depth for the first strike in the channel. As indicated before and in Table II, MCPs with the first two mentioned improvements have been fabricated and are incorporated into developmental image intensifiers. As will be discussed later in more detail, a

TABLE II

NOISE FIGURE OF MCP FOR DIFFERENT SECONDARY EMISSION COEFFICIENT γ FOR FIRST STRIKE IN THE CHANNEL

γ	OAR	N_F measured	N_F Eq. (4.6)	N_F Eq. (4.7)	N_F Eq. (4.8)
2.5	0.6	2.1	2.08	1.74	2.01
6 (MgO)	0.6	1.7	1.68	1.49	1.91
	0.85	1.4	1.44	1.30	1.60
10 (CsI)	0.6	1.5	1.56	1.43	1.88

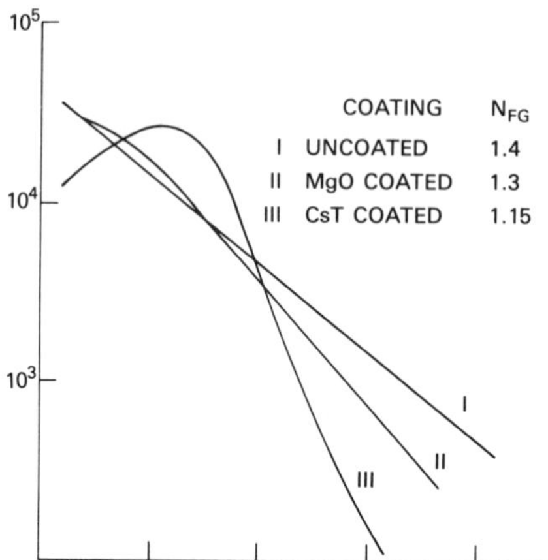

FIG. 19. Pulse-height distributions of MCP, normalized to same average energy and total number of pulses.

decrease in noise figure not only increases the probability of detecting targets at low light levels but also, often more importantly, gives a more favorable pulse height distribution. It decreases "scintillation noise." Figure 19 shows pulse height distributions for an uncoated MCP and for MCPs coated at the input with MgO or CsI. The third factor for improving N_F was to decrease the first strike depth. An obvious solution seems to be to increase the bias angle, but for bias angles above 10° distortions in the focusing of the output electrons onto the phosphor screen start to become a major problem. Another approach is to fabricate curved MCPs, as reported recently.[17]

Significantly improved noise figures and pulse height distribution can be obtained when the MCP is operated at charge saturation. Saturation takes place when the density of the electrons within a channel becomes so large that the created space-charge reduces the electrostatic field significantly. This limits the multiplication process. The pulse height distribution becomes peaked (Fig. 20) and gain noise figures as low as 1.05 have been obtained. For image intensifiers the required gain of 10^4–10^5 is too high for practical use. In addition, ion feedback significantly increases with gain. But for pulse-counting applications, operation in the charge saturated mode offers significant benefits.

E. MCP FABRICATION

The material used to manufacture microchannel plates is glass, which is basically silica, containing a certain amount of lead and/or bismuth oxides. Alkali ions are added to give the glass the desired softening and annealing

[17] J. Boutot *et al.*, *Adv. Elec. Electron. Phys.* **40A**, 103 (1976).

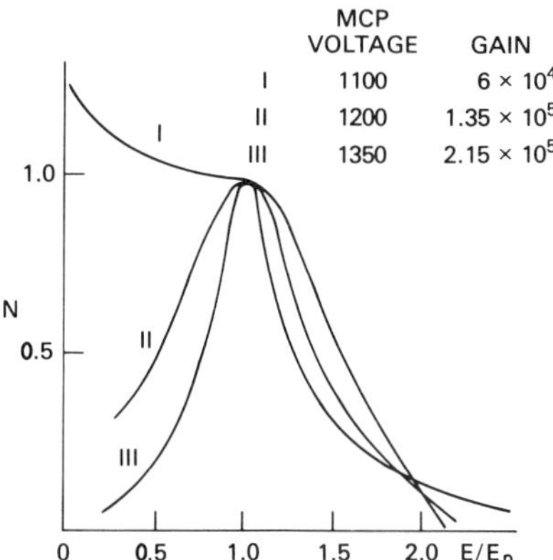

FIG. 20. Pulse-height distributions for pulse-saturated MCP, normalized for energy at the peak E_p and the number of pulses at this energy.

temperatures. Two basic techniques are currently used in the fabrication of MCPs—the "etch core" and the hollow draw technique. By far the largest number of MCPs are fabricated by the etch core method. The fabrication of a boule is very similar to the fabrication of a fiber-optics boule. First, single fibers, consisting of a tubing of the MCP glass with a solid core, are drawn. The single fibers are stacked, normally in a hexagonal pattern, fixed, and drawn again. The resulting multifiber bundles are again stacked and fused into a boule. For better handling and mounting of MCPs, a border of solid glass is normally attached around the fibers. After cutting the boule into plates of the desired thickness, grinding, polishing, and cleaning of the plates, the core is removed by a chemical etch process. If high open area ratios are to be obtained, only part of the core is removed at the input. With a different type of etch the desired shape of the core is obtained.[18] Open area ratios close to 100% have been achieved. After all the core is removed, the MCP is activated by hydrogen firing. In this process some of the lead and/or bismuth oxides are reduced. The amount of oxides reduced, and therefore the conductivity, can be adjusted by the time–temperature profile of the firing schedule. During this process the glass turns black and becomes highly absorbant. After the activation, a suitable metal (inconel) is evaporated onto the back and front surfaces. The MCP is now ready to be incorporated into current second generation image intensifier tubes. For improved second or third generation image intensifiers currently under development, two additional steps are performed. The input of the channel is coated with magnesium oxide to increase the secondary emission coefficient for the first strike. In addition, a thin film of Al_2O_3, about 50 Å, is suspended over the entire input of the MCP to prevent ion feedback.

[18] E. V. Patrick and A. R. Asam, *Proc. Elec. Opt. Syst. Design Conf.* p. 629 (September 1976).

V. IMAGE INTENSIFIER TUBES

A. Types of Image Intensifiers

Image intensifier tubes are manufactured in various configurations, with differences in sizes, photocathodes, electron focusing, gain mechanism, phosphor screens, magnification, etc. For any specific application, each of these factors must be considered carefully. By far, the largest number of image intensifier tubes have been fabricated for night vision applications. They are labeled as zero, first, second, and third generation tubes. The distinguishing factors are the type of photocathode and the gain mechanism. S-1 photocathodes have been used in zero generation tubes, sometimes called infrared image converters. First and second generation intensifiers incorporate the multialkalide S-20 photocathode; third generation tubes the negative-affinity single crystal GaAs cathode. Third generation intensifiers are still under development. The distinction between first and second generation is made by the gain mechanism. Generally, second generation intensifiers incorporate a microchannel plate and first generation tubes do not. For the purpose of the following discussion, all tubes without an MCP will be labeled first generation, independent of the cathode, if not stated otherwise.

B. Gain of Image Intensifier Tubes

In first generation image intensifiers the photoelectrons are focused and accelerated toward a phosphor screen. Three types of focusing mechanisms are used: proximity, electrostatic, and magnetic. The relative merit of the different focus mechanism will be discussed under "image transfer characteristics." The gain of an image intensifier is commonly expressed by the ratio of the input illuminance in footcandles (lumens per square foot) or lux (lumens per square meter) to the output brightness in foot lamberts ($1/\pi$ candelas per square foot) or nits (candelas per square meter). For the measurement, the intensifier is illuminated with light at a color temperature of 2854°K and the output brightness measured with a photometer. With these definitions, the gain of a one stage image intensifier can be calculated easily,

$$G = \Theta V\eta/m^2 \tag{5.1}$$

where Θ is the cathode sensitivity in amperes per lumen, V the applied acceleration voltage, η the phosphor efficiency in lumens per watt, and m the magnification of the intensifier tube. The gain as defined above is also referred to as white-light gain. The phosphor efficiency is a function of the applied voltage, and the cathode sensitivity is a function of the field strength which, in turn, is dependent on the voltage. In a typical first generation electrostatically

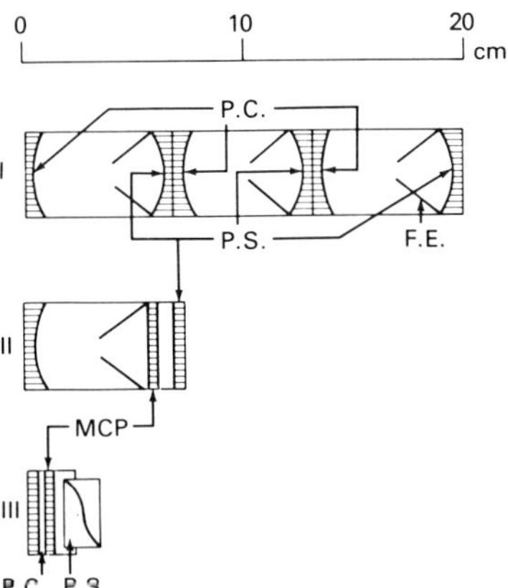

FIG. 21. Image intensifier tubes, 25 mm format. (I) Three-stage, first generation. (II) Second generation electrostatic tube—inverter. (III) Second generation proximity focused tube—wafer. P.C. is the photocathode on fiber optic; P.S. the phosphor screen on fiber optic; and F.E. the focusing electrode.

focused inverted tube $\Theta = 300\ \mu\text{A/lm}$, $V = 15\ \text{kV}$, $\eta = 25\ \text{lm/W}$, and with $m = 1$, the gain is 112 fc/fL. This gain is too low to be useful for night vision applications. To increase the gain, only the applied voltage and the magnification are free design parameters. Higher cathode sensitivities for S-20 type photocathodes have been achieved, but if specified at higher values, the low fabrication yield will increase the cost. Tubes with a demagnification ratio of 2 : 1 and somewhat higher voltage, useful down to half moonlight illumination level, are currently in production. The main advantage of this tube is its low production cost. To achieve higher gain, two or more stages are normally coupled by means of fiber optics. Figure 21 shows the widely used three-stage first generation intensifier, together with a second generation electrostatically and a second generation proximity focused intensifier. Under equal conditions, the gain of the second or third stage is always smaller than the gain of the first stage. Assuming the same photocathode, a "lumen from a 2854°K light" will generate more photoelectrons than a lumen from a green phosphor screen. So called "green light" gains of around 30 are common for good second or third stages. For a three stage tube, a gain of around 100,000 is obtained. With selected parts, higher gains are achievable, but for its main use in the small starlight scope a gain of 100,000 is more than sufficient.

The above definitions of the gain, even generally used, quite often lead to some confusion. The expressions "white light," "green light," or "starlight gain" are frequently used. These can easily be understood from the more general expression for the gain

$$G = \frac{\int_0^\infty H(\lambda)\Theta(\lambda)\, d\lambda}{680 \int_0^\infty H(\lambda)V(\lambda)\, d\lambda} V\eta \tag{5.2}$$

$H(\lambda)$ is the irradiance at the photocathode in watts per square meter or at the retina of the eye if placed in the photocathode plane. $\Theta(\lambda)$ is the cathode sensitivity and $V(\lambda)$ the relative luminosity curve for photopic vision. The above expression determines the increase in illuminance to the human eye—assuming the same optics—by the use of an image intensifier. This is obviously dependent on the spectral distribution of the input light and therefore not dependent only on tube parameters. The above-mentioned different "gains" now refer to different spectral distributions of the input light.

For night vision applications a gain definition as given by Eqs. (5.1) or (5.2) is very useful, but for almost any other use of image intensifiers it is of no value. In all cases, the output brightness or radiance $N(\lambda)$ is of interest:

$$N(\lambda) = V\eta_{\max}\eta(\lambda) \int_0^\infty H(\lambda)\Theta(\lambda)\, d\lambda \tag{5.3a}$$

with $\eta_{\max}$ the maximum efficiency and $\eta(\lambda)$ the relative output distribution. If the intensifier is viewed by the human eye the total output luminance is of more interest:

$$B = \eta V \int_0^\infty H(\lambda)\Theta(\lambda)\, d\lambda \tag{5.3b}$$

To obtain the gain, output radiance, or luminance for microchannel plate tubes, the above expressions have only to be multiplied by the electron gain.

All currently produced image intensifiers used for night vision applications are provided with an automatic brightness control (ABC). Incorporated into the power supply is a mechanism that senses the output current. If the output reaches a preset value, the gain of the tube is dropped. In second generation intensifiers, this is accomplished by lowering the voltage applied to the microchannel plate. Figure 22 shows a typical gain characteristic of the 18-mm wafer intensifier used in the Night Vision Goggle System. As can be seen, the output current is held constant over several orders of input light levels. Then at very high input levels, the output current is decreased. This is done by lowering the voltage applied between the photocathode and the MCP to prevent a destruction of the photocathode.

The ABC circuitry senses the total output current and is therefore not effective for bright point sources. Bright point sources are of frequent occurance in typical night vision scenes and if not suppressed to some degree will seriously reduce the usability of the device. As shown in Fig. 18 and described in the text, the gain of an individual channel of a microchannel plate starts to drop if the output current reaches about 10% of the strip current, and continues to drop with increasing input current. These localized saturation characteristics are one of the most significant performance improvements obtained from second generation technology. In first generation type intensifiers, no suppression of bright point sources is obtainable.

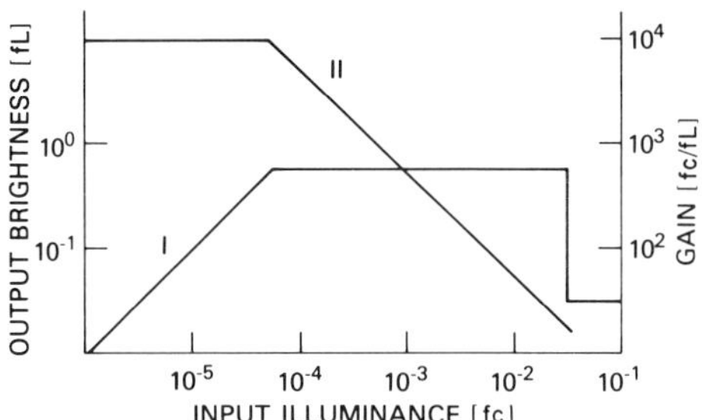

FIG. 22. Typical output brightness (I) and gain (II) as a function of average input illuminance, for image intensifiers with automatic brightness control.

C. NOISE CHARACTERISTICS

In Section II the importance of the signal-to-noise ratio for the performance of an image intensifier system has been discussed. The noise at the output of the intensifier is, first, caused by the statistical arrival of photons at the photocathode, as explained before. Additional noise is due to the less-than-ideal gain mechanism of the intensifier. The additional noise is quite often expressed by a single number, the noise figure N_F. The formula given in Section IV also applies to first generation type image intensifiers

$$N_F = D^{-1/2}[1 + (\sigma/m)^2]^{1/2}$$

with the same definitions given before.

Pulse height distributions (measuring the probability of a single scintillation having a certain energy) of three-stage first generation image intensifiers have been measured carefully[19]. From these distributions a gain noise figure of 1.2 was obtained (Fig. 23). From the total noise figure of about 1.3 obtained through a separate measurement, D can be estimated to 85%.

Noise figures of microchannel plates have been discussed before. If incorporated into an image tube, some additional factors have to be considered. Electrons hitting the closed area of an MCP will also generate secondary electrons or can be reflected by elastic or inelastic scattering. These electrons may return back to the channel plate, enter a channel, and produce an output pulse. This will increase the measured detection probability if the output pulse is within the area over which D is measured. In currently produced second generation wafer tubes, the spacing between the MCP and the cathode is around 150 μm. For an elastically reflected electron, this is also half the maximum distance between the point where it first hits the MCP and the point whence it

[19] H. K. Pollehn and J. C. Solberg, *Proc. SPIE* **21**, 31 (1970). Also H. K. Pollehn, *Elec. Opt. Syst. Design Conf.* p. 190 (September 1972).

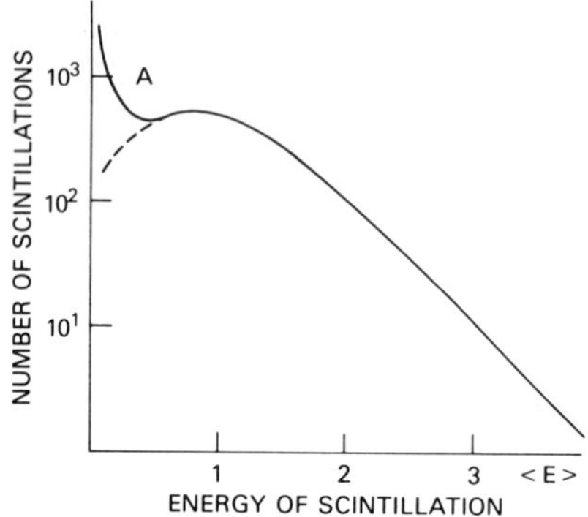

FIG. 23. Pulse-height distribution of first generation image intensifier. $\langle E \rangle$ is the average energy, A the single photon pileup.

returns back. Any electrons with less energy will travel a shorter distance. In the standard measurement of the S/N ratio and therefore D, a spot of light of 0.2-mm diameter is projected onto the photocathode and the output analyzed over a spot of about 0.3-mm diameter. Therefore, most of the backscattered or secondary electrons return to the MCP within the analyzed area, increasing the detection efficiency to around 80% (Table III) for a geometrical open area ratio of only 60%. In an inverter tube, the field strength at the input of the MCP is much lower, and most of the electrons from the closed area will not return to the MCP within the analyzed area. The detection probability is even lower than the geo-

TABLE III

AVERAGE PARAMETERS OF IMAGE INTENSIFIER TUBES

	N_{FG}	N_F	Resolution lp/mm	Gain
One stage magnetically focused tube	1.2	1.3	75	150
One-stage demagnification tube (2 : 1)	1.2	1.3	40	800
x_1 Three-stage first generation tube	1.2	1.3	36	10^5
x_1 Second generation electrostatically focused tube	1.5	2.0	32	$3 \times 10^4_{x2}$
x_1 Second generation proximity focused tube	1.55	1.7	28	10^4_{x2}
Experimental wafer tube (10 μm c-c MCP)	1.55	1.7	45	Variable
Wafer tube, Al_2O_3 filmed MCP, heavy scrub	1.5	3.0	28	Variable
Wafer tube, Al_2O_3 filmed MCP, light scrub	1.5	2.1	28	Variable
Experimental wafer tube (85% OAR, MgO coated)	1.3	1.4	28	Variable

x_1 standard production
x_2 with standard power supply
Otherwise—variable

metrical open area ratio, meaning that some electrons entering a channel will not produce an output pulse. This is even more the case of a wafer tube with an Al_2O_3-filmed plate that has been heavily scrubbed. All MCPs have to be outgassed during the fabrication of the intensifier tube. This is done by a vacuum bake at about 375°C followed by electron scrubbing. The detection efficiencies given in Table III take into account all photoelectrons not generating an output pulse. N_{FG} does not take any losses into account and has been calculated from measured distributions.

Unfilmed MCPs are also heavily scrubbed, but during the fabrication of the S-20 photocathode, alkalides can enter the channel, increasing the secondary emission coefficient. The last tube mentioned in Table III contained an Al_2O_3-filmed, funneled, and MgO-coated MCP. The D, N_{FG}, and with it N_F are the best values obtained so far in an MCP-type tube.

In the section on microchannel plates, it was stated that the pulse height distribution of an MCP is exponential, giving a gain noise figure of 1.4. Except for the MgO-coated MCP, all second generation intensifiers have gain noise figures larger than 1.4. Figure 24 shows typical pulse height distributions for a second generation wafer tube incorporating a filmed and an unfilmed MCP. The tail at larger energies can be explained by feedback processes. During the electron multiplication and acceleration processes, ions are generated mainly from the channel walls and the phosphor screen. The ions are accelerated towards the input of the channel. The ions have sufficient energy to release electrons from the channel walls. In wafer tubes with unfilmed MCPs, the ions will be further accelerated toward the photocathode. The released electrons are again multiplied by the MCP and might again release ions. The time in which all these

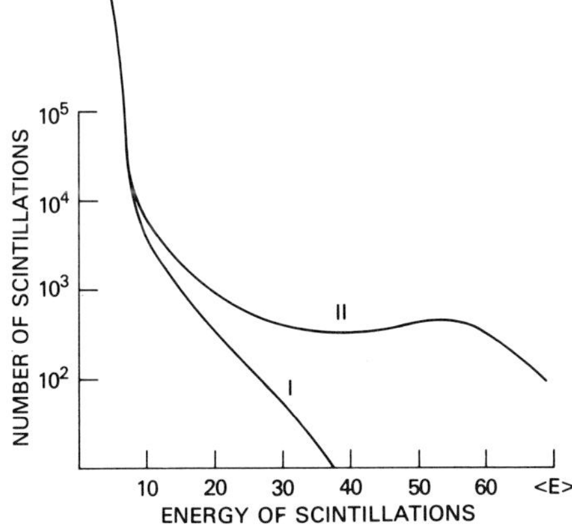

FIG. 24. Typical pulse-height distributions of second generation intensifiers showing ion feedback. (I) Wafer tube with Al_2O_3-filmed MCP or electrostatic focused tube. (II) Wafer tube with unfilmed MCP. ⟨E⟩ is the average energy of scintillation.

feedback processes occur is much shorter than the integration time of the normally used phosphors, or the integration time of the eye. A much brighter scintillation, up to 150 times the average, will be detected by the measuring equipment and the observer. Besides increasing the gain noise figure to about 1.6 in wafer tubes, these bright scintillations are responsible for the destructive, noisy appearance of the output image. Even worse, the ions hitting the photocathode degrade the sensitivity, thus shortening the lifetime of the intensifier. With Al_2O_3-filmed MCPs, and also in electrostatically focused intensifiers, ions are prevented from hitting the cathode, improving the pulse height distribution and the lifetime but at the expense of the collection efficiency. This again can be improved by coating the input of the MCP with a material of a high secondary-emission coefficient.

D. Image Transfer Characteristics

1. *Definitions*

The imaging characteristics of image intensifiers are normally described by the modulation transfer function (MTF). The veiling glare is the low-frequency part of the MTF, but since it is determined in a separate measurement and is partly caused by different processes, it is discussed separately.

The MTF is strictly defined only for linear and spatially invariant systems. First generation tubes are linear but in intensifiers incorporating MCPs, non linearities occur when the output current reaches about 10% of the strip current. If not especially mentioned, it is assumed that the intensifier operates in its linear range. All intensifiers incorporate spatially noninvariant components, fiber optics, and microchannel plates. In these components the MTF is dependent on the location and direction of the measurement pattern (mostly slit). Neglecting theoretical and mathematical purity, an average MTF can be defined for these components and is always assumed in the following discussion.

Practically all components of the intensifier and the transport of the electrons from one component to the other influence the MTF. When transported from one component to the other, the electrons are either proximity, electrostatically, or magnetically focused.

2. *Focusing*

The least image-degrading transport is obtained by magnetic focusing. In this type of focusing, a uniform magnetic field is superimposed on the accelerating electrostatic field with both field gradients parallel. The electrons released from the photocathode oscillate around a line parallel to the applied field. The locations of the nodal points are dependent on both the magnetic and electrostatic fields. Therefore, for good focusing, the relationship between the strength of the two fields is fixed and very critical. Due to the small tolerances allowable for the two field strengths and the bulkiness of the system, magnetic focusing is applied only where very high resolution (~ 80 lp/mm) over the whole field is required such as in astronomy or high quality photographic applications.

An almost equal image quality on axis can be obtained by electrostatic focusing. But since the MTF falls off toward the edge, this high quality is compromised to provide reasonable resolution over a larger format. In most electrostatically focused tubes, the input and output surfaces are curved to obtain a good MTF with a relatively short lens. Where this is not possible, a good quality image can be obtained by the use of field flatteners—as in the second generation inverter tubes where the output surface is the MCP. In all electrostatically focused intensifiers, the image is inverted. A large number of image intensifiers utilize electrostatic lenses as a focusing mechanism: the three-stage first generation intensifier, the second generation inverter tube, the zoom tubes, and the display tubes. In the demagnifier tube, the electron image from the photocathode is focused onto a smaller phosphor screen and, as mentioned before, gives the additional gain determined by the square of the ratio of the radii. All currently fabricated x-ray image intensifiers are demagnifier tubes. In zoom tubes the magnification can be varied normally between 1 and 0.3. In display tubes the electrons are focused from the photocathode onto a larger phosphor screen to be viewed without an eyepiece.

The main advantage of proximity focusing is that very short image intensifiers can be produced. Ezard[20] has derived a formula for the MTF as a function of the spacing and applied voltage.

$$\mathrm{MTF}(f) = \left| \frac{2J_1(4\pi f\, d(\phi V)^{1/2}}{4\pi f\, d(\phi V)^{1/2}} \right| \tag{5.4}$$

where J_1 is the first-order Bessel function, f the spatial frequency (lp/mm), d the distance (mm), V the applied voltage, and ϕ the "effective" energy of the photoelectrons (eV).

For S-20 photocathodes ϕ was experimentally determined to be between 0.1 and 0.2 eV. The smaller value corresponds to light at the long wavelength cutoff.

Using the above formulas and formula (5.2) for the gain, it can be shown that first generation type proximity focused intensifiers are of only marginal value for night vision applications, either due to unsufficient gain at close spacings or severe MTF degradation at wider spacings.

---

[20] L. A. Ezard, The MTF Applied to Electron-optical Systems of Image Tubes, Thesis, Univ. of Pennsylvania (1970).

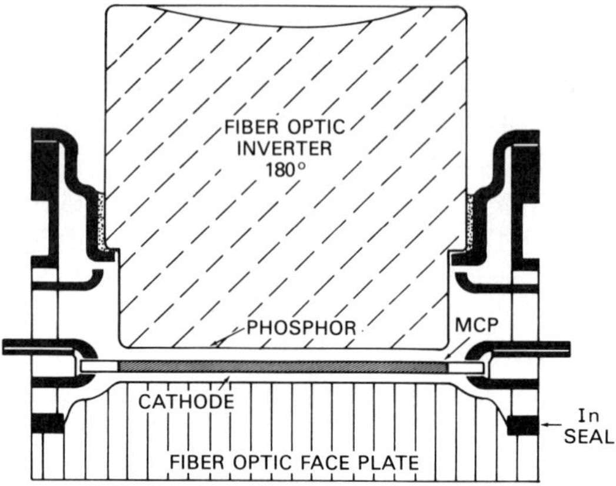

FIG. 25. Second generation wafer tube. Shaded parts metal; light parts ceramic.

The only, but very important, completely proximity focused image intensifier tube currently in production is the 18-mm second generation wafer tube (Fig. 25). For tubes with Al_2O_3-filmed MCPs, the spacing between the cathode and the MCP is around 0.3 mm and the applied voltage is 800 V; 0.15 mm and 200 V for tubes with unfilmed plates. By the above formula, the MTF should be equal as shown in Fig. 26. In all second generation image intensifiers, electrons leaving the MCP are proximity focused onto the phosphor screen. This spacing is normally around 0.9 mm and the applied voltage is 6000 V. The MTF is not well known and the curve in Fig. 26 should be taken as very approximate. But definitely the MTF degradations for this transfer are more severe than the proximity focus between the cathode and the MCP.

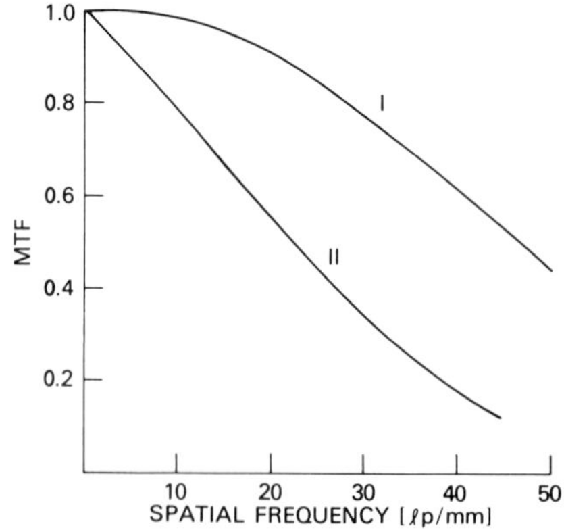

FIG. 26. MTF of proximity focus between photocathode and MCP (I) and between MCP and phosphor screen (II).

3. *Components*

MTF degradations caused by the components are negligibly small for the cathode and the fiber–optic substrates. S-20 cathodes have a maximum thickness of about 1000 Å, and single crystal III–V cathodes do not exceed a thickness of 2 μm. The cathode fiber-optic substrate and the phosphor substrate generally have a center-to-center spacing of 6 μm with the exception of the fiber optics inverter used in the 18-mm wafer tube. The fiber-optics inverter has a center-to-center spacing of 10 μm. If the cathode or the phosphor screen is deposited on a glass substrate, the MTF degradation at higher spatial frequencies, which is caused by light reflections, is covered under the veiling glare effects. Major MTF degradations are caused by the MCP and the phosphor screen. The degradiations caused by the MCP have been covered before.

4. *Phosphor Screens*

Figure 27 shows the structure of a phosphor screen. The screen consist of a thin phosphor layer on a glass or a fiber-optics substrate. The other side of the phosphor is covered with a thin (around 1000 Å) aluminum film. The aluminum film serves several purposes. It is an electrode which provides a uniform field, it prevents light generated in the screen to feed back to the cathode, and it reflects the light to increase the effective light output.

The two most important parameters of a phosphor screen are its conversion efficiency, expressed in lumens per watt, and its imaging performance expressed through the modulation transfer function. For use in image intensifiers, a high efficiency and a high MTF are desirable. It should be emphasized that for low-light-level applications the MTF at low spatial frequency is of main importance. In MCP tubes the limiting resolution (see Fig. 26) is generally below 50 line pairs/mm. The MTF values of the phosphor screen at higher spatial frequencies is therefore of no interest.

The MTF of the phosphor screen is mainly dependent on the deposition technique, the thickness of the phosphor layer, and the grain size. The luminous efficiency is, first, dependent on the type of phosphor and also on the phosphor thickness and grain size, but unfortunately in the opposite direction to the MTF. Larger grain sizes always give higher efficiencies and lower MTF values.

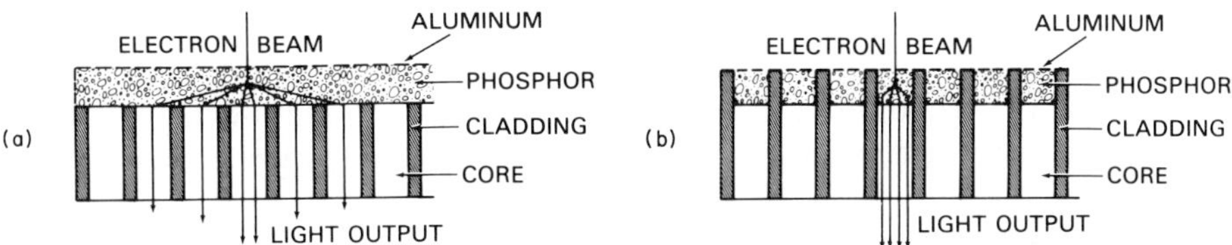

FIG. 27. Phosphor screens on fiber optics. (a) Regular screen. (b) Intagliated screen.

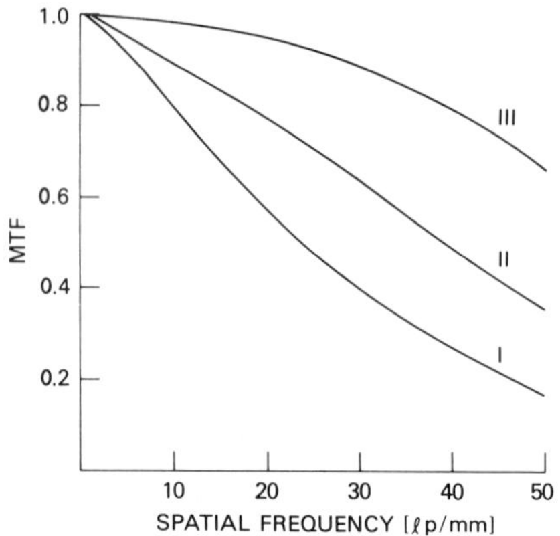

FIG. 28. MTF of phosphor screens used in image intensifier tubes. (I) Settled phosphor screen. (II) Brushed phosphor screen. (III) Intagliated phosphor screen.

With increased thickness, lower MTF values are always obtained, but higher efficiencies require higher accelerating voltages for the electrons.

Deposition techniques have been described by Diakides[21] in some detail. The most common deposition method currently used is the so-called "brush technique," which is similar to the method described by Diakides for the monolayer screen. Details of this technique are generally kept as company secrets. The MTF of these screens has significantly improved with only a slight decrease in efficiency, compared to phosphor screens fabricated by the settling technique.

Figure 28 shows the MTF of a settled and a brushed phosphor screen as currently used in the 18-mm wafer tube, together with a phosphor screen prepared by the recently developed intagliation method.[22] In this method, the core of a fiber-optics substrate is removed to a depth of a few microns by a selective etch that leaves the cladding intact. To prevent crosstalk and increase efficiency, the walls of the cavities are metallized and the phosphor deposited into the holes (Fig. 27). Measurements (Fig. 28) have shown the MTF of the phosphor screen to be equal to the MTF of the substrate.

The most commonly used phosphor types in wafer image intensifiers are the sulfides P-20 and 1052 with a peak frequency of 0.56 μm. A mixture of the silicates P1 and P39 is used in all second generation 25 mm electrostatically focused inverter tubes. The efficiencies are around 15 lm/W for 6 keV electrons.

5. *MTF of Intensifier Tubes*

The MTF of a compound system can be calculated by multiplying the MTFs of its components provided there is diffusing layer in between. By taking only

---

[21] N. A. Diakides, *Proc. SPIE* **42**, 83 (1973).

[22] J. Piedmont and H. K. Pollehn, *Proc. Eur. Elec. Opt. Conf, Geneva* (1976).

the main degrading effects into account, the MTF of a one-stage first generation type image intensifier tube can be written:

$$\text{MTF (1 gen, 1 stage)} = \text{MTF (focus)} \times \text{MTF (screen)} \qquad (5.5)$$

For low-light-level applications where a good phosphor efficiency is required, MTF (focus) is substantially unity in magnetic and electrostatic focused tubes. The MTF of these tubes is practically the MTF of the phosphor screen. For a three-stage first generation tube, the MTF is then given by

$$\text{MTF (3 stage)} = [\text{MTF (screen)}]^3 \qquad (5.6)$$

In second generation intensifiers, again taking only the main degrading effects into account

$$\text{MTF (2nd gen)} = \text{MTF (MCP)} \times \text{MTF (focus MCP} - \text{screen)} \times \text{MTF (screen)} \qquad (5.7)$$

Figure 29 shows typical MTF curves for various types of image intensifiers.

6. *Veiling Glare*

The MTF curves shown in Fig. 29 have been obtained by analyzing the image of a slit, 10 μm in width, over a 3-mm aperture. The image spread is generally larger than 1.5 mm to each side. The intensity of the light at larger distances is quite small, but can significantly degrade the performance of an image intensifier if part of a scene is at a significantly higher illumination level. This is very often the case for typical night vision scenes where the sky fills part of the field.

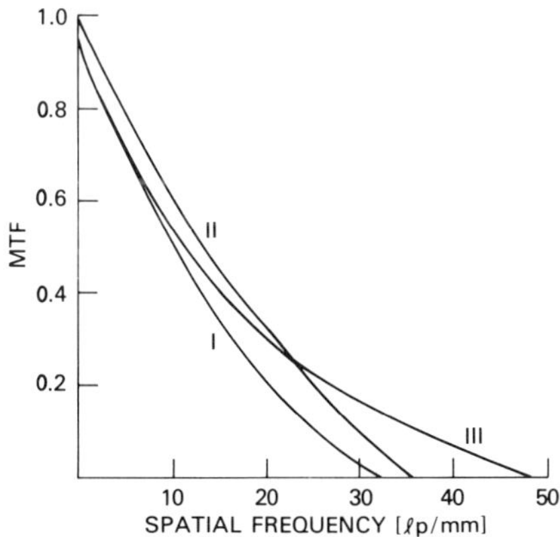

FIG. 29. MTF of image intensifier tubes. (I) Second generation wafer, 15 μm center-to-center spacing of MCP. (II) First generation three-stage and second generation electrostatic focused tube. (III) Second generation wafer, 10 μm center-to-center spacing of MCP, 0.5 mm spacing between MCP and phosphor screen.

The large area spread of the image is called veiling glare. A generally accepted method does not exist to characterize the veiling glare. A widely used method is to illuminate one-half of the intensifier and determine the output intensity as a function of the distance from the edge (half-moon method). Figure 30 shows the veiling glare determined in this way for different types of image intensifiers. Wide variations in the amount of veiling glare are experienced for individual intensifiers. But generally, first generation type magnetically and proximity focused intensifiers show the largest amount of veiling glare, followed by first and second generation electrostatic focused intensifiers, and the second generation wafer tube with the least amount of veiling glare, at least in the far field, several millimeters away from the edge.

The large area spread of the image is mainly caused by light transmitted through the photocathode and then reflected back, by electrons scattered at the various surfaces, and by feedback from the phosphor screen due to pinholes in the aluminum film. In second generation intensifiers the feedback from the phosphor screen is small since the microchannel plate acts as a very effective neutral-density filter. In any electrostatic focused intensifier, the focusing electrode significantly reduces the amount of light that can travel from the photocathode to the phosphor screen or the MCP and back from these components to the photocathode. The electrodes themselves are blackened to reduce their reflectivity. In a proximity focused second generation image intensifier, the maximum lateral distance that backscattered electrons can travel before reaching the surface from which they originated is $2d$, which is 0.3 to 0.6 mm at the

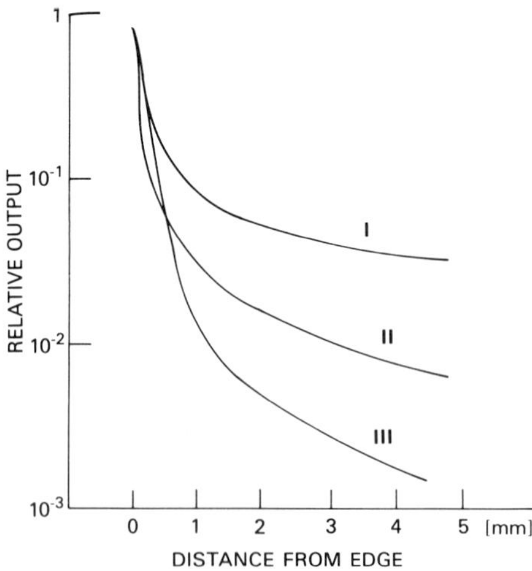

FIG. 30. Veiling glare. (I) Diode wafer tube. (II) Second generation inverter tube. (III) Second generation wafer tube.

input of the microchannel plate and about 1.5 mm at the phosphor screen. The electrons backscattered from the phosphor screen do not give a large contribution, since the accelerating voltage (6 kV) is not large against the "dead" voltage (2 kV), the minimum energy for an electron to penetrate the aluminum layer. This means that a large amount of backscattered electrons will not be able to penetrate the film or at least lose a significant amount of their energy. For light transmitted through the photocathode and reflected from the MCP, the maximum distance from the point of transmission to the first reflection is also on the order of 0.5 mm. Therefore, in the second generation wafer tube the largest contributions to the veiling glare are restricted to an area of about 0.5 mm around the image point.

The above-described method of characterizing the veiling glare has been chosen for the simplicity of the measurement technique, but also to simulate closely actual field conditions. Another frequently occurring scenario is the occurrence of a small bright-light source in an otherwise dark scene. The "half-moon" method is of little value in determining the performance of the intensifier in this situation since the automatic brightness control will lower the gain of the tube if the input intensity is raised. The automatic brightness control in second generation tubes (Fig. 22) is set so that the gain starts to decrease well before the saturation effects of the MCP occur. For uniform large-area illumination, the MCP will never operate in a saturation mode at any light level. This is different if only a small area of the intensifier is illuminated at a high light level. Then the output current is not sufficient to generate a gain reduction. Within the illuminated area, the gain will be reduced due to the saturation of the MCP, whereas the rest of the intensifier remains at its high gain. Figure 31 shows the

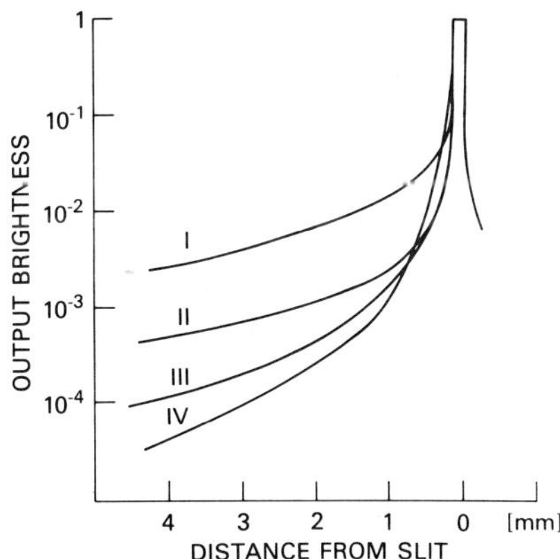

FIG. 31. Output brightness as a function of distance from slit (0.2 × 0.5 mm). Input illuminance: 2×10^{-5} fc. (I) Proximity focused diode. (II) Second generation inverter. (III) Three-stage first generation tube. (IV) Second generation wafer tube.

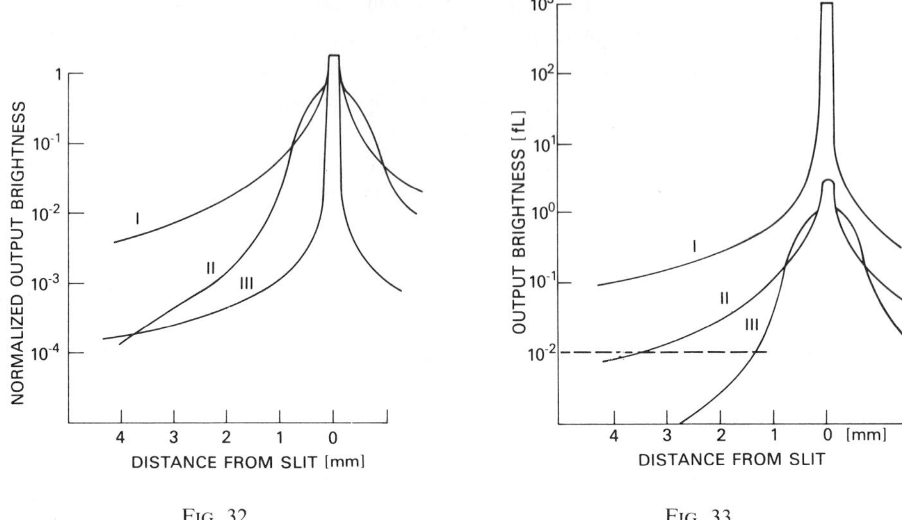

FIG. 32.

FIG. 33.

FIG. 32. Normalized output as a function of distance from slit (0.2 × 0.5 mm). Input illuminance: 10^{-1} fc. (I) Second generation inverter. (II) Second generation wafer. (III) Three stage first generation tube.

FIG. 33. Output brightness as a function of distance from slit (0.2 × 0.5 mm). Input illuminance: 10^{-1} fc. Average output brightness for scene at starlight. (I) Three-stage first generation. (II) Second generation inverter. (III) Second generation wafer tube.

output intensity distribution where the input is illuminated through a slit of 0.2 mm in width and 0.5 mm in length.

The input light level was 5×10^{-5} fc and the gain of the tubes (except for the first generation wafer diode) was set at 10,000. The distribution of the diode has been normalized to the other tubes. These data verify the results obtained with the half-moon method. Figure 32 shows the output distribution for the same tubes for an input light level of 10^{-1} fc normalized at the peak output brightness and in Fig. 33 with the actual output brightness. The distributions for the second generation intensifiers have changed drastically, whereas only slight changes can be noticed for the first generation tubes. Now assuming a scene at starlight, giving an illuminance of 10^{-6} fc at the photocathode, and within this scene a bright light giving an illuminance of 10^{-1} fc over an area of 0.2×0.5 mm, the luminous output from the bright light will be equal to the distributions shown in Fig. 33, and the luminous output due the scene will be 1.1×10^{-2} fc on the average. It is obvious that no details of the scene will be visible with first generation type intensifiers. Even 6 mm (on the screen) away from the bright light, the luminance caused by the bright light is still a factor 10 higher than the luminance due to the scene. With second generation intensifiers, especially with

(a)

(b)

FIG. 34. (a) First generation starlight scope. (b) Second generation pocketscope.

the wafer tube, scene details will be visible with the exception of the direct vicinity of the bright source. These factors have been verified in various field tests and are the main and most significant performance improvements of second generation compared to first generation intensifiers. An example can be seen in Fig. 34.

CHAPTER 11

Fiber Optics for Communications

D. B. KECK and R. E. LOVE

Corning Glass Works, Corning, New York

I. INTRODUCTION

Optical communications using glass optical waveguides is one of the most rapidly growing technologies in the world today. Indeed, all major telecommunications firms worldwide have installed systems operating with optical fibers. In addition, a host of other applications are being developed because of other beneficial features of the waveguide such as small size, dielectric properties, potential security, radiation resistance, and light weight. Included in this list are: broadband networks, military applications, process control, computer interconnection, and sensor applications. This trend is expected to accelerate in the future.

The development of the optical waveguide rests firmly on three attributes of glass: transparency, formability, and intrinsic strength. Historically, glass transparency and its associated dispersion characteristics have been the driving force behind waveguide technology. This attribute will continue to be a key

ISBN 0-12-408606-3

characteristic. The other glass attributes, formability and strength, affect the more practical aspects of the technology such as splicing, cabling, and manufacturing economics, and are becoming and will continue to become important in the future.

In the following sections these glass attributes will be explored in detail, as they affect overall waveguide communication system performance. Additionally, an overview of the key considerations and present status of individual waveguide system components will be explored. In Section II, a basic description of light propagation in the cylindrical waveguide is considered. This includes a discussion of waveguide theory and a description of basic material considerations and their effects on attenuation and dispersion, as well as waveguide emission characteristics. Waveguide strength is emerging as a key consideration affecting the general utility and overall economics of optical cables. This is discussed in Section II. Cabling of the optical waveguide is essential for its practical application. The design and construction of optical cables, as well as the factors that affect their performance and deployment, are addressed in Section III. Interconnection considerations of the waveguide to itself and to other system components, such as sources and detectors, are covered in Section IV. The theoretical discussions in Sections II and IV are generally applicable to passive components such as directional and STAR couplers. Finally, the characteristics of available light sources, detectors, and overall system design are presented in Section V.

II. OPTICAL FIBER WAVEGUIDES

A. Light Propagation in the Ideal Optical Waveguide

1. *General Description*

The propagation of light in the general dielectric waveguide is, of course, a very complex subject. Many authors have treated various aspects of this subject in considerable detail. This section summarizes many of the key concepts, results, and terminology in their papers. The interested reader may refer to the original works if more detail is desired.

A number of novel types of structures for guiding optical energy have been reported in the literature such as the "*W*-type,"[1] the "single material,"[2] and the Bragg[3] optical waveguide. However, the most practical for near-term ap-

[1] S. Kawakami and S. Nishida, *IEEE J. Quantum Electron.* **QE-10** (1974).

[2] P. Kaiser, E. A. J. Marcatili, and S. E. Miller, *Bell Syst. Tech. J.* **52**, 265 (1973).

[3] P. Yeh and A. Yariv, paper WD7-1, OSA and IEEE Topical Meeting on Integrated and Guided Wave Optics, Salt Lake City, Utah (1978).

plications is one with a core, whose refractive index is either homogeneous or radially graded, and is surrounded by a lower refractive index material. Initially, one assumes the waveguide axis is perfectly straight, and that the core diameter and all material properties are constant with length. Light propagation in this ideal structure is then considered.

Conceptually, a ray model is easiest to understand. Shown schematically in Figs. 1a and 1b are the ideal homogeneous core or "step" profile and the radially graded or "graded" profile waveguides. These structures are characterized by three quantities: the core radius a; the peak axial refractive index $n(0)$; and the surrounding index $n(0)[1 - \Delta]$, which, for the moment, is considered infinite in extent. Two types of rays propagate through these structures: "meridional" rays which pass through the axis of the waveguide and the much larger class called "skew" rays, which do not.

As shown in Fig. 1a, meridional rays in the step waveguide are totally internally reflected at the core–cladding interface and follow a zigzag path through the guide. From simple application of Snell's law at the core–cladding interface, total internal reflection for meridional rays will occur as long as the external angle $\Theta \leq \Theta_c$, where $\sin \Theta_c = n(0)\sqrt{2\Delta}$. This defines another characteristic parameter of the waveguide; its numerical aperture,

$$\mathrm{NA} = \sin \Theta_c = n(0)\sqrt{2\Delta} \tag{2.1}$$

For most waveguides considered for telecommunications, $\Delta \approx 0.01$–0.033, thus giving NA ≈ 0.2–0.35.

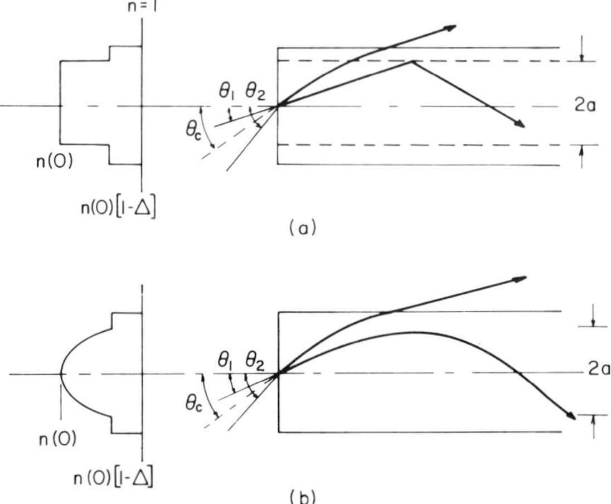

FIG. 1. Schematic diagrams of (a) a step-index waveguide and (b) a graded-index waveguide. The paths of meridional rays are indicated for each. The plot at the left indicates the radial refractive index distribution for each.

Note that since the meridional rays propagating at $\Theta = 0$ and $\Theta = \Theta_c$ travel different optical path lengths, a temporal phase delay difference, $\delta t = Ln(0) \cdot \Delta/c$, is built up in which L is the length of the waveguide. This will be related to the bandwidth of the waveguide later. By radially grading the core refractive index, propagation is no longer by reflection but by refraction. It can be shown that if the radial refractive index gradient decreases monotonically, the ray paths are sinusoidal as shown in Fig. 1b. As with the step profile waveguide, as long as the external angle at the ray entry point r is $\Theta \leq \Theta_c(r)$, the ray will be bound to the core. If the radial refractive index has the rather general form

$$n(r)^2 = n(0)^2[1 - 2\Delta f(r)] \tag{2.2}$$

where $f(0) = 0$ and $f(a) = 1$, one can show that

$$\mathrm{NA}(r) = \sin \Theta_c(r) = n(0)\sqrt{2\Delta[1 - f(r)]} \tag{2.3}$$

Thus, each radial point on the end face has a different light-acceptance property. This will affect the light-gathering properties of the graded profile waveguide as will be seen in a later section. A major benefit of the radially graded core is that temporal delay differences between various rays can be made to very nearly vanish, thus increasing the information-carrying capacity by several orders of magnitude. This is explained by the fact that the ray propagating at $\Theta = \Theta_c(r)$ spends much of its time in a lower index medium and hence, by proper choice of $f(r)$, can be made to have nearly the same optical path length as the ray at $\Theta = 0$. It will be shown later that the "parabolic" or square-law profile $f(r) = (r/a)^2$ is nearly the proper choice.

Thus far, the skew rays have only been mentioned. These rays follow spiral paths, either reflecting at the core–cladding boundary in the step profile or refracting at "critical radii" in the graded profile. Figure 2 shows schematic end-view projections of skew ray paths in the step and graded profile guides. In general, for any given skew ray there are two "turning" radii, r_1 and r_2, between

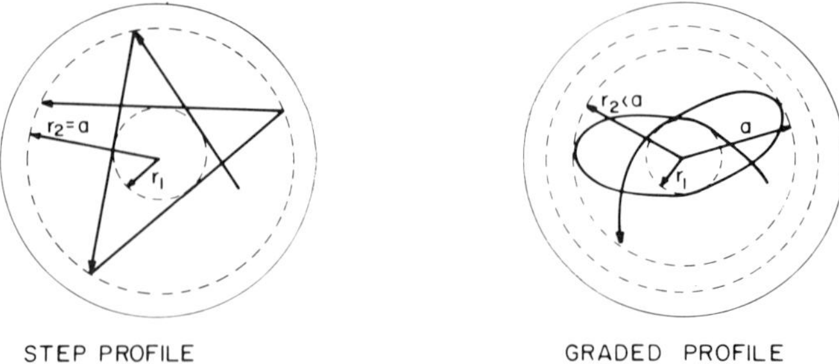

FIG. 2. End view of the ray trajectory for the step and graded profile waveguide. Rays spiral between the caustics at r_1 and r_2.

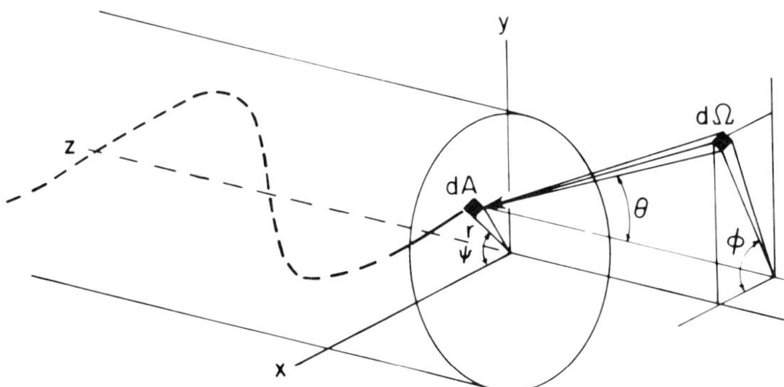

FIG. 3. Ray coordinate system used for the cylindrical fiber. A ray is characterized by its angles θ and ϕ and by its cylindrical coordinates r and ψ.

which the spiral trajectory is maintained. These radii are determined by the input angle and the position of the ray. The exact values for these radii are beyond the scope of the present work. However, the general model is useful in picturing waveguide propagation. A skew ray cannot be described solely by the angle with respect to the waveguide axis but requires a second angle ϕ, as defined in Fig. 3. This angle characterizes the angular momentum of the ray. Generally, the larger the angle ϕ, the higher the angular momentum, and the greater the turning radii. Even for skew rays, however, the condition $\Theta \leq \Theta_c(r)$ must be obeyed for bound rays to propagate. A more detailed analysis shows that for $\Theta > \Theta_c(r)$ and for sufficiently large ϕ, a "refracting"[4] or "leaky" class of rays will propagate over some length. This greatly complicates the analysis. For many problems, however, this segment of the skew ray class has sufficiently high attenuation that it may be ignored. The preceding results are obtained by solution of the Eikonal equation for the ray path in the waveguide structure. This has been discussed by many authors[5-9] both in general and for specific cases.

While the ray picture is easiest to visualize, not all problems have yet been solved using its formalism. For more detailed understanding, Maxwell's equations must be solved for the particular waveguide structure being considered. This reduces to solving the scalar wave equation in cylindrical coordinates,[5]

$$[\nabla^2 + k^2 n(r)^2]\Psi(r, \phi, z) = 0 \tag{2.4}$$

---

[4] A. W. Snyder and D. J. Mitchell, *Electron. Lett.* **9**, 437 (1973).

[5] D. Marcuse, "Light Transmission Optics." Van Nostrand-Reinhold, Princeton, New Jersey, 1972.

[6] E. G. Rawson, D. R. Herriott, and J. McKenna, *Appl. Opt.* **9**, 753 (1970).

[7] W. Streifer and K. B. Paxton, *Appl. Opt.* **10**, 769 (1971).

[8] K. B. Paxton and W. Streifer, *Appl. Opt.* **10**, 1164 (1971).

[9] A. Fletcher, T. Murphy, and A. Young, *Proc. Roy. Soc.* **223**, 216 (1954).

where $k = 2\pi/\lambda$ is the free-space propagation constant and Ψ represents the z component of either the electric or magnetic field intensity distribution. Of course, this is done subject to the boundary conditions of the specific waveguide configuration. For a purely radial index distribution, which is the sole consideration of this work, the solution can be separated into axial, azimuthal, and radial components:

$$\Psi_{\nu\mu}(r, \phi, z) = R_{\nu\mu}(r)e^{i\nu\phi}e^{i\beta_{\nu\mu}z} \tag{2.5}$$

The time dependence $e^{i\omega t}$ has been suppressed in Eqs. (2.4) and (2.5). Here $R_{\nu\mu}(r)$ is the solution to

$$\left[\frac{1}{r}\frac{d}{dr}r\frac{d}{dr} + k^2n(r)^2 - \frac{\nu^2}{r^2}\right]R_{\nu\mu}(r) = \beta_{\nu\mu}^2 R_{\nu\mu}(r) \tag{2.6}$$

Only discrete field distributions or modes will propagate in structures such as in Figs. 1a and 1b, and these are identified by the integers μ and ν. Equation (2.5) shows that the integer ν characterizes the ϕ dependence or the "angular momentum" of the mode fields. The integer μ counts the number of radial nodes in the field distribution. The quantity $\beta_{\nu\mu}$ is called the axial propagation constant and also assumes only discrete values. The axial propagation constant is a conserved quantity and is thus most important for describing light propagation in the waveguide. Equation (2.6) can be solved analytically for only the step and parabolic radial index profile classes. In those two cases, the field distributions $R_{\nu\mu}(r)$ are found to be Bessel and Laguerre–Gauss functions, respectively.

Although Eq. (2.6) can only be solved rigorously for certain forms of $n(r)$ (in particular the step and parabolic profiles) its similarity to the Klein–Gordon equation from quantum mechanics makes approximate solutions from that field applicable. This approach also results in a model for the waveguide which assists in understanding its operation. One can associate an eigenvalue or "energy level" for a mode with $\beta_{\nu\mu}$. This mode propagates in a potential well composed of two parts as deduced from Eq. (2.6). The first is merely the real refractive index $k^2n(r)^2$, which tends to bind the mode to the waveguide core. The second is a centrifugal potential $-\nu^2/r^2$, which is associated with the angular momentum of the mode and tends to unbind it from the core. This potential-well model is schematically illustrated in Fig. 4 for a given value of ν. The allowed energy levels $\beta_{\nu\mu}$ in this inverted well, exist only at ordinate values such that $R_{\nu\mu}(r)$ has an integral number of oscillations between the turning radii r_1 and r_2. These are the same turning radii described earlier in the ray model and shown in Fig. 2. Outside the turning radii, the electric and magnetic fields exhibit an exponential decrease. For bound or guided modes there is a restriction that is described by:

$$k^2n(0)^2 \geq \beta_{\nu\mu}^2 \geq k^2n(0)^2[1 - 2\Delta] \qquad \text{(bound mode condition)} \tag{2.7}$$

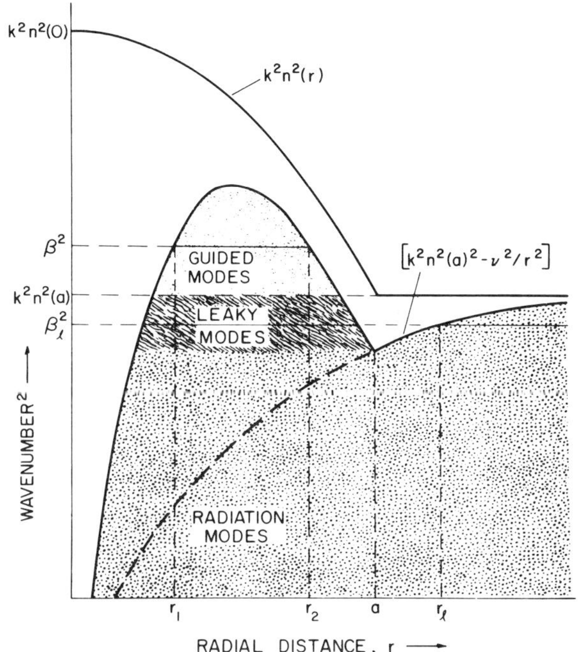

FIG. 4. Potential well schematic for a graded-index waveguide. Three regions are identified; guided modes, leaky modes, and radiation modes. The only allowed β values are those for which the mode field has an integral number of phase changes between radii r_1 and r_2.

For a given value of ν, the integer μ counts the number of discrete values of β that exist between these limits. It can be seen from this potential-well model that as ν increases, the well shifts downward, the number of μ values decreases, and the turning radii move to larger r. Thus, the large ν value or high-angular-momentum modes have their maximum field strength at large r as already indicated in the ray model.

It is also seen from the picture that energy is still "quasi-bound" to the core region for $\beta_{\nu\mu}$ values

$$k^2n(a)^2 \geq \beta_{\nu\mu}^2 \geq k^2n(a)^2 - \frac{\nu^2}{a^2} \qquad \text{(leaky mode condition)} \tag{2.8}$$

These modes are referred to as leaky in that the mode field can "tunnel" through to the point r_l and then radiate. For $\beta_{\nu\mu}$ near $kn(a)$, the tunneling barrier is very large and some of these modes can propagate with losses of less than 1 dB/km.[10,11] Generally, this number is found to be less than 10% of the number

[10] D. Gloge, *IEEE Trans. Microwave Theory Tech.* **MTT-23**, 106 (1975).

[11] R. Olshansky, *Appl. Opt.* **15**, 2773 (1976).

of bound modes; therefore, leaky modes do not substantially contribute to long-distance (1 km) propagation effects. However, over short distances (~1 meter), they definitely alter the modal power distribution. For completeness, it is also seen from Fig. 4 that light propagation with a value

$$\beta_{\nu\mu} \leq k^2 n(a)^2 - \nu^2/a^2 \qquad \text{(radiation mode condition)} \tag{2.9}$$

is not bound to the waveguide core, but becomes part of the radiation mode continuum.

Using the potential well model, the WKB approximation from quantum mechanics, and assuming the rather general index profile class [a subclass of Eq. (2.2)]

$$n(r)^2 = \begin{cases} n(0)^2[1 - 2\Delta(r/a)^\alpha] & r \leq a \\ n(0)^2[1 - 2\Delta] & r > a \end{cases} \tag{2.10}$$

one can solve Eq. (2.6) analytically.[12] Here, α is a quantity characterizing the shape of the radial index profile. For $\alpha = \infty$ and $\alpha = 2$, Eq. (2.10) reduces to the step and parabolic profile, respectively. With this profile class, the energy levels are found to form degenerate mode groups, where

$$\beta_m{}^2 = k^2 n(0)^2 \left[1 - 2\Delta\left(\frac{m}{M}\right)^{2\alpha/(\alpha+2)}\right] \tag{2.11}$$

Here, $m = 2\mu + |\nu|$ is an integer enumerating the mode group number, and the quantity

$$N = M^2 = \left(\frac{\alpha}{\alpha + 2}\right) a^2 k^2 n(0)^2 \Delta \tag{2.12}$$

is the total number of bound modes or the mode volume of the waveguide. For a uniform radiance source, Eq. (2.12) specifies the relative light-gathering ability of the α-profile waveguide. This will be discussed further in Section IV,B.

One other quantity that can be defined from Eq. (2.12) is the characteristic waveguide parameter or V value:

$$V^2 = k^2 a^2 [n(0)^2 - n(a)^2] = k^2 a^2 n(0)^2 2\Delta = 2N \tag{2.13}$$

A rigorous modal analysis of the step profile waveguide shows that if the waveguide parameters a, $n(0)$, Δ, and the propagating wavelength are chosen so that $V \leq 2.405$, only the single mode, designated HE_{11}, will propagate. This is the basis for fabricating the single-mode optical waveguide.

It is worth considering further the correspondence between the ray and modal parameters. This can be done with the WKB solution. The modal propa-

---

[12] D. Gloge and E. A. J. Marcatili, *Bell Syst. Tech. J.* **52**, 1563 (1973).

gation constant $\beta_{\nu\mu}$ is merely the projection on the waveguide axis of the wave vector k, which describes a ray as:

$$\beta_{\nu\mu} = |\mathbf{k}| \cos \theta_{\nu\mu} \tag{2.14}$$

This is shown schematically in Fig. 5. The angle $\theta_{\nu\mu}$ is taken internal to the waveguide and is simply the angle between the ray vector and the guide axis, as shown in Fig. 3. Further, the azimuthal mode number ν is related to θ and to the ray entry point r by:

$$\nu = r|\mathbf{k}| \sin \theta_{\nu\mu} \sin \phi_\nu \tag{2.15}$$

Thus, it may be seen from these two equations that only discrete rays are allowed to propagate in the waveguide. A given mode ν,μ is simply the superposition of all possible rays satisfying Eqs. (2.14) and (2.15). For the bound modes, one can obtain the relationship between the mode group number, the axial ray angle, and the radial ray entry point, from Eqs. (2.11) and (2.14):

$$\left(\frac{m}{M}\right)^{2\alpha/(\alpha+2)} = \frac{\sin^2 \theta_m}{2n(0)^2\Delta} + \left(\frac{r}{a}\right)^\alpha \tag{2.16}$$

From this one concludes that no simple relationship exists between a mode group number and its entry point and angle. The situation for leaky modes is still more complex. However, for the case $\alpha = \infty$ (step guide), the mode group number is directly proportional to the far-field angle. Thus, a plane wave at an angle θ_m will excite mode group m in this case.

2. *Attenuation*

Thus far, the ideal lossless waveguide has been considered. We now consider sources of loss which can be broadly grouped as either waveguide structure- or material-related and may be either intrinsic to the guide or induced. It is found that these losses can affect different modes differently, giving rise to the concept

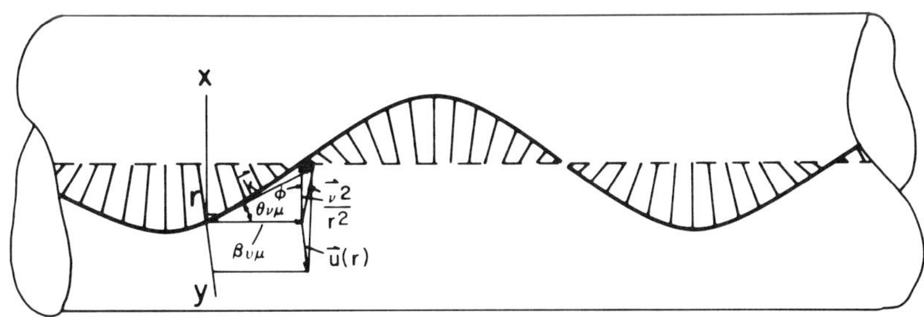

FIG. 5. Schematic diagram of a ray spiraling along a waveguide. The ray vector **k** has an axial component $\beta_{\nu\mu}$, a radial component u, and an azimuthal component ν/r. Arrows over symbols indicate vector quantities.

of differential mode attenuation. This obviously complicates any simple analysis of propagation. The following sections identify these sources and indicate their magnitude.

Since most current waveguide structure-related losses are very small, that loss source will be ignored here. Material-related losses can be broken into absorptive and radiative sources. The ensuing remarks will be primarily confined to high-silica guides made by the doped deposited silica process.[13]

Absorption loss stems from any or a combination of three mechanisms: intrinsic absorption of the basic material, impurity absorption, and induced absorption.

Intrinsic absorption from the ultraviolet band edge for high-silica waveguides is negligible beyond ~700 nm[14], which is the spectral region of interest for optical communications. Absorption from the intrinsic vibration bands in the infrared can, however, be a problem with some materials. This vibration edge is shown by the dotted curve in Fig. 6, and generates the steady increase in attenuation beginning at ~1200 nm and extending to longer wavelengths.

For most telecommunication applications, induced absorption from environmental radiation[15] is negligibly small. Because of an intrinsic distillation that takes place in the processing of high-silica waveguides, the traditional metal impurity ion absorbers such as Fe, Cu, Co, V, and Cr, appear to be a small problem. This means their concentration level is below about 10 ppb. Glasses melted by more conventional techniques, however, still appear to suffer from these impurities. The OH radical is the only impurity known to cause absorption in high-silica waveguides, and atomic vibrational bands at 725, 825, 880, and 950 nm are clearly visible in the dashed curve of Fig. 6. The level of this absorption can be controlled by processing parameters, however, and brought below the 1 dB/km level at 950 nm as demonstrated by the solid curve of Fig. 6.

Radiation or scattering losses are generated by the same three mechanisms as absorption losses: intrinsic material scattering, impurity scattering, and induced scattering.

Intrinsic material scattering represents the ultimate loss limitation. All glassy materials suffer density and compositional concentration variations, which result in refractive index fluctuations that are frozen-in at the time of manufacture. This loss source generates the familiar $1/\lambda^4$ Rayleigh scattering. Fused silica, with a value of about 1.6 dB/km at 820 nm, has one of the lowest Rayleigh scattering levels of all glasses. However, as dopant material such as germanium is added to the waveguide core to raise its refractive index, the scattering level increases. For a waveguide with NA = 0.2, the scattering loss

---

[13] R. D. Maurer, *Proc. IEEE* **123**, 581 (1976).

[14] D. B. Keck, R. D. Maurer, and P. C. Schultz, *Appl. Phys. Lett.* **22**, 307 (1973).

[15] R. D. Maurer, E. J. Schiel, S. Kronenberg, and R. A. Lux, *Appl. Opt.* **12**, 2024 (1973).

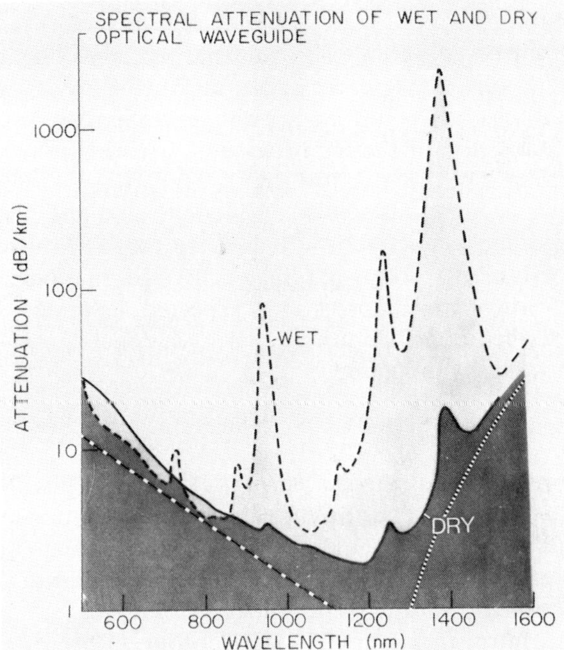

FIG. 6. Spectral attenuation of a typical "wet" and "dry" high-silica waveguide: (· · ·) infrared vibration edge, (-·-) intrinsic Rayleigh scattering.

level for a germanium–borosilicate lightguide may be as much as 1.7 times that of fused silica.[16,17]

This intrinsic scattering level is shown by the decreasing dash–dot curve in Fig. 6 as the wavelength is increased. Impurity and induced scattering losses do not appear to be a problem for most high-silica waveguides and their applications. Induced Raman and Brillouin scattering have been observed in both single- and multimode waveguides where the confined guidance and long interaction length have permitted sufficient power densities.[18,19] Generally, however, induced scattering is not a problem for typical multimode waveguides.

It is seen that the intrinsic infrared vibration bands and the instrinsic scattering curves form an attenuation minimum near about ~1200 nm. This is also evident in the spectrum of a typical "dry" waveguide shown by the solid curve

[16] P. B. O'Conner *et al.*, *Eur. Conf. Opt. Fiber Commun., Paris* Paper II.3 (1976).

[17] D. B. Keck, unpublished work.

[18] R. H. Stolen, E. P. Ippen, and A. R. Tynes, *Appl. Phys. Lett.* **20**, 62 (1972).

[19] S. M. Jensen and M. K. Barnoski, *OSA IEEE Meeting Optical Fiber Transmission II, Williamsburg, Virginia* Paper TuD7 (1977).

in Fig. 6. Attenuation values as low as 0.5 dB/km have been reported[20] at this wavelength for certain compositions. Changes in waveguide composition and parameters, as well as differences in manufacturing process and attenuation measurement, can easily account for the difference between this level and that shown in Fig. 6. Because of the importance of attenuation for system design, work in the future will be oriented toward providing light sources and detectors at these longer wavelengths, as well as optimally trading off material compositions and waveguide parameters to attain these values. Much effort will also be spent in understanding process parameters in order to economically produce waveguides with low intrinsic losses. It is expected, however, that the general mechanisms described above and the wavelength dependency shown in Fig. 6 will remain substantially unaltered.

3. *Information-Carrying Capacity*

With the low losses indicated in Section II,A,2, it is possible that the differing group velocities for the various modes, rather than attenuation, may limit the long-distance usage of waveguides. Generally, to evaluate the information-carrying capacity, the degree to which an infinitely short pulse of light broadens during transmission is considered. Pulse broadening is mainly the result of two factors: the intrinsic material, which produces intramodal dispersion; and the differing group velocities of the various modes, which produce intermodal dispersion.

Intramodal dispersion, alternatively known as material or chromatic dispersion, represents a fundamental limitation and depends on the spectral bandwidth of the light source being transmitted.[21] This broadening is due to the glass having different group refractive indices for the different wavelengths being emitted by the source. The resulting difference in group velocity gives rise to a rms intramodal pulse broadening which is simply given by

$$\sigma_{\text{intra}} = \frac{L}{c}\left(\frac{\sigma_\lambda}{\lambda}\right)\left[-\lambda^2\frac{d^2n}{d\lambda^2}\right] \tag{2.17}$$

where L is the guide length, c is the speed of light, σ_λ the rms source spectral bandwidth, λ the mean transmitted wavelength, and $n(\lambda)$ the core glass refractive index. For high-silica waveguides, $-\lambda^2(d^2n/d\lambda^2) \approx 0.021$–$0.026$. Gallium arsenide LEDs and injection lasers typically have fractional spectral bandwidths of ~ 0.02 and ~ 0.002, respectively. Thus, the pure material pulse broadening is approximately 1.5 and 0.15 nsec/km, respectively. It is noted that this broadening increases directly with the guide length.

---

[20] H. Osanai, T. Shioda, T. Moriyama, S. Araki, M. Hariguchi, T. Izawa, and H. Takata, *Electron. Lett.* **12**, 549 (1976).

[21] C. G. B. Garrett and D. E. McCumber, *Phys. Rev.* III, **1**, 305 (1970).

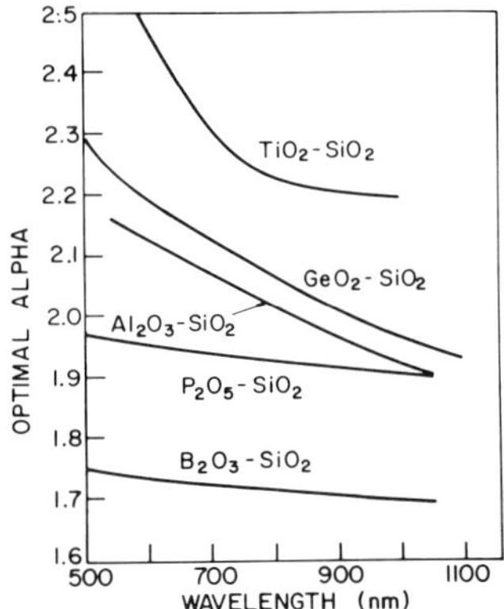

FIG. 7. Plots of the optical profile parameter as a function of wavelength. At 900 nm the germanium–silicate profile is exactly parabolic.

The second source of pulse broadening may be viewed as the effect of differing optical path lengths of transmitted rays. This is the same as differing propagation constants for the modes. Using the propagation constant from Eq. (2.11), the group delay time $\tau_m = (L/c)(d\beta_m/dk)$ for each mode group may be obtained. Having the group delay time and assuming equal mode excitation, the rms pulse broadening may be obtained to first order in Δ by[22]:

$$\sigma_{\text{inter}} = \frac{LN\Delta}{2c}\left(\frac{\alpha}{\alpha+1}\right)\left(\frac{\alpha+2}{3\alpha+2}\right)^{1/2}\left|\frac{\alpha-2-\varepsilon}{\alpha+2}\right| \tag{2.18}$$

Here, N is the axial group refractive index, α is the profile parameter [Eq. (2.10)], and $\varepsilon = (2n/N)(\lambda/\Delta)(d\Delta/d\lambda)$ is a correction factor, which accounts for the fact that $n(\lambda)$ may depend on radius. The total pulse broadening for the waveguide is simply the rms sum of these two components:

$$\sigma_{\text{tot}}{}^2 = \sigma_{\text{intra}}{}^2 + \sigma_{\text{inter}}{}^2 \tag{2.19}$$

As seen by Eq. (2.18), if $\alpha = \alpha_{\text{opt}} - 2 + \varepsilon$, the intermodal component may be made smaller than the pure material dispersion component. This requires proper choice of α for a given operating wavelength and glass composition, because $\varepsilon(\lambda)$ depends strongly on both. This dependence has been measured for several binary silicate glasses[23] and is shown in Fig. 7. By using multicomponent glasses it may be possible to make ε independent of wavelength.[24] For a germanium–silicate waveguide the value $\varepsilon(\lambda)$ is zero near 900 nm, thereby making the parabolic index profile nearly optimal. The combined effect of the

[22] R. Olshansky and D. B. Keck, *Appl. Opt.* **15**, 483 (1976).
[23] I. P. Kaminow and H. M. Presby, *Appl. Opt.* **15**, 3029 (1976).
[24] I. P. Kaminow and H. M. Presby, *Appl. Opt.* **16**, 108 (1977).

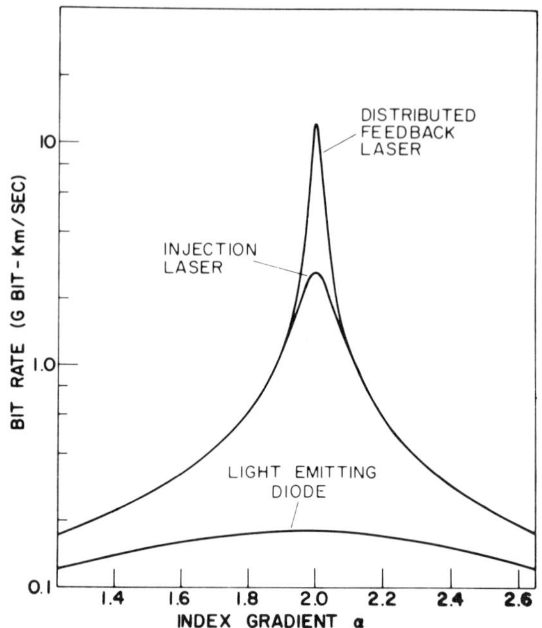

FIG. 8. Plot of waveguide information-carrying capacity as a function of the profile parameter. In each of the three curves shown, the bit rate is limited by material dispersion associated with the source spectral width.

material and intermodal dispersion is shown in Fig. 8. Here, the bit rate (Gbit · km/sec) is plotted as a function of the profile parameter α for three different types of semiconductor light sources. It is to be noted from Eq. (2.18) that the intermodal dispersion is also directly proportional to the waveguide length as well as the fractional index difference Δ. This latter fact means that more careful control of the profile parameter α is required as Δ is increased. Thus, a tradeoff exists between increasing Δ for better light-gathering ability and decreasing it to more easily achieve high bandwidths.[25]

For many systems considerations, it is the frequency response not the rms pulse broadening, which is desired. If the transmitted pulse shape is known, the frequency response can be obtained by a direct Fourier transform. For the case of the equal mode excitation considered above, however, no analytic form exists for the Fourier transform. However, the calculated -3 dB optical power frequency spectrum is bounded by the transform of a Gaussian and a rectangular pulse, on the upper and lower sides, respectively. For both of these, the -3 dB frequency can be given simply in terms of the rms pulse broadening:

$$f_{-3} = 0.187/\sigma_{\text{tot}} \qquad \text{(Gaussian)} \tag{2.20}$$

[25] R. E. Love and D. B. Keck, *Proc. SPIE* **122**, 120 (1977).

and

$$f_{-3} = 0.174/\sigma_{\text{tot}} \qquad \text{(gate)} \tag{2.21}$$

These are not unrealistic shapes to consider for the pulses transmitted by a waveguide. One expects the ideal transmitted step profile pulse to be rectangular. Similarly, for the case of mode coupling (considered in Section II,B), the transmitted pulse shape is Gaussian. Although additional complexities must be included for actual system design, the two expressions in Eqs. (2.20) and (2.21) give a simple conversion between the rms pulse broadening and bandwidth for the optical waveguide. The previous discussion has been for analog transmission. For digital transmission, knowledge of the coding scheme is required. Generally, however, a relationship similar to those in Eqs. (2.20) and (2.21) may be written between the bit rate and the rms pulse broadening.[26] For example, for an error rate of 10^{-9} and a 1% power penalty, this relationship is:

$$N \quad \text{(bits/sec)} = 0.2/\sigma_{\text{tot}} \tag{2.22}$$

This relationship was used in plotting Fig. 8 and, for this power penalty level, it is fairly insensitive to the actual pulse shape being transmitted.

B. Light Propagation in Practical Waveguides

1. *General Description*

In the practical utilization and manufacture of waveguides, several modifications of the above ideal-waveguide description are required. One finds that the axis is no longer necessarily straight and/or that variations in the fiber parameters occur along the axis of the waveguide. Physically, rays are refracted or reflected from these perturbations so that their characteristic angles change; that is, they propagate in a different electromagnetic mode. A frequency spectrum of this curvature and/or parameter variation may be defined as:

$$F(\omega) = \frac{1}{\sqrt{L}} \int_0^L f(z) e^{i\omega z} \, dz = \frac{1}{\sqrt{L}} \frac{1}{\omega^2} \int_0^L C(z) e^{i\omega z} \, dz \tag{2.23}$$

Here, $f(z)$ is the deformation function,[27] which includes not only diameter but also index variations with axial position, and $C(z)$ is the waveguide curvature. The integration is over the length of the perturbation region. From Eq. (2.11), the energy level spacing between modes may be obtained by:

$$\Delta\beta_m = \beta_{m+1} - \beta_m = \left(\frac{\alpha}{\alpha + 2}\right)^{1/2} \frac{2\sqrt{\Delta}}{a} \left(\frac{m}{M}\right)^{(\alpha - 2)/(\alpha + 2)} \tag{2.24}$$

---

[26] S. D. Personick, *Bell Syst. Tech. J.* **52**, 843 (1973).

[27] D. Marcuse, "Theory of Dielectric Optical Waveguides." Academic Press, New York, 1974.

If the frequency spectrum in Eq. (2.23) contains a component so that $\omega = \Delta\beta_m$, then energy will couple between these modes. It is generally found that appreciable coupling exists only between adjacent modes. From Eq. (2.24), a minimum coupling frequency $\Delta\beta_{\min}$ can be defined. Two regimes may then be defined. If $F(\omega)$ has components $\omega > \Delta\beta_{\min}$, mode coupling will occur and power will be transferred between modes. For the components $\omega < \Delta\beta_{\min}$, an adiabatic change of the propagation characteristics of a given mode will occur with no coupling of power between modes. The effect of both cases on attenuation and dispersion will be qualitatively considered.

2. *Attenuation*

Consider first the case of mode coupling where energy is randomly coupled between adjacent modes. This coupling is not restricted to bound modes but can also exist between bound and lossy leaky or radiation modes and vice versa. This results in excess attenuation. Eventually, a balance is achieved between energy being coupled to high-order, high-loss modes and that coupled to low-order, low-loss modes. There are two components to this attenuation. First, a transient loss is suffered in setting up this steady state condition. Second, the steady state attenuation is also increased as power leaks away from the high-order modes. For the case of random microbending in the α-class waveguide, this excess loss has been shown to be[28]

$$\gamma_1 = G(\alpha, p)(a^{2p}/\Delta^{p+1}) \tag{2.25}$$

where $0 < p \leq 3$ describes the shape of $F(\omega)$, and $G(\alpha, p)$ depends on p, on the profile shape, and assumes values between 0.1 and 1.0. This component of loss can be quite large, reaching several decibels per kilometer simply due to wrapping a waveguide on a "rough" reel. Equation (2.25) indicates that this loss is reduced by minimizing the core radius and maximizing Δ. However, additional tradeoffs arising from cabling and system considerations enter through $G(\alpha, p)$[29,30]. In general, this form of loss can be made negligibly small by careful cable design and deployment.

Even in the case where fiber parameter variation or curvature occur at a sufficiently low axial frequency so as not to result in mode coupling, excess attenuation can nevertheless occur. This may result, for example, from a diameter decrease with an attendant bound mode volume decrease. Then, some modes propagating near cutoff (near their maximum angle) may suddenly find themselves in the "nonbound" or leaky region and are momentarily lossy. If a sufficiently large number of these variations exist, all modes in excess of some

---

[28] R. Olshansky, *Appl. Opt.* **14**, 935 (1975).
[29] D. Gloge, *Bell Syst. Tech. J.* **54**, 245 (1975).
[30] R. Olshansky, *Appl. Opt.* **14**, 20 (1975).

m value will eventually be lost, thereby reducing the total waveguide mode volume and producing another form of transient loss. Olshansky and Nolan[31] have calculated the loss due to a Gaussian diameter distribution about a nominal value. They find that a relative diameter variation as small as 2% rms will produce a 0.56 dB transient loss in 1 km. As manufacturing tolerances are improved, this effect is not expected to be a problem.

Another mechanism for nonmode coupling loss is bending. It has been shown that the mode volume of a bent waveguide is less than that of a corresponding straight lightguide.[32] Thus, again, a transient loss will occur. Experimentally this has been found to be small in single- and multimode waveguides for bend radii larger than about 10 cm.[33] However, further correlation between theory and experiment is required in this area.

3. *Dispersion*

While the effect of mode coupling perturbations on attenuation is detrimental, the effect on pulse broadening is beneficial.[34] The coupling of power between modes propagating with different group velocities tends to average out the group delay differences between modes. The situation is perfectly analogous to the "random walk" problem, and one finds that in the steady state coupling limit, the rms pulse broadening increases only as the square root of length rather than linearly, as indicated by Eqs. (2.17) and (2.18). The length-dependent behavior of the rms pulse broadening is schematically shown in Fig. 9 for step and parabolic guides with excess mode coupling losses of 2 and 10 dB/km. Behavior similar to this has been experimentally observed in many laboratories.[35] Two regions exist for which the rms pulse broadening length dependence may be mathematically specified:

$$\sigma(z) = \sigma_u z, \qquad z \ll L_c \tag{2.26}$$

or

$$\sigma(z) = \sigma_c(\sqrt{z/\gamma_1}), \quad z \gg L_c \tag{2.27}$$

where σ_u and σ_c are constants characteristic of the waveguide, γ_1 the excess mode coupling loss rate, and L_c the coupling length as indicated in Fig. 9. Mathematically, the coupling length is given by:

$$L_c = \frac{1}{\gamma_1}\left(\frac{\sigma_c}{\sigma_u}\right)^2 \tag{2.28}$$

---

[31] R. Olshansky and D. A. Nolan, *Appl. Opt.* **15**, 1045 (1976).
[32] N. S. Kapany, "Fiber Optics." Academic Press, New York, 1967.
[33] F. P. Kapron, D. B. Keck, and R. D. Maurer, *Appl. Phys. Lett.* **17**, 423 (1970).
[34] S. D. Personick, *Bell Syst. Tech. J.* **50**, 843 (1971).
[35] E. L. Chinnock *et al.*, *Proc. IEEE* **61**, 1499 (1973).

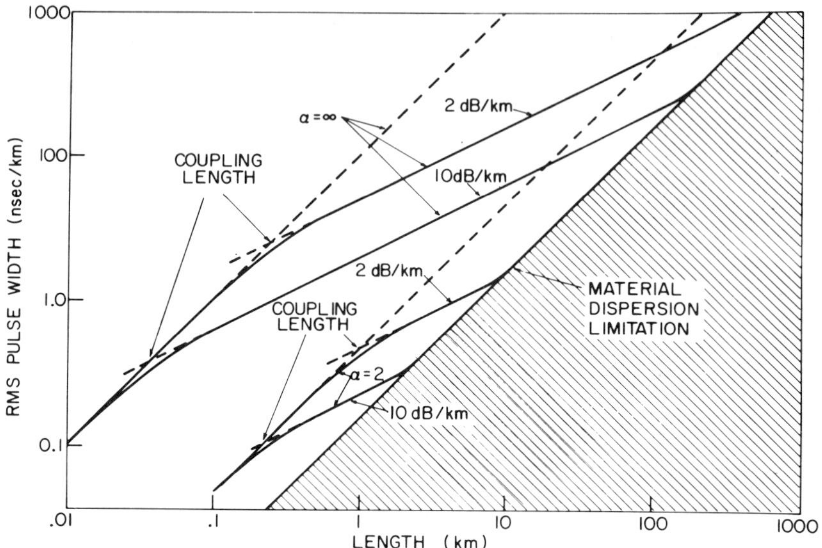

FIG. 9. Plot of the rms pulse broadening as a function of fiber length for both step and parabolic profile waveguides. Two mode coupling losses are assumed for each. A $(\text{length})^{1/2}$ dependence is predicted after the coupling length.

By taking the ratio of the square of the coupled and uncoupled rms pulse widths from Eqs. (2.26) and (2.27), a quantity independent of the guide length may be defined. This quantity specifies the attenuation–dispersion trade off and is commonly written as:

$$R^2\beta = 4.3L_c\gamma_1 = 4.3\left(\frac{\sigma_c}{\sigma_u}\right)^2 = \text{const} \tag{2.29}$$

Here, β is the excess mode coupling loss and R is the ratio of mode coupled to uncoupled rms pulse width.

The constant ranges from approximately 0.5 to 2 for the step and parabolic waveguide profiles, respectively. Thus, as shown in Fig. 9, for a fivefold increase in mode coupling loss, a $\sqrt{5}$ benefit in pulse-broadening rate in the steady state is obtained. Thus, one can approximately calculate the relative merits of mode coupling for a particular system.

Other than variations in the shape of the radial index profile with length, the general low-frequency (non-mode coupling) fiber parameter variations that cause attenuation are not expected to affect dispersion. Fluctuations and/or constant deviations from the optimal $\alpha = 2 + \varepsilon$ profile shape given in Fig. 7 will, however, cause pulse broadening. From first-order perturbation theory, Olshansky[36] has shown that the mean square deviation from the optimal

[36] R. Olshansky, *Appl. Opt.* **15**, 782 (1976).

shape averaged over the guide length determines the magnitude of the excess pulse broadening. Calculations have been made for various perturbations from the optimal profile.[36,37] The effect can be severe. For example, for a 1% amplitude sinusoidal variation about the optimal profile at the worst possible frequency, a 2 nsec/km excess broadening will result. Fortunately, these perturbations appear to be small in present waveguides because no effects have been observed.

4. *Measurement of Fiber Waveguide Properties*

As might be expected from the above discussion of factors affecting light propagation in optical waveguides, the measurement of their properties is becoming more difficult as improved performance and higher precision is required. Waveguide attenuation is one parameter that has received considerable attention. One finds, for example, that the attenuation rate differs for the various propagating modes.[10,38] This may occur, for example, from material effects, waveguide-dimensional variations, or external effects such as bending. This differential attenuation affects any simple measurement of either attenuation or pulse broadening because the relative power excited in various modes must now be specified in order to generate a meaningful measurement. Similarly, if mode coupling occurs because of an external perturbation, as might occur by tightly wrapping the waveguide on a rough surface, then both the attenuation and pulse broadening will be functions of the waveguide length until the steady state condition is achieved, $L \gg L_c$. These effects were demonstrated by measuring the transmitted power through 4 km of step waveguide.[39] The power as a function of length is shown in Fig. 10a. For the first 2 km, the attenuation rate for a 0.32NA launch is seen to be decreasing with length. However, a constant rate was observed beyond 2 km. The difference between the extrapolation of the steady loss rate to $L = 0$ and the initial power is the transient loss. It is seen that the transient loss depends on the numerical aperture of the excitation beam. For a launching NA of 0.32, a transient of approximately 0.6 dB is observed. However, for a launch NA of 0.1, a negative transient of about 0.3 dB is observed, indicating that not all modes of the steady state power distribution are initially excited. Similar behavior is shown in Fig. 10b for a graded-index waveguide with a launch NA = 0.1. Here, the transient loss is about 1.2 dB. This is not unexpected, due to the larger fraction of leaky modes excited in the graded-index compared to the step-index waveguides for this launch NA.

---

[37] J. A. Arnaud and W. Mammel, *Electron. Lett.* **12**, 7 (1976).

[38] R. Olshansky, S. M. Oaks, and D. B. Keck, *OSA IEEE Topical Meeting Opt. Fiber Transmission II, Williamsburg, Virginia* (1977).

[39] R. Olshansky, M. G. Blankenship, and D. B. Keck, *Eur. Conf. Opt. Fiber Commun., 2nd, Paris, France* (1976).

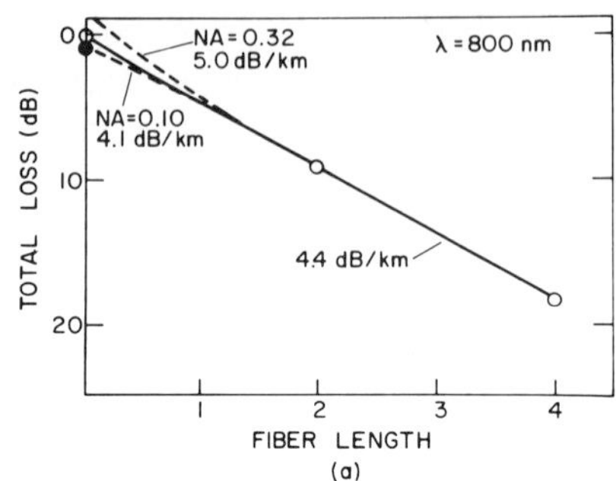

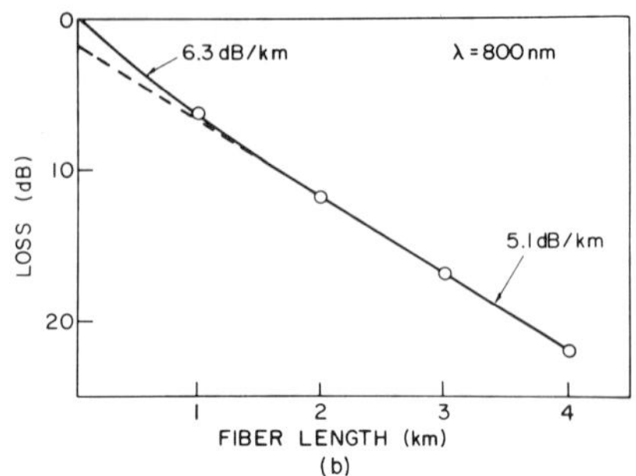

FIG. 10. Plots of total loss as a function of length for two 4-km waveguides. In (a) the loss at two launch numerical apertures is shown for a step-index waveguide. In (b) a similar curve is shown for a 0.1NA launch into a graded-index waveguide. A steady state loss occurs after about 2 km.

This same behavior obviously affects to some degree the measurement of other waveguide parameters such as pulse broadening and numerical aperture. Currently, work is under way to devise techniques for more reproducibly specifying waveguide parameters. However, from this discussion it is apparent that any quotation of fiber parameters should specify the method of excitation and measurement conditions, as well as the degree of possible mode coupling present. The user may then meaningfully utilize this information in the design of a particular optical waveguide system.

C. Fiber Emission Characteristics

For many purposes, knowledge of the waveguide emission characteristics is required. Indeed, the waveguide far- and near-field radiation patterns contain much information about the waveguide parameters as well as the degree of external influences on the waveguide. To calculate these patterns, one begins by considering the waveguide phase-space or mode volume. Quite generally, one can consider a collection of rays propagating with respect to the waveguide optic axis. As described earlier in Section II,A,1, each ray may be instantaneously characterized by two spatial and two angular coordinates r, ψ and θ, ϕ, respectively, which define a phase space as shown in Fig. 3. The elemental phase-space volume is given by:

$$\Delta v = r\,dr\,d\psi \sin\theta\,d\theta\,d\phi \tag{2.30}$$

The total phase-space or waveguide-mode volume is derived simply by integrating Eq. (2.30) over the limits imposed on each coordinate by the propagation characteristics of the particular waveguide:

$$v = \int_0^a \int_0^{2\pi} \int_0^{\theta_c(r)} \int_0^{2\pi} r\,dr\,d\psi \sin\theta \cos\theta\,d\theta\,d\phi \tag{2.31}$$

Considering only bound mode propagation, for example, $\theta_c(r)$ is the maximum acceptance angle specified in Eq. (2.3); the $\cos\theta$ factor accounts for the reduction of the effective core area for nonnormal incidence. Assuming equal excitation of all waveguide modes, Eq. (2.31) may be used to obtain the near- and far-field radiation patterns of the waveguide. If nonconstant excitation of the modes occurs either during light input or through differential attenuation, the waveguide radiation patterns cannot generally be calculated.

1. *Near-Field Radiation Pattern*

Consider first the light distribution which would be observed upon microscopic examination of the waveguide end. This is obtained by differentiating Eq. (2.31) with respect to r and performing the integration over all solid angles. This, of course, is also the same as the amount of light from a constant radiance source coupled into a differential area $r\,dr\,d\psi$ of the guide at radius r. Assuming equal excitation of all modes (or equivalently constant source radiance), the integration may be performed for the α-profile class of Eq. (2.10). The result can be divided into bound and leaky mode contributions by using appropriate limits on the integration as obtained from Eqs. (2.7) and (2.8):

$$\left.\begin{array}{r} 0 \le \phi \le \pi/2 \\ 0 \le \eta \le (1 - R^{\alpha})^{1/2} \end{array}\right\} \quad \begin{pmatrix}\text{bound}\\ \text{modes}\end{pmatrix} \tag{2.32}$$

and

$$\left.\begin{array}{r} 0 \le \phi \le \pi/2 \\ (1 - R^{\alpha})^{1/2} \le \eta \le \left[\dfrac{1 - R^{\alpha}}{1 - R^2 \sin^2 \phi}\right]^{1/2} \end{array}\right\} \quad \begin{pmatrix}\text{leaky}\\ \text{modes}\end{pmatrix} \tag{2.33}$$

where $\eta = \sin\theta(r)/n(0)\sqrt{2\Delta}$ and $R = r/a$. The resulting near-field pattern is then:

$$\frac{1}{2\pi R}\frac{dP}{dR} = \pi a^2 \sin^2\theta_c(1 - R^{\alpha}) \qquad \begin{pmatrix}\text{bound}\\ \text{modes}\end{pmatrix} \tag{2.34}$$

or

$$= \pi a^2 \sin^2\theta_c(1 - R^{\alpha})\left[\frac{1}{\sqrt{1 - R^2}} - 1\right] \qquad \begin{pmatrix}\text{leaky}\\ \text{modes}\end{pmatrix} \tag{2.35}$$

The total (bound + leaky) and bound mode power as a function of radius for $\alpha = 2$ and $\alpha = \infty$ are shown as the solid and dashed curves of Fig. 11, respectively.

It is found that for very short waveguide lengths (< 1 m), the dashed curves most closely resemble the coupled and emitted radial power distributions. For longer lengths, differential mode attenuation usually alters the radial power distribution. The solid curves are then a closer approximation to the true

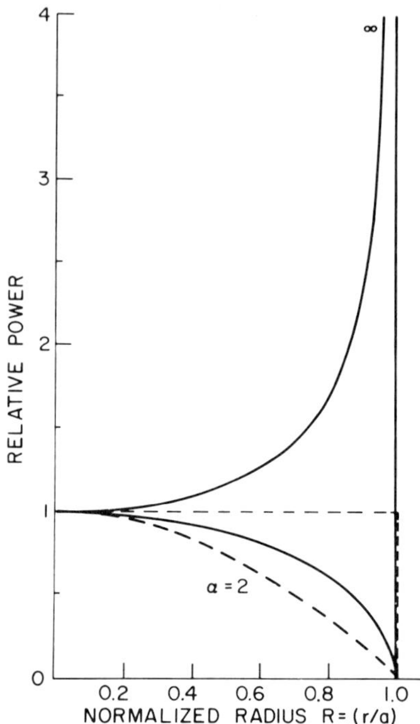

FIG. 11. Calculated near-field radiation patterns for a step ($\alpha = \infty$) and parabolic ($\alpha = 2$) waveguide. The dashed curve represents only bound modes, while the solid curve includes both bound and leaky modes.

situation. It is noted that Eq. (2.35) and the resulting short-length near-field power distribution offers a simple technique for measuring the radial profile parameter. This is currently used by many laboratories. If very detailed knowledge is required, then Eq. (2.35) is obviously no longer valid, and one must have exact knowledge of the differential mode attenuation characteristics of the specific fiber.

2. *Far-Field Radiation Pattern*

A similar analysis can be done for the far-field radiation pattern by differentiating Eq. (2.31) with respect to the axial angle θ, again assuming constant modal excitation, and integrating over ϕ and R.

The integration can be done for bound modes for any α value. However, when leaky modes are included, it can be done analytically only for certain profile shapes, specifically $\alpha = 2$ and $\alpha = \infty$. For equal excitation of the bound modes, the far-field pattern is simply:

$$\frac{1}{2\pi\eta}\frac{dP}{d\eta} = \pi a^2 \sin^2 \theta_c (1 - \eta^2)^{2/\alpha} \qquad \text{(bound modes } \eta \leq 1) \tag{2.36}$$

If leaky modes are included for the cases $\alpha = 2$ and $\alpha = \infty$, the results are:

$\alpha = 2$

$$\frac{1}{2\pi\eta}\frac{dP}{d\eta} = \pi a^2 \sin^2\theta_c(1-\eta^2)^{1/2}, \qquad (\eta \leq 1) \qquad \begin{pmatrix}\text{including}\\ \text{leaky}\\ \text{modes}\end{pmatrix} \tag{2.37}$$

$\alpha = \infty$

$$\frac{1}{2\pi n}\frac{dP}{d\eta} = \pi a^2 \sin^2\theta_c \begin{cases} 1 & (\eta \leq 1) \\ \left\{1 - \dfrac{2}{\pi}\left[\sin^{-1}\left(\dfrac{\eta^2-1}{\eta^2}\right)^{1/2} + \dfrac{1}{\eta}\left(\dfrac{\eta^2-1}{\eta^2}\right)^{1/2}\right]\right\} & (\eta \geq 1) \end{cases}$$

(including leaky modes)

(2.38)

From the above η dependence, it is seen that the far-field pattern contains information about the axial numerical aperture of the waveguide. Leaky modes and differential mode attenuation, however, complicate this in the same manner as described for the near-field pattern. Using Eqs. (2.36)–(2.38), normalized far-field radiation patterns with and without leaky modes for both parabolic and step waveguides can be calculated and are shown as the solid and dashed curves in Fig. 12, respectively. It is noted that for the parabolic guide, both leaky and bound modes have the same maximum angle, $\sin\theta = \sin\theta_c = n(0)\sqrt{2\Delta}$. Thus, important information can be obtained fairly unambiguously. For the step profile, leaky mode power exists out to $\theta = \pi/2$. Differential attenuation very rapidly removes most of this, but exact interpretation is very

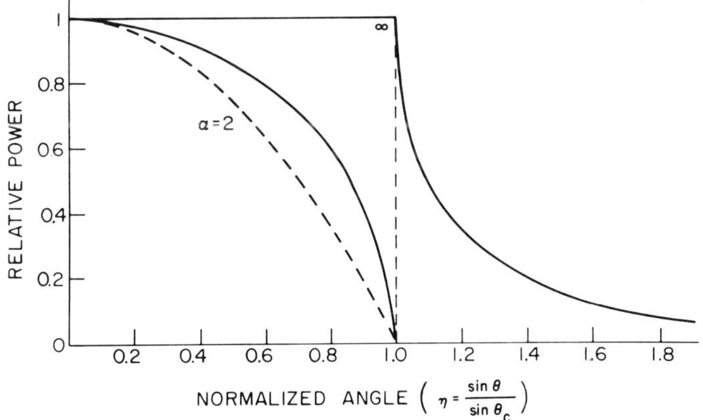

FIG. 12. Calculated far-field radiation pattern for a step ($\alpha = \infty$) and parabolic ($\alpha = 2$) waveguide. The dashed curve represents only bound modes while the solid curve includes both bound and leaky modes.

difficult. Over lengths of a few meters, the solid curve is quite representative of the far-field radiation pattern. Using the far-field radiation pattern, various definitions of the waveguide numerical aperture are being extracted. These include the sine of the angle below which 90% or 100% of the total power is contained, or for which the radiation pattern is at 10% of its peak value. These are measured on both long (> 500 m) and short (< 2 m) lengths of waveguide. Experimentally, it is found[39] that the numerical aperture measured by any of these methods decreases about 10% for a high-silica waveguide in a 1-km length due to differential mode attenuation. Each of these definitions has its own merits for a particular application. The 100% short-length numerical aperture has the virtue that it can be related to material constants and thus may ultimately prove most useful for process as well as systems considerations.

D. Waveguide Strength

The behavior of glass waveguides under stress is qualitatively different from that of ductile materials[40] such as copper wire. The latter exhibits a well-defined yield point; considerable irreversible plastic deformation occurs prior to fracture. Glass fibers, on the other hand, are perfectly elastic. The relationship between stress and strain is linear up to the point of fracture and there is no residual permanent deformation when the load is removed.

The probability of fracture under load is governed solely by the depth and spatial distribution of surface flaws. Under tensile loading, the applied stress is intensified at the crack tip and the resulting crack growth eventually leads to failure. Failure always involves two independent processes, flaw initiation and flaw propagation. The tensile stress acting along the length of a fiber resulting from an applied load cannot initiate a surface flaw. Flaws must be present before the fiber is stressed for fracture to occur at stress levels less than the intrinsic material strength. They may be caused by abrasion in either the fiber or cable manufacturing processes or they may be present as structural or compositional imperfections in the as-drawn fiber.

1. *The Nature of Waveguide Fiber Strength*

The velocity of crack propagation is given by an empirical relationship:

$$\delta = AK^n \tag{2.39}$$

where δ is the flaw depth; K the stress intensity factor, which is a measure of the relative stress at the flaw tip; n the stress corrosion constant that ranges from

[40] C. J. Phillips, *Am. Sci.* **53**, 20 (1965).

15 for soda lime glasses[41,42] to 50 for fused silica; and A a material constant. Equation (2.39) is characteristic of a wide range of brittle materials. Flaw growth, or stress corrosion, depends upon the chemical interaction of water with the glass surface at the flaw tip. The nature of the surface chemistry that controls flaw growth is not understood quantitatively. However, it appears to be relatively independent of water vapor concentration in the environment above a few percent relative humidity, at least for high-silica glass waveguides.

The applied stress σ_a and the flaw size are related to the stress intensity factor by the Griffith equation[40]:

$$K = Y\sigma_a \delta^{1/2} \tag{2.40}$$

where Y is a geometrical factor that depends upon the shape of the specimen.[43] Fiber fracture results if K exceeds K_c, the critical stress intensity factor, which is a material constant. Note that K_c may be reached by increasing σ_a alone, δ alone, or both simultaneously. In general, fiber failure cannot be characterized by a particular value of applied stress or flaw depth.

A qualitative understanding of the nature of waveguide failure probability may be obtained by examining the implications of Eqs. (2.39) and (2.40).

(a) Equation (2.40) implies two failure probability regimes. If the applied stress and the initial flaw size δ_i are such that

$$Y\sigma_a \delta_i > K_c \tag{2.41}$$

fracture occurs immediately. On the other hand, if

$$Y\sigma_a \delta_i < K_c \tag{2.42}$$

the fiber will not fracture immediately, but may at a later time as determined by δ_i and σ_a. These two cases are commonly referred to as fast fracture and fatigue fracture, respectively.

(b) A given length of fiber can be identified by its weakest flaw. Depending on the applied stress level, either fast fracture or fatigue fracture (if it occurs) will take place at that flaw.

(c) Due to the random nature of the flaw severity distribution with length, failure probability must increase monotonically as the fiber length under stress is increased.

(d) Similarly, two nominally identical fibers, as might be obtained by breaking a fiber into two equal lengths, would be expected to fail at significantly different stress levels (fast fracture) or at significantly different times under the fatigue fracture condition expressed by Eq. (2.42).

---

[41] A. G. Evans and S. M. Weiderhorn, *Int. J. Fracture* **10**, 379 (1974).

[42] S. M. Weiderhorn, *J. Am. Ceram. Soc.* **50**, 407 (1967).

[43] ASTM Special Tech. Publ. No. 381, Fracture Toughness Testing and its Applications (1964).

2. *Failure Probability*

It is evident from the foregoing discussion that the strength of fiber waveguides cannot be readily characterized by a single engineering parameter such as "tensile strength" or "yield point." In fact, it is best not to regard tensile strength as being a property of waveguide fibers; one must learn instead to think of waveguide fiber strength in terms of failure probability. As pointed out above, a fiber may fail immediately or at a later time, depending upon the applied stress level and the depth of the weakest flaw. In the case of rapid fracture, it can generally be shown that the cumulative failure probability is given by[44,45]:

$$F = 1 - \exp[-N(\sigma)l] \tag{2.43}$$

where l is the fiber length and $N(\sigma)$ the number of flaws per unit length that fail at a stress level less than σ, the fast-fracture stress. Setting $N(\sigma) = (\sigma/\sigma_0)^m(l/l_0)$ yields the well-known Weibull relationship[46]:

$$F = 1 - \exp\left[-\left(\frac{\sigma}{\sigma_0}\right)^m \frac{l}{l_0}\right] \tag{2.44}$$

which is often used to characterize the failure probability of waveguide fibers. The constants σ_0, l_0, and m are determined experimentally by measuring the stress level at which a large number of fibers, selected from the same flaw population, fail. The fracture stress values are ranked in ascending order and $F(\sigma)$ is given by the fraction that fail below σ. If a plot of $\log[\ln[1/(1 - F)]]$ versus $\log(\sigma)$ yields a reasonably good straight line, as is often the case, then one is justified in determining m from its slope. The value of $(\sigma_0)^m l_0$ is found by determining σ_1; the value of σ at which $F = 0.632$. [For this value of F, the exponent in Eq. (2.44) is equal unity.] Therefore:

$$(\sigma_1)^m l = (\sigma_0)^m l_0 \tag{2.45}$$

Figure 13 is an example of a Weibull plot for which this method of analysis is applicable. The slope is 4.33. If we set l equal to the reference l_0, then σ_0 is equal to 5.6 lb, that is, σ_0 is equal to σ_1.

If different fiber lengths, l_1 and l_2, from the same flaw population are characterized by the same value m, then:

$$\sigma_2 = \sigma_1(l_1/l_2)^{1/m} \qquad \text{for equal } F \tag{2.46}$$

This is a convenient means of deriving F for long fiber lengths from more easily obtained short-length data. Fortunately, there is evidence that m is independent of length for l greater than 10 m.[47]

---

[44] R. D. Maurer, *Appl. Phy. Lett.* **10**, 220 (1975).

[45] R. Olshansky and R. D. Maurer, *J. Appl. Phys.* **47**, 4497 (1976).

[46] W. Weibull, *Trans. R. Inst. Technol., Stockholm* No. 27 (1949).

[47] B. Justice, *Fiber Integrated Opt.* **1**, 115 (1977).

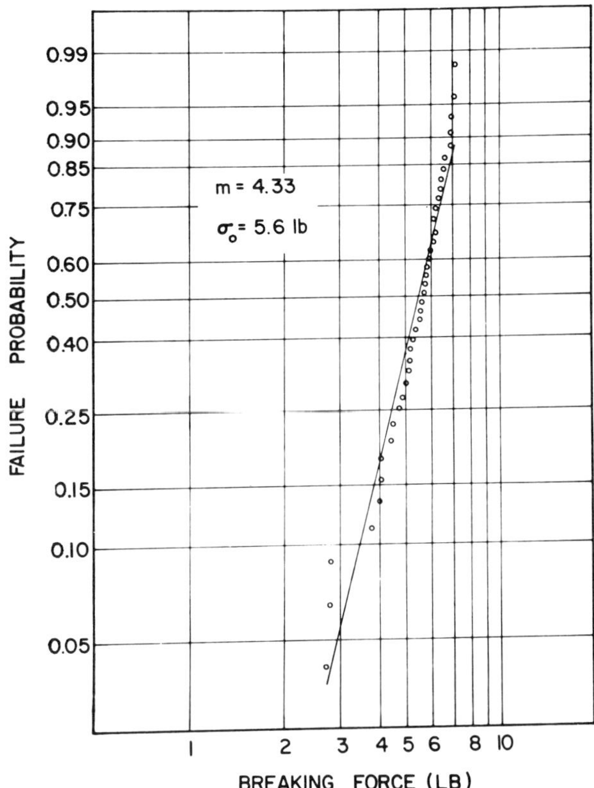

FIG. 13. Weibull plot of fast-fracture strength distribution for 104-μm-diameter, 60-cm-gauge-length waveguide fibers.

A Weibull-type relation can be obtained for fatigue fracture by combining Eqs. (2.39) and (2.40) to derive a time-to-failure expression in terms of σ, the applied stress σ_a, and other constants.[45,48] Substituting the resulting expression for σ in Eq. (2.44) yields:

$$F = 1 - \exp\left[-\left(\frac{t}{t_0}\right)^{m/(n-2)}\left(\frac{\sigma_a}{\sigma_0}\right)^{nm/(n-2)}\frac{l}{l_0}\right] \tag{2.47}$$

where t is the time to failure. As in the case for fast fracture, a large number of fibers from the same population may be placed under load and the time to failure ranked in order of increasing time. If a Weibull plot of $\log[\ln[1/(1 - F)]]$ versus $\log t$ yields a straight line, the value of $m/(n - 2)$ may be determined

[48] R. E. Love, *Proc. SPIE* **77**, 69 (1976).

from the slope. The value of n may then be obtained if m is known from fast-fracture measurements on fibers from the same population. For equal F it also follows that:

$$\sigma_{a_1} = \sigma_{a_2}\left(\frac{t_2}{t_1}\right)^{1/n}\left(\frac{l_1}{l_2}\right)^{(n-2)/mn} \tag{2.48}$$

where the subscripts 1 and 2 denote two combinations of applied stress, gauge length, and time.

3. *Screen Testing and Zero Failure Probability*

Screen testing entails stressing a waveguide fiber for a short period of time prior to the final winding step in the manufacturing process.[49] The stress is usually applied on-line and the level is chosen to ensure a reasonable yield through the apparatus. Thus, the entire length of fiber that survives this test does not, by definition, contain any flaws when:

$$Y\sigma_s\delta^{1/2} \geq K_c \tag{2.49}$$

where σ_s is the screen stress; otherwise it would have failed. Stated another way, the remaining flaws are such that:

$$\delta \leq (K_c/Y_{\sigma_s})^2 \tag{2.50}$$

Screen testing, therefore, sets an upper bound on initial flaw size. Given this initial condition on flaw size, Eqs. (2.39) and (2.40) can be combined and a solution in closed form obtained for the minimum time to failure at stress levels less than σ_s. The result is[50]:

$$t_{min} = 10^{-6}\left(\frac{\sigma_s}{\sigma_a}\right)^n \frac{B}{\sigma_s{}^2} \tag{2.51}$$

where all the stresses are in psi and constant $B = 0.0036$. The relationship between screen stress and applied stress to assure zero failure probability in 20 years is shown in Fig. 14. The stress corrosion parameter n is approximately 23 for fiber waveguides manufactured with the doped deposited silica process.

Before concluding this section, it should be emphasized that in practical situations and applications, the flaw or failure stress distribution *per se* is a characteristic never actually observed by the end user. As far as fiber strength requirements are concerned, there is only one operational strength parameter intrinsic to a given length of fiber; namely, the strength of the weakest flaw. Herein lies the value of the screen test; it sets the maximum flaw depth and, in conjunction with Eq. (2.51), allows a realistic assessment of the in-use stress under which zero failure probability can be guaranteed.

---

[49] B. K. Tariyal, D. Kalish, and M. R. Santana, *Ceram. Bull.* **56**, 204 (1977).

[50] S. T. Gulati, *Horizon House Int. Telecommun. Exposition, 1st, Atlanta, Georgia*, p. 702 (1977).

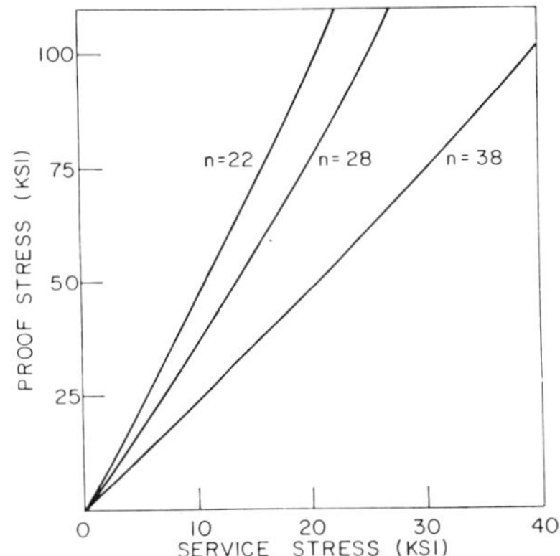

FIG. 14. The relation between screen stress and in-use stress to guarantee zero failure in 20 years. For high-silica content fibers, n is approximately 23.

III. OPTICAL CABLES

A. INTRODUCTION

Immediately after the fiber drawing operation and before any physical contact with the fiber surface occurs, an organic coating is applied to the fiber. This coating may serve several purposes, one of which must be to prevent abrasion damage from occurring throughout the subsequent pulling, screen testing, and winding steps of the manufacturing process. Coating materials that have been used to provide this function exclusively include Kynar®,[51] cellulose acetate lacquer,[52] and epoxy acrylate.[53] All of these materials are characterized by high elastic modulus and Knoop hardness. It is important to realize that bare, uncoated fibers could not survive the handling and tensioning operations normally associated with fiber and cable manufacture.

Coated fibers are packaged, or cabled, to provide a practical degree of protection from mechanical forces and the environmental conditions likely to be encountered during installation and normal usage. Each cable design and configuration, as well as the materials selected, must be consistent with particular installation conditions, as well as system and application requirements.[54] The

---

[51] Corning Glass Works Product Bulletin No 1-R, Multimode Corguide™ Fibers.

[52] Corning Glass Works Product Information, Optical Waveguides for Light Communication (1977).

[53] H. N. Vazorani, H. Schonhorn, and T. T. Wang, *OSA IEEE Meeting Optical Fiber Transmission II, Williamsburg, Virginia*, paper Tu B3-1 (1977).

[54] S. G. Frood and J. Lees, *Proc. IEEE* **123**, 597 (1976).

challenge of optical cable design and manufacture is to satisfy the mechanical and environmental specifications at minimum cost without degrading the signal transmission characteristics of the optical fiber.

B. Cable Design Principles

The transmission characteristics of cabled fibers are governed by three principal considerations: in situ microbends, temperature, and elongation.

1. *In Situ Microbends*

In addition to protecting optical fibers from external forces, the arrangement of jackets, strength members, armoring, etc., is designed to minimize microbend effects. The term "microbend" refers to small bends (axial distortions) of the waveguide axis that may occur whenever there is contact with adjacent structural elements in the cable. Depending upon the surface roughness of the materials in contact with the fiber, the extent of contact, and the contact pressure, a significant increase in attenuation may result. Theoretical calculations show that bends only 3 μm in amplitude with an average spacing of 10 m result in light being scattered out of the waveguide and lead to a 24 dB/km increase in attenuation.[28]

Commercially available cables, as well as experimental cables described in the literature, overcome this effect by surrounding the fiber with one or more plastic jackets, referred to collectively as "buffers." The buffer jacket system is designed to mechanically isolate and decouple the fiber from geometrical irregularities within the cable structure. Theoretical analysis[54] shows high modulus (100 kg/mm^2) jackets are more effective than soft (1 kg/mm^2) jackets and hybrid jackets consisting of either a soft shell with a hard interior or vice versa are more effective than single-layer jackets.[29] Also, all things being equal, microbend losses decrease with increasing jacket radius. Experimental results on cabled fiber attenuation confirm that microbend effects are a serious consideration and that theoretical predictions are qualitatively correct.[55,56] It has proven difficult to control microbend losses by placing buffer materials in intimate contact with the fiber. Depending on the materials used, the degree of contact, and manufacturing method employed, the so-called "tight buffer" approach can result in an additional, or excess, attenuation of between 1 and 5 dB/km.[55]

Experience has shown, on the other hand, that entubulating fibers inside a very loosely fitting jacket (buffer) results in less than 1 dB/km excess attenuation.

---

[55] R. A. Miller and M. Pomerantz, *Proc. Int. Wire Cable Symp., 23rd, Atlantic City, New Jersey* p. 266 (1974).

[56] J. C. Smith and M. Pomerantz, *Proc. Int. Wire Cable Symp., 25th, Cherry Hill, New Jersey* p. 226 (1976).

Many optical cable manufacturers have adopted this approach, commonly referred to as a "loose tube" construction. It is more difficult to terminate cables with this loose buffer construction than those with tight buffers since, in the latter case, fibers can be positioned more precisely relative to each other. It is still too early in the development of fiber splice and connector hardware, however, to judge if this will be a consideration affecting buffer design.

2. *Attenuation as a Function of Temperature*

The attenuation of all-glass waveguides is unaffected by temperature changes in the -55–$+80°C$ range.[57] Nevertheless, cabled fibers often exhibit a strong temperature-dependent attenuation. This effect, which is more pronounced at low temperatures, is due to thermally induced microbends resulting from the inevitably large mismatch in termal expansion coefficients between the waveguide and the surrounding plastic buffers, jackets, and strength members. As the temperature is decreased, the cable materials contract more than the fiber and there is a greater tendency for axial distortions of the fiber axis to occur.

This effect tends to be more pronounced in cables that utilize a tight buffer construction and, hence, buffer geometry in these cables (as well as material homogenieties) must be precisely controlled. The loose tube buffer design tends to isolate fibers more completely from contractions of the buffer itself, as well as from the surrounding cable elements, and is therefore generally more effective in preventing microbends of the fiber axis under changes in temperature.

3. *Waveguide Elongation*

Because of their small cross-sectional area, waveguide fibers cannot normally bear a significant fraction of a cable's tensile stress rating. Optical cables always include strength members to reinforce the cable and to provide practical levels of load-bearing capability.

For cable design purposes, it is convenient to regard the reinforcement elements as limiting fiber elongation at the rated cable tensile stress.[58] Waveguide elongation S_f is usually not allowed to exceed σ_s/E_f, where σ_s is the fiber screen test stress and E_f the Young's modulus of the fiber. In order to illustrate some elementary cable design principles, consider a single-fiber, single-strength member cable, both being of equal length in the unstressed condition. The cable tension is given by:

$$T = S_c E_c A_c + S_f E_f A_f \tag{3.1}$$

[57] R. D. Maurer, *Proc. IEEE* **123**, 581 (1976).

[58] S. G. Frood, W. E. Simpson, and A. Cook, *Proc. Int. Wire Cable Symp., 23rd, Atlantic City, New Jersey* p. 276 (1974).

where the subscript c refers to the cable strength member and A denotes cross sectional area. To a first approximation, the tension is given by:

$$T = S_c E_c A_c \tag{3.2}$$

provided the fiber term in Eq. (3.1) is small compared to $S_c F_c A_c$, which is almost always the case. But since the initial length and the length under load of both the fiber and strength member are equal, S_f is equal to S_c and therefore:

$$T = S_f E_c A_c \tag{3.3}$$

But $S_f \leq S_f(\text{max}) = \sigma_s/E_f$. Hence, the tensile strength rating T_R must be such that:

$$T_R \leq S_f(\text{max}) E_c A_c \tag{3.4}$$

It is important to note that although the fiber does not contribute a significant fraction of the load-bearing capability, T_R is nevertheless a linear function of $S_f(\text{max})$. The greater σ_s is, the smaller E_c and A_c need be to satisfy Eq. (3.4). The implication of this is that cable material costs may be less for fibers that are proof tested at higher stress levels.

By designing the cable to include excess initial fiber length, the tension rating of the cable may be increased relative to that indicated in Eq. (3.4) for a given σ_s, E_c, and A_c. This situation is illustrated in Fig. 15. The new tensile strength rating is given by:

$$T_R \leq \{S_f(\text{max}) + S_e[1 + S_f(\text{max})]\} E_c A_c \tag{3.5}$$

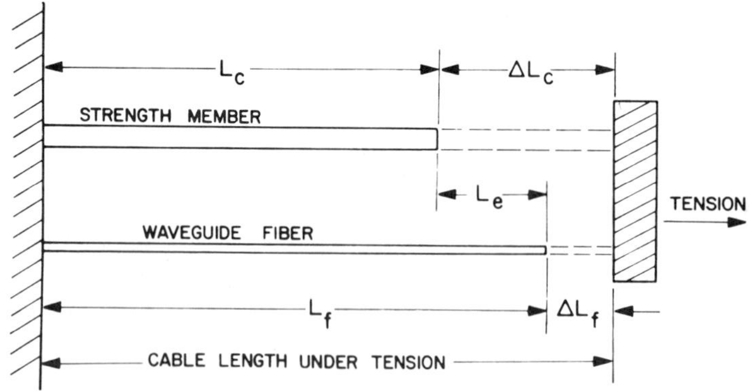

FIG. 15. Elementary optical cable consisting of a single fiber and a single strength member. L_c and L_f denote, respectively, the initial length of the strength member and fiber in the unstressed condition. The initial excess fiber length is L_e. This illustration shows the cable under a tension T, which is of sufficient magnitude that the excess length has been taken up and both the fiber and strength member are of equal length.

where S_e is the initial fractional difference in length between the fiber and the reinforcement element. Omitting the second order term yields,

$$T_R \le [S_f(\max) + S_e]E_c A_c \tag{3.6}$$

The temperature dependence of T_R may also be treated within the context of this simple model. We make the approximation that thermal expansion coefficient of the fiber ($\approx 10^{-6}/°C$) may be neglected compared to that of the reinforcement element ($\approx 10^{-5}/°C$ for metals and $\approx 5 \times 10^{-5}$ for plastics). Initially, before loading, the length of the strength member is given by:

$$L_c(t) = L_c(t_0)[1 + \alpha(t - t_0)] \tag{3.7}$$

where α is the thermal expansion coefficient of the strength element, t is the ambient temperature, and t_0 a reference temperature. Referring to Fig. 15 again, we have:

$$S_e(t) = \frac{L_f - L_c(t)}{L_c(t)} \tag{3.8}$$

Substituting for $L_c(t)$ from Eq. (3.7) yields:

$$S_e(t) = \frac{S_e(t_0) + 1}{1 + \alpha(t - t_0)} - 1 \tag{3.9}$$

and, from Eq. (3.5), we find:

$$T_R(t) \le \left(\frac{[S_f(\max) + 1][S_e(t_0) + 1]}{1 + \alpha(t - t_0)} - 1\right)A_c E_c \tag{3.10}$$

which is the maximum tensile strength rating of the cable as a function of temperature. Equation (3.10) may be rewritten in the form of a simple derating relationship:

$$T_R(t) = T_R(t_0)\left(1 - \frac{\alpha(t - t_0)}{S_e(t_0) + S_f(\max)}\right) \tag{3.11}$$

provided $\alpha(t - t_0)$ and S_f are much less than unity. Note: The safe tensile strength rating decreases with increasing temperature and the derating factor decreases as either $S_e(t_0)$ or α_s increase. Consider a numerical example: the thermal expansion coefficient of most plastics suitable for use in optical cables is the order of $5 \times 10^{-5}/°C$ and, depending upon cable design and permissible fiber elongation, $S_e(t_0) + S_f$ lies between 0.001 and 0.01. Let $t - t_0$ equal 10°C. Therefore, for this example:

$$5 \times 10^{-2} \le \frac{\alpha(t - t_0)}{S_e(t_0) + S_f} \le 5 \times 10^{-1} \tag{3.12}$$

and hence:

$$50\% \leq \frac{T_R(t)}{T_R(t_0)} \leq 95\% \tag{3.13}$$

This is a realistic situation. Hence, careful attention should be paid to the ambient temperature during cable installation to insure that $T_R(t)$ is not exceeded.

C. Cable Constructions

The practical implementation of the cable design principles discussed above can be illustrated by considering two cable configurations. Both were manufactured using conventional cabling equipment suitably modified to process waveguide fibers; both have been installed using techniques common to electrical cables of comparable size, weight, and flexibility; and both have operated successfully in field trial communication systems.

The cable shown in Fig. 16 is an example of a tightly coupled structure.[55] Six urethane-buffered waveguide fibers are stranded around a seventh. To minimize microbend losses, each fiber is lubricated with a thin film of silicone

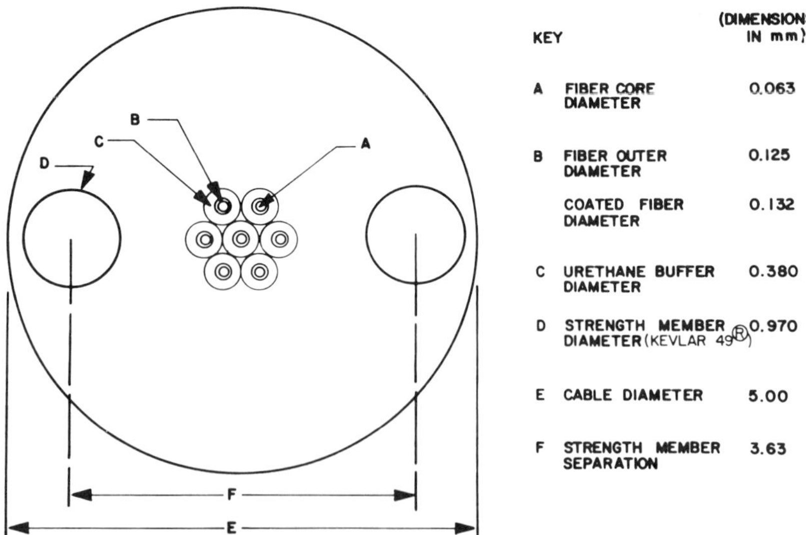

Fig. 16. Cross-sectional view of a cable structure representative of a tightly coupled construction [J. F. Frazier and R. A. Miller, Research and Development Technical Rep., ECOM-73-0348-F (1975)].

TABLE I

TYPICAL MICROBEND-INDUCED LOSSES FOR EACH FIBER IN THE CABLE STRUCTURE SHOWN IN FIG. 3.2c[a]

Initial waveguide attenuation (dB/km)	Final (cabled) waveguide attenuation (dB/km)	Excess attenuation (dB/km)
4.0	4.7	0.7
4.4	5.2	0.8
4.4	5.4	1.0
5.0	5.4	0.4
6.0	6.2	0.2
6.2	9.0	2.8
7.6	9.0	1.4

[a] J. F. Frazier and R. A. Miller, Research and Development Technical Rep., ECOM-73-0348-F (1975).

fluid prior to extruding the buffer. This serves to decouple the fibers from axial motion of the urethane buffer. A urethane jacket, which encapsulates the buffers, is then extruded and drawn down over the seven buffered fibers. At the extrusion orifice, the urethane flows freely over and around each buffer. As it cools, each buffer jacket is molded firmly in position relative to the others. This prevents the shifting and bunching of fibers under the bending and twisting of the cable. To further protect the fibers from external forces and to provide tensile reinforcement, an outer urethane jacket containing strength members positioned 180° apart is extruded over the core assembly. The strength members are Kevlar 49®, a high-tensile-strength aramid fiber. At the rated tensile strength of 50 kg force, the elastic modulus times the cross-sectional area of the Kevlar 49® limits fiber elongation to about 0.1%. The lay of the six fibers around the central member does not result in appreciable excess fiber length. Hence, the fibers are subjected to a tensile stress immediately as tension is applied to the cable. A measure of the excess attenuation, due to microbend-induced scattering before and after cabling is shown in Table I. The temperature coefficient of attenuation is shown in Fig. 17.

A six-fiber cable illustrating the loose buffer design is shown in Fig. 18.[59] Each waveguide is contained within a loosely fitting Halar 300® jacket that is reinforced with a thin layer of polyester. The inner diameter of the jacket is about eight times the diameter of the fiber. These buffer jackets are laid around a polyurethane member which contains a high-modulus carbon steel wire. Two concentric layers of Kevlar 49®, separated by an extruded polyurethane jacket,

[59] Siecor Optical Cable Product Bull. (1977).

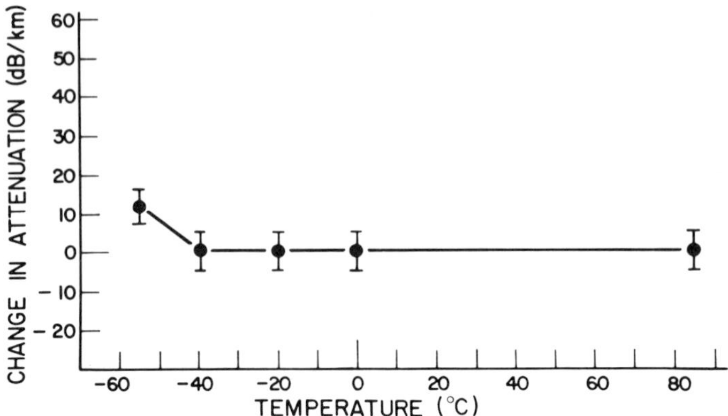

FIG. 17. Typical attenuation vs temperature data for the cable design shown in Fig. 16 [J. F. Frazier and R. A. Miller, Research and Development Technical Rep., ECOM-73-0348-F (1975)].

protect the core assembly from lateral forces and impact. The outer jacket is a weather-resistant polyethylene formulation. The spun layers of Kevlar 49®, as well as the polyurethane and polyethylene jackets, are loosely coupled mechanically to maximize the degree of mechanical isolation between the waveguides and the rest of the cable structure. The polyurethane-encapsulated steel wire, which is not a strength member *per se*, limits the bend radius and prevents the cable structure from buckling under bending. The excess fiber length within the buffer jackets is optimized to minimize the cross-section area of the Kevlar 49® required to achieve the tensile load rating and to minimize microbend losses. Excess attenuation is less than one dB/km.

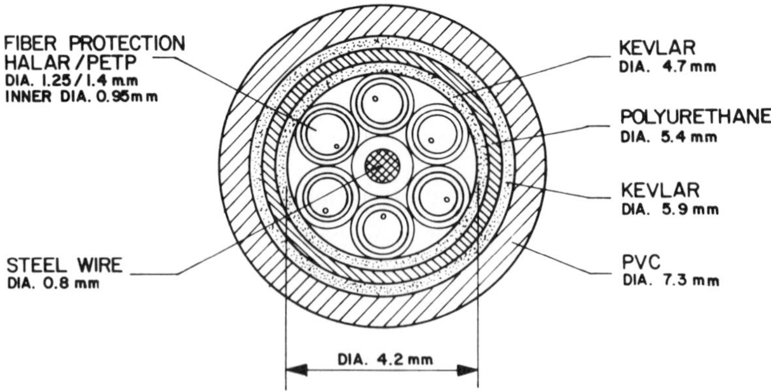

FIG. 18. Six- and eight-fiber cable design representative of the "loose tube" (Siecor) construction [Siecor Optical Cable Product Bulletin (1978)].

IV. OPTICAL WAVEGUIDE COUPLING

A. Theory

To utilize the tremendous potential of optical waveguides in practical systems, one must consider, in addition to attenuation and bandwidth, the coupling of light into, out of, and between waveguide sections connecting the system. Coupling light out of a waveguide and detecting it presents no particular difficulty, since most detectors are of sufficient size to gather all emerging radiation. However, since input coupling and interconnection depend strongly on waveguide parameters, they deserve more detailed discussion.

Since most present systems use multimode waveguides, this discussion will be restricted to them. As in Section II,C, the most useful quantity to consider is the waveguide mode volume. We consider a collection of rays propagating in the phase space defined earlier and characterized by a density function $\rho(r, \psi; \theta, \phi)$. If the average power for each ray is δP, then the optical power contained within an elemental volume of phase space, defined by Eq. (2.30), is

$$\Delta P = \delta P \rho(r, \psi; \theta, \phi)\, r\, dr\, d\psi \sin\theta\, d\theta\, d\phi \tag{4.1}$$

The quantity $\delta P \rho$ is the radiance $B(r, \psi; \theta, \phi)$ of the optical field. From Liouville's theorem, one can derive the well-known result that the radiance or, alternatively, the density of rays, is conserved as they move through phase space. This, of course, assumes no absorption in the transmission medium or reflection at discontinuous boundaries. Liouville's theorem also shows that the volume surrounding a fixed number of rays (points) in phase space is conserved although the surrounding boundary may, in fact, deform. Thus, once the volume and density in phase space is defined (e.g., by the radiation source), no optical system can increase this density of points. The most that an optical system can do is redistribute them within the phase-space volume. Although this conservation principle is very well known there are omnipresent attempts to violate it. This has been particularly true with the increasing use of optical waveguides.

The preceding discussion has important implications for optical waveguides. With proper usage it allows us to calculate the amount of power coupling into and between waveguides.

B. Source-to-Fibre Coupling

The significance of the preceding section on source-to-fiber coupling is that if the phase space volume of the fiber (v_{f}) is smaller than that of the source (v_{s}), no optical system will enable all the light to be coupled into the fiber. Of course, in the reverse case proper lensing will allow the fiber to collect all the source light. If we define the source radiance as $B(r, \psi; \theta, \phi)$, then the total light coupled into the bound modes of an optical waveguide will be obtained by integrating Eq. (4.1) over the phase-space volume of the waveguide:

$$P = \int_0^a \int_0^{2\pi} \int_0^{\theta_c(r)} \int_0^{2\pi} B(r, \phi; \theta, \phi) r \, dr \, d\psi \sin\theta \cos\theta \, d\theta \, d\phi \qquad (4.2)$$

Here again, $\theta_c(r)$ is the maximum bound mode acceptance angle given in Eq. (2.3), and a is the core radius of the guide. The $\cos\theta$ factor accounts for the reduction in core area due to non-normal incidence of the light.

Equation (4.2) assumes that the number of modes is sufficiently great that all waveguide phase-space points uniformly accept light.* For present multimode waveguides this is a sufficiently good approximation. It should also be emphasized that the limits of integration in Eq. (4.2) have considered only the bound waveguide modes. If the limits are extended to include the refracting or leaky rays, then considerably greater coupled power will be calculated. However, due to strong attenuation, as discussed previously, most of these rays are lost within the first few meters.[11] Therefore, Eq. (4.2) should give a value for the coupled power which is within 25% of the true value.

Obviously, without specific knowledge of the source radiance, the coupled power cannot be evaluated. Fortunately, many light-emitting diodes have a nearly Lambertian emission in which $B(r, \psi; \theta, \phi) = B_0 = \text{const}$. For this case and considering the radial refractive index profiles in Eq. (2.10), the total power coupled into bound modes may be evaluated by:

$$P = \left(\frac{\alpha}{\alpha + 2}\right) B_0 \pi^2 a^2 n^2(0) \Delta = \left(\frac{\alpha}{\alpha + 2}\right) B_0 \pi A_f (\text{NA}_f)^2 \qquad (4.3)$$

where A_f and NA_f are the core area and fiber numerical aperture, respectively.

It is seen from Eq. (4.3) that the coupling efficiency depends on the radial refractive index profile, being maximum for the step profile ($\alpha = \infty$) and decreasing to $\frac{1}{2}$ that for the important parabolic profile ($\alpha = 2$). As physically expected, the coupling is directly proportional to the fiber core cross-sectional area. Also, as with optical systems in general, the coupling efficiency depends on the square of the numerical aperture, which, in turn, depends on the fractional index difference between the waveguide core and cladding. These two dependences prompt interest in large-core, high-numerical aperture waveguides for some applications.

* Strictly speaking, one must calculate the modal coupling efficiency $C_{\nu\mu}$ as the overlap integral between each waveguide mode and the source radiation field:

$$C_{\nu\mu} = \frac{\int B(r, \psi; \theta, \phi) E_{\nu\mu}(r, \psi; \theta, \phi) \, dv}{\int |E_{\nu\mu}(r, \psi; \theta, \phi|^2 \, dv}$$

where $E_{\nu\mu}$ is the electric field distribution for the mode ν, μ. A summation over all coupling coefficients then gives the desired coupled power.

C. FIBER-TO-FIBER COUPLING

A means of jointing or splicing waveguide fibers is necessary to extend fiber length during cable installation, as well as to repair damaged cables. In these situations, a permanent splice is usually employed. On the other hand, connectors are needed to interface with fibers (pigtails) permanently attached to solid state sources or detectors inside transmitter and receiver models. Bulkhead-type in-line connectors are used for this purpose. Freely suspended in-line connectors are needed principally for testing or experimental purposes. In all cases, the most important concern is light loss at the joint. Analogous to electrical connectors, this light loss at the joint is called insertion loss or coupling loss. This loss is generally on the order of 1 dB and 0.5 dB, respectively, for state-of-the-art connectors and splice fixtures.

1. *Theory*

Generally, joint losses are either intrinsic or extrinsic. Intrinsic losses originate from variations in fiber parameters which affect mode volume as given by Eqs. (2.12) or (4.3). If the mode volume of the receiving waveguide is less than that of the source waveguide, loss results. Extrinsic losses are due to lack of precision in the splice and connector elements used for aligning and retaining the fibers being joined.

Assuming perfect jointing hardware, consider first the magnitude of intrinsic losses. To do this, one uses Eq. (2.12) to define the coupling ratio $\rho = N_R/N_S$ (for the case $\rho \leq 1$), where N_R and N_S are the mode volumes of the receiving and source waveguide, respectively. Then, for given values of the fiber parameters, the coupling loss in decibels is given by[60]

$$L_c = -10\left[\log\frac{\alpha_S(\alpha_R + 2)}{\alpha_R(\alpha_S + 2)} + \log\left(\frac{a_R}{a_S}\right)^2 + \log\left(\frac{\mathrm{NA}_R}{\mathrm{NA}_S}\right)^2\right] \tag{4.4}$$

where α, a, and NA, respectively, are the waveguide's profile parameter, core radius, and numerical aperture. As an example, a 1% difference in these three waveguide parameters between source and receiver fibers would result in a 0.17-dB loss due to radius and NA variations, and a 0.025-dB loss due to profile variations.

Of course, one must be careful in applying Eq. (4.4) to the maximum statistical variations in each of these parameters for production waveguides. Worst-case losses can be obtained from such analysis. However, this is hardly representative of the loss to be expected from mating randomly selected waveguides. To quantify this more practical case, a Monte Carlo computer calculation has been made of the insertion loss, based on computing the overlapping

[60] F. L. Thiel and R. M. Hawk, *Appl. Opt.* **15**, 2785 (1976).

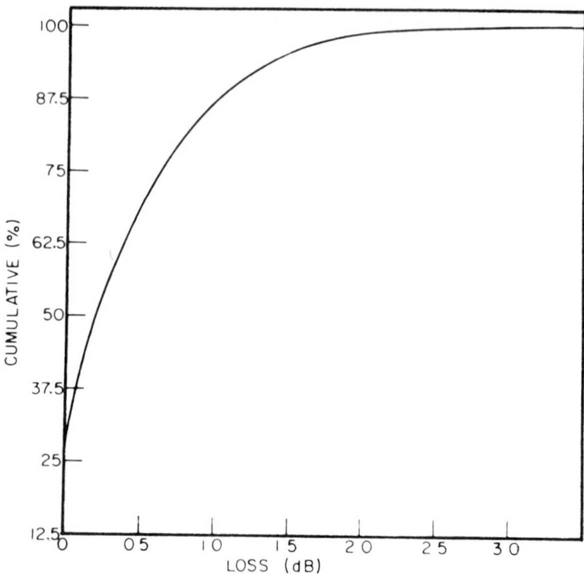

FIG. 19. Plot of the cumulative splice-loss distribution for random matings of fibers having a Gaussian distribution of core radius, numerical aperture, and α profile. The fiber parameter distributions correspond to typical manufacturing tolerances. Half the matings exhibit insertion losses less than 0.2 dB.

mode volume between thousands of pairs of waveguides having typical "production" distributions of core radius, numerical aperture, and profile parameter.[61] The resulting probability plot in Fig. 19 more nearly represents the practical case. Here it is seen that 50% of the interconnections will have losses of less than 0.2 dB and 90% less than 1.15 dB.

Extrinsic losses arise from any of six sources: lateral misalignment of the waveguide axes, angular misalignment, end separation, mode coupling distortions, fiber end finish, and Fresnel reflections. The magnitude of these losses can be calculated by again considering the overlapping mode volumes.[62,63] Experimental results, based on the work of Chu *et al.*,[64] for the first three sources of loss are shown in Fig. 20. The normalized end separation and offset (lateral displacement) are defined as s/R and d/R, respectively, where s is the axial separation in μm, d is the offset in micrometers, and R is the fiber core radius in micrometers. The angular misalignment α was normalized to the angle associated with the meridional numerical aperture, $\sin^{-1}(\mathrm{NA}_0)$. It is not

---

[61] F. L. Thiel and D. H. Davis, *Elec. Lett.* **12**, 340 (1976).

[62] C. M. Miller, *Bell Syst. Tech. J.* **55**, 917 (1976).

[63] D. Gloge, *Bell Syst. Tech. J.* **55**, 905 (1976).

[64] T. C. Chu and A. R. McCormick, *Bell Syst. Tech. J.* **57**, 595 (1978).

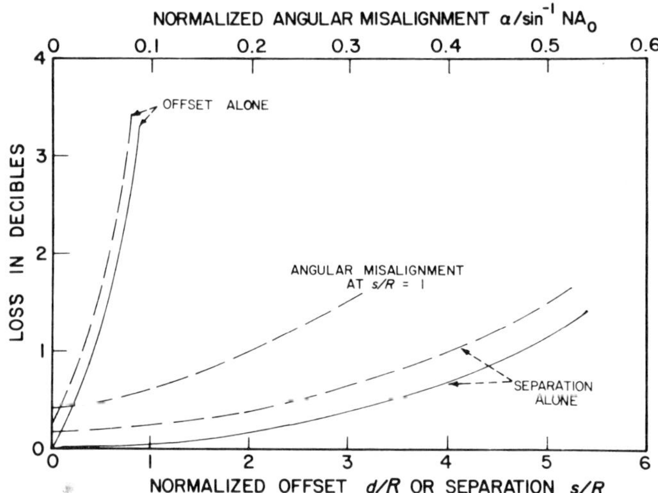

FIG. 20. Loss in decibels as a function of normalized offset d/R, separation s/R, and angular misalignment $\alpha/\sin^{-1}(NA_0)$, for graded-index fibers. These curves are based on the results of two experiments. The first (——) was done with fibers having a 50-μm core diameter and a 100-μm outside diameter; the second (– – –) with fibers having 55-μm core diameter and 110-μm outside diameter [T. C. Chu and A. R. McCormick, *Bell Syst. Tech. J.* **57**, 55 (1978)].

exactly correct to consider each loss mechanism as acting independently, but it is useful as a guideline in designing splice fixtures and connector hardware to regard the separate contributions as if they were independent. Lateral misalignment is usually the major cause of insertion loss. Unfortunately, it is also the most difficult and costly to control. Note, however, that this source of insertion loss, as well as that due to end separation, can be reduced in inverse proportion to the core diameter for fixed values of s and d.

Any calculated value of either intrinsic or extrinsic losses may be an overestimate. Due to differential mode attenuation and/or mode coupling, the mode volume of the source waveguide that is actually excited, assuming uniform excitation, may be less than the calculated mode volume. That is, the receiving mode volume may be able to accept light from the source waveguide with much smaller loss than a calculation might indicate. The magnitude of this effect has been calculated.[63] It may decrease the expected loss by a factor of 3 or 4 if steady state light propagation conditions have been achieved in the source waveguide. Many connector prototypes are available and much loss data has been reported. It now seems reasonable to expect that practical connectors can be made with losses less than 1.0 dB.

2. *Fiber End Preparation*

To obtain low insertion loss, the fiber ends to be joined must be flat, smooth, and reasonably perpendicular ($<2°$) to the fiber axis. Also, since fiber coatings

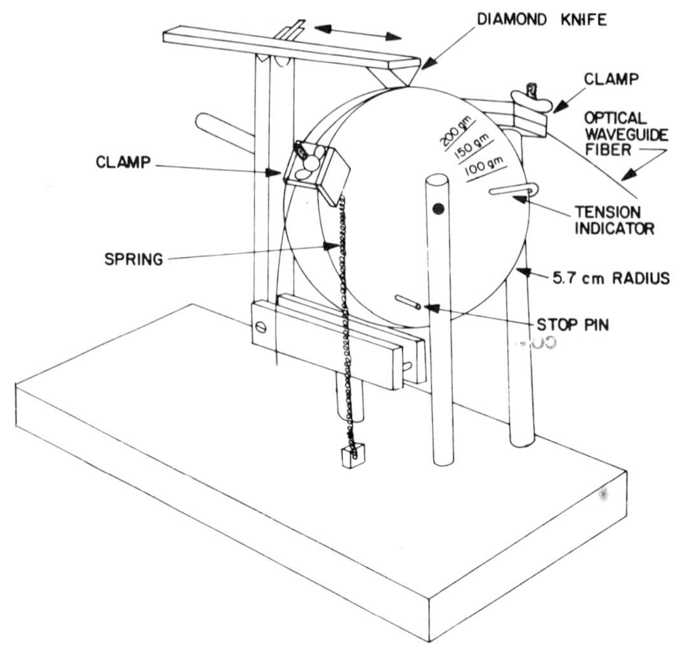

FIG. 21. Fiber breaking apparatus (Corning Glass Works Application Note No. 21).

and buffers are usually not sufficiently concentric or uniform in thickness to serve as the referencing means for fiber-core alignment, such coatings must be removed prior to making a splice joint or attaching a connector end. In removing the coating, care should be exercised not to damage the fiber surface. Depending on the coating material and its degree of adhesion to the fiber, manufacturers recommend removal with solvents or with a mechanical stripping tool.

Fiber ends are prepared for termination by lightly scribing the fiber surface with a tungsten carbide or diamond-edged blade, bending the fiber at that point over a radius of several inches with the scribe mark located on the outer side of the radius, and applying a slight tension. This causes the fiber to fracture at the point of the flaw.[65] Satisfactory ends for many purposes may be prepared manually by scribing the fiber while it is held against one's thumb and then applying a slight tension as it is bent over the thumb. Experimentation[66] with different radii of curvature has shown that a radius of about 6 cm produces a mirror surface across the entire waveguide for a wide range of tensions and scribe pressures. A semiautomatic apparatus like the one shown in Fig. 21

[65] DOD-STD-1678, Test Method 8040.

[66] D. Gloge, P. W. Smith, D. L. Bisbee, and E. L. Chinnock, *Bell Syst. Tech. J.* **52**, 1579 (1973).

produces reproducibly good ends regardless of operator skill. For 125-μm diameter fibers, a tension between 125 and 175 gm and a 5-gm force on the scribe tool produces satisfactory ends. A number of hand tools for preparing fiber ends are also available.[67,68]

3. *Jointing Techniques and Fixtures*

Perhaps the simplest fixture for splicing two fiber ends consists of an accurately machined *V* groove in a block of metal or plastic and a flat cover plate. The fibers are laid in the groove, butted together, and then either clamped in position with the cover plate or cemented in place with a suitable adhesive. Often, a small amount of transparent epoxy resin is placed between the fiber ends to minimize Fresnel reflections. This also reduces end-separation losses. Elaborations and modifications of this basic approach have been employed successfully to slice optical cables in a number of experimental optical communication systems. Average splice loss is reported to be about 0.4 dB.[69,70]

Additional grooves may be formed in the block to splice multiple fiber cables.[71,72] To provide environmental protection and isolation from external forces, the entire fixture may be mounted in a conventional electrical cable splice box or cable sleeve.

Another *V*-groove-type fixture, the "loose tube" splice,[73] is illustrated in Fig. 22. A fiber splice is made by first filling a short length of rectangular tubing with index-matching epoxy. The two fiber ends are then inserted into opposite ends and flexed in the same plane, causing the tube to rotate, and forcing the fiber against a corner. The fibers are then butted together and held firmly in place until the epoxy cures.

Three-point support arrangements constitute another basic approach to fiber splices. An example of such a fixture is shown in Fig. 23.[74] Here, a cylindrical section of thin-wall tubing is indented (crimped) at three points to form the fiber positioning mechanism. The cylinder may be filled with index-matching epoxy if a permanent splice is desired. The fiber ends, from which a short section of

---

[67] Thomas and Betts Corp., Optical Connector Product Bull. (1977).

[68] Deutsch Corp., Optical Connector Product Bulletin (1978).

[69] H. Pascher, *OSA IEEE Meeting Opt. Fiber Transmission, Williamsburg, Virginia* Paper ThA2-1 (1977).

[70] F. Aoki, K. Ando, Y. Ueno, M. Kajitani, K. Tsukada, and S. Shiraishi, *OSA IEEE Meeting Fiber Transmission, Williamsburg, Virginia* Paper ThB4-1 (1977).

[71] A. H. Cherin and P. J. Rich, *OSA IEEE Topical Meeting Opt. Fiber Transmission, Williamsburg, Virginia* p. 6 (1975).

[72] C. M. Miller and C. M. Schroder, *OSA IEEE Conf. Laser Electroopt. Syst.*, San Diego, California p. 82 (1976).

[73] C. M. Miller, *Bell Syst. Tech. J.* **54**, 1215 (1975).

[74] J. F. Dalgleish, H. H. Lakas, and J. D. Lee, *Eur. Conf. Opt. Fiber Commun., 1st, London* p. 87 (1975).

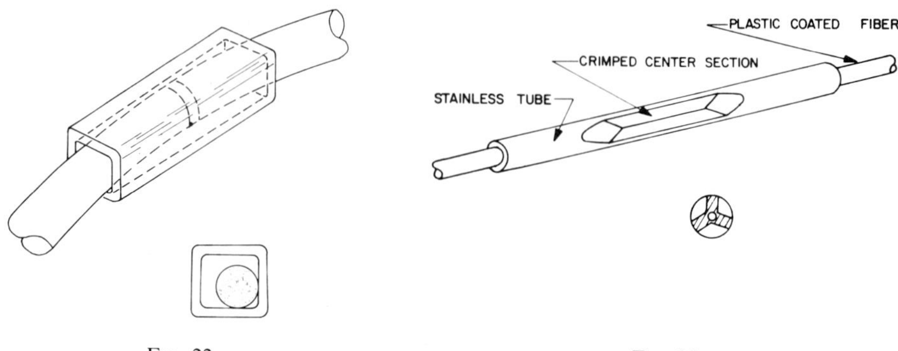

FIG. 22. FIG. 23.

FIG. 22. Loose-tube splice [C. M. Miller, *Bell Syst. Tech. J.* **54**, 1215 (1978)].
FIG. 23. Three-point support splice (Bell Northern: Single Optical Fiber Splice BNRS-10).

the buffer jacket has been removed, are inserted into the cylinder and butted together; then the cylinder ends are crimped firmly around the buffer jacket. This design has the practical advantage of holding the fiber rigidly in place while the epoxy cures. Also, the crimp around the jacket provides strain relief for the joint, as well as a good measure of protection from external environmental and mechanical forces.

The three-rod splice is a type of three-point support fixture which has the advantage of using readily available materials.[60] As shown in Fig. 24, the fibers are butted together inside the interstitial channel between three sections of drill rod. For the range of fiber diameters that are of practical interest, it is always possible to select a suitable standard drill rod diameter. The drill rod sections are held together with a short section of shrink tubing. As before, the channel containing the fibers may be filled with a transparent index-matching epoxy.

The insertion loss associated with all splicing techniques is highly dependent upon care in fixture assembly, fixture dimensional tolerances, insertion loss measurement methods, and the uniformity of fiber-core diameter, NA, and

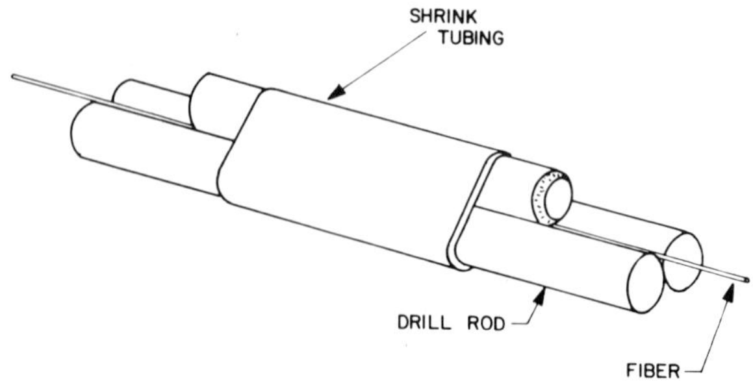

FIG. 24. Three-rod splice arrangement [Corning Glass Works Application Note No. 6].

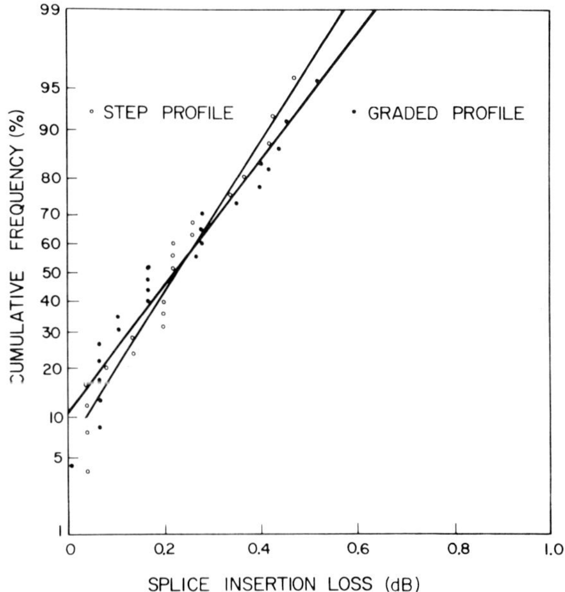

FIG. 25. Siecor optical cables *V*-groove splice-insertion loss data for Corning Glass Works waveguide fibers.

the α profile. Unfortunately, without detailed knowledge of these considerations as they apply to a particular type of splice fixture or fiber population, it is difficult to compare the various insertion-loss measurements reported in the literature. It appears, however, that less than 0.1-dB insertion loss can be achieved between two precisely aligned fiber ends resulting from the same fracture. For randomly selected ends, insertion loss ranges between 0.1 and 1.5 dB. Typical insertion loss data for a *V*-groove fixture is shown in Fig. 25.[75]

In addition to tolerance-related insertion-loss considerations, fiber end faces should be maintained free of dust and other foreign contaminants.[60] Thus, those surfaces against which sharp fiber edges make sliding contact during the insertion operation should be hard and free of contaminants. Any opaque particles scraped off such surfaces may become lodged between fiber cores and lead to additional insertion loss. Also, fixture design and assembly procedures should not generate microbends in the fiber axis, which may also contribute to insertion loss. This could occur, for example, if the adhesives used to hold the fibers in position are not cured uniformly or if the surface of any clamping assembly in contact with the fibers is not smooth and uniform.

Finally, before a fixture design is specified, insertion loss as well as overall splice integrity should be evaluated under anticipated environmental and

[75] Siecor Optical Cables, Inc., Private communication (1978).

mechanical use conditions. For many applications and situations, temperature and humidity cycling tests may be the most critical in view of the large difference in thermal expansion coefficients between waveguide fibers, metals, and plastics and the relatively poor moisture resistance of many organic compounds.

Fusing fibers together by means of an electric arc may be practical for certain applications. Several workers have demonstrated the feasibility of this technique in the laboratory.[76,77] A considerable amount of operator skill is required, but, with practice, splice joints averaging about 0.2 dB insertion loss can be obtained. A coating to protect the fiber from abrasion damage is applied after the splice is formed.

4. *Optical Waveguide Connectors*

At the time of this writing, optical connector technology is evolving rapidly. Prototype connectors are now available from several manufacturers.[67,68,78] Insertion-loss specifications range from 1 to 3 dB, depending upon fiber type and connector design. The basic design approach is the same as for fiber splices; namely, to accurately align and position two fiber ends together using connectors. However, a convenient means of mating and demating connector ends must be provided. To this end, many of the mechanical arrangements common to electrical connectors for mating, sealing, and locking two connector ends together are used. Also, optical connector ends are designed to protect fibers from mechanical damage and environmental contaminants in both the mated and unmated condition.

The principal design difficulty with optical connectors is the requirement to accurately control both lateral misalignment between fibers and fiber end separation. In order to achieve less than 1.0 dB insertion loss, which will be the requirement in the future for many optical communication applications, fiber cores (62.5-μm diameter) must be aligned to better than 8 μm and end separation should be less than 15 μm, assuming two identical fibers, a dry joint, and no other source of loss. It is difficult, at reasonable cost, to achieve those tolerances by simply centering fibers within mating ferrules or sleeves. The cumulative effect of concentricity and diameter tolerances constitute a serious limitation of this approach. It may, however, be viable only in situations where fiber diameter tolerances are the order of ± 1 μm and where individual fibers are centered by optical means relative to an independent referencing system and then fixed in place with an adhesive or mechanical clamp.

A very promising design approach is the overlap-style connector that avoids tolerance accumulation effects by spatially separating the fiber mating plane

---

[76] Y. Kohanzadeh, *Appl. Opt.* **15**, 793 (1976).

[77] D. L. Bisbee, *Appl. Opt.* **15**, 796 (1976).

[78] AMP Optical Connector Product Bull. (1978).

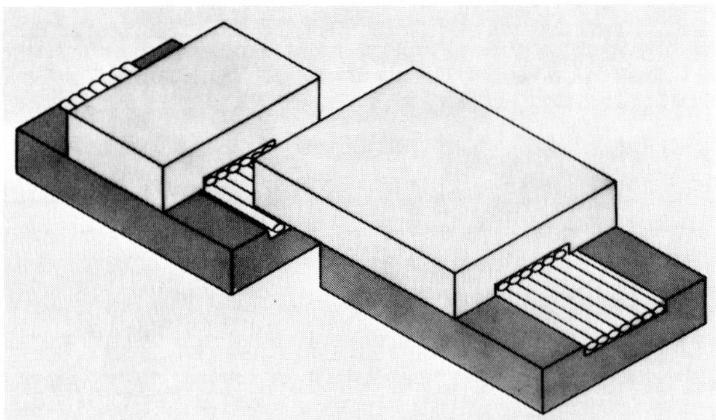

FIG. 26. Illustration of the overlap concept. Linear alignment rod arrays (multiple three-rod splices) are located in the left- and right-hand sections. These arrays overlap, thereby ensuring alignment in the region of the fiber joints. The extension of the lower arrays facilitates fiber insertion [F. L. Thiel and R. M. Hawk, *Appl. Opt.* **15**, 2785 (1976)].

from the mechanical mating plane of the connector halves. Connectors of this type have been fabricated by several manufacturers[67,78,79] and by other workers in the field.[60] The various embodiments of the design concept are based on the splice fixture designs discussed earlier. An example of the alignment means characteristic of multiple fiber overlap-style connectors is illustrated in Fig. 26.[60] Here, the basic three-rod splice approach has been extended to an array of rods that have been separated into two halves. Note that the two mechanical mating planes are not coincident with the fiber mating plane. In the mated condition, each of the two fiber ends are aligned as in the three-rod splice fixture To compensate for fiber diameter variations, a self-centering mechanism may be provided by coating the rods with a resilient plastic material. In this embodiment, the coated diameter of the rod is designed to make the interstitial channel slightly smaller than the smallest expected fiber diameter. Laboratory prototype connectors based on the overlapping three-rod splice fixture exhibit an average insertion loss of 0.57 ± 0.2 dB for identical step-index fibers of 125-μm core diameter,[60] and 85-μm core diameter.

A fiber jointing technique, developed by the Deutsch Corporation, that is not based on accurately aligning and butting two fiber ends together, is illustrated in Fig. 27.[80] The fiber ends are automatically aligned, centered, and positioned relative to each other as they are inserted into opposing conical sections that form a double-lens system in a transparent plastic medium.

[79] W. L. Schumacker, Electronic connector study group, *Connector Symp., 10th, Cherry Hill, New Jersey* p. 193 (1977).

[80] M. A. Holzman, *OSA IEEE Conf. Laser Electro-Opt. Syst., San Diego, California* Paper WAA4 (1978).

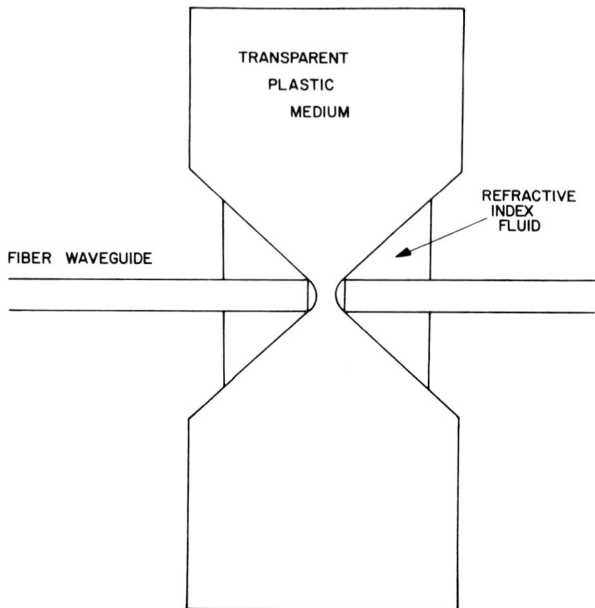

FIG. 27. The double-lens fiber-jointing technique (developed by Deutsch Corporation). A viscous refractive index fluid is self-contained within two opposed conical cavities. The prescription for the lens formed by the fluid at the fiber ends is a function of waveguide optical and mechanical characteristics.

The lens prescription is determined by the NA, cladding and core diameters, and by the index profile of the fibers. A viscous index-of-refraction fluid is permanently self-contained in the reservoirs formed by the lens cavities. In practice, the plastic lens medium is situated within a socket to which two connector ends may be attached.[68] As the two halves are mated, a means of moving the fiber ends into position inside the conical lens cavities is provided.

V. COMMUNICATION SYSTEMS

A. INTRODUCTION

The essential elements of an optical communication link are illustrated in the block diagram shown in Fig. 28. An input electrical signal is amplified or conditioned as necessary to drive a light-emitting diode. Light emitted from the diode is thus intensity modulated so as to reproduce variations in the input electrical signal level, and is focused or otherwise directed into the optical waveguide. Upon leaving the waveguide, the light signal is detected by a photo-

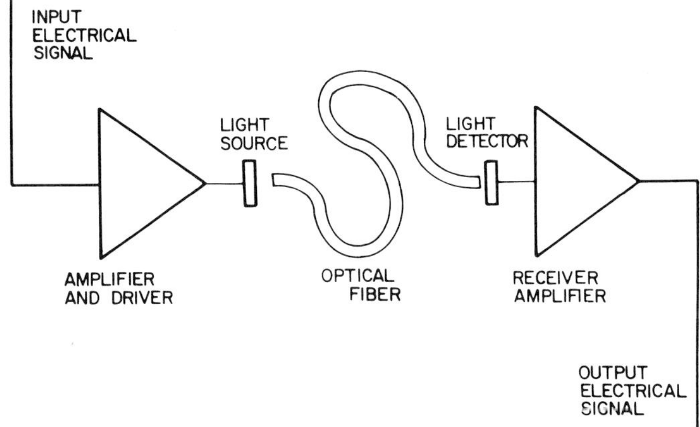

FIG. 28. The basic elements of an optical communication link.

voltaic diode and converted back to an electrical signal. The electrical signal is then amplified and processed as required by the application.

The design of an optical link and the selection of the key components; namely, the source, detector, and optical waveguide, is an iterative process. It generally begins by establishing the optical signal quality and level needed at the receiver, amplified to meet signal-to-noise ratio or bit-error-rate requirements. Selection of the light source and detector depend upon considerations such as cost, modulation rate, source-to-fiber coupling efficiency, optical output power level, and detector sensitivity. The basis for selecting a particular optical waveguide is, of course, the attenuation and bandwidth needed, which, in turn, is determined by the signal level quality required at the receiver detector.

This section discusses the principle of operation and the characteristics of sources and detectors suitable for use in optical communication links. In addition a brief outline of link design considerations is also given. In-depth analysis of system design tradeoffs may be found in the extensive literature on this subject.[81–84]

B. Diode Light Sources

It is a general characteristic of semiconductor *p–n* junctions that electrons and holes injected into the junction region under forward bias can recombine

[81] S. D. Personick, *Bell Syst. Tech. J.* **52**, 843 (1973).

[82] M. Barnoski (ed.), "Fundamentals of Optical Fiber Communications." Academic Press, New York, 1976.

[83] S. E. Miller, T. Li, and E. A. J. Marcatili, *Proc. IEEE* **61**, 1726 (1973).

[84] G. R. Elion and H. A. Elion, "Fiber Optics in Communication Systems." Deckker, New York and Basel, 1978.

to emit light under certain conditions. The intensity of the emitted radiation and its angular distribution, as well as the source-to-fiber coupling efficiency, depend upon junction design and geometry. The wavelength of the recombination radiation is equal to ch/E, where c is the velocity of light in free space, h is Planck's constant, and E the band-gap energy.

A number of alloys from the III and V families of elements have band-gap energies in the range of interest for optical fiber communications; namely between 1.5 and 1.0 eV, which corresponds to emission wavelengths between 800 and 1300 nm. For example, GaAs emits at approximately 900 nm, and $Al_xGa_{1-x}As$ can be tailored to emit between 780 and 900 nm by varying the concentration of aluminum. Other III–V systems from which room-temperature lasers have been fabricated include the ternary alloys GaAsSb[85,86] and the quaternary alloy GaInAsP/InP.[87] As discussed in Section II, there is an intense interest in devices fabricated from these alloys because they can be tailored to emit radiation between 1.0 and 1.4 μm, the spectral region of lowest fiber waveguide attenuation.

Light-emitting diode sources are designed so that light may be emitted from either the edge of the junction or through one of the surfaces.[88–90] One of the most effective surface emitters is the Burrus high-radiance diode.[89] The arrangement of the alloy layers that form the junction of this device is shown in Fig. 29. The potential well formed by the different band energies of GaAs and GaAlAs tends to confine the injected electrons to the GaAs p side of the p–n junction. Hence, there is a high probability that the recombination radiation will be generated here. The recombination region is laterally confined by placing the positive electrode directly beneath the etched well in the top GaAs n layer. This design enhances the probability that the emitted radiation will be captured by the optical fiber located at the bottom of the well and eliminates the possibility of light reabsorption in the GaAs n layer before it reaches the fiber. Junctions of this kind between two dissimilar semiconductors with different energy gaps are called heterojunctions.

To a good approximation, high-radiance diodes are Lambertian emitters; therefore, the light-coupling efficiency E_f into the fiber is given by:

$$E_{\mathrm{f}} = \frac{\alpha}{\alpha + 2} \frac{A_{\mathrm{f}}(\mathrm{NA}_{\mathrm{f}})^2}{\mathrm{A}_{\mathrm{s}} n_0{}^2} \tag{5.1}$$

---

[85] C. J. Nuese, G. H. Olsen, M. Ettenberg, J. J. Gannon, and T. J. Zamerowski, *Appl. Phys. Lett.* **29**, 807 (1976).

[86] R. F. Nahory, M. A. Pollack, F. D. Beebe, and J. C. DeWinter, *Appl. Phys. Lett.* **28**, 19 (1976).

[87] J. J. Hsieh, J. A. Rossi, and J. P. Donnelly, *Appl. Phys. Lett.* **28**, 709 (1976).

[88] J. K. Butler, H. Kressel, and I. Ladany, *IEEE J. Quantum Electron.* **11**, 402 (1975).

[89] C. A. Burrus and B. I. Miller, *Opt. Commun.* **41**, 307 (1971).

[90] H. Kressel and M. Ettenberg, *Proc. IEEE* **63**, 1360 (1975).

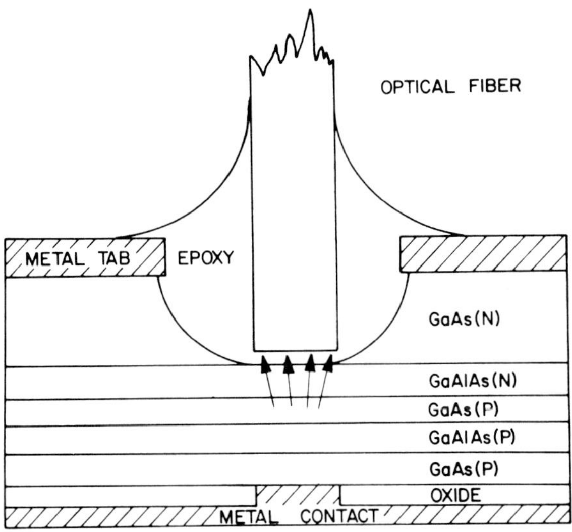

FIG. 29. The Burrus high-radiance diode structure. Recombination radiation is generated, as indicated by the arrows, on the GaAs p side of the p–n junction.

where A_s is the area at the base of the etched well, n_0 the refractive index of the bonding material between the fiber and the bottom surface of the well, and NA_f the peak (or meridional) numerical aperture of the fiber. A typical plot of the optical power coupled into a fiber from a high-radiance diode as a function of drive current is shown in Fig. 30.

The semiconductor diode laser is superficially similar in design to a high-radiance diode. Layers of GaAs and GaAlAs are successively grown using liquid-phase epitaxial techniques to obtain the structure shown in Fig. 31. When a forward voltage of about 1.5 V is applied, either holes or electrons (depending on the dopant levels) are "injected" across the junction where they recombine to emit light of a characteristic wavelength. By cleaving the chip on both ends to obtain mirrorlike reflecting surfaces, an optical cavity that acts as a light resonator is formed. Providing, of course, that the electron and hole densities are sufficiently high, the partial reflection of light back into the junction region at the GaAs p-layer-to-air interface initiates laser action. As in the case of the high-radiance diode, the potential well formed by the GaAlAs and GaAs layers tends to confine the injected electrons or holes to the GaAs p layer where recombination radiation is generated. Lateral confinement of the injected carriers is achieved by means of a stripe electrode. The lower refractive index of the GaAlAs layers on both sides of the GaAs p layer forms a planar optical waveguide and serves to confine the recombination radiation within the GaAs p layer. These radiation and current confinement techniques enhance the

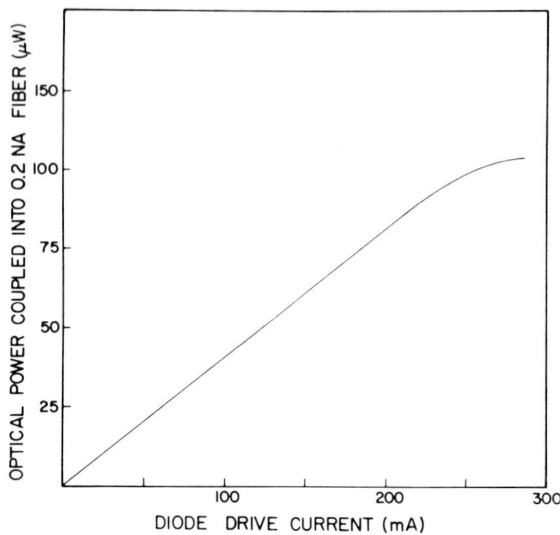

FIG. 30. Optical power coupled into a fiber from a typical high-radiance diode as a function of drive current.

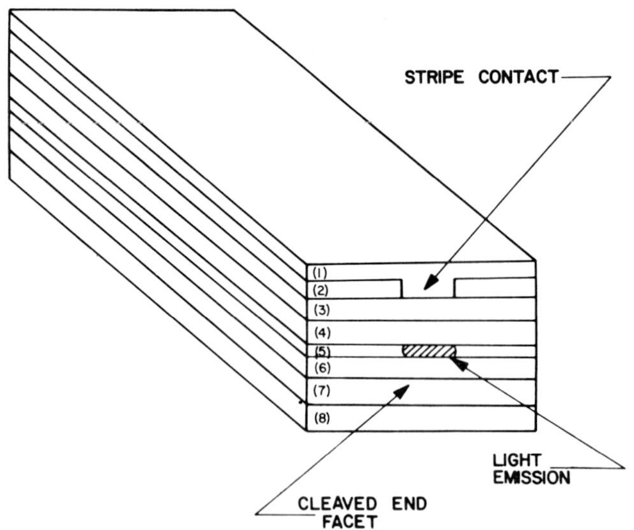

FIG. 31. Stripe geometry double heterostructure GaAs/GaAlAs laser. Carriers are injected into the active GaAs p layer beneath the stripe contact (≈ 10-μm wide). Light is emitted through the cleaved end facets. (1) Metal contact. (2) Insulating oxide. (3) GaAs (P). (4) GaAlAs (P). (5) GaAs (P) (active layer). (6) GaAlAs (N). (7) GaAs (N). (8) Metal contact.

probability of stimulated recombination radiation. The end result is the emission of a relatively coherent ($\approx$2-nm spectral width) light beam from the end face of the GaAs *p* layer as indicated in Fig. 31.

Since the area of the end face is small compared to the core area of most optical waveguides, a large fraction of the emitted radiation may be coupled into the waveguide. Typically, a 3–6 dB coupling loss is associated with injection laser diodes, compared to a 17 dB typical input coupling loss for high-radiance diodes. Laser diodes characteristically yield higher output optical power; 10–20 mW as compared to a few milliwatts from a high-radiance diode. Also, the narrower spectral width (1–2 nm compared to 30–40 nm for the high-radiance diodes) leads to less material dispersion (higher bandwidth) per unit length in the optical waveguide. High-radiance diodes, on the other hand, have longer life; about 100,000 h compared to about 10,000 h for the best commerically available laser diodes. Much progress, however, is being made in extending laser diode lifetime.

Edge-emitting LEDs are similar in structure to the laser geometry shown in Fig. 31. In these devices, the lateral width of the emitting region is matched to the core size of waveguide fibers, typically 50–100 μm. There is partial internal guiding of the spontaneous radiation by the heterojunctions, which improves the directionality of the junction, but the coupling loss is still considerably greater (12–16 dB) than that of laser diodes (3–6 dB). The spectral width of the emitted radiation is approximately 40 nm. Thus, due to material dispersion in the fiber, these devices are not used in applications requiring high bandwidth. In addition, the modulation rate of the device itself is limited to about a few hundred megahertz since it is difficult to decrease injected carrier lifetime without sacrificing internal quantum efficiency.[91] They are best suited for short-distance, moderate-data-rate applications where cost considerations are of paramount importance. On the other hand, these devices are more reliable (long life) than laser diodes and are less expensive to manufacture.

C. Diode Light Detectors

It is now generally accepted that silicon *p–i–n* photodiodes or avalanche photodiodes (APD) are correct choices for detecting the light in an optical communication system. These devices are most sensitive to radiation at wavelengths between 800 and 900 nm, corresponding to the emission wavelength of GaAlAs diode sources. In both devices, the absorption of a photon produces an electron hole pair that results in a net charge flowing through an external

---

[91] M. Ettenberg, H. F. Lockwood, J. Wittke, and H. Kressel, Technical digest, *Int. Electron Devices Meeting, Washington, D.C.* p. 317 (1973).

load. The responsiveness of p–i–n diodes is on the order of 0.5 A/W of incident optical power. In APDs, however, the initial electron charge is multiplied by the mechanism of collision ionization within the device; consequently, larger values of responsivity may be realized. It is possible in practical applications to realize current gains up 100 times greater than with p–i–n diodes, but this potential advantage of APDs must be weighed against their higher operating voltage—several hundred volts compared to 20–40 V for p–i–n diodes. This is an extremely critical consideration because the multiplication factor is a strong function of the back bias voltage, as well as temperature. APDs also exhibit a somewhat lower response time than p–i–n diodes, but, in almost all cases, overall system bandwidth is limited by other considerations.

The recent interest in operating optical communication systems at wavelengths greater than 1 μm, coupled with the poor sensitivity of silicon photodiode detectors in this region of the spectrum, has resulted in the development of experimental p–i–n diodes and APDs fabricated from the III–V system of alloys. The same ternary and quaternary alloys that have direct band gaps suitable for source devices in the 1.0–1.4 μm region are, of course, candidate materials for diode detectors.

Both single heterostructure p–i–n photodiodes and double-heterostructure APDs have been successfully fabricated in the GaInAsP/InP materials system.[91–93] Response times for experimental diodes are comparable to those of silicon diodes. For example, at a wavelength of 1.22 μm and with the device reverse-biased to yield a gain of 10–12, the response time was measured at 150 psec, a value that may actually have been due to the rise time of the laser used to make the measurement.[94] Much more work is needed to establish the commercial feasibility of these devices but the experimental work carried out to date suggests that fiber optic communication systems operating at wavelengths between 1.1 and 1.4 μm will be technically feasible in the future.

D. System Design

Once the signal to be transmitted, the link length, the bit-error rate, and the rise time (or bandwidth) are given, the basic approach used to design optical communication systems is quite straightforward. The first step involves constructing a plot of the mean optical power required at the receiver input to obtain the desired bit rate and bit-error rate. Figure 32 is an example of such a plot for state-of-the-art APD and p–i–n diode receivers and a bit-error rate of

---

[92] H. H. Wieder, A. R. Clawson, and G. E. McWilliams, *Appl. Phys. Lett.* **31**, 468 (1977).

[93] R. L. Davies and F. E. Gentry, *IEEE Trans. Electron Devices* **ED-11**, 313 (1964).

[94] J. J. Hsieh, C. E. Hurwitz, and C. C. Shen, *IEEE ERA Electro-78 Convent., Boston, Massachusetts* Session 29 (May 1978).

10^{-9}. Similar curves would result for other bit-error rates. Typical values for the optical power that may be coupled into an optical fiber from a LED or laser diode are also shown. For convenience in calculating power loss budgets, the ordinate in Fig. 32 is given in terms of dBm, that is, 10 log (power in milliwatts/one milliwatt). In practice, a plot such as this is constructed for a number of detectors and sources from the information given in the manufacturers' data sheets. It is then possible to determine the maximum fiber attenuation permissible in order to transmit a certain distance at a specified bit rate. Allowance for splice or connector losses, as well as source degradation over the expected life of the system, must also be taken into account. In most system designs, an additional margin of 3–6 dBm is usually factored in for contingencies. Signal amplitude is often the limiting factor in determining the maximum distance between source and receiver. However, for high-bit-rate applications such as CATV, either signal dispersion or bandwidth may be the critical consideration governing the choice of system components, as well as the limiting factor in determining the maximum permissible distance between source and detector. In a cascade-connected system, the overall rise time must

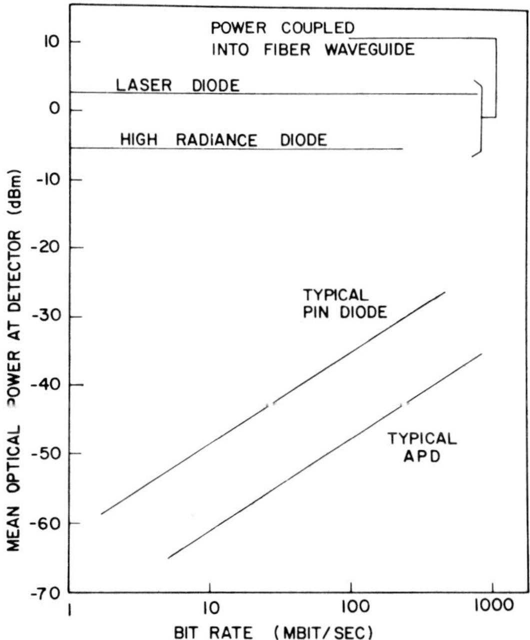

FIG. 32. Typical mean optical power required to be incident on p–i–n diodes and APDs to realize a bit error rate of 10^{-9} as a function of bit rate. Representative values for the optical power coupled into a fiber from high-radiance and laser diodes are also indicated. The link decibel budget is given by the difference between the optical power coupled into the fiber and the required power output from the fiber.

be consistent with the bit-rate requirement. As a rough approximation, the overall rise time is 1.1 times the square root of the sum of the squares of the rise time of the individual components. The rise time for a fiber may be taken to be the rms pulse dispersion, which for Gaussian pulses is equal to 0.187 divided by the bandwidth as given in Eq. (2.20).

Author Index

Numbers in parentheses are references numbers and are included to assist in locating references in which authors' names are not mentioned in the text.

F

G

H

Subject Index

T

U

V

W

X

Y

Z